SPECTRAL LINE SHAPES

AIP CONFERENCE PROCEEDINGS 328

SPECTRAL LINE SHAPES
Volume 8
12th ICSLS
TORONTO, CANADA JUNE 1994

EDITORS: **A. DAVID MAY**
J. R. DRUMMOND
UNIVERSITY OF TORONTO

EUGENE OKS
AUBURN UNIVERSITY

American Institute of Physics   New York

Authorization to photocopy items for internal or personal use, beyond the free copying permitted under the 1978 U.S. Copyright Law (see statement below), is granted by the American Institute of Physics for users registered with the Copyright Clearance Center (CCC) Transactional Reporting Service, provided that the base fee of $2.00 per copy is paid directly to CCC, 27 Congress St., Salem, MA 01970. For those organizations that have been granted a photocopy license by CCC, a separate system of payment has been arranged. The fee code for users of the Transactional Reporting Service is: 0094-243X/ 87 $2.00.

© 1995 American Institute of Physics.

Individual readers of this volume and nonprofit libraries, acting for them, are permitted to make fair use of the material in it, such as copying an article for use in teaching or research. Permission is granted to quote from this volume in scientific work with the customary acknowledgment of the source. To reprint a figure, table, or other excerpt requires the consent of one of the original authors and notification to AIP. Republication or systematic or multiple reproduction of any material in this volume is permitted only under license from AIP. Address inquiries to Series Editor, AIP Conference Proceedings, AIP, 500 Sunnyside Boulevard, Woodbury, NY 11797-2999.

L.C. Catalog Card No. 94-74309
ISBN 1-56396-326-4
DOE CONF-9406286

Printed in the United States of America.

Contents

Preface xiii

LOW DENSITY PLASMAS: HYDROGEN LIKE AND HELIUM LINES

Some Novel Concepts for Spectroscopic Diagnostics in Tokamaks 3
    D. Voslamber*

Line Profile Measurements in Microwave Produced Discharges 28
    H. Schlüter, S. Hirsch, and G. Himmel

Experimental Study of Two HeI Lines with Forbidden Components 30
    C. Pérez, I. de la Rosa, J. A. Aparicio, S. Mar, and M. A. Gigosos

Fine-Structure Asymmetry of Doppler-Free Stark-Broadened Two-
Photon Polarization Line Profiles of Hydrogen Lyman-α 32
    J. Seidel, A. Steiger, and K. Grützmacher

Improved Theory of Ion Impact Broadening in Magnetized Plasmas
and its Diagnostic Applications 34
    E. Oks and A. Derevianko

Line Shapes in Astrophysics 36
    C. Stehlé*

Experimental Study of the He II P_β Line Shape 58
    N. Konjević, I. Stefanović, and M. Ivković

Influence of Ion-Dynamics on the Shape of the He I 4713 Å and
7065 Å Lines 60
    Z. Mijatović, N. Konjević, R. Kobilarov, and M. Ivković

Experimental Study of the Temperature in a Helium Plasma 62
    J. A. Aparicio, M. A. Gigosos, S. Mar, C. Pérez, and M. I. de la Rosa

Strong Collision Contributions to Shift and Width of Hydrogen
Spectral Lines 64
    A. Könies and S. Günter

Stark Broadening of He I Lines 66
    T. Schöning

Broadening and Shift of the Paschen Alpha Line 68
    A. Döhrn, V. Helbig, S. Günter, and A. Könies

Co-operative Collision Processes for the Population of Ionic Excited
States 70
    S. Bliman, M. Cornille, and K. Katsonis

LOW DENSITY PLASMAS: COMPLEX RADIATORS AND GENERAL RESULTS

Temperature Dependence of the Triply Ionized Oxygen Stark Widths 75
    N. Konjević, B. Blagojević, M. V. Popović, and M. S. Dimitrijević

Influence of Ion-Dynamics on the Shift of C I 5052.17-Å Spectral Line in Plasma ... 77
    Z. Mijatović, N. Konjević, R. Kobilarov, and S. Djurović

Stark Parameters Regularities within Spectral Series of Several Multicharged Ions ... 79
    J. Purić and M. Ćuk

"Exact" Analytic Formulas for the Stark Broadening of Isolated Ion Lines ... 81
    S. Alexiou and Y. Maron

Fermi Method of Equivalent Photons for Stimulated Emission of Polarization-Induced Multiphoton Radiation by Multiply Charged Ions ... 83
    V. A. Astapenko and A. B. Kukushkin

Lasing on a Weak Intercombination Transition ($4p^4S_{3/2} \rightarrow 4s^2P_{3/2}$) in ArII Plasma ... 85
    D. A. Shapiro, S. I. Kablukov, S. A. Babin, and S. V. Khorev

Influence of the Oscillator Strengths on the Stark Broadening of Rb I Lines ... 87
    M. S. Dimitrijević and S. Sahal-Bréchot

Effect of Dissociative Recombination on Spectral Line Profiles in Neon Glow Discharge ... 89
    J. Szudy, R. Ciuryło, A. Bielski, J. Domysławska, and R. S. Trawiński

Ion Dynamics and Microfield Smoothness ... 91
    S. Alexiou

Spectra Deconvolution Using Biraud's Method ... 93
    A. Lesage, M. Depiesse, and J. Richou

Study of Profiles of Some CI and NI Multiplets ... 95
    A. Goly and T. Wujec

Dense Arc Electrode Plasma Instability and Line Broadening ... 97
    P. Serapinas

Anomalous Broadening—Anomalous Electric Fields? ... 98
    E. Oks (Discussion Leader)

HIGH TO ULTRA-HIGH DENSITY PLASMAS

Application of Spectral Line Shapes to the Study of High Density ICF Plasmas ... 105
    C. J. Keane,[*] B. A. Hammel, S. H. Langer, R. W. Lee, A. Calisti, L. Godbert, R. Stamm, and B. Talin

Spatially Resolved X-ray Emission Spectroscopy from Dense Plasmas ... 122
    E. Leboucher-Dalimier,[*] P. Angelo, H. Derfoul, P. Gauthier, and A. Poquérusse

Line Broadening of Nonhydrogenic Ions in Plasmas ... 134
    S. Glenzer

Width of the Hydrogen Lyman-α Line within an Approach Based
on a Green's Function Technique for Electrons and Computer
Simulations for the Ions 151
    W. Olchawa, J. Halenka, and S. Günter

Analysis of Lithiumlike Line Shapes in a Gas-Liner Pinch 153
    R. Stamm, A. Calisti, L. Godbert, T. Meftah, C. Mossé, B. Talin,
    and S. Glenzer

Analysis of K-Shell Line Emission Spectra Emitted by Short-Pulse
Laser-Produced Plasmas 155
    O. Peyrusse, D. Gilles, J. C. Kieffer, C. Y. Coté, and Z. Jiang

Ab Initio Non Uniform Microfield Joint Distributions in Plasmas 156
    A. V. Demura, D. Gilles, B. C. Huynh, and C. Stehlé

Line Shape Diagnostics for Solid Density Plasmas Produced
by Ultra Intense Subpicosecond Laser 158
    Z. Jiang, J. C. Kieffer,* M. Chaker, G. Korn, S. Coe, G. Mourou,
    O. Peyrusse, and D. Gilles

Dicke Narrowing in Dense Plasmas 160
    D. A. Shapiro* and E. V. Podivilov

Effects of Microfield Nonuniformity in Dense Plasmas 177
    A. V. Demura* and C. Stehlé

Stark Widths of Hydrogen Spectral Lines in One- and Twofold
Ionized Helium Plasmas 209
    B. Grabowski, J. Halenka, and W. Olchawa

BV, CVI and NVII Lyman Line Profiles from 10 PS KrF-Laser
Produced Plasmas... 211
    H. R. Griem, Y. Leng, J. Goldhar, and R. W. Lee

Stark Broadening Calculations of $3d$-$5f$ Transition in Al XI 213
    N. Ben Nessib, Z. Ben Lakhdar, H. Nguyen, and J. P. Arranz

X Emission from Dense Plasma Created by Colliding Foils 215
    P. Angelo, P. Gauthier, E. Leboucher-Dalimier, A. Poquérusse,
    and C. Back

Shift and Width of He II Lines 217
    S. Günter, M. Stobbe, A. Könies, and J. Halenka

COLLISION INDUCED SPECTRA

Rototranslational Collision-Induced Absorption by H_2-H_2 Pairs
at Temperatures from 600 to 7,000 K.. 221
    C. Zheng and A. Borysow

The Effect of Rotational Level Mixing in Far Wings Collision
Induced Spectra 223
    W. Glaz and G. C. Tabisz

The Induced Dipole Moment and Collisional Interference
in the Rotational Spectrum of HD-He and HD-Ar 225
    G. C. Tabisz, B. McQuarrie, B. Gao, and J. Cooper
The Influence of the Anisotropic Potential on the Spectral Moments
of Collision-Induced Absorption of CO_2 and N_2 Pairs 227
    M. Gruszka and A. Borysow
Investigation of Collision Induced Line Shape of a Satellite Associated
with $4s^2\ ^1S_0$-$4s3d\ ^1D_2$ Transition of Calcium............................ 229
    M. A. Gondal, M. A. Khan, and M. H. Rais
Effect of the Atomic Polarization on the Spectral Line Far Wings
Induced by Anisotropic Collisions 231
    A. L. Zagrebin and M. G. Lednev
Effect of Collisions on the Atomic Forbidden Transitions:
$He(2\ ^1S, 2\ ^3S) + Ne \rightarrow He(1\ ^1S) + Ne + \hbar\omega$ 233
    A. L. Zagrebin and S. I. Tserkovnyi
Collision Induced Spectra: $Hg(6\ ^3P_2)$+He, Ne, Ar, Kr,
Xe$\rightarrow Hg(6\ ^1S_0)$ +He, Ne, Ar, Kr, Xe+$\hbar\omega$ 235
    A. Z. Devdariani, A. L. Zagrebin, and M. G. Lednev
Spectral Invariants for the Absorption Coefficient of CO_2-Ar Pairs 237
    M. Moraldi and A. Borysow

SELECTIVE REFLECTION AND OTHER LINE SHAPES

Atomic Spectral Line Shapes from Selective Reflection Laser
Spectroscopy 241
    M. Fichet,* N. Papagiorgiou, F. Schuller, D. Bloch, and M. Ducloy
Blue Shift Paradox in Selective Reflection 249
    T. A. Vartanyan, D. Bloch, and M. Ducloy
Subnatural Linewidth in Large Optical Fields 251
    L. M. Narducci,* C. H. Keitel, G. L. Oppo, and M. O. Scully
Precision Measurement of Transition Linewidths 252
    W. A. van Wijngaarden and J. Li
Accurate Profiles of Solar Infrared OI Triplet Lines 254
    A. A. Galal, N. H. Youssef, and M. M. Behery
Energy Transfer and Energy Pooling Collisions in Li-Cd System 256
    G. Pichler, D. Azinović, and S. Milošević
Dynamic Measurement of Gas Temperature and Pressure Using
Infrared Spectroscopy 258
    R. Berman, P. Duggan, M. P. Le Flohic, A. D. May, and J. R. Drummond
Lineshape of Light Absorption by Excitons Under Band-to-Band
Transitions in Low Dimensional Molecular Structure 260
    N. I. Grigorchuk

LINE MIXING AND FAR WING LINE SHAPES

Line Mixing Effects in the Impact Limit: ECS Analysis of Various Experiments Made on Some IR Bands of CO_2 in Helium 265
    C. Boulet* and J. Boissoles

Line Shapes in the Far Wings of Allowed Transitions 274
    R. H. Tipping* and Q. Ma

The Wings of Pressure Broadened Molecular Bands 288
    G. Birnbaum*

Semiclassical Analysis of the Interbranch Line Coupling in the Infrared Band Shapes of Linear Molecules 298
    N. N. Filippov and M. V. Tonkov

Close-Coupling Calculation of Line-Mixing Parameters for D_2-He 300
    R. Brezina, W.-K. Liu, and S. Green

Experimental HF-Ar Lineshape Parameters in Far Infrared: Broadening, Shifts, and Line Mixing 302
    I. M. Grigoriev, N. N. Filippov, A. V. Rozanov, and M. V. Tonkov

Line Mixing in Q-Branches of Π-Δ Transitions of CO_2 304
    M. V. Tonkov, F. Thibault, and R. Le Doucen

Line-Mixing and Duration-of-Collision Effects in the ν_3 R-Branch Band Head of CO_2 306
    D. Tobin, L. Strow, S. Hannon, and J. W. C. Johns

Line Mixing and State-to-State Rates in D_2 Determined from the Raman Q Branch 308
    A. D. May, P. M. Sinclair, J. W. Forsman, and J. R. Drummond

Water Vapor Absorption in MMW Atmospheric Windows. Continuums 310
    A. Bauer, M. Godon, J. Carlier, and Q. Ma

Close Examination of the Line Mixing Concept in the Line Wing Absorption 312
    S. D. Tvorogov, O. B. Rodimova, and L. I. Nesmelova

Extension of the Quasistatic Far-Wing Line Shape Theory to Multi-Component Anisotropic Potentials 314
    Q. Ma and R. H. Tipping

Photoabsorption Studies of Quasimolecules of Alkaline-Earth and Related Metal Atoms 316
    Y. Sato*

Line Shapes in the Far Wings of $Hg\,{}^3P_1$-${}^1S_0$ Resonance-Line Broadened Due to the $Hg\,{}^3P_1 \to {}^3P_0$ Fine-Structure Transitions in Collisions with N_2 and CO 341
    Y. Sato, K. Ohmori, T. Kurosawa, H. Chiba, M. Okunishi, and K. Ueda

Line Shapes in the Far Wings of Hg^3P_1-$^1S_0$ Resonance Line
Broadened Due to the Chemical Reactions: $Hg^*(^3P_1)+H_2, D_2 \rightarrow HgH(X^2\Sigma^+, v, j)$
+H, D 343
  K. Ohmori, H. Chiba, T. Kurosawa, M. Okunishi, K. Ueda,
  and Y. Sato

Interatomic Potentials for Cd-Xe and Cd-Ar from the Cd 326.1 nm
Line Wings Measurements 345
  G. D. Roston and T. Grycuk

Semiclassical Coherences in Collisional Redistribution 347
  R. J. Bieniek and I. M. Bell

Effects of Interaction Hg-C_{60} Observed on the Far Red Wing of the
Hg 253.7 nm Line... 349
  T. Grycuk, M. Tchaplyguine, E. Czerwosz, and P. Byszewski

LINE SHAPES IN DENSE MEDIA

Raman Spectra of Formamide in DMSO 353
  A. Mortensen, O. F. Nielsen, J. Yarwood, and V. Shelley

Field Ionization of High Rydberg States of CH_3I in Liquid Argon 355
  R. Reininger and A. Al-Omari

Line Shapes of the Fundamental Vibration-Rotation-Phonon and
Pure Rotation-Phonon Spectra in HD..................................... 357
  R. M. Herman, B. Weiner, and P. B. Shaw

Many-Body Correlations in the Far Infrared Absorption in
Liquid OCS... 359
  H. Stassen and W. A. Steele

An Analysis of the Radiation Line Shape from a Multiple Quantum
Well Structure in Four-Wave-Mixing 361
  C. J. Hsu

Absorption Spectra Shapes of Silver Colloid Aggregates 363
  V. P. Safonov, Yu. E. Danilova, and V. A. Markel

BROADENING AND SHIFTING IN NEUTRALS AND IONS

Ability to Predict Molecular Rotation-Vibration Line Shapes................. 367
  S. Green*

Collisional Broadening and Coupling in the Rotational Spectrum
of the Asymmetric Rotor CHF_2Cl 372
  G. Buffa, O. Tarrini, G. Cazzoli, L. Cludi, G. Cotti, and C. Degli Esposti

A Study of Collisional Lineshapes of Ammonia Transitions 374
  A. Ciucci, G. Baldacchini, F. D'Amato, G. Buffa, and O. Tarrini

Foreign-Gas Broadening of Stark-Tuned IR Resonances in Ammonia
at Low Pressures 376
  R. A. Gordon

Lineshape Analysis of Speed Dependent Collisional Width
Inhomogeneities in CO Broadened by He, N_2, and Xe 378
  P. Duggan, P. M. Sinclair, A. D. May, and J. R. Drummond

Broadening and Shifting of the Raman Q Branch in Pure D_2 and
D_2-He Mixtures 380
  P. M. Sinclair, P. Duggan, M. P. Le Flohic, J. W. Forsman,
  J. R. Drummond, and A. D. May

Broadening and Shift of the Lines of Molecular Ions by Collisions
with Neutral Perturbers 382
  O. Tarrini, G. Buffa, G. Cazzoli, and L. Dore

Non-Adiabatic Effects in the Broadening and Shift of the
K $7s\ ^2S_{1/2}-4p\ ^2P_{3/2}$ Transition Perturbed by Ar 384
  W. Kreye and J. Kielkopf

Intensities and N_2-Broadened Half-Widths in the ν_3-Fundamental
Band of CO_2 at Atmospheric Temperatures 386
  Z. Li, P. Varanasi, and M. Weber

Line Shapes and Collisional Effects in Resonant Degenerate
Four-Wave Mixing 388
  L. A. Rahn,* S. Williams, and R. N. Zare

EXCIMERS

Diffuse Bands in Intermetallic Excimers 391
  S. Milošević*

Theoretical and Experimental Studies of the LiHg-Blue
Green Bands 406
  D. Gruber, L. Windholz, X. Li, M. Gleichmann, and B. Heβ

Temperature Dependence of the Kr*Ar Exciplex Emissions
Lineshape 408
  R. Reininger, J. L. Subtil, C. Vincent-Donnet, P. Laporte,
  and P. Gürtler

ULTRACOLD ATOMS

Slow Atom Collisions: A Review 413
  P. S. Julienne*

Theory of Line Shapes for Cold Atom Collisions 415
  C. J. Williams*

Ultracold Collision Studies: A New Physical Regime Where
Lineshape Analysis Informs Dynamics 417
    J. Weiner, R. Napolitano, V. Bagnato, L. Marcassa, and P. S. Julienne
Photoassociation of Ultracold Atoms 420
    D. J. Heinzen, J. D. Miller, and R. A. Cline
Liquid Helium Line Broadening and Shifts of Alkaline Earth
and Alkali Metal Atoms 422
    Y. Takahashi, K. Fukuda, T. Kinoshita, and T. Yabuzaki
"Slow" Molecules Ensemble Choice in Room Temperature Gas
for High Resolution Coherent Transient Spectroscopy 424
    N. N. Rubtsova, L. S. Vasilenko, and E. B. Hvorostov

APPENDIX

Minutes of the Meeting of the International Committee 427
Author Index 428
Subject Index 431

*Invited speaker

PREFACE

The 12th International Conference on Spectral Line Shapes was held from June 13 to 17, 1994 in the Physics Department of the University of Toronto, Toronto, Canada. The spectral line shape conferences are held biannually and alternate between Europe and North America. This year's conference was attended by 100 participants from 15 countries. The proceedings of the last seven conferences have been published under the title Spectral Line Shapes by de Gruyter, Berlin (volumes 1–3); A. Deepak Publishing, Hampton, VA (volume 4); Ossolineum Publishing, Warsaw, Poland (volume 5); American Institute of Physics, NY (volume 6); and Nova Science Publishers, Commack, NY (volume 7).

The conference format consisted of five invited papers in the morning, a two-hour session of contributed posters in the mid-afternoon, followed by a discussion period. The evenings were left free for informal discussions. The program of invited speakers was suggested primarily by the International Committee. The program emphasized those physical processes associated with the formation of spectral lines as observed in absorption and scattering of radiation by plasmas and neutral gases. The program also included papers in related fields. These proceedings include the full length text of many of the 25 invited speakers and all of the extended two-page abstracts of the approximately 80 contributed papers. Included in the Appendix is the minutes of the meeting of the International Committee.

Line profiles are used to study the fundamental physics of atomic and molecular interactions and are also employed as powerful diagnostic tools for many media (e.g., gas discharges, flames, plasmas, and planetary and astrophysical objects). These research areas have seen a resurgence of interest with the enabling advances made in laser and electronic technology. These proceedings are intended to record the latest advances in the field, to display the vitality of the subject, and to encourage young investigators to participate in the international community involved in line shape studies.

The organizers and participants would like to thank the following for financial support:

The International Union of Pure and Applied Physics
The International Science Foundation
The Office of the Vice President and Provost, University of Toronto
The Department of Physics, University of Toronto

INTERNATIONAL COMMITTEE

K. Burnett	Oxford University	U.K.
A. Devdariani	Saint Petersburg University	Russia
N. Feautrier	Observatoire de Paris–Meudon	France
C. Iglesias	LLNL (Livermore)	U.S.A.
D. Kelleher	NIST (Gaithersburg)	U.S.A.
J. Kielkopf	University of Louisville	U.S.A.
N. Konjevic	University of Belgrade	Yugoslavia
H. Nguyen	Université P. & M. Curie	France
G. Pichler	Institute of Physics (Zagreb)	Croatia
J. Seidel	PTB–Institute (Berlin)	Germany
R. Stamm	Université de Provence	France
R. Tipping	University of Alabama	U.S.A.
E. Yukov	Lebedev Physical Institute	Russian
M. Zoppi	C. N. R. (Firenze)	Italy

LOCAL COMMITTEE

David May	University of Toronto	Canada
Jim Drummond	University of Toronto	Canada

SPONSORS

The conference was sponsored by the University of Toronto, by the Department of Physics, and by the International Union of Pure and Applied Physics. In accordance with IUPAP rules, no bona fide scientist was excluded from participation on the grounds of national origin or political considerations unrelated to science.

LOW DENSITY PLASMAS: HYDROGEN LIKE AND HELIUM LINES

SOME NOVEL CONCEPTS FOR SPECTROSCOPIC DIAGNOSTICS IN TOKAMAKS

D. Voslamber

Ass. EURATOM-CEA, DRFC-STPF; CEN Cadarache

F-13108 St.Paul-lez-Durance

ABSTRACT

In recent years, new concepts for spectroscopic diagnostics in tokamaks have been developed with the emphasis being laid on local measurements. Localization can be achieved by using neutral particle beams or laser beams. While the former are now currently used for spectroscopic purposes, application of the latter is still in an evolutionary stage. One of the most promising concepts using laser beams is Doppler-free two-photon excited fluorescence, particularly in the Lyman-α lines of the hydrogen isotopes H, D and T. A detailed theoretical study has shown that this method should allow local and isotope-specific measurements of atomic hydrogen groundstate densities; further, in favorable situations, this technique should also permit to measure atomic velocity distributions and macroscopic flows (plasma rotation), as well as the effective ion charge number Z_{eff} and the magnetic field direction.

INTRODUCTION

Tokamak plasmas are quite different from the ICF plasmas treated in the preceding contribution[1]. The main differences lie in the much lower electron density (which is typically ten orders of magnitude smaller), and in the much larger size and lifetime (the former measuring in meters, the latter in seconds if not minutes). In a tokamak (see Fig.1) the plasma is contained in a toroidal vacuum vessel which is surrounded by coils to produce a toroidal magnetic field of 3-5 T. In addition to the toroidal field, a toroidal current is induced by using the torus as the second circuit of a transformer. The current produces a poloidal magnetic field which adds to the toroidal one, so that the resulting magnetic field lines show a helical structure. The magnetic field configuration thus produced is essential for the confinement and stability of the plasma. Heating of the plasma is achieved by the induced current as well as by additional heating mechanisms, such as neutral beam injection and high frequency electro-magnetic waves. In the center of the plasma the temperature reaches typically 10^7-$10^8\,K$, the electron density 10^{19}-$10^{20}\,m^{-3}$. Towards the edge, these quantities decrease by one or more orders of magnitude. The neutral particle density is typically 10^{12}-$10^{14}\,m^{-3}$ in the center and increases by 4 or 5 orders of magnitude towards the edge.

4 Spectroscopic Diagnostics in Tokamaks

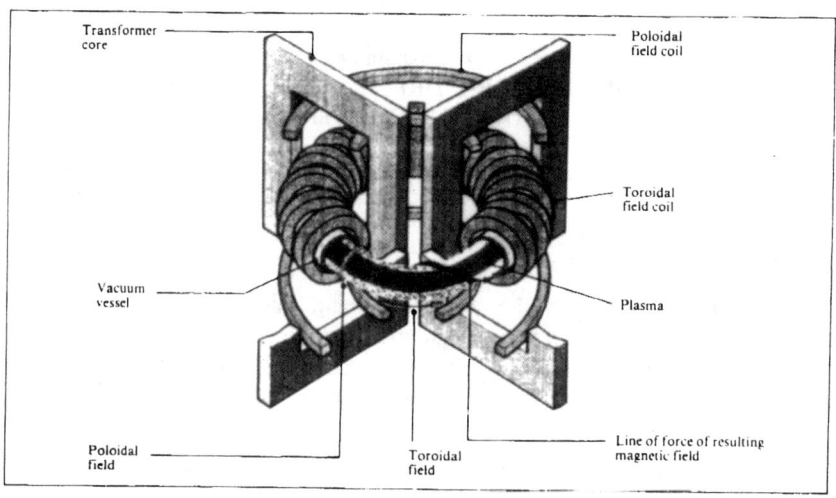

Figure 1. Schematic drawing of a tokamak.

Among the various diagnostic techniques employed to measure the plasma parameters, spectroscopy plays a prominent role. It exploits virtually all charged and uncharged particle species present in the plasma, in particular the hydrogen isotopes H or D, which are the main plasma components (in a fusion reactor there will also be T), sometimes He (which is artificially introduced and studied because of its importance as a fusion product in a future reactor), and the impurities (e.g. C, O and others) which are always present in their various ionization stages. The spectroscopical investigations cover wavelength regions ranging from the visible to the soft X-ray domain and include the continuous (e.g. bremsstrahlung) spectrum as well as the numerous spectral lines emitted from the different particle species. Conventional measurements of the latter concentrate on line intensities, intensity ratios and line shapes, to obtain information mainly on the densities and temperatures of the emitter species and, if possible, also on their macroscopic velocity (e.g. to measure plasma rotation). In connection with numerical plasma modeling, these data provide information on such properties as impurity transport, radiation losses and deviations from ionization equilibrium; they may also serve to adjust the electron density and temperature results obtained from other diagnostics, because the electrons are decisive for the upper level populations of the spectral lines observed. Besides line intensity and line shape measurements, polarization measurements of Zeeman components have also been suggested[2] to determine the magnetic field direction and thus the current distribution in the tokamak plasma.

As a consequence of the large temperatures and small densities in tokamaks spectral line shapes are in most cases largely dominated by Doppler broadening, at least in conventional emission spectroscopy. However, the line shapes are usually no pure Gauss profiles as they are distorted by various additional effects, such as the fine structure, the overlapping of lines originating from different isotopes (e.g. in the case of hydrogen), the Zeeman effect, the motional Stark effect, and non-thermal upper level populations. These effects do not always appear simultaneously, but some of them mostly occur. Unfortunately, they are usually too small to yield valuable information on their causes, but they are often too large to be neglected.

Conventional spectroscopy in tokamaks has been passive, in the sense that it has consisted of just analyzing the radiation which is emitted by the plasma in any case, without any further provision. Such techniques suffer from bad localization, because the signals measured are integrated over the line of sight. Also, the radiation recorded this way comes mainly from the plasma edge (except the one emitted from high-Z impurities), so it usually yields little information on the plasma center. More recent concepts, in contrast, are based on active spectroscopy, which consists of using neutral particle or laser beams. These techniques have the advantage that they localize a point on the sightline (its crossing point with the beam) so that the plasma parameters are measured for a given location in space. There are additional advantages, depending on the method used, which will be described in the forthcoming two chapters. The first of these will give a short overview on active spectroscopy using neutral particle beams, the second will present a more detailed description of the recently developed concept of Doppler-free two-photon excitation of hydrogen Lyman-α.

ACTIVE NEUTRAL BEAM SPECTROSCOPY

There are two main motivations for injecting fast neutral particle beams into tokamak plasmas: first they constitute an efficient means of heating the plasma (and also of driving additional currents and flows), second they provide a variety of diagnostic facilities. The short overview presented here will be restricted to hydrogen (H or D) beams, although purely diagnostic beams of e.g. helium[3], lithium[4,5] and barium[6] atoms have also been used in the past.

Active hydrogen beam spectroscopy has been employed in a variety of tokamaks (JET, DIII, PBX, TFTR) [7-16]. The current practice consists of analyzing spectral lines (often in the visible, e.g. hydrogen H_α or impurity lines between Rydberg states) emitted from a given point in the beam. As already mentioned, the local character of the information obtained this way constitutes an important advantage over conventional (passive) spectroscopy. Further significant advantages arise from the abundance of information that measurements of this kind are able to provide. Indeed, since the fast (typically 50-100 keV/amu) beam atoms penetrate deeply into the plasma and interact with the plasma ions via charge exchange (CX), hydrogen, helium and light impurity spectral line radiation can also be generated in those inner plasma regions where these elements are normally completely ionized. This is substantial for exploring the important regions around the magnetic axis which otherwise are hardly accessible to spectroscopic investigation. Furthermore, as part of the injected beam atoms are excited by the plasma, these emit line radiation as well and yield further valuable information on the plasma properties.

An eloquent example of the above is the H_α spectrum induced by the beam. The two mentioned effects of the plasma-beam interaction (charge exchange and excitation of beam atoms) entail two entirely different contributions to the spectrum, one of them being emitted by the plasma, the other one by the beam (see Fig.2 for a calculated example).

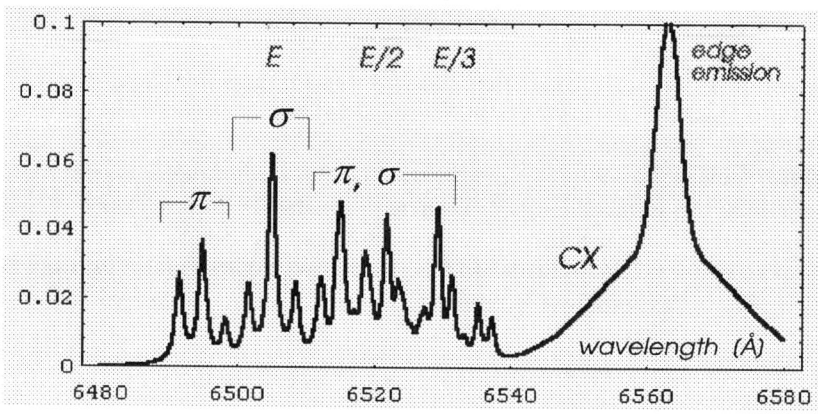

Figure 2. Example of a calculated H_α spectrum showing the Doppler-shifted Stark patterns of the three energy groups E, $E/2$, $E/3$ ($E=100$ keV) of a neutral deuterium beam and the CX- and cold edge line of the plasma.

The beam contribution, emitted by the injected fast atoms after excitation by the plasma, is composed of three spectral features originating from the three velocity classes ($v = v_0$, $v_0/\sqrt{2}$, $v_0/\sqrt{3}$ corresponding to accelerated H^+, H_2^+, H_3^+) present in the beam. These large velocities have two effects on each of the features, namely a very large Doppler shift (typically several tens of Å's) and a large Stark splitting (of about 4 Å). The latter is due to the electric $\vec{v} \times \vec{B}$ field which acts on the atoms when these move across the magnetic field $\vec{B}$. It is noteworthy that the Zeeman effect is usually negligible against this large Stark effect. In most cases the three Stark multiplets partially overlap, so that diagnostic information is best extracted from the outest one (left side in the example of Fig.2).

The Stark components are of particular interest for magnetic field measurements. Indeed, from the distance between neighboring components (which is proportional to $|\vec{v} \times \vec{B}|$) and from their polarization (which is partly σ, partly π with respect to $\vec{v} \times \vec{B}$) one gets valuable local information on the strength and the direction of the magnetic field component $B_\perp$ perpendicular to $\vec{v}$. This, in turn, helps to determine the current distribution which is of great importance for the confinement and the stability of the plasma. Moreover, the intensity of the beam emission provides information on the mean ion charge number Z_{eff} (which is a measure for the average impurity content in the plasma, see its definition in Eq.(7) below) because the impurity ions participate efficiently in the beam attenuation as well as in the excitation of the beam atoms[10]. Further, fluctuations in the intensity of the emitted radiation can be traced to fluctuations of the charged particle densities (turbulence) and thus provide a means of diagnosing the latter[16].

The plasma contribution to the H_α spectrum is emitted by atoms which are created from hydrogen ions of the plasma by charge exchange interaction with the beam atoms. This contribution (right side of Fig.2) is therefore a spectral line which is essentially Doppler broadened, with the total intensity, width and shift reflecting, respectively, the density, temperature and macroscopic velocity (plasma rotation) of the hydrogen ions. Since a possible rotation velocity is very much smaller than the beam velocities, the center of this line is close to the unshifted H_α position, implying that the line is usually well separated from the three Stark multiplets emitted from the beam atoms. It generally overlaps with a narrower H_α line emitted from the cold plasma edge. Difficulties in the determination of ion density, temperature and plasma rotation from the line shape arise from this latter fact as well as from the various distorting effects mentioned in the introduction. A further problem, especially for high temperatures, comes from the velocity-dependence of the charge exchange cross section, which may cause the velocity distribution of the atoms resulting from charge exchange to be different from that of the ions. A comprehensive discussion of all these perturbing effects has been given in Ref.8 in the general context of charge exchange recombination spectroscopy.

As already mentioned, the beam also induces charge exchange reactions with the impurity ions to which the above considerations on hydrogen can thus be readily extended. By exploring as systematically as possible all impurities present in the plasma (which is commonly done with the hydrogenic ion species using visible transitions between Rydberg states), the local densities, temperatures and rotation velocities of the different ion species can be determined individually. Gathering all the densities, in particular, allows one to reconstruct the radial profile of the effective ion charge number Z_{eff}. Of course, the various perturbing effects mentioned above are found here again. One of the major difficulties in the case of higher Z impurities is that the charge exchange processes do not lead to a thermal population of the excited states of the recombining ions. The reason for this is that the recombination occurs preferentially into certain highly excited states with high orbital quantum number l. In a simplified picture, the subsequent de-excitation then follows a cascade of $\Delta l = \mp 1$ transitions and thus results in an overpopulation of the high orbital quantum numbers in all levels. In a more realistic picture, additional complications must be taken into account. One of these is that neighboring levels are mixed by the motional Stark and Zeeman effect so that the $\Delta l = \mp 1$ selection rule is no longer applicable. Another complication arises from collisional redistribution between neighboring levels. Both effects not only modify the line under consideration but also generate new lines which can be observed.

DOPPLER-FREE TWO-PHOTON INDUCED FLUORESCENCE IN LYMAN-α

1. INTRODUCTORY REMARKS

One of the limitations of active beam spectroscopy is the large Doppler broadening of the spectral lines investigated. Although, as has been illustrated above, Doppler profiles provide valuable data on the properties of the plasma, they unfortunately also hide much of the information contained in the spectra. For common hydrogen lines e.g. they cover several Angstroms and thus obscure any structure which is below the Angstrom scale. This includes the isotopic differences between the spectral positions of lines emitted by hydrogen, deuterium and tritium (which might be exploited for isotope-specific density determinations) as well as any structure associated with sub-Doppler line profiles (as e.g. Stark profiles and other interesting features which will be discussed below).

In high resolution laser spectroscopy, several techniques have been developed which allow the elimination of linear Doppler broadening[17]. As has been suggested previously[18-20], one of these techniques, namely Doppler-free two-photon excitation (and detection of the subsequent fluorescence) may also be applied to the hydrogen isotopes in tokamak plasmas, though with more difficulty and, as will be outlined below, under rather severe conditions imposed on the performances of the laser spectrometer used.

The spectral line of choice for the two-photon transition is Lyman-α, in spite of the fact that the fluorescence occurs in the VUV. One reason is that excitation in this line occurs from the groundstate; this is indeed the only state which has still some population in the deeper plasma regions and, moreover, is the state which is really representative of the neutral atom density. A further reason is that Lyman-α is the only transition resisting collisional depolarization[21] which is an important aspect when one wants to determine the magnetic field direction (q-profile).

Finally, Lyman-α has a simple structure and is readily accessible to theoretical interpretation.

A possible alternative is Doppler-free two-photon excitation in Lyman-β and detection of the fluorescence in Balmer-α. This method keeps the advantage of giving direct access to the groundstate density and avoids the experimental difficulties encountered in the VUV. However, it suffers from almost complete collisional depolarization, implies a lower fluorescence yield and is theoretically more involved. Moreover, sufficient laser pulse energies at the wavelength required ($\lambda \approx 2051$ Å) do not seem to be available at present.

Doppler-free two-photon excitation in the *1S-2S* transition of Lyman-α has been employed in high resolution spectroscopy with regard to fundamental atomic physics[22]. When applying this technique to tokamak diagnostics, however, the physical situation is entirely different and requires complete reconsideration of the method in all its theoretical and experimental details. The most obvious new features are the much lower neutral atom densities, the much higher temperatures and, above all, the fact that the optically active atoms are embedded in a plasma and subject to external magnetic and electric fields. A detailed analysis of the situation has been given in Ref.19, mainly with regard to the determination of neutral atom densities and of the magnetic field direction. The calculations presented there include estimates of the expected experimental errors and an optimization of the physical and instrumental parameters to be chosen for the experiment. An important ingredient in the formalism used is the two-photon absorption profile; in Ref.19 this has been assumed to be mainly shaped by Stark broadening due to atom-ion collisions. A complementary analysis[20] has shown that there are also important cases where the dominant broadening arises from the relativistic transverse Doppler effect and from the motional Stark shifts of the fine structure components. Also the hyperfine structure has been found to be non-negligible in certain cases. A summary of the main results obtained in this study will be given in the following sections.

2. PRINCIPLE OF THE PROCESS

The elimination of linear Doppler broadening in two-photon absorption is achieved by using two counter-propagating laser beams with wave vectors $\vec{k}$ and $-\vec{k}$ and with angular frequency $c|\vec{k}| = \omega/2$, where ω is close to the resonance frequency of the transition[17]. In the rest frame of an atom moving with velocity $\vec{v}$, the Doppler shifted frequencies of the two lasers appear as $(\omega/2 \mp \vec{k} \cdot \vec{v})/\sqrt{1-(v/c)^2}$. If each beam provides one of the two photons absorbed, the sum frequency is $\omega/\sqrt{1-(v/c)^2}$, which is close to ω to the extent that the relativistic correction is small. As a consequence, all atoms, regardless of their velocity, are in resonance with the two laser beams and have an equal chance to be excited; this implies both a substantial increase of the upper level population relative to ordinary two-photon excitation processes and a considerable narrowing of the absorption profile, in which the linear Doppler broadening is now eliminated. In the case of hydrogen Lyman-α in a tokamak plasma, the reduction of the line width amounts to three orders of magnitude and entails complete separation of the three two-photon absorption lines of the isotopes H, D and T. By tuning the laser frequency over the three resonances, the shapes of these lines can, in principle, be measured too, provided the laser bandwidth is sufficiently small.

Two-photon induced Lyman-α fluorescence in a tokamak plasma is conveniently looked at as an atomic three-photon process (two photons absorbed, one emitted) occurring under the influence of the surrounding plasma and the external fields. For a qualitative understanding of the various intervening effects, these are best introduced by steps, as is schematically done in Fig.3. The left side shows the first two ($n=1$ and $n=2$) energy levels of an unperturbed hydrogen atom at rest, as described by the non-relativistic Schrödinger equation. Associated with the upper level are the four spherical eigenstates $|l,m>$, where l and m denote, as usual, the orbital and magnetic quantum numbers. (The principal quantum number n has been omitted for notational simplification.) Introducing the strong magnetic field of a tokamak (3-5 T) leads to a strong Zeeman splitting in the Paschen-Back limit. The two central levels (corresponding to states $|0,0>$ and $|1,0>$) are unshifted and coincide in reality; they have been separated for clarity in the figure. In accordance with the high temperatures prevailing in the tokamak plasma ($10^7 - 10^8 K$), the atom is then given a large velocity $\vec{v}$; this induces a strong electric Lorentz field (equal to $\vec{v} \times \vec{B}$) in the atom's rest frame and leads to a motional Stark effect in addition to the Paschen-Back effect. As a consequence, the outer levels spread further out by an amount which is comparable in order of magnitude to the original Zeeman splitting. Further, with the exception of state $|1,0>$ which remains unaffected, the new eigenstates are all superpositions of the three states $|0,0>$, $|1,\pm 1>$, implying that they contain both S and P admixtures. Hence they can correspond with the groundstate through both two-photon and one-photon transitions, i.e. they can be excited by absorption of two photons and subsequently decay by emission of one fluorescence photon.

After two-photon excitation into one of the three states mentioned has taken place, the atom can either emit a Lyman-α photon immediately, or it can first undergo collisional transitions to the other levels and radiate from one of these. According to the selection rules $\Delta m = \pm 1$ for σ-polarization and $\Delta m = 0$ for π-polarization (referred to the magnetic field direction), the fluorescence is π-polarized when arising predominantly from the $|1,0>$ state and σ-polarized when arising predominantly from the other three states. Both cases can occur and may, in principle, be exploited for measurements of the magnetic field direction.

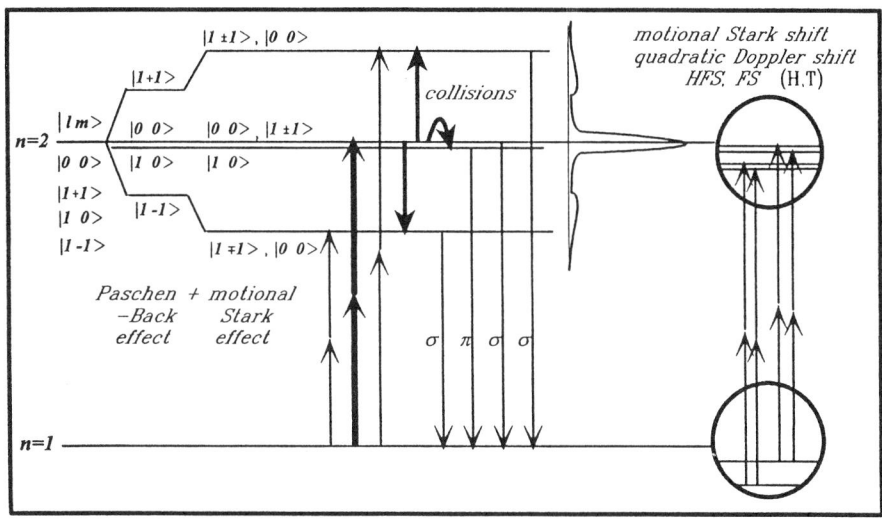

Figure 3. Schematic illustration of the physical processes characterizing the generation of two-photon induced fluorescence in a tokamak plasma

While all the three states containing an S-admixture can in principle be reached by the two-photon transition, there is actually only one of them, namely the unshifted one, for which this is possible in practice. The reason is that the positions of the outer levels are strongly velocity-dependent (as a consequence of the velocity-dependence of the electric $\vec{v} \times \vec{B}$ field) and give thus rise to a large inhomogeneous broadening of the corresponding side components of the two-photon absorption profile. The unshifted level, in contrast, is not affected by this broadening; as sketched in Fig.3, it gives rise to a very sharp peak in the middle of the two-photon absorption profile and is thus accessible to efficient excitation.

In the framework of the non-relativistic Schrödinger equation and the non-relativistic atomic motion considered so far, the line shape of the narrow central peak is uniquely determined by natural broadening, Stark broadening (mainly caused by collisions with the plasma ions), and, in case of very strong laser fields, ionization broadening (due to ionization from the excited state by a third photon). As already mentioned, however, there are situations where the broadening mechanisms originating from relativistic effects on the dynamics of both the bound electron and the atomic center of gravity are comparable or even more important than those mentioned above. For an accurate evaluation of the line shape it is thus indispensable to take these effects into account. Furthermore, besides the fine structure thus included, also hyperfine structure turns out to be non-negligible; this is in fact larger than the residual spin-orbit splitting left over from the Paschen-

Back effect. The right side of Fig.3 symbolizes the additional effects mentioned above for the isotopes H and T (nuclear spin $I = 1/2$).

3. VELOCITY-DEPENDENCE AND SPLITTING OF ENERGY LEVELS

For general values of I it can be shown that the two energy levels of interest (the ground level and the one excited in the two-photon transition) are each globally shifted and split into $2(2I+1)$ sublevels. Taken in the atom's rest frame and to a precision better than $10^8\,Hz$, these sublevels are given by the expressions

$$E_1(\beta;m_s,m_I) = -E_{Ry} - B(g_e\mu_e m_s + g_n\mu_p m_I)$$
$$-\alpha^2 E_{Ry}\left(\frac{1}{4} - 6.357\alpha - \frac{8M_e}{3M_p}g_n m_s m_I\right), \qquad (1)$$

$$E_2(\beta;m_s,m_I) = -\frac{1}{4}E_{Ry} - B(g_e\mu_e m_s + g_n\mu_p m_I)$$
$$-\alpha^2 E_{Ry}\left(\frac{7}{192} + \frac{1/24 - 0.813\alpha}{1+\beta^2} - \frac{M_e}{3M_p}\frac{1-\beta^2/20}{1+\beta^2}g_n m_s m_I\right)$$
$$-\frac{\alpha^4 E_{Ry}^2}{\hbar\omega_L}\frac{\beta^2(0.00299 - 0.00043(1+\beta^2))}{(1+\beta^2)^2}m_s. \qquad (2)$$

Here, the electronic and nuclear spin quantum numbers m_s and m_I take the values $m_s = \pm 1/2$, $m_I = -I, -I+1, \ldots, +I$. In the further quantities defined in the above equations subscripts e, p, n refer to the electron, proton and nucleus, respectively. α denotes the fine structure constant, $E_{Ry} = e^4 M_r / 2(4\pi\varepsilon_0\hbar)^2$ the Rydberg energy, where $M_r = M_e M_n / (M_e + M_n)$ is the reduced mass of the electron-nucleus pair. B is the magnetic field strength and $\omega_L = eB/2M_e$ the Larmor frequency. $\beta = 6v_\perp / \alpha c$ is a measure for the atomic velocity component $v_\perp$ perpendicular to the magnetic field. $\mu_e = e\hbar/2M_e$ and $\mu_p = e\hbar/2M_p$ are the electron and proton magnetons, g_e and g_n the electron and nucleus g-factors. We have $g_e = -2.0023$ for the electron, $I = 1/2$ and $g_n = 5.586$ for the proton, $I = 1$ and $g_n = 0.8576$ for the deuteron, $I = 1/2$ and $g_n = 5.958$ for the triton.

The second terms in the first lines of Eqs.(1) and (2) (those proportional to B) represent the coupling of the magnetic moments of the electronic and nuclear spins with the external magnetic field. Since these are identical for the upper and the lower level and do thus not enter the frequencies of the various line components, they have been omitted in the schematic at the right side of Fig.3. This is the reason why all sublevels under the "magnifiers" are shifted downward and why there are only two (instead of four) sublevels under the lower "magnifier".

The second lines in Eqs.(1) and (2) include the relativistic and Lamb shift corrections to the Schrödinger equation as well as the coupling between the electronic and nuclear magnetic moments. The last line in Eq.(3) represents the residual spin-orbit coupling left over from the (incomplete) Paschen-Back effect.

There are $2(2I+1)$ possible transitions between the lower and upper sublevels (hence four in the example of Fig.3), namely those conserving the spin quantum numbers m_s and m_I. The transition frequencies depend all on the variable β and thus on the atomic velocity $\vec{v}$. The essential cause for this dependence is the S-P fine structure mixing induced by the electric $\vec{v} \times \vec{B}$ field; this is again a motional Stark effect, though much smaller than the one discussed before in the framework of the non-relativistic Schrödinger equation. For β varying from 0 to ∞, the transition frequencies increase from a given value (depending on the isotope considered) to another value lying about 6 GHz higher. The upper set of curves in Fig.4 demonstrates this behavior for hydrogen ^{1}H; it is quite similar for tritium. For deuterium one has three (instead of two) doublet curves, which however lie much closer together. Note that the curves refer to the transition frequencies

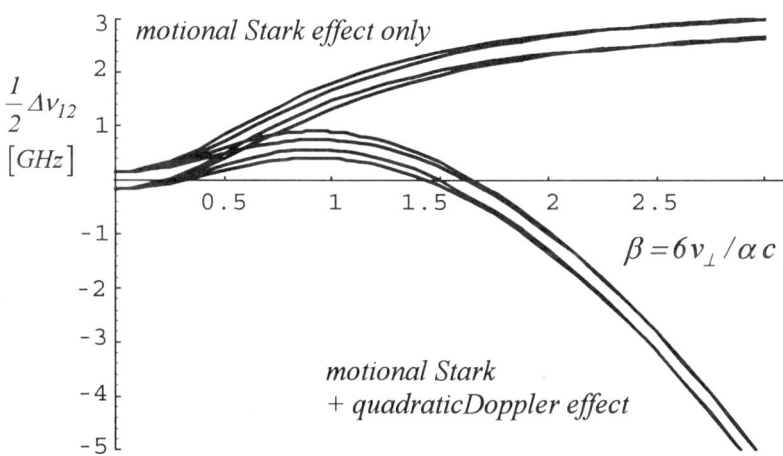

Figure 4. Velocity-dependence of the central two-photon transition frequencies of hydrogen (^{1}H) Lyman-α due to motional Stark and quadratic Doppler effect.

divided by two (in order to correspond to the frequency domain of the laser). They actually represent $\Delta v_{12}/2$, where $\Delta v_{12} = v_{12} - v_0$ with v_{12} being the transition frequency defined by $v_{12}(\beta;m_s,m_I) = (E_2(\beta;m_s,m_I) - E_1(\beta;m_s,m_I))/h$ and v_0 being a reference frequency which is formally obtained from v_{12} by putting $\beta = m_s = m_I = 0$; we have $v_0 = 2.4660563 \; 10^{15} Hz$ for H, $v_0 = 2.4667273 \; 10^{15} Hz$ for D, and $v_0 = 2.4669541 \; 10^{15} Hz$ for T. The main splitting of the curves in Fig.4 is due to hyperfine structure, the smaller splitting to residual spin-orbit coupling (assuming $B = 4\,T$).

The lower set of curves in Fig.4 represents the transition frequencies in the laboratory frame, i.e. includes the relativistic (quadratic) Doppler effect from the atomic motion in addition to the motional Stark effect. For simplicity the velocity has, in the figure, been chosen perpendicular to $\vec{B}$. It is seen that the velocity-dependence due to the quadratic Doppler effect goes into the direction opposite to that of the motional Stark effect: while the latter tends to raise the curves, the quadratic Doppler effect tends to lower them. This is a very fortunate circumstance, as it entails a considerable mutual cancellation of the two motional level shifts and thus a substantial reduction of the total inhomogeneous line broadening (typically from more than *3GHz* when either effect acts alone to less than *1GHz* when both effects act together, as can be seen from inspection of Fig.4). As a consequence, the two-photon excitation probability is significantly larger than expected from either the motional Stark broadening or the quadratic Doppler broadening.

4. TWO-PHOTON ABSORPTION PROFILE FOR A GIVEN VELOCITY

The total broadening (and shift) of the two-photon absorption profile results from a combination of all the broadening mechanisms mentioned above. For not too strong laser intensities, ionization broadening can be neglected. Hence, in addition to the inhomogeneous broadening mechanisms discussed in the preceding paragraph, one has essentially natural and Stark broadening. For all details in the theoretical treatment of the latter we refer to Refs. 19 and 21, where it has been shown that Stark broadening in tokamak plasmas arises predominantly from atom-ion interactions and can be treated in the impact theory. It is noteworthy that neither Stark broadening (which in usual cases is approximately[25] homogeneous) nor natural broadening can be considered as homogeneous in the case investigated here. The main reason is that the Stark width of the two-photon excited state is a velocity-dependent mixture of *S*-widths and *P*-widths, the latter being by about a factor of three smaller than the former[19,21]. Similarly, the natural width is a velocity-dependent mixture of the finite natural *P*-width and the negligibly small *S*-width. In spite of this the term "inhomogeneous broadening" will, for notational simplicity, in the following be reserved to motional Stark and Doppler broadening.

The natural and Stark two-photon absorption profile for a given atomic velocity $\vec{v}$ is a sum of Lorentzians, corresponding to the set of two-photon transitions involved. In the laboratory frame it appears as

$$L_c(\vec{v}, v) = \frac{1}{2(2I+1)} \sum_{m_s=-1/2}^{+1/2} \sum_{m_I=-I}^{+I} \frac{W(\beta)/\pi}{\left(v - \frac{1}{2} v_{12}(\beta; m_s, m_I)\sqrt{1-(v/c)^2}\right)^2 + W(\beta)^2} \quad (3)$$

where subscript c indicates that the profile is restricted to the central transitions (thus excluding the two very broad side components associated with the outer levels). The independent variable v denotes the frequency of the two photons if it is common to both; in case that the two wave vectors do not exactly cancel each other, one has $v = (v_1 + v_2)/2 - (\vec{k}_1 + \vec{k}_2) \cdot \vec{v}/4\pi$.

As the transitions differ only by their electronic and nuclear spin variables, the Lorentzians have all the same width. (Note that the shift vanishes for the transitions considered here.) In the following we indicate an expression for the halfwidth which is slightly more general and accurate than the one given in Refs. 19,21. The numerical factors in the "Coulomb logarithms" have indeed been determined here according to Refs. 26,27, implying an exact solution of the Schrödinger equation for strong collisions[26] and a treatment of screening based on the dielectric function of the plasma[27,28]. In addition, the presence of impurity ions has been accounted for in a more consistent way. Under the assumption that there is a main ion species (e.g. hydrogen) which is distinctly more abundant than the others, a simplified expression for the halfwidth (HWHM) can be written as

$$W(\beta) = \frac{1}{4\pi(1+\beta^2)}\left[\beta^2 R + Z_{eff}^* \frac{12 n_e \hbar^2}{M_e^2 Q_m}\sqrt{\frac{2\pi M_A}{k_B T_A}}\left(\ln\Lambda_1 + \beta^2 \ln\Lambda_2\right)\right] \quad (4)$$

where

$$\Lambda_1 = \frac{0.0468 Q_m^5 M_e^5 \varepsilon_0^{1/2}(k_B T_A)^3}{B^2 M_A^{5/2}(e\hbar Z_{eff})^3 [n_e(1+Z_{eff})]^{1/2}(1+\beta^2)}, \quad \Lambda_2 = \frac{1.56 Q_m^2 M_e^2 k_B T_A}{B M_A e\hbar Z_{eff}(1+\beta^2)^{1/2}} \quad (5)$$

$$Q_m = \sqrt{1 + \frac{T_m M_A}{T_A M_m}}, \quad Z_{eff}^* = Z_{eff} + (Q_m - 1)\frac{(Z_{eff}-1)Z_{eff}^{imp}}{Z_{eff}^{imp}-1} \quad (6)$$

$$Z_{eff} = \frac{1}{n_e}\sum_i n_i Z_i^2, \quad Z_{eff}^{imp} = \frac{\sum_{i\neq m} n_i Z_i^2}{\sum_{i\neq m} n_i Z_i}. \quad (7)$$

16   Spectroscopic Diagnostics in Tokamaks

Here, $R = 3.13 \ 10^8 \ rad/s$ is the natural halfwidth of a pure P-state, n_e the electron density; M_A, M_m and T_A, T_m are the masses and temperatures of the optically active atom species and the main ion species, respectively.

The electron density n_e being usually known from other diagnostics, the quantity to be diagnosed from the Stark width is the effective ion charge number Z_{eff}, as defined in the first of Eqs.(7). As can be seen from Eq.(4), the Stark contribution to W is not directly proportional to Z_{eff}, but rather to Z_{eff}^* which is related to Z_{eff} by the second of Eqs.(6). This relation requires, in principle, knowledge of a further quantity, namely the mean charge number Z_{eff}^{imp} of all ions except the main species. However, since this number frequently refers to the impurities and is thus usually large, it has little influence in the expression for Z_{eff}^*. It is therefore legitimate to write the simpler relation

$$Z_{eff}^* = Q_m(Z_{eff} - 1) + 1 \qquad (8)$$

where, furthermore, Q_m can often be replaced with $\sqrt{2}$ (see the first of Eqs.(6)).

5. FLUORESCENCE YIELD AS A FUNCTION OF LASER DETUNING

The quantity of interest for the diagnostics under consideration is the total (i.e. frequency-integrated) fluorescence yield as a function of the laser frequency, observed in a given solid angle and direction and possibly analyzed for a given polarization. Assuming the process to be stationary (which is of course an approximation for a pulsed experiment) and far from saturation, it follows from the formalism of Ref.19 and from the above equations that the number of detected photons is given by

$$N = \frac{n_A E_{L1} E_{L2} \Omega l}{4\pi \Delta t r^2 (\pi c h \nu_0)^2} \mu \varepsilon S \left\langle |\vec{e}_1 \cdot \vec{e}_2|^2 F_c\left(\frac{1}{2}(\nu_1 + \nu_2) - \frac{1}{4\pi}(\vec{k}_1 + \vec{k}_2) \cdot \vec{v}\right)\right\rangle \qquad (9)$$

with

$$F_c(\nu) = 3\int f(\vec{v})\left(X(\beta) + Y(\beta)\sin^2 \vartheta \cos^2 \psi\right) L_c(\vec{v}, \nu) d\vec{v}. \qquad (10)$$

In Eq.(9), n_A is the atomic groundstate density, E_{L1} and E_{L2} are the pulse energies of the two laser beams, Δt is the pulse duration and observation time (both assumed equal), r and l are the radius and length of the observed volume (assumed cylindrical) of the overlapping beams, and Ω is the solid angle of the observed light cone. Further, μ and ε are loss factors; the first includes losses of the excited atoms due to ionization by a third photon, to stimulated two-photon emission and to escape from the observed volume; the second represents all transmission losses of the optical detection system, including the quantum efficiency of the detector. (Note that owing to the large Doppler width of the emitted fluorescence, losses due to radiation trapping in the plasma are usually negligible.) The quantity S in the above equation measures the *1S - 2S* transition strength[19]; it equals $8.07\,10^{-26}\,s^{-2}$. $\langle\cdots\rangle$ denotes an average over the modes of the two laser beams, characterized by the frequencies v_1, v_2, wave vectors $\vec{k}_1$, $\vec{k}_2$ and polarization vectors $\vec{e}_1$, $\vec{e}_2$. The function F_c occurring in the average is defined in Eq.(10); it represents an average over atomic velocities, $f(\vec{v})$ denoting the velocity distribution and $L_c(\vec{v},v)$ being the velocity-dependent two-photon absorption profile defined in Eq.(3). The quantities $X(\beta)$ and $Y(\beta)$ are complicated but weakly dependent functions of β, whose explicit expressions can be found in Ref.19; physically, they arise from collisional redistribution among the excited levels; they have indeed been calculated using the general theory of the collisional redistribution of light[23,24]. ϑ is the angle between the line of sight and the magnetic field direction; both span the "plane of sight" with respect to which ψ denotes the angle of a linear polarization analyzer.

When the spread of the laser modes can be neglected and the scalar product of the polarization vectors has the maximal value one (e.g. linear polarization in the same plane or circular polarization in the opposite sense), the average in Eq.(9) simply yields $F_c(v_L)$, where v_L is the mean laser frequency. This function then represents the characteristic line shape exhibiting the dependence of the fluorescence on the detuning of the laser. The dependence of this line shape on ϑ and ψ (see Eq.(10)), reflects respectively, the anisotropy and polarization induced by the magnetic field. Summing over two orthogonal polarizations and averaging (or integrating) over the directions of observation leads to a considerable simplification of Eq.(10). Indeed, it can be shown that

$$\sum_{\psi}\left(X+Y\overline{\sin^2\vartheta}\cos^2\psi\right)=2X+\frac{2}{3}Y=\frac{1}{3(1+\beta^2)} \qquad (11)$$

where the bar denotes averaging over ϑ. Note that in the limit of very high charged particle densities this result is also obtained without the above averaging, because the collisional transitions among the excited levels then lead to a complete equipartition of their populations, with the consequence that the fluorescence is isotropic and unpolarized from the outset. This implies[29] $X\to 1/6(1+\beta^2)$, $Y\to 0$.

The factor $1/(1+\beta^2)$ in the rhs. of Eq.(11) is nothing but the fraction of S-admixture to the central excited states considered for the two-photon excitation. Since the two outer states share the remaining S-contribution, corresponding to a fraction of $\beta^2 / 2(1+\beta^2)$ for each of them, the profile in Eq.(10) acquires a simple normalization rule if the two broad side components are included. Indeed, on summing over the polarizations, averaging over the directions of observation and integrating over the laser frequency one obtains

$$\sum_\psi \int \overline{F(v_L)} dv_L = 1 \qquad (12)$$

where subscript c has now been omitted to indicate that the side components due the outer levels are included. This normalization plays no role in practice, however, because it will neither be possible to collect the fluorescence over all directions where it is emitted, nor to observe it for laser detunings in the broad side components. It is more realistic, therefore, to evaluate Eqs.(9,10) for given values of the angle ϑ, possibly summed over the polarizations ψ to account for the absence of a polarization analyzer.

Figures 5 and 6 show various results for the function $F_c(v_L)/6$, expressed in terms of the laser detuning $\Delta v_L = v_L - v_0$. Except for Fig.5 (c), all of them are calculated for $\vartheta = \pi/2$ (observation perpendicular to the magnetic field) and except for Fig.5 (d), all of them are summed over the two polarizations. The electron density and the magnetic field have the fixed values $n_e = 3\,10^{19} m^{-3}$ and $B = 4T$, respectively. Protons are taken as the main ion species causing Stark broadening when the atomic species is hydrogen ^1H, and deuterium ions are taken when the atomic species is deuterium or tritium. In Fig.5 the effective ion charge number is throughout $Z_{eff} = 1$ and each case is presented for the two temperatures $T = 10^6 K$ and $T = 3\,10^7 K$. Figs.5 (a) and (b) demonstrate the influence of the splitting due to hyperfine and fine structure (both combined), the former being by far the most important. While the effect is negligibly small in the case of deuterium (owing to the small nuclear magnetic moment), it is seen to be considerable in the case of tritium and can there even be resolved if the laser bandwidth is sufficiently small. It is noteworthy that it persists even at temperatures as high as $T = 3\,10^7 K$.

Figs.5 (c) and (d) show the dependence of the fluorescence on the direction of observation (c) and on the direction of polarization (d). Both dependences have the same physical origin, namely π or σ dipole radiation, the dipoles being oriented with respect to the magnetic field. The solid and dotted lines correspond, respectively, to observation perpendicular and parallel to the magnetic field in graph (c) and to a linear polarizer turned parallel and perpendicular to the magnetic field in graph (d). As an interesting fact we note that for low temperatures the solid curves lie above the dotted ones while for high temperatures they lie below them. The reason is that for low temperatures the electric $\vec{v} \times \vec{B}$ field is, on the average, very small, implying that the central two-photon excited states have only small P-admixtures. Hence they are almost metastable, with the consequence that the fluorescence is mainly induced by collisions to the neighboring states. The nearest

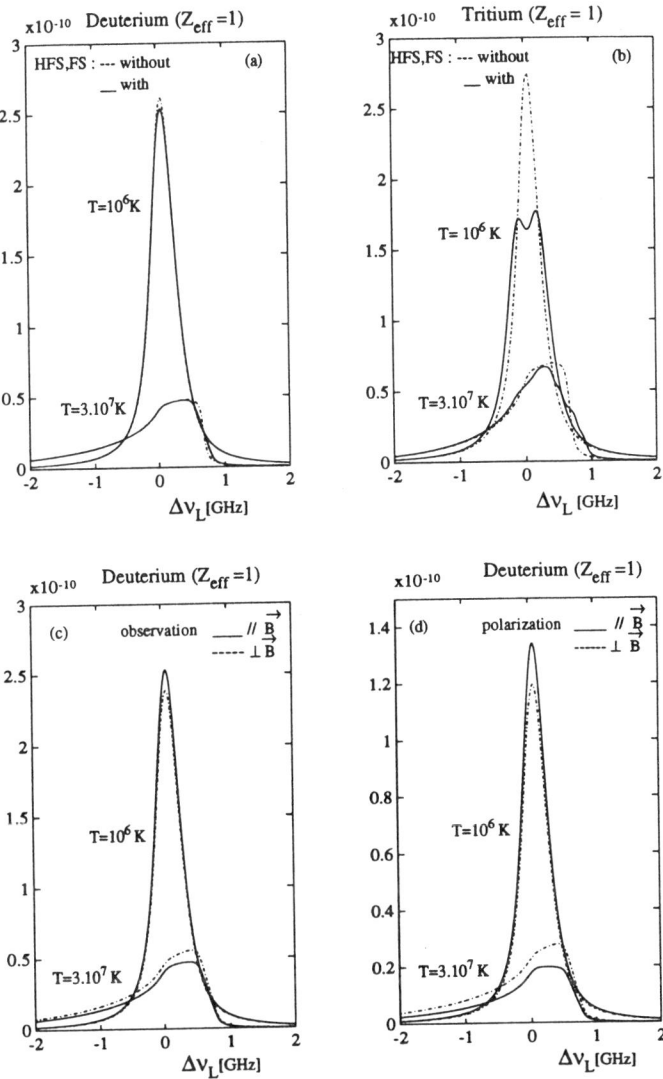

Figure 5. Line shape factor $F_c(\nu_L)/6$ of the fluorescence yield as a function of the laser detuning. Graphs (a) and (b) show the combined effect of hyperfine and fine structure splitting for the cases of deuterium and tritium. Graphs (c) and (d) display, for the case of deuterium, the anisotropy of the fluorescence and its dependence on the direction of linear polarization.

and thus favored one is the unperturbed central P-state which gives rise to π-polarization. Photons polarized parallel to the magnetic field direction therefore predominate over those polarized perpendicular to this direction. For high temperatures the situation is reversed, because the two-photon excited central states have then large P-admixtures which enable them to radiate preferably a σ-photon by direct decay to the groundstates instead of first undergoing collisional transitions. Photons polarized perpendicular to the magnetic field direction therefore prevail. The above explanation includes the situation in Fig.5 (c) because this simply reflects the directional characteristics of π or σ dipole radiation.

The dependences displayed in Figs.5 (c) and (d) are moderate, though distinct; they would still be larger for lower electron densities. The dependence on the direction of polarization can, in principle, be exploited for measurements of the magnetic field direction[19].

Figure 6 illustrates the sensitivity of the fluorescence to changes of the temperature and of the effective ion charge number Z_{eff}. The profiles in graphs (a) and (b) have been calculated for hydrogen ^{1}H, those in graphs (c) and (d) for tritium. The effective ion charge number is $Z_{eff} = 1$ in (a) and (c) and $Z_{eff} = 3$ in (b) and (d). In each graph the temperature changes by a factor of about three from one curve to the other. The strong dependence on the temperature in (a) and (c) is due to the predominance of inhomogeneous (motional Stark and quadratic Doppler) broadening over plasma Stark broadening; it can in such cases be used for measurements of the temperature of the atomic species under consideration. The dependence is weaker in (b) and (d), because plasma Stark broadening here provides a significant contribution and prevails for the lower temperatures. As is demonstrated in graph (d), there is no clear tendency of the temperature-dependence in cases where the contribution of inhomogeneous broadening is small: increasing the temperature monotonously from $10^6 K$ to $10^7 K$ leads to first raising (see the curve for $T = 3\ 10^6 K$) and then lowering the fluorescence. The reason for this behavior lies in the somewhat complicated manner in which the cancellation between motional Stark and quadratic Doppler effect takes place for small velocities (see Fig.3).

The sensitivity to changes of Z_{eff} can, to a certain extent, be seen by confronting the left ($Z_{eff} = 1$) with the right ($Z_{eff} = 3$) graphs and comparing curves with the same temperature. As expected, there is a strong Z_{eff}-dependence for the relatively low temperatures (below $10^7 K$, say) where Stark broadening prevails. These cases are therefore suitable for measurements of the effective ion charge number Z_{eff}. Whether a given situation is more suitable for temperature measurements or for measurements of Z_{eff} can to some extent be inferred from the form of the line shapes; these are strongly asymmetric when inhomogeneous

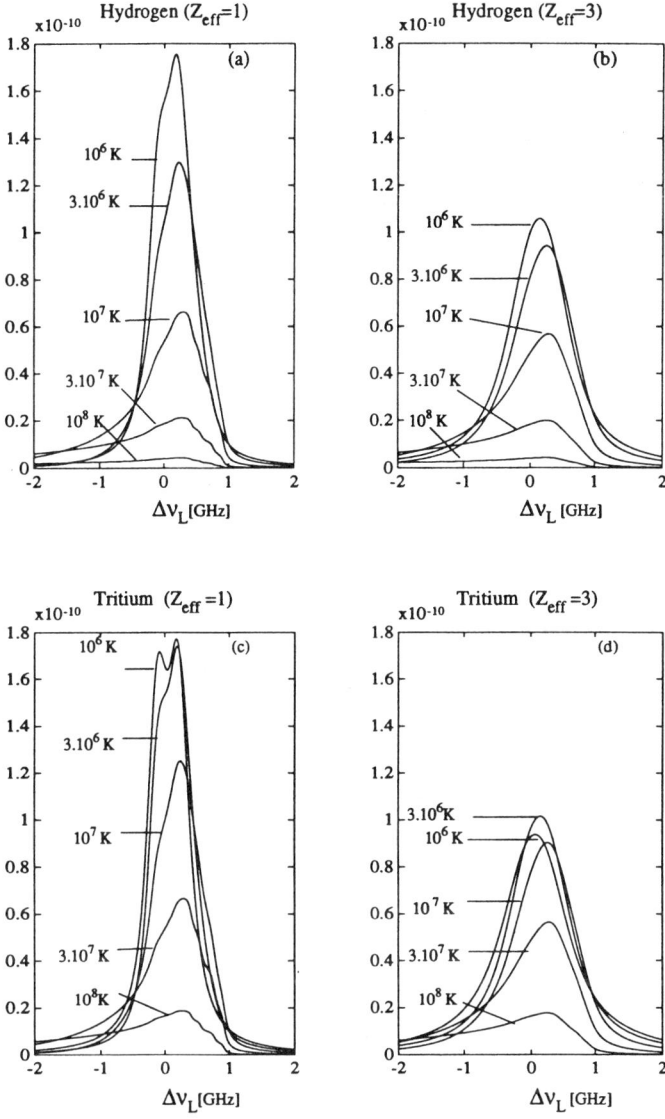

Figure 6. Line shape factor $F_c(\nu_L)/6$ of the fluorescence yield as a function of the laser detuning. The influence of the temperature and of the effective ion charge number Z_{eff} is displayed for the cases of hydrogen ^{1}H (graphs (a) and (b)) and tritium (graphs (c) and (d)).

broadening prevails and resemble Lorentzians (like the low-temperature curves in the right graphs) when plasma Stark broadening is dominant.

A general feature of all line shapes presented in Figs. 5 and 6 is the drastic decrease of the peak intensity when the temperatures become high. This is only partly due to the larger inhomogeneous broadening. A second reason is that in the limit $T \to \infty$ the central excited states gradually loose their S-contribution (which is picked up by the outer levels) and tend to pure P-states. Hence the two-photon absorption rate and thus the total fluorescence yield tend to vanish in this limit.

6. REQUIREMENTS FOR DIAGNOSTIC APPLICATION

In order for the diagnostic to be reliable with regard to signal-to-noise ratio and spectral resolution, the number of detected photons (as given by Eq.(9)) should be as large as possible and its dependence on the laser detuning should reveal, as precisely as possible, the structure of the two-photon absorption profile. As one of the prerequisites for achieving this, the average line shape in Eq.(9) should be as narrow as possible, which means that the spread of radiation modes in the laser beams should, if possible, be kept so small that no appreciable broadening is added to that of the line shape $F_c(\nu_L)$. The width of the latter being typically on the order of $1\,GHz$ (see Figs.5 and 6), it is thus desirable that both the laser bandwidth and the broadening introduced by the beam divergence be kept below this value.

The broadening from the beam divergence is due to the residual linear Doppler effects which occur because the wave vectors $\vec{k}_1$, $\vec{k}_2$ of any pair of photons taken from the two beams do not cancel exactly. Under the simplifying assumption that the sums $\vec{k}_1 + \vec{k}_2$ are in the plane perpendicular to the beams and obey a Gaussian distribution with variance $(\pi \nu_0 \gamma / c)^2$ (γ being the angle characterizing the divergence), the residual Doppler profile is easily shown to be

$$L_D(\nu) = \frac{2c}{\gamma \nu_0 \sqrt{k_B T_A / M_A}} \exp\left(-\frac{4c|\nu - \nu_0/2|}{\gamma \nu_0 \sqrt{k_B T_A / M_A}}\right). \qquad (13)$$

Choosing hydrogen ^1H at $T = 10^7 K$ (or deuterium at $T = 2\,10^7 K$) as a typical example, the width (FWHM) of this profile stays below $1\,GHz$ for divergence angles $\gamma < 1\,mrad$.

If the widths introduced by the laser are not distinctly narrower than the width of $F_c(\nu_L)$, they may be accounted for by folding F_c with both the frequency profile of the laser and (as an approximation) the profile given in Eq.(13).

Finally, inspection of Eq.(9) shows that it is also desirable to have a small spread of polarizations. Indeed, to avoid a noticeable decrease of the peak intensity, the scalar products of the polarization vectors should remain close to unity.

As for the first factor in the rhs. of Eq.(9), its maximization is limited by various physical and technical constraints; for a detailed discussion we refer to Ref.19. One of the important points is that the most restrictive lower limit for the focal radius r arises from the ionization of the excited atom by a third photon. Indeed, for steadily decreasing r the laser intensity (for given pulse energy) eventually increases to values where the probability for ionization exceeds the one for spontaneous decay. Denoting by r_{ci} the critical focal radius where both are equal and assuming the same pulse energy $E_{L1} = E_{L2} = E_L$ for the two beams, we have approximately[19] $\Delta t\, r_{ci}^2 = 7.8\, 10^{-13} E_L [SI]$. Putting this into Eq.(9) causes the photon number to be directly proportional to the laser pulse energy E_L.

For an evaluation of the pulse energies which are necessary to make the measurements sufficiently precise, we write the expected experimental error (noise-to-signal) for the observed photon numbers as

$$\frac{\Delta N}{N} = \frac{\sqrt{N + N_p}}{N} \qquad (14)$$

where N_p is the number of background photons entering the measurement. This expression assumes that the shot noise of the detector is the main error source and obeys Poisson statistics. Essentially in accordance with Ref.19 we evaluate the error for $N_p = 2 r l \Omega \Delta t \cdot 10^{15} m^{-2} sr^{-1} s^{-1}$, $r = r_{ci}$, $l = 0.15 m$, $\Omega = 10^{-2} sr$, $\Delta t = 3\, ns$, $\mu = 0.5$, $\varepsilon = 0.13$. The photon number N is calculated from Eq.(9) on assuming the line shape (right factor) to be unaffected by laser broadening; it is taken at peak intensity. Table 1 presents the results obtained for deuterium atoms. The relative error $\Delta N / N$ is indicated there for $n_e = 3 \cdot 10^{19} m^{-3}$, $Z_{eff} = 1$ and for different values of the neutral particle density n_A and the temperature T, as well as for different laser pulse energies E_L (the first column, 30 mJ, corresponding to energies presently available, the second one, 100 mJ, to energies announced for the near future[30]).

It is seen from these results that for not too high temperatures ($T < 3 \cdot 10^7 K$, say) laser pulse energies of 30-100 mJ should already be sufficient for measurements of neutral particle density variations down to densities on the order of $10^{14} m^{-3}$. An excellent precision is attained for higher neutral particle densities or/and lower temperatures (say $n_A \geq 10^{15} m^{-3}$, $T \leq 10^7 K$), such as are prevailing in the plasma edge or in divertor plasmas. Here, absolute density measurements should also be possible (from the combined knowledge of the area and the shape of F_c). Comparison with the numbers in parentheses, which have been calculated for

$N_p = 0$, shows that considerably lower values of the neutral particle density could be measured if there were some way to suppress the Lyman-α background radiation. Since this is primarily emitted from the cold plasma edge, implying a spectral width smaller than that of the fluorescence line, a filter with a bandwidth adapted to this narrow line might improve significantly the signal-to-background ratio.

$n_A [\text{m}^{-3}]$	$T[\text{K}]$	laser pulse energy E_L		
		30 mJ	100 mJ	300 mJ
10^{13}	10^8	80 (0.4)	30 (0.2)	10 (0.1)
	10^7	6 (0.1)	2 (0.06)	1 (0.03)
	10^6	3 (0.1)	1 (0.04)	0.6 (0.02)
10^{14}	10^8	8 (0.1)	3 (0.06)	1 (0.04)
	10^7	0.6 (0.03)	0.2 (0.02)	0.1 (0.01)
	10^6	0.3 (0.02)	0.1 (0.01)	0.06 (0.01)
10^{15}	10^8	0.8 (0.04)	0.3 (0.02)	0.1 (0.01)
	10^7	0.06 (0.01)	0.02 (0.006)	0.01 (0.003)
	10^6	0.03 (0.01)	0.01 (0.004)	0.006 (0.003)

Table 1. Relative error (noise-to-signal) in the case of deuterium for different plasma parameters and laser energies. The numbers in parentheses give the error without background radiation.

Finally it is worth mentioning that in those situations where the precision is very high, it is also of interest to consider the use of lasers whose spectral bandwidth or beam divergence is not small in the sense specified above. While loosing the ability of resolving the atomic line shape F_c and exploring the plasma environment this way, such laser systems would have other advantages which we briefly discuss in the following.

When the laser bandwidth is large compared to the width of F_c, the average line shape in the rhs. of Eq.(9) essentially represents the frequency profile of the laser (because for small beam divergence the average is essentially the convolution of the laser profile with F_c). Measurements of the neutral density, for instance, then require only knowledge of this laser profile and of the normalization constant of F_c, without involving the detailed line shape of the latter. This signifies a substantial simplification of the calibration problems, as has been discussed in Ref. 19.

Further, when the beam divergence is so large that the residual (linear) Doppler width significantly exceeds the width of F_c, the average line shape in the rhs. of Eq.(9) essentially represents the residual Doppler profile, as given approximately in Eq.(13). Whether or not this situation prevails can be judged from the particular form of the profile which is characterized by a sharp central maximum (unsteady derivative in the approximation (13)). If it prevails, the residual Doppler profile can be exploited for a measurement of the neutrals' temperature in a more comfortable way than by the use of the profile of F_c (which possibly necessitates a disentanglement of Stark and quadratic Doppler broadening).

A still different way of measuring the temperature (and even velocity distribution) of the neutrals would consist of introducing artificially a modest amount of linear Doppler broadening by imposing a small angle between the two beam directions. The width of the resulting linear Doppler profile could then be monitored and adapted to the actual situation by varying the angle between the beams (which should have small divergence in this case). The profile would be a direct image of the distribution of the velocity component pointing perpendicular to the beams and lying in the plane spanned by them. By turning this plane, the velocity distribution could be exploited in all directions perpendicular to the beam.

7. PRESENT EXPERIMENTAL SITUATION

Until recently, the applicability of Doppler-free two-photon induced Lyman-α fluorescence was rather limited in practice because the existing light sources did not yet have all the characteristics required for performing the diagnostics efficiently. In particular, the weak laser pulse energies available so far would have confined the diagnostics to regions with high neutral particle densities, such as prevailing in the plasma edge of a tokamak. Recent progress in laser technology achieved at PTB in Berlin, however, has led to a significantly more optimistic situation[30]. A high quality solid-state laser spectrometer has been developed there, which is tunable at 243 nm, provides pulses of 3 ns duration with energies of 30 mJ (very probably extensible to 100 mJ in the near future), has a bandwidth smaller than 500 MHz and a beam divergence less than 1 mrad. As follows from the discussion in Sec.6, these characteristics should allow one to perform at least part of the measurements outlined in this paper. As a preparing step for future application in tokamak diagnostics, the spectrometer has very recently been installed on the PSI-1 plasma generator (a magnetically stabilized high-current arc) at the Berlin division of IPP Garching[31], where the generation of two-photon excited Lyman-α fluorescence is presently being tested.

8. CONCLUDING REMARKS

In the preceding sections we have outlined the great diagnostic potential which (linear) Doppler-free two-photon excitation of Lyman-α fluorescence has for tokamak plasmas. We have shown that a wealth of information can be deduced from the characteristics of the fluorescence, such as its strength, its functional dependence on the laser detuning and its polarization. Examples are the isotope-specific measurement of the hydrogen groundstate densities from the respective strengths of the fluorescence lines, the determination of the neutrals' temperature from inhomogeneous (essentially quadratic Doppler) line shapes or the determination of the effective ion charge number Z_{eff} from Stark profiles, and finally the determination of the magnetic field direction from the polarization. In addition, by using slightly crossing (instead of exactly counter-propagating) laser beams, measurements of the temperature and, moreover, of the atomic velocity distribution, can be based on the "artificial" linear Doppler profiles thus introduced. The present status of laser technology should allow application of at least part of these diagnostics in the near future.

REFERENCES

1. C.Keane, 12th ICSLS, these Proceedings
2. U.Feldman, J.F.Seely, N.R.Sheeley Jr., S.Suckewer and A.M.Title, J.Appl.Phys. 56, 2512 (1984)
3. Nobuhiro Nishino, Rev.Sci.Instrum. 62, 2695 (1991)
4. K.McCormick et al.(20 authors), Phys.Rev.Lett. 58, 491 (1987)
5. W.P.West, D.M.Thomas, J.S.de Grassie and S.B.Zheng, Phys.Rev.Lett. 58, 2758 (1987)
6. M.Wickham, S.Fornaca, N.H.Lazar and N.Rynn, Rev.Sci.Instrum. 55, 1748, (1984)
7. A.Boileau, M.G.von Hellermann, L.D.Horton, J.Spence and H.P.Summers, Plasma Phys.Control.Fusion 31, 779 (1989)
8. R.J.Fonck, in *Spectral Line* Shapes, Vol.6 (10th ICSLS, Ed. L.Frommhold and J.W.Keto), AIP, New York 1990
9. A.Boileau, M.G.von Hellermann, W.Mandl, H.P.Summers, H.Weisen and A.Zinoviev, J.Phys.B, 22, L145 (1989)
10. W.Mandl, R.C.Wolf, M.G.von Hellermann and H.P.Summers, Plasma Phys.Control.Fusion 35, 1373 (1993)
11. D.Wroblewski, K.H.Burrel, L.L.Lao, P.Politzer and W.P.West, Rev.Sci.Instrum. 61, 3552 (1990)
12. D.Wroblewski and L.L.Lao, Rev.Sci.Instrum. 63, 5140 (1992)
13. F.M.Levinton, R.J.Fonck, G.M.Gammel, R.Kaita, H.W.Kugel, E.T.Powel and D.W.Roberts, Phys. Rev. Lett. 63, 2060 (1989)
14. F.M.Levinton, Rev.Sci.Instrum. 63, 5157 (1992)

17. M.D.Levenson and S.S.Kano, *Introduction to Nonlinear Laser Spectroscopy*, Academic Press, San Diego 1988
18. D.Voslamber, *9th ICSLS - Abstracts of Contributed Papers* (Ed. J.Szudy), Nicholas Copernicus University Press, Torun 1988; and Verhandl. DPG (VI) 23, P 5.25 (1988) and 27, P 19.47 (1992)
19. D.Voslamber, Report EUR-CEA-FC-1387 (1990)
20. D.Voslamber, Verhandl. DPG (VI) 28, P 15.66 (1993)
21. D.Voslamber, Report EUR-CEA-FC-1342 (1988)
22. T.W.Hänsch, S.A.Lee, R.Wallenstein and C.Wieman, Phys.Rev.Lett. 34, 307 (1975)
23. D.Voslamber and J.-B.Yelnik, Phys.Rev.Lett. 41, 1233 (1978)
24. D.Voslamber, "Multiphoton processes in the presence of collisions" in *Quantum Optics and Spectroscopy* (Ed. J.Fiutak, J.Mizerski, M.Zukowski), Nova Science Publishers, Commack, N.Y. 1993
25. J.Seidel, Z.Naturforsch. 34a, 1385 (1979)
26. H.Pfennig, Z.Naturforsch. 26a, 1071 (1971)
27. H.Capes, Thèse d'État, Université Paris-Sud (1980)
28. H.Capes and D.Voslamber, Phys.Rev. A5, 2528 (1972)
29. The corresponding limit of X in Ref.19 is erroneous and should be replaced with the expression given here.
30. K.Grützmacher, I.de la Rosa, J.Seidel, A.Steiger and D.Voslamber, in *Proceedings of the 21st EPS Conference on Controlled Fusion and Plasma Physics* (Montpellier 1994), Vol.18 B, part III, p.1290
31. H.Behrendt, W.Bohmeyer, L.Dietrich, G.Fußmann, H.Greuner, H.Grote, M.Kammeyer, P.Kornejew, E.Pasch, in *Proceedings of the 21st EPS Conference on Controlled Fusion and Plasma Physics* (Montpellier 1994), Vol.18 B, part III, p.1328

LINE PROFILE MEASUREMENTS IN MICROWAVE PRODUCED DISCHARGES

H. Schlüter, S. Hirsch, G. Himmel

Ruhr-University of Bochum, Institute of Experimental Physics II
44780 Bochum, Germany

Pulsed high-power microwave discharges up to atmospheric pressure represent suitable media for coherent and non-coherent light sources. The calibration of these discharges is hampered by the use of closed microwave propagation structures. In fact, material probes and antennas fed through the wall of the waveguide may strongly interfere with the electric field distribution of the propagating mode. This suggests the application of emission spectroscopic diagnostics. Nevertheless, even at medium pressures and in the absence of any considerable influence of collisions between neutral atoms on the line shape (e.g. van-der-Waals broadening, Dicke-narrowing), it does not suffice to use standard techniques for the analysis of the Stark broadening. There are cases where the effects of the fluctuating plasma microfield are noticeably superimposed by the dynamic Stark effect due to the oscillating microwave field.[1] At the highest electron densities achieved in the present study, however, this dynamic Stark effect is masked by the Stark broadening so that its influence proves negligible indeed.

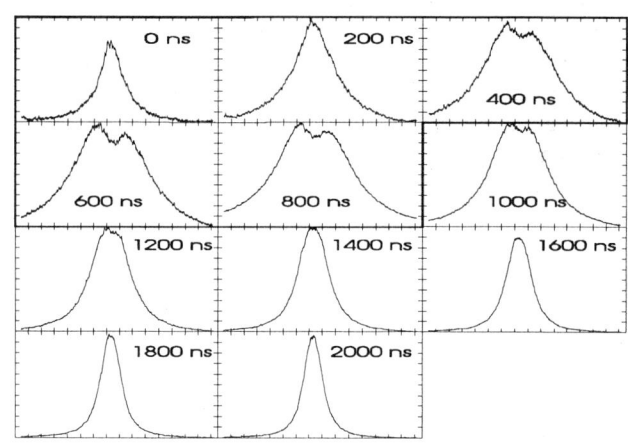

Fig.1. Time-resolved measurements of the Balmer line Dβ.
p[He] = 500 Torr;
Time resolution is 180 ns;
"0 ns" refers to the first 180 ns after ignition of the discharge.
The wavelength scale is given in 0.1 nm ; the wavelength increases to the right side.

Aiming at maximal electron density, a 250 kW magnetron (f = 9.4 GHz, pulse duration = 10^{-6} s, repetition rate = 500 s^{-1}) was used as a microwave source. Up to 95% of the incident microwave power was coupled into the discharge by means of a plasma microwave resonator.[2] The working gas was helium (10 Torr < p[He] < 500 Torr) with a small amount of deuterium (always less than 1%) for diagnostic purposes. The line spectra were taken side-on from the capillary discharge (inner tube diameter = 4 mm) which was orientated transversely to the axis of the waveguide and was observed through a small hole in the wall of the waveguide. Time resolved line spectra (see Fig.1) were obtained with a multichannel spectrum analyzer or alternatively with a gated photon counter attached to the exit slit of the spectrometer. Due to the skin effect a nearly rectangular radial density profile was expected. This assumption was supported by pictures taken end-on with a gated CCD-camera showing a nearly homogeneous distribution of plasma luminosity.

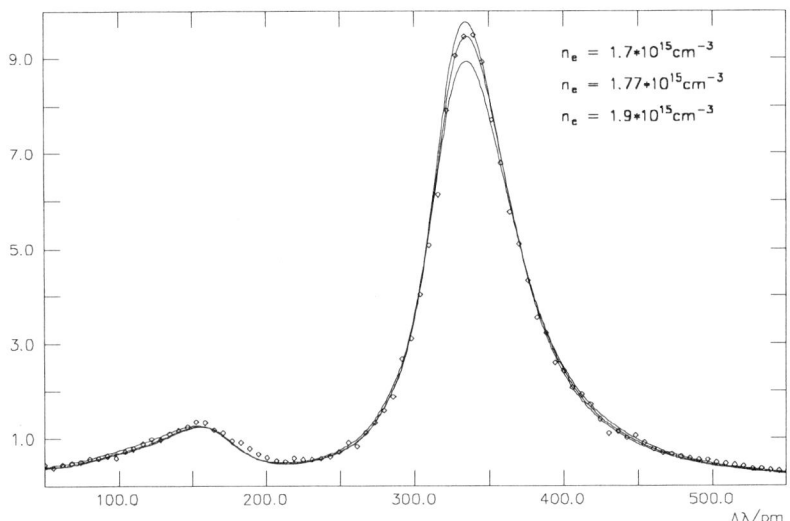

Fig.2. Measured profile of the HeI-line at 447.1 nm ($4^3D, 4^3F - 2^3P$) compared with MMM-profiles for three different electron densities; $T_e = 1$ eV, $T_{He} = 500$ °K, p[He] = 30 Torr.

Figure 2 depicts a measured HeI line profile recorded just at the beginning of the postdischarge when the microwave pulse elapsed. For the theoretical description including ion dynamic corrections the model microfield method[3] (MMM) was chosen. The electron temperature was determined from the intensity ratio of the Balmer lines $D\alpha$ and $D\beta$. The instrumental line profile and the Doppler profile were incorporated through the convolution with the corresponding MMM-profile. Fitting theoretical profiles to the measurements, an accuracy better than 15% is achievable in the determination of the electron density. Thus the maximum electron densities ranged between 10^{15} cm^{-3} and 10^{16} cm^{-3}, depending on the pressure and the microwave power. This shows that highly overdense plasmas may be obtained with densities being about 5000 times higher than the critical density.

1. G. Boehm, Microwave Discharges, Fundamentals and Applications (NATO ASI series B: Physics, vol. 302, Plenum Press, N.Y., 1992), p. 215.
2. S. Hirsch, G. Himmel, G. Ströhlein, Proc. XXIth Int. Conf. on Phen. in Ion. Gases (Bochum), Contr. Papers 1, 59 (1993).
3. A. Mazure, C. Goldbach, G.Z. Nollez, Z.Naturforsch. 34a, 773 (1979).

EXPERIMENTAL STUDY OF TWO HEI LINES WITH FORBIDDEN COMPONENTS.

C. Pérez, I. de la Rosa, J.A. Aparicio, S. Mar and M.A. Gigosos
Departamento de Optica. Universidad de Valladolid (Spain).

Plasma diagnostic methods based on lines with forbidden components are widely used at present. This kind of lines have a complicated structure sensitive to electron density variations. In this work, two HeI lines with forbidden components, 447.1 and 492.2 nm, have been investigated with a pulsed discharge in pure Helium gas. Electron density, accurately measured by interferometry at two wavelengths, 632.8 and 488.0 nm, ranges from 1.54×10^{22} to 1.56×10^{23} m^{-3}. The HeI temperatures were estimated to be in the interval 20000 to 25000 K. For both spectral lines, different profile parameters are examined in order to be used for plasma diagnostic purposes. Available experimental data [1-9] are included for comparisons.

The experimentally determined line parameters, which give information about these complex profiles are: $\lambda_{1/2}$, the total half width; S, the separation between peaks of both lines; F/A, the ratio of the intensities of forbidden to allowed component and, D/A, the ratio of the deep between the two lines and the intensity of the allowed component.

The pulse arc, that was used as a plasma source, is described elsewhere[10]. Experimental results for the 447.1 nm line are in the following four figures. In them, the above mentioned parameters are plotted versus the electron density, and they are shown in comparison with other experimental data. Some of them have been taken from the figures in their references, therefore they may exhibit sligth inxactness.

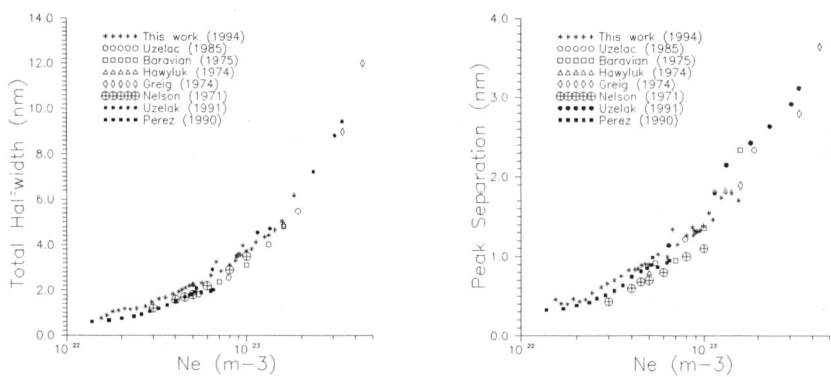

Fig. 1 and 2. Total Halfwidth and Peak separation for the 447.1 nm line.

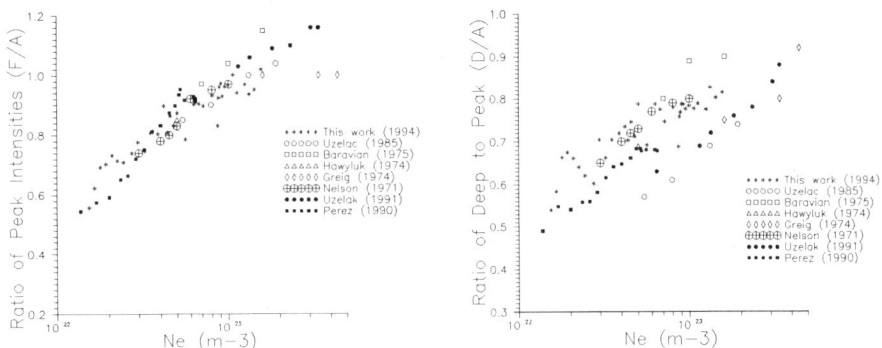

Fig. 3 and 4. Ratio of Peak intensities and Deep to Peak for the 447.1 nm line.

The most remarkable fact in these figures is that all the experimental results, from the different authors, show the same functional dependency of the parameter under study with the electron density. Nevertheless, each group of data shows some kind of relative displacement referred to others. This displacement seems be due to something inherent to each experiment, the reason for that is unknown up to now. These figures suggest that all the line parameters seem to be suitable for Helium plasma diagnostics, bar the D parameter that exhibits a considerable scatter of the data. Results for the other line provide similar features, but with less previous experimental data for comparisons.

Authors thank the Dirección General de Investigación Científica y Técnica for its finantial support under contract No.Pb/90/0353.

References.
(1) N.I. Uzelac, I. Stefanovic and N. Konjevic, J. Quant. Spectrosc. Radiat. Transfer 46, 5, 447 (1991).
(2) N.I. Uzelac and N. Konjevic, Phys. Rev. A, 33, 1349 (1985).
(3) G. Baravian, J. Bretagne, J. Godart and G. Sultan, Z. Physik. B 20, 260 (1975).
(4) R.J. Hawryluk, G. Bekefi and E.V. George, Phys. Rev. A 10, 265 (1974).
(5) J.G. Greig, L.A. Jones and R.W. Lee, Phys. Rev. A 9, 44 (1974).
(6) R.H. Nelson and A.J. Barnard, J. Quant. Spectrosc. Radiat. Transfer 11, 161 (1971).
(7) B.T. Vujicic, XVI International Conference on Phenomena in Ionized Gases, 4, Düsseldorf (1983).
(8) C.S. Diatta, Thèse, Université d'Orleans (1977).
(9) C. Pérez, Tesis, Universidad de Valladolid (1990).
(10) M.A. Gigosos, S. Mar, C. Pérez and I. de la Rosa, Phys. Rev. E, 449, 2, 1575 (1994).

Fine-Structure Asymmetry of Doppler-Free Stark-Broadened Two-Photon Polarization Line Profiles of Hydrogen Lyman-α

J. Seidel, A. Steiger, and K. Grützmacher

Physikalisch-Technische Bundesanstalt, Institut Berlin, D-10587 Berlin, Germany

With Doppler-free two-photon polarization spectroscopy[1], the Stark-broadened line profiles of the hydrogen and deuterium 1s– 2s transitions could recently be measured[2,3] for the first time in arc plasmas with electron densities of 10^{22} m^{-3} and below, where the profile of the corresponding one-photon Lyman-α line is completely determined by Doppler broadening. These measurements were done to investigate the ion dynamic effects, which are theoretically predicted to dominate the Stark broadening at low plasma densities[4]. In fact, the deep depression in the very line centre of two-photon Lyman-α, which is characteristic for static ion broadening (cf. Fig. 1), is absent in the measured profiles. (Two-photon Lyman-α has no central Stark component and resembles one-photon Lyman-β.)

However, the measured profiles did show a distinctly asymmetric depression in their central region (with the red peak higher than the blue one), or at least a "shoulder" on the blue side of the central peak (Figs. 1 and 2). The physical reason for this was unclear at first, but reconsideration[3] of calculations[5] for the static Stark components of one-photon Lyman-α indicated that the asymmetry might be due to fine-structure splitting, which still produces a distinct intensity asymmetry of the Stark components even if field strengths are already high enough to result in symmetric linear Stark shifts large compared to the fine-structure splitting.

Therefore, Stark-broadening computer simulations of the Doppler-free two-photon polarization line profile of Lyman-α have now been carried out which take account of fine-structure splitting. The simulations employed the collision-time technique, which yields the correct ion impact limit at low plasma densities[6]. Electron broadening was taken in the impact limit of the unified theory. Results are compared to experimental data in Figs. 1 and 2.

The theoretical results are rather insensitive to temperature, but do depend on the electron density (Fig. 2). In the experiment[3], the electron density was determined from the (static-ion) Stark broadening[7] (excluding the line centres) of the Balmer-β and -γ lines (estimated uncertainty 15 %, including plasma inhomogeneity). The electron densities which are to be inferred from Figs. 1 and 2 are about 10 to 25 % higher than these values. This indicates that the uncertainty of low electron densities determined from Balmer-line Stark profiles is considerably larger than the uncertainty of higher electron densities[8].

In summary, our combined experimental-theoretical work demonstrates that both the measurement and the computer-simulation technique are well understood and yield reliable results, if care is taken to account for all effects which may be essential under the plasma conditions considered in each case.

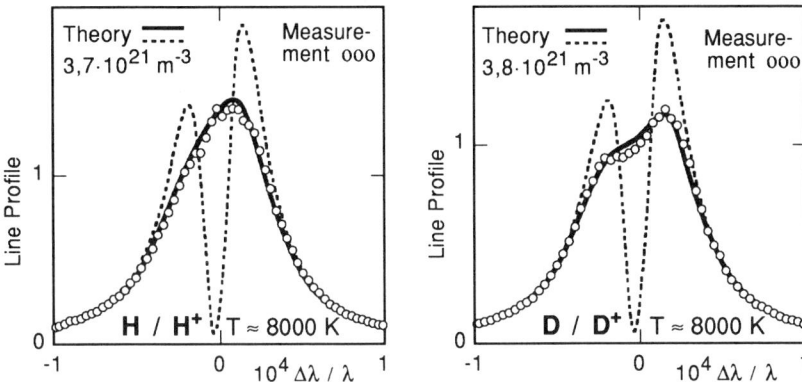

FIG.1. Stark broadening of the Doppler-free two-photon polarization line shape of the 1s – 2s Lyman-α transition. Experimental and theoretical results for a hydrogen plasma and a deuterium plasma. *Full lines*: computer simulation including ion dynamic effects; *broken lines*: simulation for static ions (the same for H and D). The simulations account for the fine structure splitting of the upper level. Their results are normalized to the same area as the experimental data; only the line centre is shown in the figure.

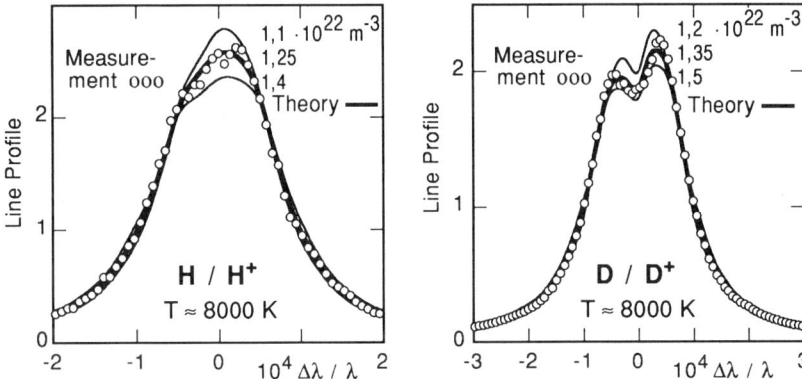

FIG.2. Same as Fig. 1, but at higher electron density and shown for extended wavelength ranges. Instead of the static ions result, simulation results including ion dynamic effects are shown for different electron densities

References
1. K. Danzmann, K. Grützmacher, B. Wende, Phys. Rev. Lett. **57**, 2151 (1986)
2. A. Steiger, K. Grützmacher, *Spectral Line Shapes, Vol. 7*, eds. R. Stamm and B. Talin (Nova Science Publ. 1993), p. 141
3. A. Steiger, Dissertation (Technische Universität Berlin 1993)
4. J. Seidel, Z. Naturforsch. **32a**, 1207 (1977)
   R. Stamm, E.W. Smith, B. Talin, Phys. Rev. A **30**, 2039 (1984)
5. G. Lüders, Ann. Phys. [6] **8**, 301 (1951)
6. J. Seidel, *Spectral Line Shapes, Vol. 6*, eds. L. Frommhold and J.W. Keto (AIP Conf. Proc. 216, New York 1990), p. 98
7. C.R. Vidal, J. Cooper, E.W. Smith, Astrophys. J. Suppl. **25**, 37 (1973)
8. V. Helbig, Contrib. Plasma Phys. **31**, 183 (1991)

IMPROVED THEORY OF ION IMPACT BROADENING IN MAGNETIZED PLASMAS AND ITS DIAGNOSTIC APPLICATIONS

E.Oks and A.Derevianko
Physics Dept., Auburn University, Auburn, AL 36849, USA

A new spectroscopic method for <u>local</u> measurements of an effective charge $Z_{eff}=\Sigma Z_i^2 N_i/N_e$ of high temperature plasmas (e.g., in tokamaks) was proposed[1] and recently implemented[2]. The idea of the method is the following. Under conditions typical of tokamaks ($N_e \sim 10^{13} cm^{-3}$, $T_a \geq 10^2 eV$) Stark broadening of hydrogen spectral lines (SL) is controlled by <u>ion impact broadening (IIB)</u>. Then a resulting homogeneous Stark width $\gamma_s \propto \Sigma Z_i^2 N_i/\langle V_i \rangle$ is a linear function of Z_{eff} (summation includes both the major plasma component (H^+ or D^+) and impurity ions). A homogeneous width may be determined experimentally using a saturation of an optical transition n'-n in hydrogen by laser light with a small spectral width. Indeed, in this case the observed fluorescence SL shape has the Voigt profile with a dispersion component of a halfwidth $\Gamma_B \approx (d_{12}E_0/\hbar)[(\gamma_{r,n}^{-1}+\gamma_{r,n'}^{-1})\gamma_{s,nn'}]^{1/2} \gg \gamma_{s,nn'}$, where $\gamma_{r,n}$ and $\gamma_{r,n'}$ are the radiative widths of the levels n and n', $\gamma_{s,nn'} \gg \gamma_{r,n}+\gamma_{r,n'}$, and E_0 is the laser amplitude. This significant enhancement of the width of dispersive component makes it comparable to the Doppler width and allows to extract it from the observed Voigt profile and thus to measure Z_{eff}.

Whether or not this method will be of a broad practical use is contingent upon developing a <u>detailed, consistent theory of IIB</u> as opposed to rough estimates of IIB employed in [1,2]. This development constitutes a subject of the present paper.

Our theory of IIB has two distinctive features. First, we take into account a strong magnetic field $B \geq 1T$ characteristic for tokamaks. This magnetic field splits up a hydrogen level of a principal quantum number n into 2n-1 equidistant multiplets of degeneracy g ranging from 1 to 2n-1. The point is that the states inside each degenerate multiplet are interconnected by nondiagonal matrix elements $\Phi^{n.d.}$ of the IIB-operator what can result in drastic modifications of impact widths of SL components[3].

Second, our theory of IIB is a <u>generalized</u> semiclassical theory that is free from a shortcoming of the standard semiclassical theories of Stark broadening which were intrinsically divergent at small impact parameters. A <u>convergency</u> of our theory is achieved in spirit of paper[4]: an influence on a radiating atom of the projection $[\mathbf{E}(t)]_B$ of an ion-produced field $\mathbf{E}(t)$ onto the direction of the magnetic field $\mathbf{B}$ is taken into account <u>exactly</u> (on equal footing with $\mathbf{B}$) while two other components of the field $\mathbf{E}(t)$ are allowed for in the impact approximation. The property making it possible is that interactions with both a magnetic field **B** and an electric field **E**∥**B** can be diagonalized in parabolic quantization.

Our impact width γ_s consists of two physically different parts. An adiabatic contribution $\gamma_s^{a.}$ proportional to matrix elements of z^2 ($Oz \parallel \mathbf{B}$) is calculated along the lines of the "old" adiabatic theory[5]. Our result for nonadiabatic contribution (proportional to matrix elements of x^2+y^2) may be presented as a substitution of the standard broadening function a ($Re\Phi \propto a$) by a generalized (yet elementary) function a(Y) depending on a new dimensionless parameter $Y=0.160 n Z_i B(T)/T_i(eV)$. Physically, Y is the ratio of a magnetic splitting $\Delta\omega_B$ to the ion Weisskopf frequency Ω_W. Thus γ_s in the reality has the form $\gamma_s = \gamma_s^{n.a.} + \gamma_s^{a.}$, $\gamma_s^{n.a.} \propto \sum Z_i^2 N_i a(Y(Z_i))/<V_a>$, $\gamma_s^{a.} \propto \sum Z_i^2 N_i/<V_a>$. The point is that perturbers of different $Z_i=1-30$ result in a <u>significantly different $a(Y(Z_i))$</u> making it impossible to express $\gamma_s^{n.a.}$ as a function of Z_{eff}. In other words, the same width γ_s may correspond to values Z_{eff} differing by 30-40% (contingent upon relative contributions of light and heavy impurities). Thus practically all hydrogen SL <u>cannot</u> be employed for the method[1,2] (particularly, Z_{eff} inferred in[2] by using SL H_α is regrettably irrelevant).

Fortunately we have found one exception from this rule. For the unshifted (π) component of SL L_α corresponding to transition from the doublet (100), (010) to the ground state (000), <u>the nonadiabatic width vanishes</u> as a result of a mutual cancellation of diagonal and nondiagonal matrix elements of x^2+y^2 of the doublet. Apart from the general physical interest, this <u>unique phenomenon allows indeed to measure locally Z_{eff}</u> by employing the unshifted (π) component of SL L_α whose impact width is purely adiabatic:

$$\gamma_s(L_\alpha^\pi) = \gamma_s^{a.}(L_\alpha^\pi) = (Z_{eff}-1+2^{-1/2})\gamma_0, \quad (1)$$
$$\gamma_0 \equiv 72(\hbar/m_e)^2 N_e (2\pi M_H/T_H)^{1/2} I(R) \text{ or}$$
$$\gamma_0(s^{-1}) \approx 2.45 \times 10^{-4} N_e (cm^{-3}) [T_H(eV)]^{-1/2} I(R), \quad I(R) \approx 0.209 + 6^{-1}\ln R,$$
$$R \equiv (m_e V_H/6\hbar Z_i)[T_e/4\pi e^2 Z_{eff} N_e]^{1/2} \approx 1.1 \times 10^8 T(eV)[Z_{eff}^3 N_e(cm^{-3})]^{-1/2}.$$

For $N_e=10^{13} cm^{-3}$, $T_H=10^2 eV$, $Z_{eff}=4$ the homogeneous width from (1) is $\gamma_s=1.2\times 10^9 s^{-1}$. To observe the Voigt profile with equal dispersive and Doppler components it should be used a laser field $E_0 \sim 10^2 kV/cm$ what is feasible.

REFERENCES

1. V.A.Abramov,V.S.Lisitsa,Sov.J.Plasma Phys.**3**,451(1977).
2. S.S.Bychkov, R.S.Ivanov, G.I Stotskii, Sov.J.Plasma Phys.**13**, 769 (1987).
3. M.L.Strekalov,A.I.Burshtein,Sov.Phys.JETP **34**,53(1972).
4. Ya.Ispolatov,E.Oks, JQSRT **51**, 129 (1994).
5. V.S.Lisitsa, Sov.Phys.Usp. **122**, 449 (1977).

LINE SHAPES IN ASTROPHYSICS

C.Stehlé

DARC et UPR 176 du CNRS
Observatoire de Paris, 5 Place J. Janssen, 92195 Meudon, France

ABSTRACT

Some applications of the spectral line shapes theory to astrophysics are briefly rewieved. Among them is the determination of the stellar parameters using hydrogen line profiles, the description of stellar winds in terms of line asymetry, the computation of radiative acceleration for broad lines. Recent theoretical results concerning the line widths of hydrogen lines as also synthetic spectra of hydrogen and ionic radiators are presented.

INTRODUCTION

The connection of the spectral line shapes to astrophysics has been reviewed in many text books especially for the application to stellar atmospheres[1-3]. Line shapes are indeed an important tool for astrophysicists[4]. They are used for interpreting the emergent fluxes or as input physics for modelling the composition of the stellar matter. The subject is very large. Due to wide range of gas composition in the different parts of the universe, one may find molecules, even clusters, or neutrals or ions. Molecules dominate the cold interstellar medium and the planetary atmospheres, whereas ions are the most abundant in the star interiors. Molecules may affect significantly the spectrum of stars and sun in particular, and their spectra are often included in the modelling of stellar atmospheres. It is well known for example that the pressure induced broadening modifies dramatically the opacity of cold atmospheres of brown dwarfs and giant planets[5].

It is vain to pretend to discuss all the line shapes applications to astrophysics. The present paper is restricted to the spectra emitted by atoms (with emphasis on hydrogen) or ions in stars. We will restrict ourselves to plasma temperature and density conditions where the broadening is essentially attributed to the interactions between the radiating species and the free plasma charges. Even with such restrictions, the problem is wide due to the variety of plasma conditions and of radiating species, which may be found. The plasma conditions may vary strongly from a star to an other one, and inside each star from the atmosphere to the interior[6].

In the first part of this paper, we shall report selected examples where accurate line shapes are needed. Afterwards we shall focus our attention on data which might be useful for astrophysics. Section 2 presents new results concerning estimations of the hydrogen line widths at low densities. Section 3 is devoted to recent results on hydrogen, hydrogenic and helium like lines. The wide problem of radiation redistribution will be not discussed.

I : ASTROPHYSICAL CONTEXT

From the applications areas of the line shapes to stellar astrophysics, one may distinguish two classes. One is the use of observed spectras as a diagnostic of stellar parameters. The second, which may be correlated to the first one, is the introduction of line shapes together with a considerable amount of other physics ingredients to model some region of the star, which may be for example the envelope or the atmosphere.

I.1/ Interpretation of the observations.

The progresses in the modern high resolution spectroscopy allow a detailed interpretation of the observed spectra. With satellites the observations have been extended to the X-UV and IR regions. These observational progresses should take benefit from the recent theoretical progresses in atomic and molecular physics. The emerging line shapes reflect the history of the light travel through the stellar atmosphere. Thus the radiative transfer allows to deduce informations about the atmosphere layers.

From the applications of the line shapes in this context, we shall retain the caracterisation of stellar winds using the line asymmetry and also the determination of the surface gravity [1] g which is necessary for stellar classification and building of atmosphere models. This surface gravity is defined in terms of the stellar radius R and mass M by

$$g = GM / R^2 \qquad (1)$$

It is generally determined from the whole shape of an hydrogen line which is emitted in a large depth in the atmosphere. The shape of the emergent line is a signature of the plasma composition (i.e. electronic density and temperature) and of the radiative transfer through the atmosphere. Assuming a given surface gravity g, a chemical composition and an effective temperature T_{eff} defined in terms of the intrinsic star luminosity L by

$$T_{eff}^4 = \frac{L}{4 \pi \sigma_R R^2}, \qquad (2)$$

---

[1] *Typical values of g (in cm s^{-2}) are 2 10^4 for the main sequence stars, 300 for giants, 20 for supergiants and 10^8 for White Dwarfs.*

it is possible to model the atmospheres and to deduce the emerging lines shapes of hydrogen. Such tables of line shapes have been calculated extensively[8]. Inversely the knowledge of the stellar emergent line shapes allows to deduce the gravity, if the other parameters are known. Difficulties appear when the line are partly blended together as it occurs in the spectra of White Dwarfs, because the theoretical estimation of the intensity between the lines must include line interference and field ionisation effects. Figure 1 illustrates the sensitivity of the corresponding synthetic spectrum[10] to the details of the wings of Hydrogen Balmer lines calculated in various theoretical approaches[11,12].

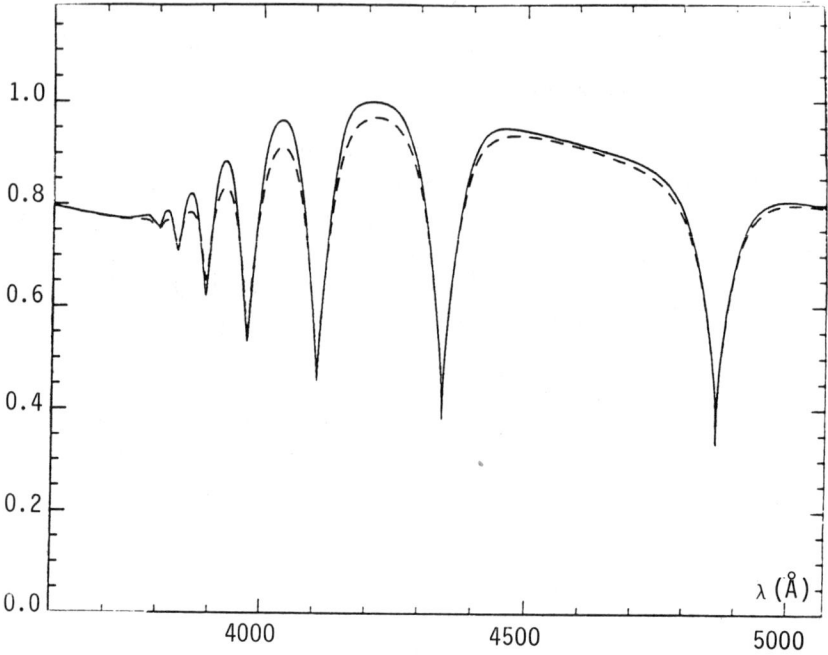

Figure 1: Comparison of a synthetic Balmer spectrum of a of DA White Dwarf[10] calculated with the Stark profiles of Schöning and Buttler [12] (based on the Unified Theory [13], dashed lines) and Seaton[11] (solid line), for log g=8 and T_{eff}=20000K. The increased flux in the line wings with Seaton's profiles is a direct result of the reduced wing opacity.

The application of the line shapes to the problem of the gravity determination requires a correct description of both the line centers and wings. For this application, astrophysicists use the tables of line shapes published by Vidal et al.[13]. These authors use the *static approximation* for the ionic contribution to the line profile. This approximation means that the plasma ions, which perturbe the radiator, can be considered as fixed in space during the radiative process. This approximation fails in the line centers at low electronic densities (i.e. less than 10^{16} cm^{-3} for Hα). Thus, it may be interesting to test the incidence of the *ion dynamics effects* on the gravity determination. The contribution of the fine structure effects, which are also currently neglected in theoretical calculations, could be also inspected. Both effects modify the line centres. Ion dynamics effects lead to a Stark width of the Hα line which exceeds those obtained in the static approximation. An attempt to include these ion dynamics effects has been done recently[7,14]. The conclusion of both papers indicate only small effects of ion dynamics on the emerging shapes of Hα, Hγ of hydrogen and Pα of He$^+$ for hot stars. This conclusion might by at least revisited, because in these work the profiles are given by empirical expressions which, to our oppinion, underestimate the halfwidth value(2). The recent tables of H profiles in the MMM method may be used alternatively for such applications[15]. These two last calculations unfortunately neglect the fine structure effects which can be included in the line shape in the qualitative manner[19].

An other application is the interpretation of asymmetric lines associated to stellar winds. Extreme cases of such profiles are observed in the UV spectra of hot stars and are attributed to expending optically thick envelopes[20]. The corresponding typical P-Cygni profiles consist in a shortward displaced absorption component and a longward displaced emission component. The mass ejection rate can reach 10^{-5} M$_0$/year in M supergiants. The asymmetry is caused by the conjunction of Doppler effect and radiative transfer. In the case of weak winds(3), the asymmetry due to the wind is small and may be, in extreme cases, on the same order than the natural asymmetry of the line. It is important to have some indications on these line asymmetries[21]. Using the usual Hα line, Lanz and Catala[22] expect to detect winds corresponding to 10^{-11} MO/yr in the main sequence A type stars with the modern high resolution detectors. In this context, the eventual effect in the line shape of the correlation between the Doppler and the Stark effect may be addressed, which is induced by the anisotropy of velocity for light propagation parallel or perpendicular to the direction of the wind [23,24].

---

2 Hence the values adopted there are deduced from the work of Oza & al.[16] in the relaxation theory which differ by a factor of two from those obtained by Stehlé and Feautrier[18] in the impact approximation. But the calculations of Seidel[17] using accurate numerical simulations are in agreement with the impact results, which are independently confirmed by MMM calculations (see Figure 3 below).

3 For example, the solar wind has a terminal velocity of 300 km/s at the earth orbit and a mass loss rate of 10^{-14} MO/yr (ie 10^{12} g/s)

Koestler & Allard[25] found a different application of the line shapes in the observation of molecular satellites in the Lyman α lines in White Dwarfs. These satellites reflect the density of neutrals Hydrogen and of protons along the atmosphere.

I.2/ Envelopes and atmospheres modelisation.

Line shapes enter in the models of radiative envelopes by the estimation of the Rosseland opacity k_R, defined in terms of the monochromatic opacity k_ν and of the Planck function B_ν by

$$\frac{1}{k_R} = \frac{\pi}{4\sigma_R T^3} \int_0^\infty d\nu \, \frac{1}{k_\nu} \frac{dB_\nu}{dT} \qquad (3)$$

The absorption coefficient k_ν is the sum of the various absorption mechanisms of the different plasma constituents. The line shape intensity ϕ_ν enters in the expression of the contribution $k_{\nu,ij}$ of the transition between bound states i,j of the radiator A, initialy in state i, through

$$k_\nu = N(A,i) \, \phi_\nu \, \frac{\pi e^2}{m_e c} f_{ij} \qquad (4)$$

where N(A,i) is the volumic density of radiators A in state i, f_{ij} is the absorption oscillator strength and m_e the electron mass

Recent improvements of the averaged opacity calculation are achieved by including a considerable amount of lines [26,27] which where missing in the previous tabulations[28]. In this context, the line shapes can not be ignored, because the Rosseland mean is included by the inverse of the absorption coefficient. Lee[29] discussed recently the line shapes effects in the averaged opacity calculations. He indicated that, even in the case of a spectrum containing a huge number of lines, the wings opacity may increase dramatically the Rosseland mean opacity. But such calculation is difficult to carry exactly and the results are only indicative. Hydrogen opacity is affected by the broad hydrogen lines, but accurate values are probably unnecessary for the determination of k_R, except for the contribution of the strong red Lyman α wing which lies in a transparency window between the Balmer discontinuity and the Lyman α line.

The line shapes play an important role in the calculation of the radiative acceleration[30-32]. This diffusion mechanism may contribute to interpret the surface abundances of various elements. It occurs in the stellar envelopes and/or atmospheres. Various anomalous abundances have been reported[31] by comparing with stars of similar atmospheric parameters or with typical solar system abundances[33]. The radiative diffusion mechanism is due to the net driving radiative force acting on an absorbing particle which pushes it towards the star surface, in the case where the radiative acceleration is larger than the gravity. The radiative acceleration g_r at ν, in the frequency interval $d\nu$, acting on the element A (density N_A, mass m_A) is

$$m_A \, g_r = \frac{k(\nu,A)}{N_A} \, \phi_\nu \, \frac{d\nu}{c} \tag{5}$$

where $k(\nu,A)$ is the contribution of A to the monochromatic absorption coefficient. In the opaque envelope (radius r), the radiative flux ϕ_ν is approximately equal to

$$\phi_\nu = \frac{4\pi}{3} \frac{1}{\rho k_\nu} \frac{\partial B_\nu}{\partial T} \left[-\frac{\partial T}{\partial r} \right] \tag{6}$$

where k_ν ($= k(\nu,A) + k_{rest}$) is the total absorption coefficient, ρ the volumic mass, T the temperature and B_ν the usual Planck function.

Details of the absorption coefficient are needed, in particular in the case of blended lines[34]. As the calculation of radiative acceleration among the various layers of the stellar envelopes is difficult and time consuming, it may be interesting to use some parametrized expressions [35] to calculate the "background" opacity (i.e. k_{rest}) (see equations 5 and 6).

Finally we point out an important application of the lines shapes of hydrogen in construction of stellar atmospheres models. Although the details of the lines may not influence dramatically the structure of the atmosphere model, accurate profiles are needed to model the emerging flux. Great numerical and physical advances have been made in the construction of non LTE model atmospheres[36]. Although the main interest is usually concentrated in the intense hydrogen lines, the lines of helium play a similar role for extremely rich helium stars[37].

I.3/ Astrophysical requests.

For some problems, like for the contribution of weak lines to the Rosseland mean, only the halfwidth values are needed. In other cases, as for the gravity determination or for the radiative forces in the case of blends, one needs the details of the line shapes from the centre to the line wings. This concerns intense lines of H, He+ and the most abundant elements.

Let us take the example of the radiative diffusion calculation. It will be in general necessary to compute a series of lines for a large number of plasma compositions characterising the different star layers. As the number of elements of interest for such application is very large, it is convenient to dispose either of accurate tables of line shapes, or from fast codes giving results with a reasonable accuracy. Special care must be paid to the plasma description. Hence the line shapes of hydrogen and of ions are sensitive to the field distribution function P(E)

(Section 3). This quantity varies with the radiator charge, the charges of the perturbing plasma ions, the electronic density and temperature and the ratio of the ionic to electronic temperature. In the case of cold stars the contribution to the broadening by neutrals and the resonance broadening must be carefully checked [38,39].

Which are the available data?

For the hydrogen lines, astrophysicists use traditionally the tables of Griem[40] or of Vidal et al.[13]. The plasma is there composed on electrons and ions with equal concentrations (i.e. Z=1). Ion dynamics effects are neglected. This may be justified depending on the conditions and on the use. Hence dynamics effects affect line centres of the lowest lines of each series at moderate and low density whereas the static limit is valid for the line wing, in a first approximation *(4)*. All these theoretical results are based on the following approximations. The plasma-radiator interaction is described in terms of plasma electric microfield (this corresponds to the dipolar interaction). Only the states belonging to the same principal quantum number n are mixed by this electric field (no quenching approximation). Fine structure (fs) effects are neglected. These (fs) effects should be introduced for the Hα line at electronic densities smaller than 10^{14} cm^{-3}. They are significant as long as the Doppler linewidth is smaller than the typical fine structure separtion. As a result of these three last assumptions, the line shapes are symmetrical with respect to the line centre. An improvement has been done recently in the tabulations of line shapes allowing for ion dynamics effects[15].

Schöning and Buttler [41,42] recently published tables of line shapes of He$^+$ ion using the last three approximations together with the static ion assumption. The method is based on the grounds of the Unified Theory of Vidal et al.[13] developed for the case of hydrogen and by Greene and Cooper[43] for the broadening of H-like ions. The natural broadening is not included in these tables. This broadening varies as Z^4 and is important for the ionic lines, especially at low densities. Ion dynamics effects are also neglected. Like for hydrogen, this is not important as long as the Doppler width dominates the line core. An estimation of the order of magnitude of this effects may be obtained using the new impact formulas given in section II, or the results obtained by Stehlé[44] in the Model Microfield Method.

As Hydrogen and Helium are the dominant elements of the universe, these tables are of practical interest. Some parametrization formulas have been also proposed for these lines. They give the line shapes with less accuracy but they are specially interesting for intensive calculations[35].

Astrophysicists also use for Hydrogen and H-like ions the pioneering tables of Underhill and Waddel[45], especially to describe the lines which are missing in the other tables. These authors used the Holstmark theory for the electric field

---

4 We shall neglect all the quantum effects which are important in the far wings only.

distribution functions of ions which is only valid for hydrogen at low density, due to the neglect of plasma effects and of the repulsion between the charges of the radiator and the perturbing ions. The wings are underestimated due to the missing electronic contribution. Applying a corrective multiplicative factor of two, one obtains the true wings intensity in the case where the electronic contribution can be described in the same manner that the ionic ones. This is almost exact for hydrogen but not for ionic lines due to the different coulomb interactions of ions and electrons at short distances from the radiator. These tables may be used only for qualitative variations.

For the weak lines it is often sufficient to have an estimation of the width. The reader may find in the compilation of Fuhr and Lesage[46] precious informations. In the next section we shall present the semi-classical method which has been widely used in this context and gives, on average, halfwidth values with a good accuracy.

II : LINE CENTRES AT LOW DENSITIES.

We shall not discuss the trivial broadening due to spontaneous radiative decay and concentrate ourselves to the collisional broadening. Except for the case of very broad lines the impact description in terms of collisional rates is appropriate, as long as the collisional duration t_c is small compared to the inverse of the line width $\Delta\omega_{1/2}^{-1}$. Isolated lines have a Lorentzian shape. The width $\Delta\omega_{1/2}$ is related to the usual scattering operator S. To calculate it, one may use two different approaches: one way is to use a quantum description of the collision. This method is suitable at low velocities and near the threshold.

In the various astrophysical plasmas the temperatures are large and the long range dipolar part of the coulomb interaction allows to use the semi-classical description of the collision. The motion of the perturbing charges is treated classically (rectilinear trajectories for a collision with a neutral and hyperbolic, for collision with an ion). This method is numerically fast and almost sufficient for many astrophysical applications. It has been widely used[40,47-52]. We shall here focus our attention on the broadening by ions (Z_p, N_p) and on the dipolar contribution to the cross-sections, which dominates in the case of optically allowed transitions. Thus, we put

$$V(t) = -eZ_p\, \mathbf{r}\, \mathbf{d}/r^2(t) = -\mathbf{d}\, \mathbf{E}_p(r(t)) \qquad (7)$$

The halfwidth expression is

$$\Delta\omega_{1/2} = 2\pi\, N_p \int_0^\infty v f(v) dv \int_0^\infty b\, db\, P(b,v)$$

with

$$P(b,v) = \{S_{ii}(\mathbf{b},\mathbf{v})\, S^{-1}_{ff}(\mathbf{b},\mathbf{v}) - 1\}_{Ang.\ Average} \qquad (8)$$

The angular average is done over all the directions of the relative velocity (at infinity) **v** (v, Ω_v), and b is the impact parameter. Using a perturbative treatment for the calculation of the scattering matrices, one obtains

$$P(b,v) = \left\{ \int_0^\infty dt_1 \int_0^{t_1} dt_2 \; \left[\left(\sum_i V^I_{ii'}(t_1) V^I_{i'i}(t_2) + \sum_f V^I_{ff'}(t_1) V^I_{f'f}(t_2) \right) - V^I_{ii}(t_1) V^I_{ff}(t_2) \right] \right\}_{Ang.\,Average} \quad (9)$$

where $V^I(t)$ is the interaction potential in the interaction representation. These appproximations lead to the following expression for $\Delta\omega_{1/2}$ [48]:

$$\Delta\omega_{1/2} = \frac{2\pi N_p}{3\hbar^2} \; [\,\mathbf{d}_{ii'} \cdot \mathbf{d}_{i'i}\,] \int_0^\infty v f(v) dv \int_0^\infty b\,db$$

$$\int_0^\infty dt_1 \int_0^{t_1} dt_2 \exp(-i\omega_{ii'}(t_2 - t_1)) \; \mathbf{E}_p(t_1)\,\mathbf{E}_p(t_2) + \quad similar\ terms.] \quad (10)$$

In terms of the total ionic plasma microfield $\mathbf{E}(t) = \Sigma\,\mathbf{E}_p(r(t))$, this is equivalent to

$$\Delta\omega_{1/2} = \frac{1}{3\hbar^2} \; [\,\mathbf{d}_{ii'} \cdot \mathbf{d}_{i'i}\,] \int_0^\infty dt\,\exp(-i\omega_{ii'}t)\,C_{EE}(t) + similar\ terms.] \quad (11)$$

where

$$C_{EE}(t) = \left\{ \mathbf{E}(0)\,\mathbf{E}(t) \right\}_{Average} \quad (12)$$

Thus the quantity of interest is the Fourier transform of the ionic microfield autocorrelation function in the limit of independent particles. The use of independent particles, each of them affected by the reduced mass μ, is called the μ-ion approach. It may be noticed that the crossection value is independent whatever one chooses the centre of mass or laboratory frame. The result may be expressed in terms of modified Bessel functions[47].

In the case of overlapping lines the line width expression is more complicated and we shall focus the attention to the case of lines of hydrogen or hydrogenic ions. As usually we neglect the fine structure effect and connect together only states with the same principal quantum numbers (no quenching approximation). In the impact approximation the near wings intensity of the transition $(i, i'..) - (f, f'..)$ is given by

$$I(\omega) = \frac{\Gamma}{\pi\Delta\omega^2} = \frac{\sum_{if,i'f'} \mathbf{d}_{if}\,\gamma_{if,i'f'}\,\mathbf{d}_{i'f'}}{\pi\Delta\omega^2 \sum_{if} \mathbf{d}_{if}\,\mathbf{d}_{if}} \quad (13)$$

which allows to define the mean relaxation operator Γ. The structure of $\gamma_{if,i'f'}$ is similar to those given in expression (6).

An interesting feature of the line shapes of one electron ions in the impact limit is the Lorentzian behaviour in the line core, with an halfwidth value equal to Γ. This has been demonstrated earlier[18] for the Lyman and Balmer lines. Calculations using the Model Microfield Method confirm also this behaviour and indicate that the result may be extended to all the hydrogenic transitions. The line width expression is analytical. The contributions of each charged species of perturbers are additive ($\Gamma = \sum_p \Gamma_p$). For the ions of charge Z_p (density N_p), one has

$$\Gamma_p = \Delta\omega_{p,1/2} = 2 N_p \sqrt{\frac{2\pi}{m_p kT}} \, k_{n,n'} \, a_0^2 \, I_H \, f(b_c, b_d) \quad (14)$$

with $\quad b_c^2 = \dfrac{I_H \, a_0^2 \, k_{n,n'}}{kT}, \quad b_D^2 = \dfrac{kT}{4\pi e^2 N_e} \dfrac{1+<Z>^2}{<Z>},$

In these expressions b_c and b_D are the strong collision radius and the Debye length, T is the temperature, N_e the electronic density, a_0 the Bohr radius, I_H the hydrogen ionisation energy, $<Z>$ the mean charge of the plasma ions, $<Z^2>$ the mean squared charge. The function f is given by

$$f(b_c, b_d) = (1 - \gamma + Ln\frac{b_D^2}{b_c^2}), \quad \text{for rectilinear trajectories (neutral radiator)}$$

$$f(b_c, b_d) = \exp(-\frac{a^2}{b_c^2}) - (1+\frac{a^2}{b_c^2}) E_1(\frac{a^2}{b_c^2}) + 2 \exp(-\frac{a}{b_c}) E_1(\frac{a}{b_D}),$$

for hyperbolic trajectories (ionic radiator of charge Z_t),

where $a^2 = Z_p Z_t e^2/(2kT)$, γ is the Euler constant, E_1 the exponential integral, and $k_{n,n'}$ is an intrinsic parameter of the transition n - n', those expression is given in terms of the dipolar line strengths by

$$k_{n,n'} = \frac{2}{3\sum_{if} S_{if}}$$

$$[\sum_{if} S_{if} \{\sum_{ii'} \frac{S_{ii'}}{2l_i+1} + \sum_{ff'} \frac{S_{ff'}}{2l_f+1}\} - \sum_{ii'ff'} 2\sqrt{S_{if} S_{i'f'} S_{ii'} S_{ff'}} \begin{Bmatrix} l_{i'} & l_{f'} & 1 \\ l_i & l_f & 1 \end{Bmatrix}] \quad (15)$$

In the general case, analysis of the numerical results indicate that $k_{n,n'}$ (with $n'=n+\delta n$) is *exactly* equal to

$$k_{n,n'} = 1.5 \, [\, \delta n^2 (\delta n^2 - 1) + 2 \, n \, n' \, (2\delta n^2 - 1) \,] \quad (16)$$

The asymptotical results of Griem[53] concerning the n-n+1 transitions[5] agree only qualitatively with the present results. These new expressions are valid at low densities when the halfwidth value is small compared to the plasma frequency

---

[5] *i.e.* $k_{n,n+1} = 1.5 \, n^2 [\, 1.5 + (2/e^2) \, Ln(2n/3) \,]$

(ionic or electronic, depending on the nature of the involved collisions). The total width value $\Delta\omega_{1/2}$ is obtained by adding the electronic and ionic contributions $\Delta\omega_{e,1/2}$ and $\Delta\omega_{p,1/2}$. The line shape is lorentzian for detunings $|\Delta\omega|$ smaller than the ionic plasma frequency ω_{pi}. This low density halfwidth value differs up to a factor of 10 from those obtained in the usual static approaches[13,40]. However, after convolution with the Doppler line shape, the discrepancy may be reduced in a large number of cases. The line shape recovers the static limit at large detunings allowing to get a smooth interpolation between the impact line core and the static wing results.

We want now to discuss some paradox concerning <u>only</u> the broadening of <u>ionic</u> lines. The key point of the impact halfwidth is the field autocorrelation functions $C_{EE}(t)$. The time average of $C_{EE}(t)$ in terms of collisions is always positive[44]. In fact this time integral of the autocorrelation function of the field acting on an ion is strictly equal to zero [54]. To demonstrate this last rule, one uses the proportionality between the field and the nature of the electrostatic force acting on the ion. One makes also the assumption of the diffusion limit at large times. This zero average value explains the oscillations of $C_{EE}(t)$ with the time, which are observed in molecular dynamics simulations (although the integral over the time is not calculated, even at low density where these oscillations are very smooth). The oscillatory behaviour of $C_{EE}(t)$ indicates thus departures from the collisional scheme (or μ-ion model, where the collisions are uncorrelated between themselves). The binary assumption fails at time larger than the typical collisional time. It is in fact confusing to use for the collision an history time between $-\infty$ and $+\infty$, because several collisions occur during this time'which leads to the diffusion limit at large times. On the other hand the binary picture is justified at low densities for times smaller than the collisional time (i.e. the inverse of the ionic plasma frequency, $1/\omega_{pi}$), thus it is expected that the ``true" autocorrelation function and the ``collision" field autocorrelation function agree at times smaller than $1/\omega_{pi}$, which corresponds to the relevant part of $C_{EE}(t)$ versus the time. This requires that the typical time for the diffusion regime t_{dif} [6] is larger than $1/\omega_{pi}$. This point will be illustrated in the next section. The problem of the positive value of $\int_0^\infty C_{EE}(t)$ in the collisional picture is not important for the purpose of the calculation of cross-sections. The quantity of interest for our purpose is the field autocorrelation function in the μ-ion picture, which we will note hereafter $\mathcal{C}_{EE}(t)$ and which is in good agreement at small times with the true ones in the low density limit.

---

[6] *A rough estimation of* t_{dif} *is given in terms of the diffusion constant D by* $t_{dif} = Dm/kT$.

In these computations of the halfwidth, it has been assumed implicitly that correlations between the Stark and the Doppler profiles can be neglected. This approximation justifies the use of the μ-ion picture but prevents in fact to recover the diffusion limit at large times as discussed previously. The typical time scale for the establishment of the diffusion regime t_{dif} must be long compared to the inverse of the halfwidth. This condition concerns only the case of the broadening of lines emitted by ions.

III: LINE SHAPES OF H AND H-LIKE IONS.

The variety of the line shapes requests for astrophysical applications explains the need of building codes able to give Stark broadened line shapes which are valid from line centre to the line wings, for weakly to high correlated plasmas. The Model Microfield Method developed by Brissaud and Frisch [55,56] for the case of broadening of neutrals is adapted to this purpose. This method works very well for the broadening of hydrogen and helium [57,58]. Its extension to the case of broadening of ions, proposed by Stehlé [59] is recent [60,61]. We shall present the principles of the theory for the case of neutral radiators and some results on hydrogen lines. Afterwards we shall present the extension of the method to the broadening of ions.

III.a: Hydrogen broadening.

Concerning the hydrogen lines, the theoretical and experimental results of the last decades indicate the necessity to include the ion dynamics effects. Such effects are important for the lowest lines of each series, giving, for example, broader lines for Lyα and Hα and reduced dips for Lyβ and Hβ, compared with the static results. These effects are noticeable at low and moderate densities and disappear at high densities and in the line wings. In this context the ab-initio Monte-Carlo simulations [17,62] are considered as theoretical benchmarks. Such computer experiments are rather difficult and usually restricted to moderate electronic densities ($>10^{14}$-10^{15} cm^{-3}) and to the centres of the lines. The Relaxation Theory, developed by Greene[63], is a semi-analytical theory which requires the knowledge of characteristic plasma functions which are calculated also by means of computer-simulations. In fact, it should be able to treat the line wings. This method seems to fail to recover the impact limit, which may be a consequence of the difficulty to carry Monte-Carlo simulations at low densities. The Model-Microfield has not these limitations. We shall recall briefly the principles of this method.

The electric microfield is supposed to vary stepwise with the time. In each time intervals t_i, t_{i+1}, the field value has a constant value E_i. The jumping times obey a Poisson law, with a density ν, which depends on the value of the field before the jump, i.e. $\nu(E)$. The expression of the Fourrier transform of the time evolution operator (directly connected to the line shape expression) is analytical. It is given in terms of the field distribution functions $P(E)$[64] and frequency jump. This frequency jump is a free parameter which can be chosen to describe the variations of the field autocorrelation function.

We may define two expressions for this field autocorrelation function, one is the binary correlation function $C_{EE}(t)$ in the μ-ion model presented previously and the other is the MMM field autocorrelation function $C_{M;EE}(t)$ which is

$$C_{M;EE}(t) = \int_0^\infty dE\, E^2\, P(E)\, \exp(-\nu(E)t) \qquad (17)$$

It is important to understand that the notion of field autocorrelation function is relevant for the line shapes only in the low density limit when the profile is given by the collisional limit. It may be showed that the damping operator of the impact limit, which *must* be reached at low densities[59], is given in terms of the field autocorrelation function $C_{EE}(t)$. Thus it is necessary to have a good agreement between $C_{M;EE}(t)$ and $C_{EE}(t)$ in this limit. In the high density limit there is no need to describe the exact plasma field autocorrelation function because the static limit is reached and such dynamic quantity is not of interest. We choose to determine the frequency jump $\nu(E)$ such that $C_{M;EE}(t) \approx C_{EE}(t)$.

It is expected to have in this way *a theoretical interpolation of the line shape* between the low density collisional regime described previously and the high density static limit. According to the preceding discussion, this choice of $\nu(E)$ prevents to use our $C_{M;EE}(t)$ to describe the field autocorrelation function at large times, as noticed by Alastuey [65]. Indeed, this is not our aim. In practice the jumping frequency $\nu(E)$ is choose to reproduce the time variation of $C_{EE}(t)$ [55,56]. Its expression is analytical. We have checked that the MMM gives accurate near line wings values[61].

Such MMM line shapes have been used to reproduce the experimental results of Wiese et al. [66] concerning the emissivity of the Balmer series in a plasma arc. Using the procedure of level dissolution of Däppen et al.[67] for the calculation of the level populations of Hydrogen and for the contribution of the photoionization below the threshold we found a very good agreement with the experimental results, which confirm together a correct line wing treatment by MMM and a reasonable treatment of bound-free contribution below the threshold[67] (Figure 2).

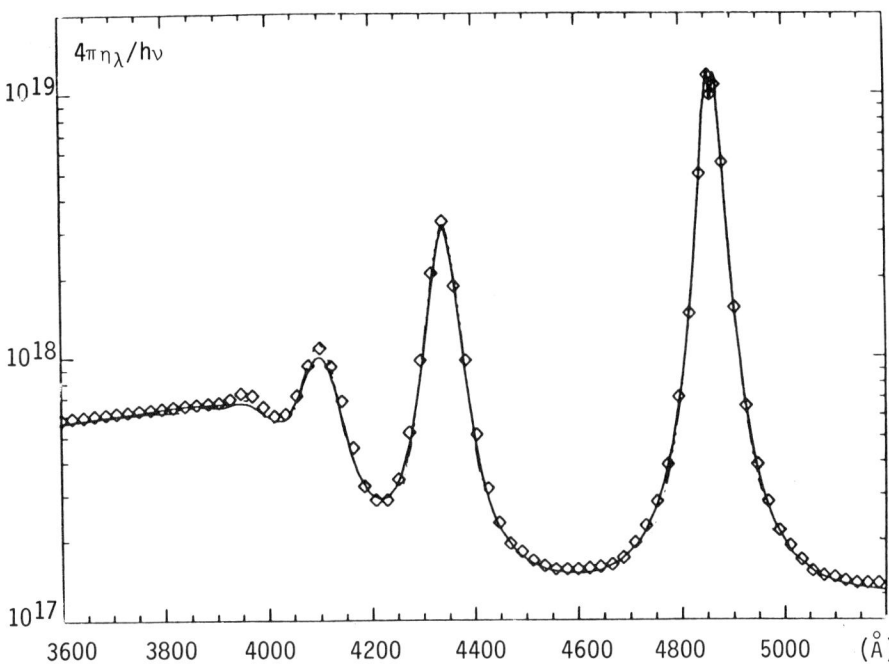

Figure 2: spectral emissivity $4\pi\eta_\lambda/h\nu$ in cm^{-3} $Å^{-1}$ s^{-1} under the conditions of the experiment of Wiese et al [66] (Ne=8.68 10^{16} cm^{-3}, T=13000K). MMM [68] and experimental results.

The ion dynamics effects are illustrated in figure 3 which shows the variations of the halfwidth value of the Hα line at 10^4 K versus the electronic density in the Relaxation Theory[16], the MMM and the Unified Theory[12] (static ions). The results obtained by Seidel[17] in Monte-Carlo simulations confirms in the line centres the impact limit results of Stehlé and Feautrier[18].

Recent tables of Balmer and Lyman lines have been published in the MMM for the conditions of the stellar envelopes[15]. An analytical parametrization has been proposed for the case of a plasma of protons by Clausset et al.[35].

In the low density limit we have checked that the impact limit is recovered in the line centre, as it must be. In the line wings the ion and electron contributions to the intensity are correctly given by the static limit, leading to the usual $|\Delta\omega|^{-5/2}$ Holstmark intensity.

To conclude this paragraph concerning hydrogen line shapes, it seems suitable to include the fine structure effects in the computations especially for the Hα line. A more accurate description of the red line wing of the Ly α line is also needed, including molecular effects in the interaction between H and H^+ or H and H (leading to the formation of molecular satellites[25]). This importance of the red wing may be understood by inspecting the opacity variations in this UV region. Progresses are suitable in the description of the plasma field ionisation effects, which should be tested by new experiments. An extension of the formalism of MHD to the NLTE case has been recently proposed by Hubeny et al.[69].

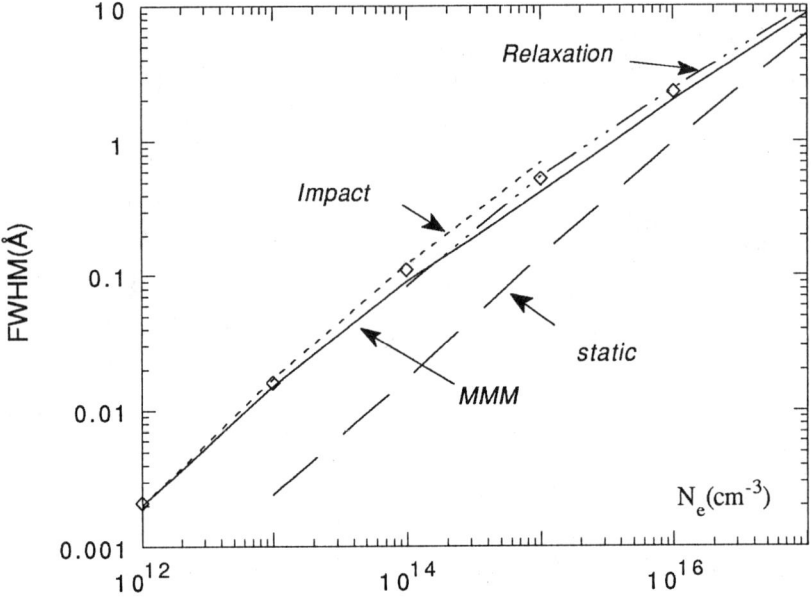

Figure 3: variations of the halfwidth value (FWHM in Å) of the hydrogen Hα line at $10^4 K$, versus the electronic density (cm^{-3}). MMM: full line, Relaxation Theory [16]: —··—; Monte-Carlo[17]: ◊ ; Static [13] : — — ; Impact [18] : -----. The Doppler broadening is not included.

III.b/ Lines of one electron ions.

The situation is slightly different for ionic radiators than for neutrals. This is a consequence of the Coulomb interactions between the radiator and the surrounding charges which modify the statistical properties of the plasma microfields. The experimental results are relatively few except for He^+.

Let us discuss briefly the contribution of electrons to the shape. It can be described in terms of absorption or emission during collisions between an electron and the radiating ions. Such approach has been also used in the semi-classical description of the collision[41-43]. It leads to the impact limit in the line centre and the static limit in the far line wings. The contribution due to spontaneous emission increases as Z^4 whereas the impact electronic contribution decreases as Z^{-2}. This indicates the important role of the radiative decay in the line shape.

The dominant broadening in the line centre results from the interactions between the radiator and the free plasma ions. The static approach underestimates the width of the lines exhibiting a central component but recovers the correct wing intensity if the electronic contribution is properly described [41-43]. The existing dynamics methods are up to now restricted to line centres. Among these methods, the molecular dynamics simulation [70,71] allows to follow the motion of each ion perturbed by all the others during the time of the computer experiment. It is expected to mimic the time dependent statistics of the plasma. The relaxation theory already used for hydrogen atom has been extended to the case of ionic radiator[16,72]. The characteristic plasma functions are calculated by Monte Carlo simulations but unfortunately the use of rectilinear trajectories in these simulations may underestimate the repulsion effects due to the Coulomb interactions. The kinetic theory[73] treats in an unified manner the problem of the radiator velocity and field correlations, which is neglected in all the other approaches. Like the model of components mixing[74], it is up to now restricted to the line centres where it has a wide range of practical applicability.

We prefered to use the MMM method in order to treat a large number of plasma conditions [44,75]. This method is rather simple to use and is able to describe the line wings if the electronic contribution is correctly included.

The principles of the method are the same than for hydrogen. The frequency jump $v(E)$ allows to describe the time variations of the field autocorrelation function $C_{EE}(t)$. Contrary to the neutral case this function is finite at time 0. We used a different method to determine $C_{EE}(t)$ than for neutrals. The details of its determination are presented elsewhere[44]. We required that the time average of the MMM field autocorrelation function is equal to the time average of $C_{EE}(t)$, i.e.

$$\int_0^\infty \frac{E^2 P(E)}{v(E)} dE = \int_0^\infty C_{EE}(t)\, dt \tag{18}$$

We emphasise again that the covariance of the microfield is adjusted to $C_{EE}(t)$ and not to the true field autocorrelation function, and that MMM is unable to reproduce the oscillations of the last ones. But like in the case discussed in the preceding section this method should allow to reproduce qualitatively the short times variations of this true function in the low density case. This is illustrated in Figure 4 which shows the variations of the field autocorrelation function calculated in the MMM and in the kinetic theory[76] for an impurity of Ar^{17+} in protons at 10^7K and an electronic density equal to $1.5\ 10^{23}$ cm^{-3}. The impact limit is recovered by the model at low densities, as demonstrated earlier [59].

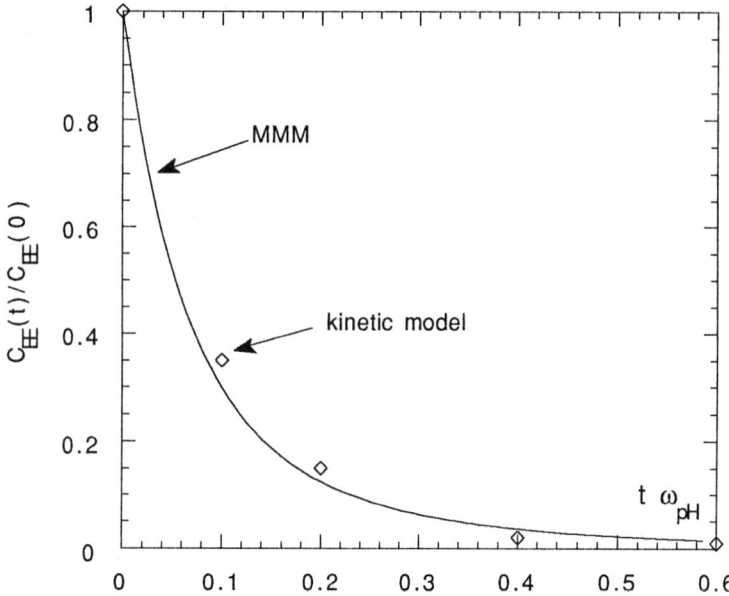

Figure 4: field autocorrelation function in the MMM compared with the field autocorrelation calculated in the kinetic model[76] for Ar^{17+} in H^+ (T=10^7 K, N_e= 1.5 10^{23} cm^{-3}). The concentration of Argon is 10^{-4}. The time is in units of the H^+ plasma frequency ω_{pH} (ω_{pH}^{-1}=2 10^{-15} s.rd^{-1}, t_{dif}=1.5 10^{-13} s, deduced from the value[90] of the diffusion coefficient D=2.9 cm^2/s)

We present new results concerning the lines of He^+ ion and of more highly ionised ions, like Carbon and Argon. We have used the dipolar and no quenching approximations. The electronic contribution is introduced in the line centres in the impact theory. This allows to test our method by comparing with other theoretical and experimental data.

i/ Helium lines.

Detailed calculations have been done for various lines of this ion. The field distribution function is taken from Hooper[64]. We present here only results concerning the Paschen α line (4860 Å), and compare with other experimental results. We note a very good agreement at low densities. We note large ion dynamics effects in the line centres at low densities. The radiative broadening operator 0.5(1/t_i+1/t_f) is equal to 4.5 10^9 and 9.2 10^8 rad.s^{-1} for the Hα and Pα lines and dominates the broadening at densities smaller than 10^{13} and 10^{12} respectively (T=10^4K) (Figure 5). It does not modify the present results.

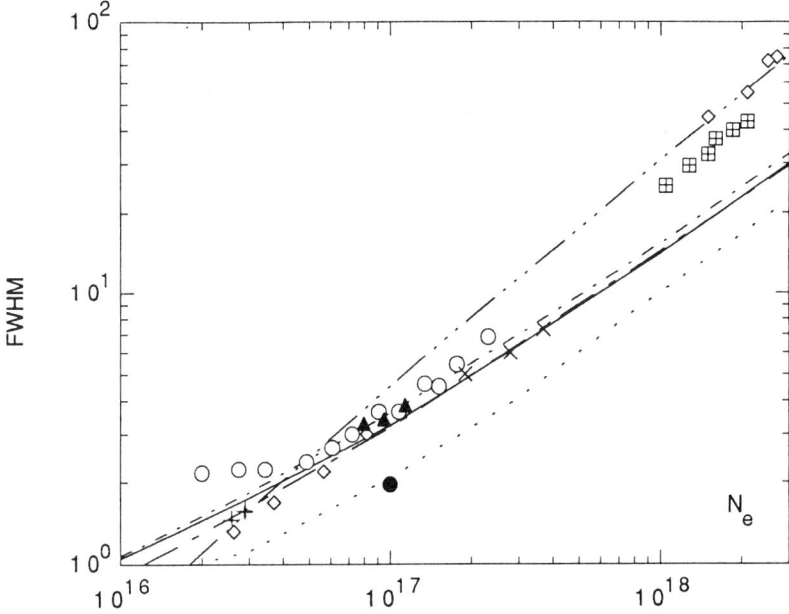

Figure 5: FWHM(Å) of the He^+ $P\alpha$ versus the electronic density (cm^{-3}) for different temperatures and perturbing ions. The Doppler effect is included. Full curve: MMM, $10^5 K$, He^+; — . — MMM, $10^5 K$, H^+; — - — MMM, 40000 K, H^+; ..... $10^5 K$ static ions[44]; —...— Kepple[77] (static); • Greene[78](static); Jones et al.[79]; ∗ :Soltwisch and Kusch[80]; × Oda and Kiriyama[81]; ◊ Pittman et al.[82], ○ Pittman and Fleurier[83]; Gawron et al.[84],+ Stefanovic et al.[89].

ii/ Carbon lines

One electron carbon ion is fund in stellar envelopes. Its lines have also been studied in the context of X-ray laser research. The differences between the astrophysical and experimental conditions concern mainly the plasma composition which is composed almost from protons in envelopes and carbon ions in the second case. We report here results concerning the halfwidth of the Hα line (180 Å) for C^{5+} perturbers. For this application we calculated the field ditribution function either using the cluster formalism of Baranger-Mozer[85-86] or Monte Carlo simulations[87] depending on the plasma conditions[75]. The radiative damping rate due to spontaneous emission decay (3.92 10^{11} rds.s^{-1}) dominates the line shapes for densities smaller than 10^{17} cm^{-3}. Figure 6 shows the variations of the width with the temperature and the density. Doppler broadening, fine structure and spontaneous emission effects are not included. This figure illustrates the ion dynamics effects. We notice some disagreement with the results of Oza et al. in the Relaxation Theory[72], which are not well understood.

54 Line Shapes in Astrophysics

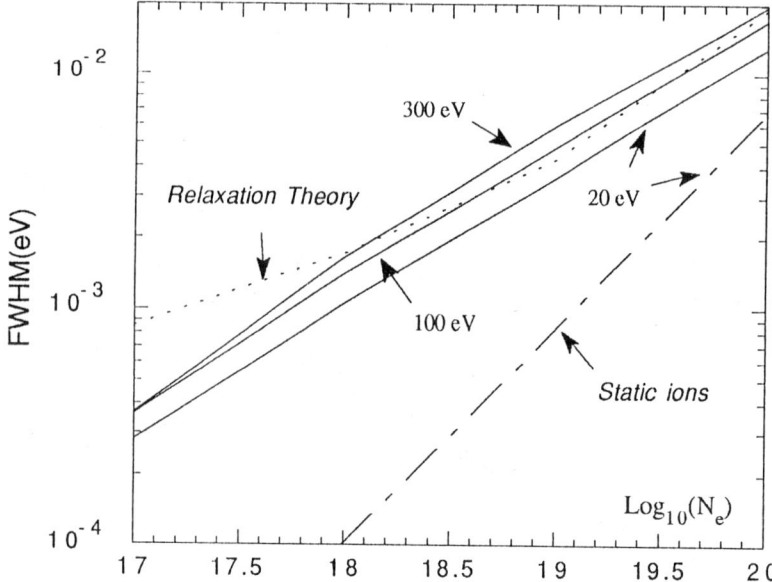

Figure 6: Halfwidth values (FWHM) of the Hα line of C^{5+} preturbed by C^{5+} versus the electronic density at different temperatures.

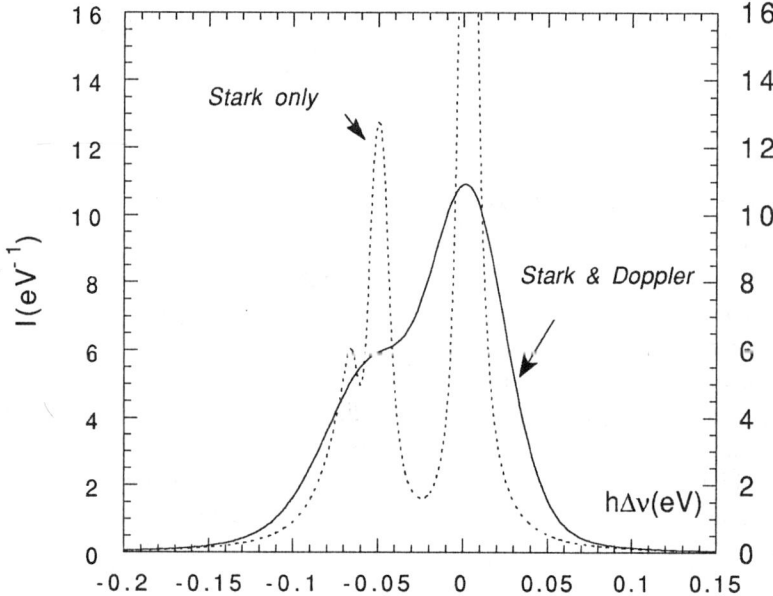

Figure 7: Ly α line of C^{5+} including Doppler, fine structure and spontaneous emission decay.

IV: CONCLUSIONS

In the results presented here the contribution of fine structure and spontaneous emission decay are neglected. The method can easily include these effects, as illustrated in Figure 7, which shows the profile of the C^{5+} Lyman α line at 6.7 10^{19} cm^{-3} and 4.1 10^{5}K (conditions of Ap stars envelopes). The methods reported in this paper are specially adapted to astrophysical applications which require the maximum accuracy compatible with extensive line shapes calculations. We have emphazised the importance of the ion dynamics effects in the line shapes.

ACKNOWLEDGEMENTS.

I would like to thank T. Lanz, M.C. Artru, R.Cayrel, C. Catala, G. Alecian, D. Gilles, J. Babel, for their suggestions and helpfull comments, J. Richer for providing some diffusion coefficients, and the Comité National Français d'Astronomie for his financial support.

REFERENCES

1. D. Mihalas, Stellar Atmospheres(W.H.Freeman & co, San Fransisco 1978)
2. G.W.Collins, The Fundamentals of Stellar Astrophysics (W.H.Freeman & co., N.Y. 1989).
3. E. Schatzman, F. Praderie, "Les Etoiles", Interéditions du CNRS, (1990).
4. B.E.J. Pagel, J. Phys. B: At. Mol. Phys, 4, 279 (1971).
5. P. Lenzuni, D.F. Chernoff, E.E.Salpeter, Astr. J. Sup. Ser. 76, 759 (1991)
6. J.E. Vernazza, E.H. Avrett, R. Loeser, Astrophys. J., 184, 605 (1973).
7. R. Napiwotzki, T. Rauch, A&A 285, 603 (1994).
8. R.L. Kurucz, Astrophys. J. Suppl. 40, 1 (1979).
9. P. Bergeron, F. Wesemael, G. Fontaine, Astrophys. J. 367, 253 (1991).
10. P. Bergeron, private communication.
11. M. Seaton, J. Phys. B: At. Mol. Opt.Phys. 23, 3255 (1990).
12. T. Schöning, K. Buttler, private communication to Bergeron (ref. 10).
13. C. Vidal, J. Cooper, E.W. Smith, Astrophys. J. Suppl., 25, 37 (1973).
14. T. Schöning. A&A 282, 994 (1994).
15. C.Stehlé, A&AS 104, 509 (1994).
16. D.H. Oza, R.L. Greene, D.E. Kelleher D.E., PRA 37, 531 (1988).
17. J. Seidel, in "Spectral Line Shapes", Vol. 6, Ed. L. Frommhold, J.W. Keto (AIP, N. Y. 1990) p.99.
18. C.Stehlé, N.Feautrier, A&A 127, 263 (1983), A&A 255, 368 (1988).
19. A. Sanchez, R.D. Fulton, H.R. Griem, PRA 35, 2596 (1987).
20. G.L. Olson, J.I. Castor, Astrophys. J., 244, 179 (1981).

21. A.V. Demura, G.V. Sholin, J. Quant. Spectrosc. Radiat. Transfer 15, 881 (1975).
22. T. Lanz, C. Catala, A&A 257, 663 (1992).
23. J. Seidel, Z. Naturforsch. 34a, 1385 (1979).
24. V. Kesting, Spectral Line shapes Vol. 11 (ed. R.Stamm and B.Talin.1993).
25. D. Koestler, N. Allard, Spectral Line Shapes Vol. 11 (ed. R.Stamm and B.Talin.1993).
26. F.J.Rogers, C.A. Iglesias, Astrophys. J. Suppl. 79, 507 (1992).
27. M. Seaton, Yu-Yan, D.Mihalas, A.K.Pradhan, Mon. Not. R. Astron. Soc. 266, 805 (1994).
28. A.N. Cox, J.N. Stewart, Astrophys. J. 67, 113 (1962).
29. R.W. Lee, Spectral Line ShapesVol. 11, ed. R.Stamm and B.Talin.1993).
30. S. Vauclair, G. Michaud G.,Y. Charland, A&A, 31, 381 (1974).
31. G. Michaud, Physica Scripta 36, 112 (1987)
32. G.Michaud, Physica Scripta T47, 143 (1993)
33. N. Grevesse, Physica Scripta T8, 49 (1984).
34. J. Babel, A&A 283, 189 (1994)
35. F.Clausset, C. Stehlé, M.C. Artru, A&A in press
36. I. Hubeny, T. Lanz, A&A 262, 501 (1992)
37. S. Dreizler et al., A&A, 235,234 (1990)
38. R. Cayrel, G.Traving, Z. Astroph., 50, 239 (1960).
39. M.C. Lortet, E. Roueff, A&A 3, 462 (1976).
40. H.R. Griem, Spectral Line Broadening by Plasmas (Ac. Press, N.Y.1974).
41. T. Schöning, K. Buttler, A&A 219, 326 (1989).
42. T. Schöning, K. Buttler, A&AS 78, 51 (1989).
43. R.L. Greene, J. Cooper, J. Quant. Spectrosc. Radiat. Transfer 15, 1045 (1975).
44. C.Stehlé, A&A, accepted for publication.
45. A.B. Underhill, J.H. Waddel , NBSC 603 (1959).
46. J.R. Fuhr, A. Lesage, NIST Special Publications, 366, Supplt 4 (1992)
47. K.Alder, A. Bohr, Huus T., Mottelson B. & Winther A., Rev. Mod. Phys. 28, 432 (1956).
48. S. Alexiou PRA 49, 106 (1994)
49. S. Sahal-Bréchot, A&A, 1,91 (1969).
50. M.S.Dimitrijevic, S. Sahal-Bréchot , V. Bommier, A&AS 89, 581 (1991)
51. M.S. Dimitrijevic, N. Konjevic, A&A 172, 345 (1987).
52. M.S. Dimitrijevic, N. Konjevic, A&A 163, 297 (1986).
53. H.R. Griem, Astroph. J. 148, 547 (1987).
54. E.L. Pollock, J.C. Weisheit , Spectral Line Shapes,Vol. 3, Ed. F. Rostas (de Gruyter, N.Y. 1985), p181
55. U.Frisch, A.Brissaud, J. Quant. Spectr. Radiat. Transfer 11, 1753 (1971).

56. A. Brissaud, U. Frisch, J. Quant. Spectr. Radiat. Transfer 11, 1767 (1971).
57. J. Seidel, Zeit. Naturforsch. 32a, 1195 (1977).
58. A. Mazure, G. Nollez G., Ann. Phys. Fr. 9, 675 (1984).
59. C. Stehlé C., Radiative Properties of Hot Dense Matter, Ed.W. Goldstein et al.(Singapore, World Sc.1990), p.91.
60. T. Schöning, J. Phys. B: At. Mol. Opt.Phys. 26, 899 (1993).
61. C. Stehlé, J. de Physique, C1, 1, 121 (1991).
62 R. Stamm, E.W. Smith, B. Talin, PRA 30, 2039 (1984).
63 R.L. Greene, J. Quant. Spectr. Radiat. Transfer, 27, 185 (1982).
64. C.F. Hooper, PR 165, 215 (1968).
65. A. Alastuey, J. Lebovitz, D. Levesque, PRA, 43, 2670 (1991).
66. W.L. Wiese, D.E. Kelleher, D.R. Paquette, PRA 6, 1132 (1972).
67. W. Däppen, L. Anderson L., D. Mihalas, Astroph. J. 319, 195 (1987).
68. C. Stehlé, S. Jacquemot, A&A, 271, 348 (1993).
69. I. Hubeny, D.G. Hummer, T. Lanz, A&A 282, 151(1994).
70. R. Stamm, B. Talin, E.L. Pollock, C.A. Iglesias, PRA 34, 4141 (1986).
71. A.V. Anufrienko, A.E. Bulyshev, A.L. Godunov, A.V. Demura, Yu K. Zemtsov, V.S. Lisitsa, A.N. Starostin , Sov. Phys. JETP 76, 219 (1993).
72. D.H. Oza, R.L. Greene, D.E. Kelleher, PRA 34, 4519 (1986).
73. D.B. Boercker, C.A.Igesias, J.W. Dufty , PRA 36, 2254 (1987).
74. A. Calisti, F. Khelfaoui, R. Stamm, B. Talin, Spectral Line Shapes, Vol. 6, ed. L. Frommhold and J.W. Keto (AIP , N.Y.1990).
75. D. Gilles, C. Stehlé, Laser And Part. Beams, to be published.
76. D. Boercker, Spectral Line Shapes,Vol.11, ed. R. Stamm and B. Talin (1993)
77. P.C. Kepple, PRA 6, 1 (1972)
78. R.L. Greene, Phys. Rev. A 14, 1447 (1976).
79. L.A. Jones, E. Kälne, D.B. Thomson, J. Quant. Spectrosc. Radiat. Transfer 17, 175 (1977).
80. H.Soltwisch, H.J. Kusch, Z. Naturforsch. 34a, 310 (1979).
81. T. Oda, S. Kiriyama, J. Phys. Soc. Japan, 49, 385 (1980).
82. T.L. Pittman, P. Voigt P., D.E. Kelleher, PRL 45, 723 (1980).
83. T.L. Pittman, C. Fleurier, PRA 33, 1291(1986).
84 A. Gawron, S. Maurmann F.Böttcher, A. Meckler, H.J. Kunze, PRA 38, 4737 (1988).
85. M. Baranger, B. Mozer, PR 115, 521 (1959), PR 118, 626 (1960)
86. A.V.Demura, Preprint 4632/6, Institute of Atomic Energy, Moscow (1988)
87. A. Angelie, D. Gilles , Annales de Physique (Fr) Col.3, 11, 157 (1986)
88. J. Richer, private communication
89. I. Stefanovic, M. Ivkovic, N. Konjevic, submitted

EXPERIMENTAL STUDY OF THE He II P$_\beta$ LINE SHAPE

N. Konjević, I. Stefanović and M. Ivković
Institute of Physics, 11080 Belgrade, P. O. Box 68,
Yugoslavia

ABSTRACT

Using a low pressure repetitively pulsed discharge in helium-hydrogen gas mixture, the Stark broadened profiles of He II 320.3 nm, P$_\beta$, line was investigated. The electron density ranging between 2.3×10^{16} cm^{-3} and 2.8×10^{16} cm^{-3} was measured with He-Ne laser interferometer at 3.39 µm. The intensity ratio of He II P$_\beta$ and He I 318.8 nm line was employed to determine the electron temperature in the range between 34000 K and 38000 K. At these experimental conditions, the profiles of investigated He II line was recorded and compared with corresponding theoretical line shapes. Our experimental P$_\beta$ half widths are in agreement with several other experiments and some theoretical calculations. The inclusion of ion-dynamics in the evaluation of the central part of P$_\beta$ considerably improves the agreement between the experiment and the theory.

INTRODUCTION

The shape of helium P$_\beta$ line is similar to the well known shape of H$_\beta$ line. The ratio of average values of two maxima to the value of minimum, defined as a "dip" turned out to be very sensible on the plasma constituents (or the reduced mass of perturbers)[1]. Measuring the value of "dip" should be a good test for existing theories[2,3] of Stark broadening of hydrogenic lines. In this work we present recently obtained set of experimental data.

EXPERIMENT

As we alredy described the plasma source and the experimental setup in previous paper[4], in this one, only a few details would be considered. A low pressure (continuous flow of the He:H$_2$=1:1 gas mixture at an initial pressure of 1 mbar) repetitively pulsed arc is used as a plasma source. It consists of a low inductance, 2.5 µF discharge capacitor and a 10 mm internal diameter quartz discharge tube. The distance between the electrodes (on both ends of the tube) is 180 mm. The pulsed arc is fired at 4 kV by a grounded grid thyratron with a repetition rate of 105 pulses per minute. The discharge current was critically dumped, with a duration of 6.3 µs and a peak value of 1.3 kA. The light from the plasma source was observed end-on, through a 1m monochromator (inverse linear dispersion of 0.43 nm/mm) equipped with a photomultiplier. In order to perform the line spectrum's recordings from the repetitively pulsed discharge, the signals from the photomultiplier were averaged by a boxcaraverager&integrator while the monochromator was scanning continuously at the speed of 0.05 nm/min.

RESULTS

The experimental results for P_β full halfwidth $\Delta\lambda_{1/2}$ are given in Table I compared with theoretically calculated data $\Delta\lambda_{1/2}^K$ (quasistatic treatment of ions) and $\Delta\lambda_{1/2}^S$ (with ion-dynamics included) performed by Kepple[2] and Stehlé[3] respectively. In the same table, the experimental line "dips" are compared with the theoretical ones. The "dip" is here defined to be the quantity (peak-minimum)/peak, where the peak is the average of two maxima of P_β line.

Table I: Values of full halfwidths ($\Delta\lambda_{1/2}$) and "dips" of P_β versus electron density (N_e): exp. is for experimental results; K and S are for theoretical values from Kepple[2] and Stehlé[3] respectively.

N_e ($\times 10^{16}$ cm^{-3})	$\Delta\lambda_{1/2}^{exp}$ (nm)	$\dfrac{\Delta\lambda_{1/2}^{exp}}{\Delta\lambda_{1/2}^{K}}$	$\dfrac{\Delta\lambda_{1/2}^{exp.}}{\Delta\lambda_{1/2}^{S}}$	"dip"$_{exp}$	$\dfrac{\text{"dip"}_{exp.}}{\text{"dip"}_{K}}$	$\dfrac{\text{"dip"}_{exp}}{\text{"dip"}_{S}}$
2.80	0.53±0.05	0.87	0.97	0.14±0.01	0.36	0.74
2.68	0.54±0.05	0.88	-	0.16±0.01	0.39	-
2.58	0.53±0.05	0.92	1.02	0.14±0.01	0.36	0.72
2.49	0.50±0.05	0.96	0.99	0.18±0.01	0.44	0.90
2.41	0.48±0.05	0.94	0.96	0.17±0.01	0.43	0.86
2.33	0.48±0.05	0.97	1.01	0.17±0.01	0.42	0.86

REFERENCES

1. W. L. Wiese, D. E. Kelleher and V. Helbig, Phys. Rev. A11, 1854 (1975).
2. P.C. Kepple, Phys. Rev. A6, 1 (1972).
3. C. Stehlé, Astron. Astrophys. to be published.
4. N. Konjević, in "Spectral Line Shapes" Vol. 6, edited by L. Frommhold and J. Keto, API Conf. Proc. 216, New York (1990), p.16.

INFLUENCE OF ION-DYNAMICS ON THE SHAPE OF THE He I 4713 Å AND 7065 Å LINES

Z. Mijatović, N. Konjević*, R. Kobilarov and M. Ivković*

Institute of Physics, Trg Dositeja Obradovića 4, 21000 Novi Sad, Yugoslavia
*Institute of Physics, P.O. Box 68, 11080 Belgrade, Yugoslavia

This paper is an extension of our first study of the influence of the ion-dynamics to the width and shift of the two $He\ I$ lines.[1] Here we report the results of the study for additional two prominent isolated $He\ I$ lines in hydrogen-helium plasma. In order to achieve better quality of the line shape and shift recordings at lower electron densities when ion-dynamics effects are of greater importance, the new repetatively pulsed plasma source is built and signal averaging technique for light detection is applied. The electron densities in the range $(2.5-5.9)10^{15}\ cm^{-3}$ are measured by 10.6 μm laser interferometry, the electron temperatures ranging from 19300 to 23600 K are determined from the ratio of H_γ line intensity to the underlying continuum while the gas temperatures from 5000 to 12600 K are measured from the Doppler component of $He\ I$ line profiles. The experimental Stark widths and shifts are compared with theoretical results evaluated from the following formulae for the width[2,3] w and shift[1,2] d at the half-width

$$w = w_e(1 + gW_j A_n)N_e 10^{-16} \quad \text{and} \quad d = (d_e \pm 3.2 A_n g_1 D_j w_e)N_e 10^{-16}$$

where $g = 1.75(1 - 0.75R)$, $R = 0.090 N_e^{1/6} T_e^{-1/2}$, $A_N = AN_e^{1/4}10^{-4}$ and $g_1 = g/1.75$. In the above equations w_e and d_e are the electron impact half halfwidth and shift in angstrom units, respectively, and A is the ion broadening parameter. All three quantities are for electron density $N_e = 10^{16}\ cm^{-3}$. The dynamic ion broadening parameters W_j and D_j are described by the following expressions:

$$W_j = \begin{cases} 1.36 B^{-1/3}/g & , \quad B < (1.36/g)^3 \\ 1 & , \quad B \geq (1.36/g)^3 \end{cases}$$

and

$$D_j = \begin{cases} (2.35 B^{-1/3} - 3A_N^{1/3} R)/2g_1 & , \quad B < 1 \\ 1 & , \quad B \geq 1 \end{cases}$$

where $B = A_N^{1/3}(0.0806 w_e/\lambda^2 N_e^{2/3})(\mu/T_g)$ with atom-ion perturber reduced mass μ in *amu* and gas temperature T_g. In case when $W_j = 1$ and/or $D_j = 1$ the influence of ions to the line shape is treated quasistatically.

For evaluation of the $He\ I$ Stark widths and shifts we used electron impact half halfwidths w_e and shifts d_e and ion broadening parameter A from several theoretical calculations. For the examples presented in Fig. 1 the data from Ref. 4 are used.

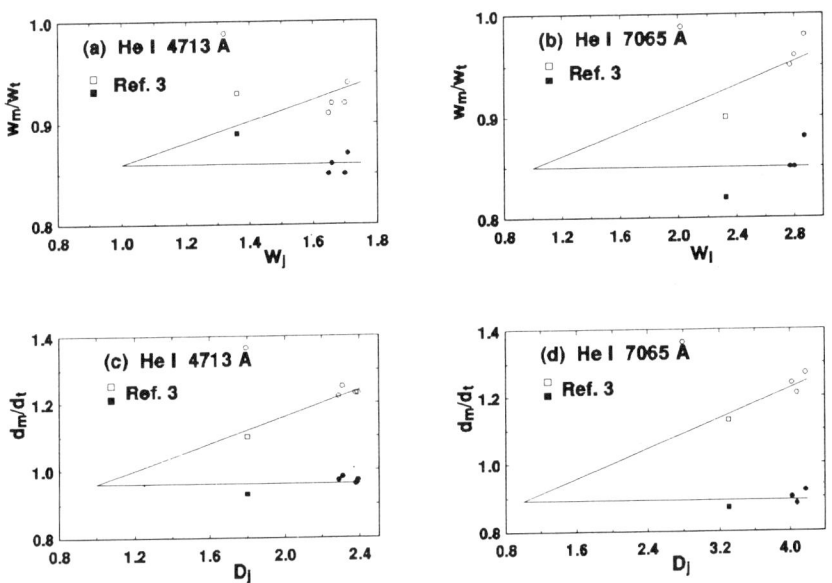

Fig. 1 w_m/w_t vs ion-dynamic parameter W_j for (a) 4713 Å and (b) 7065 Å line, and d_m/d_t vs ion-dynamic parameter D_j for (c) 4713 Å and (d) 7065 Å line.

REFERENCES

1. R. Kobilarov, N. Konjević and M. V. Popović, Phys. Rev. A **40**, 3871 (1989).

2. J. Barnard, J. Cooper and E. W. Smith, J. Quant. Spectrosc. Radiat. Transfer **14**, 1025 (1974).

3. D. E. Kelleher, J. Quant. Spectrosc. Radiat. Transfer **25**, 191 (1981).

4. H. R. Griem, *Spectral Line Broadening by Plasmas* (Academic, New York, 1974).

EXPERIMENTAL STUDY OF THE TEMPERATURE IN A HELIUM PLASMA

J.A. Aparicio, M.A. Gigosos, S. Mar, C. Pérez and M.I. de la Rosa
Departamento de Optica. Universidad de Valladolid.
47071 Valladolid, Spain

In this work, we give some evidences of the non-equilibrium between species in an Helium plasma. These evidences appear when calculating the plasma temperature on the base of the usual equilibrium hypothesis.

The plasma was created by discharging a capacitor bank of 20 μF charged up to 8 kV. The filling pressure was 30 mbar and a continuous flow of Helium gas, at a rate of 28 cm^{-3}/min, was kept through the lamp during the whole experiment.

Interferometric and spectroscopic end-on measurements were simultaneously done all over the plasma life, which is about 500 μsec. A Twyman-Green interferometer, working at two wavelengths (632.8 nm He-Ne and 488.0 nm Ar) allows us to obtain the electron density evolution. This ranges from 5×10^{21} to 1.4×10^{23} m^{-3}. Two spectrometers equipped with Optical Multichannel Analyzers were used for the plasma light studies. One of them, a Jarrell-Ash of 0.3 m focal length with low resolution, monitorizes the HeI 501.6 nm line in order to control the plasma repeatability. The other one, a Jobin-Yvon 1.5 m, was used to measure the intensities of several HeI lines 388.86, 412.08, 438.79, 471.31, 501.56, 587.56, 667.81, 706.51, 728.13 nm and the HeII 468.57 nm. More details about the experimental set-up are given elsewhere[1] and they will be provided at the Conference.

Assuming that the whole number of particles is constant, and that the number of neutral particles is always equal to the initial number minus the electron density, the temperature has been calculated from different methods. In the first one, we add the assumption that the HeI upper levels are populated according to the Boltzmann law, and with the help of the HeI lines intensities, we have obtained the temperature by a Boltzmann plot at each time of the plasma life. From this, the HeII temperatures have been also calculated. In the second method, we used Saha equation with the electron density determined experimentally. In both methods, we have observed that a relatively great variation of the particles density does not affect seriously to the obtained temperatures.

Results from both methods are given in the following figure. In this, it appears the evidence of non equilibrium between both species, at least during the emission time of HeII 468.6 nm line. Therefore, in this temporal range, it is not possible to use the intensities ratio of HeII/HeI lines. However, we think that, from the 70 μs, the population of the HeI upper levels follows the Boltzmann law, and in this way, we will be able to calculate the Boltzmann temperature.

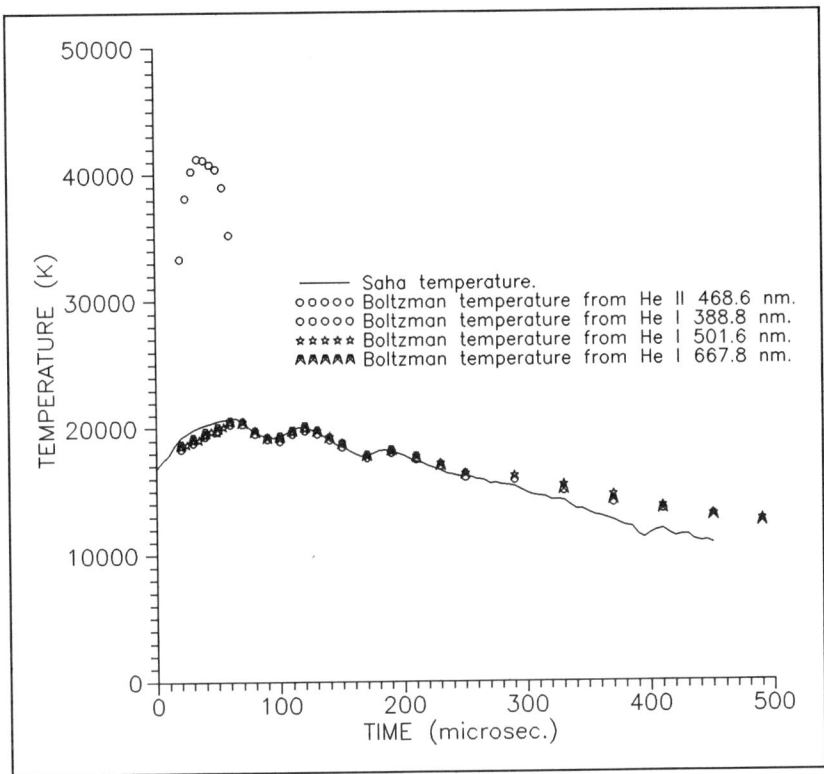

Authors will thank Santiago González for his help with the mechanical development and the Dirección General de Investigación Científica y Técnica (Ministerio de Educación y Ciencia) of Spain for its financial support under Contract No. PB-90-0353.

References

1. M. A. Gigosos, S. Mar, C. Pérez and I. de la Rosa, Phys. Rev. E, <u>49</u>, 2, 1575 (1994).

STRONG COLLISION CONTRIBUTIONS TO SHIFT AND WIDTH OF HYDROGEN SPECTRAL LINES

A. Könies and S. Günter
Fachbereich Physik, Universität Rostock, 18051 Rostock, Germany

INTRODUCTION

Whereas the electronic contributions to the width of spectral lines have been treated successfully within the unified theories (e.g. [1]), there are still open questions concerning the calculation of hydrogen line shifts. Due to the applied no-quenching approximation only shifts due to $\Delta n = 0$ transitions are included in the unified theory. Up to now, the most successful treatment of hydrogen line shifts has been reached using a second order Born approximation for the electron-atom collisions. However within a semiclassical treatment of the electron-radiator collisions, a Born approximation produces divergent integrals for shift and width. Although these divergencies may be overcome using a full quantum mechanical theory, contributions of strong electron-atom collisions to the line shift are considerably overestimated within a second order Born approximation. Therefore, up to now, a cut-off procedure for small impact parameters as proposed by Griem [3] has been applied.

THEORY

The width and shift of the two-particle energy can be obtained by the solution of the corresponding Bethe-Salpeter equation. Using a special scattering channel, namely the interaction of a plasma electron and the two-particle bound state, a Dyson equation for the T-matrix can be written:

$$\cdots + \cdots + \cdots = iT_3^{e,b} \quad (1)$$

This equation describing a three-particle scattering problem is solved approximately via partial summation (see [2]). Compared to the former second order Born approximation an additional correction factor $[1 + iA]^{-1}$ occurs in the self energy formula:

$$\Sigma(\Omega_\lambda) = -\frac{1}{e^2} \int d\vec{q} \int d\vec{p} \, \frac{1}{(2\pi)^6} \frac{M_{n\alpha}^{(o)}(-\vec{q}\,) \, V(q) \, M_{\alpha n}^{(o)}(\vec{q}\,)}{1 + iA(n, \vec{p}, \vec{q}, \Omega_\lambda, E_p)} \times$$
$$\frac{1}{(-i\beta)} \sum_\mu V^s(\vec{q}, -\omega_\mu) \, G_2(\alpha \, \Omega_\lambda - \omega_\mu) \frac{i(f_e(E_p) - f_e(E_{\vec{p}-\vec{q}}))}{E_p + \omega_\mu - E_{\vec{p}-\vec{q}}}. \quad (2)$$

For weak collisions (small momentum transfers) the correction term is small compared to unity, whereas it causes remarkable deviations from the Born approximation for large q thus accounting for strong collisions.

RESULTS

The theory which was outlined in the foregoing has been applied to several hydrogen lines. In the Table the shift of the H_α line is given in comparison with the experimental one and other theoretical results. The agreement between theory and experiment is very good.

Table I. Shift of the Hydrogen H_α line at an electron density of 10^{17} cm^{-3} and a temperature of 12000 K.

this paper	0.48 Å
Griem [3]	0.54 Å
Kelleher et al. [4]	0.43 Å
Halenka [5]	0.52 Å
Vitel [6]	0.31 Å (19000 K)

The same holds for P_α where the theoretical value for the maximum shift is 9.74 Å at an electron density of $10^{17} cm^{-3}$ compared to the experimental shift of 10.66 Å. [7].

Employing the theory, the width can be obtained in the same manner. A comparison with experimental results, however, needs the inclusion of ion dynamic effects.

It should be mentioned again that all results have been obtained within a consequent quantum mechanical many particle theory including strong collision contributions. Thus they do not depend from any arbitray parameters.

REFERENCES

[1] C.R. Vidal, J. Cooper, and E.W. Smith, J. Quant. Spectrosc. Radiat. Transfer 10, 1011 (1970), Astrophys. J. Suppl. 25, 37 (1973)

[2] S. Günter, Phys.Rev. E **44**, 500 (1993).

[3] H.R. Griem, Phys. Rev. A38, 2943 (1988)

[4] D.E. Kelleher, N. Konjevic, and W.L. Wiese, Phys. Rev. A **20**, 1195 (1979).

[5] J. Halenka, Proc. XVII ICPIG, eds. Bakes and Sörley, p. 993 (Budapest 1985).

[6] Y. Vitel, J. Phys. B: At. Mol. Phys. **20**, 2327 (1987).

[7] A. Dörn and V. Helbig, private communication

STARK BROADENING OF He I LINES

T. Schöning
Institut für Astronomie und Astrophysik der Universität München
81679 München Scheinerstraße 1 Germany

Modern techniques in astrophysical spectroscopy make it possible to observe He I absorption line spectra of hot stars with high resolution and signal to noise ratio. The analyses allow precise determinations of helium abundances and provide information of the physical conditions prevailing in the stellar atmospheres. Due to the high temperatures NLTE line formation theory is a prerequisite for these analyses and requires the use of accurate Stark broadening functions.

For isolated helium lines comprehensive data tables compiled from semiclassical[1,2] and unified theory[3] calculations are available. However, apart from the work of Gieske & Griem[4] reliable broadening data are still lacking for transitions involving quasi-degenerate levels. Hence abundance determinations of helium based on the diagnostics of overlapping He I lines could be affected by considerable errors due to theoretical uncertainties in the broadening parameters[5].

Therefore we have extended our previously developed quantum mechanical method for the computation of complete Stark profiles for multielectron ions[6] to the case of neutral radiators. First results have been obtained for selected $n = 2 \to n = 2, 3, 4$ transitions in the spectrum of neutral helium[7].

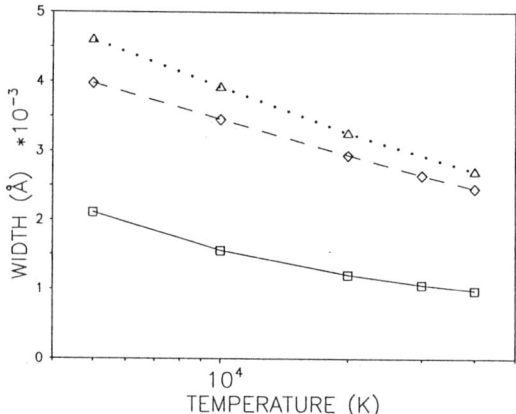

Figure 1: He I 2^1P^o–4^1D^e at 4922 Å. Electron impact line width versus electron temperature ($N_e = 1 \times 10^{13}$ cm^{-3}). Comparison of our CC calculations (full) with the semiclassical results of Dimitrijević & Sahal-Bréchot[1] (broken) and Griem[2] (dotted).

As an example we compare electron impact widths calculated according to our method using close-coupling theory (CC) with semiclassical results[1,2] for the $2^1P^o-4^1D^e$ transition (Fig. 1). The CC calculations deviate substantially from the semiclassical calculations but moderate convergence is observed with increasing electron temperature. The comparison with experimental measurements[8] reveals good agreement, especially with regard to the width of the allowed component which is overestimated by the unified theory[3] (Fig. 2).

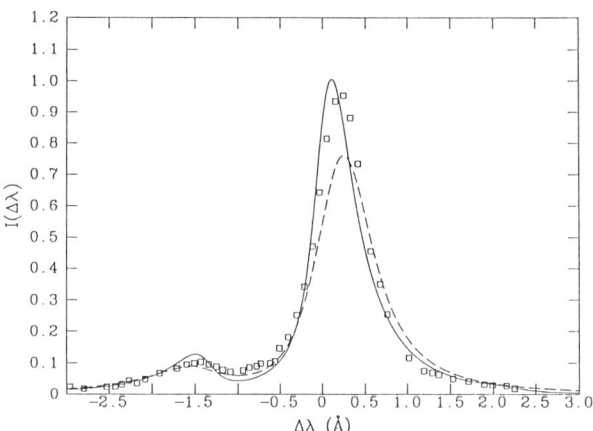

Figure 2: He I $2^1P^o-4^1D^e$ at 4922 Å. Comparison of the observed spectrum[8] (□) with our calculations (full) for the plasma parameters $N_e = 1 \times 10^{15} \text{cm}^{-3}$, $T_e = 18000$ K and gas temperature $T_0 = 13000$ K. Ion dynamic effects for the H$^+$ perturbers as well as Doppler and instrumental broadening have been included. A comparison with the unified theory[3] is also shown (broken).

REFERENCES

1. M. S. Dimitrijević, S. Sahal-Bréchot, Astron. Astrophys. Supp. Ser. 82, 519 (1990).
2. H. R. Griem, Spectral Line Broadening by Plasmas (Academic, N.Y., 1974), p. 320.
3. A. J. Barnard, J. Cooper, E. W. Smith, J. Quant. Spectrosc. Radiat. Transf. 14, 1025 (1974).
4. H. A. Gieske, H. R. Griem, Astrophys. J. 157, 963 (1969).
5. D. J. Lennon, P. L. Dufton, Astron. Astrophys. 155, 79 (1986).
6. T. Schöning, J. Phys. B: At. Mol. Opt. Phys. 26, 899 (1993).
7. T. Schöning, J. Phys. B: At. Mol. Opt. Phys. , submitted.
8. H. Richter, A. Piel, J. Quant. Spectrosc. Radiat. Transf. 33, 615 (1985).

BROADENING AND SHIFT OF THE PASCHEN ALPHA LINE

A. Döhrn and V. Helbig
Universität Kiel, D 24098 Kiel, Germany

S. Günter and A. Könies
Universität Rostock, D 18051 Rostock, Germany

The width and the shift of the plasma broadened P_α line was determined from the spectrum of a wallstabilized arc runnning in neon or argon with admixtures of hydrogen. In order to increase the electron density the arc chamber was designed to allow measurements at pressures up to 7 atm.. For the spectral investigations a 2m vacuum monochromator was used that prevented water vapour absorption to degrade the recorded line profiles. A lead-selenid element served as detector.

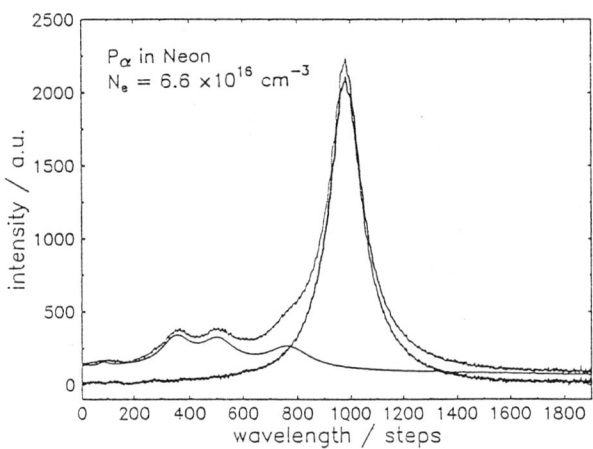

Fig. 1: Profile of the P_α line emitted from a neon plasma. The upper curve shows the recorded profile. The lower curve is the profile after subtracting the neon background shown in between obtained from a model function that was least squares fitted to the original curve.

For the plasma analysis a second monochromator was installed for simultaneously recording the profile of H_β. The halfwidth of this line was used as electron density probe using the broadening parameters of[1]. The recorded data were digitized and stored on a computer for further processing. The main problem in the data analysis of the P_α-line were neighbouring lines that partially overlapped at higher electron densities. Fig. 1 shows an example for the neon case. A model function accounting for all known spectral lines of the carrier gas in the surrounding of the hydrogen line was used in a least squares routine to fit the recorded spectrum and to extract the neon or argon background.

Full shifted and asymmetric line profiles have been calculated using a consequent quantum mechanical many-particle approach. Contributions of strong electron-atom collisions are included via partial summation of the corresponding T-matrix[2]. In Fig. 2 the calculated shifts are compared to the experimental ones. The dashed line shows the resulting shift including many-particle effects whereas the solid line represents the shifts resulting from a binary collision approximation. The increasing uncertainties at higher electron densities indicated in the figure are due to problems with the carrier gas lines.

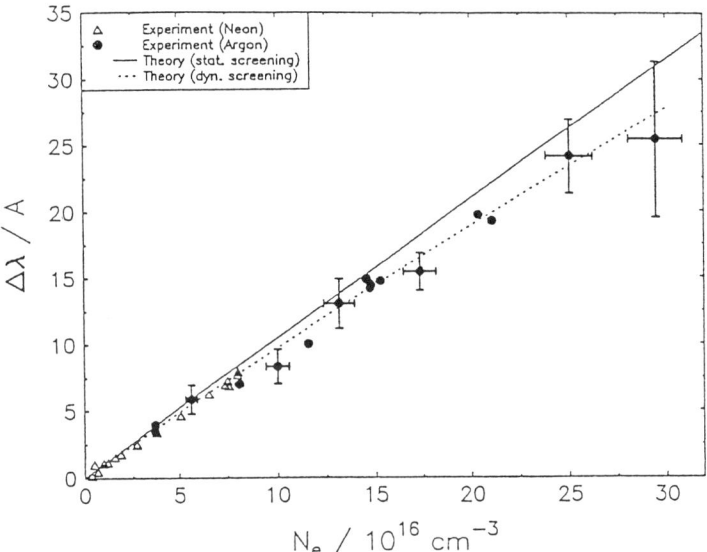

Fig. 2: Shift of the P_α line versus electron density. The experimental values from the neon and the argon measurements are compared to theoretical data.

[1] V. Helbig and K.-P. Nick, J. Phys. B **14**, 3573 (1981).
[2] A. Könies and S. Günter, this volume, PB-6.

CO-OPERATIVE COLLISION PROCESSES FOR THE POPULATION OF IONIC EXCITED STATES

S. BLIMAN

URA 775 - CNRS, LSAI, Univ. Paris-Sud, 91405 Orsay Cedex, France
and LRME, Univ. Marne la Vallée, 93166 Noisy-le-Grand, France

M. CORNILLE

UPR 176 - CNRS, DARC, Obs. de Paris, 92195 Meudon Cedex, France

K. KATSONIS

URA 073 - CNRS, LPGP, Univ. Paris-Sud, 91405 Orsay Cedex, France

In analyzing spectral lines emitted by ions embedded in hydrogen plasmas, a current modelling assumption is that the ionization equilibrium is dominated by ionization and dielectronic recombination. On the basis of experimental observation we show that in many different isoelectronic sequences where, above a ground state sits a long-lived metastable state, the charge exchange (CX) collisions populate states currently identified as resulting from dielectronic recombination (DR).

The He-like Carbon ion C^{4+} has a long-lived metastable state $1s2s\ ^3S_1$ which in case of charge exchange collision gives:

CX $\quad C^{4+}\ (1s2s)\ ^3S_1 + H \rightarrow C^{3+}\ [(1s2l)\ ^3L\ nl]\ ^{2,4}L + H^+$

(note that ^1S states decaying radiatively by two photon emission are disregarded) whereas dielectronic recombination gives:

DR $\quad C^{4+}\ (1s^2)\ ^1S_0 + e^- \rightarrow C^{3+}\ [1s2ln'l']$.

The Li-like ion $C^{3+}\ (1s^2\ 2s)\ ^2S_{1/2}$ through DR and Transfer Excitation (TE) would give:

DR $\quad C^{3+}\ (1s^2\ 2s)\ ^2S_{1/2} + e^- \rightarrow C^{2+}\ (1s^2\ 2pnl)\ ^{1,3}L$

TE $\quad C^{3+}\ (1s^2\ 2s)\ ^2S_{1/2} + H \rightarrow C^{2+}\ (1s^2\ 2pnl)\ ^{1,3}L + H^+$.

For C^{2+} there exists a long-lived metastable level C^{2+} ($1s^2\,2s2p$) $^3P_{0,2}$ above the ground state C^{2+} ($1s^2\,2s^2$) $^1S_0$. These ions give respectively:

DR $\quad\quad C^{2+}\,(1s^2\,2s^2)\,^3S_0 + e^- \rightarrow C^+\,(1s^2\,2s2pnl)$

CX $\quad\quad C^{2+}\,(1s^2\,2s2p)\,^3P_{0,2} + H \rightarrow C^+\,(1s^2\,2s2pn'l')\,^{2,4}L + H^+$.

For the case of measured or/and calculated cross sections the corresponding rate coefficients have been estimated showing the importance of CX in comparison with DR (see e.g. Bliman et al. 1994). For instance, our CTMC calculations give for the He-like Carbon CX cross section a maximum of about $25*10^{-16}$ cm^2 for a collision energy of 10 keV/amu, followed by a plateau towards smaller energies, leading to a much higher charge exchange rate than the total DR rate given in the litterature for C^{4+} (e.g. a maximum of $1.2*10^{-12}$ cm^3/s^{-1} for $10^{6.4}$ K approximatively as calculated by Badnell et al. 1990).

REFERENCES

S. Bliman, M. Cornille, K. Katsonis: Phys. Rev. A, 1994, in press.

N.R. Badnell, M.S. Pindzola, D.C. Griffin: Phys. Rev. A, 1990, 41, 2422.

LOW DENSITY PLASMAS: COMPLEX RADIATORS AND GENERAL RESULTS

TEMPERATURE DEPENDENCE OF THE TRIPLY IONIZED OXYGEN STARK WIDTHS

N.Konjević, B.Blagojević, M.V.Popović and M.S.Dimitrijević*
Institute of Physics, 11080 Belgrade, P.O.Box 68, Yugoslavia
*Astronomical Observatory, 11050 Belgrade, Volgina 7, Yugoslavia

INTRODUCTION

Broadening and shift of spectral lines in plasmas are subject of numerous experimental studies. Unfortunately, most of the reported data are from the measurements at a single electron temperature or in the best case the results are taken in a small temperature range. Tne lack of the experimental data in a wider temperature range makes a detailed test of the Stark broadening theoretical calculations unreliable. Furthermore, without the knowledge of the line width and shift dependence upon electron temperature comparison of the experimental results obtained at different plasma conditions becomes very difficult.

The aim of this paper is to supply the experimental and theoretical data for the widths of the prominent triply ionized oxygen lines in a large electron temperature range. The reported experimental results together with other experimental data will be used for the testing of various theoretical calculations.

THEORY

By using the semiclassical-perturbation formalism[1] we have calculated electron-, proton-, and ionized helium-impact line widths for O IV $3s^2S-3p^2P^0$ and $3p^2P^0-3d^2D$ multiplets and the results are given in Table 1.

TABLE 1.

| | | PERTURBER | | |
| | | Electrons | Protons | Ionized He |
Transition	T [K]	Width [A]	Width [A]	Width [A]
3s-3p	40000	1.100e-01	2.380e-02	3.250e-03
3066.4 [A]	70000	8.560e-02	3.660e-03	4.420e-03
c=0.28e21	100000	7.410e-02	4.440e-03	5.200e-03
	170000	6.100e-02	5.630e-03	5.880e-03
3p-3d	40000	1.170e-01	2.230e-03	3.080e-03
3410.9 [A]	70000	9.100e-02	3.620e-03	4.310e-03
c=0.34e21	100000	7.840e-02	4.410e-03	5.200e-03
	170000	6.420e-02	5.850e-03	6.050e-03

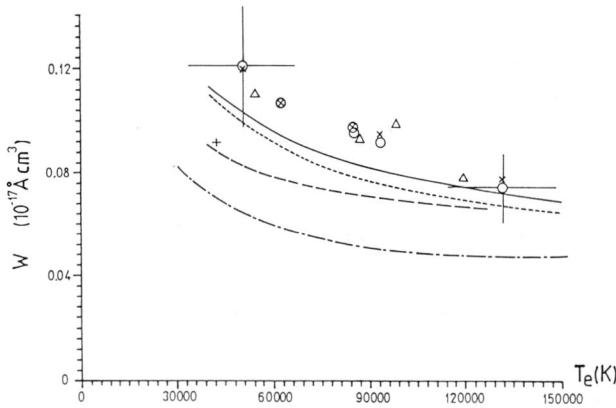

Fig 1. Full Stark widths (refered to an density of 10^{17}cm^{-3}) for the O IV 3s-3p multiplet vs electron temperature. Theory: ___, semiclassical, electrons + He$^+$ impact widths, ..., semiclasical, electrons only (see Table 1.), _ _, semiclassical aproximation[5,4], _._, modified semiempirical formula[5].

EXPERIMENT AND COMPARISONS

The Stark widths of the O IV lines belonging to the multiplets form Table 1 are measured in the plasma (initial gas mixture He:O$_2$=98.6:1.4) of a low preasure pulsed arc. Electron densities in the range (2.1-6.4)x10^{17}cm^{-3} were determined from the width of the He II P$_\alpha$ line while electron temperatures between 50800 and 131800 K are measured from the Boltzman plot of O IV line intensities. Our experimental results for 3s^{2}S-3p^{2}P^0 multiplet are compared in Fig.1 with other experiments[2,3] and our semiclassical results from Table 1. For comparison other simplified theoretical approaches[4,5] are taken as well.

REFERENCES

1. S.Sahal-Brechot, Astron.Astrophys. 1, 91 (1969); 322 (1969).
2. J.Purić, S.Đeniže, A.Srećković, M.Platiša and J.Labat, Phys.Rev. A37, 498 (1988).
3. S.Glenzer, J.D.Hey, H.J.Kunze, J. Phys. B27, 413 (1994).
4. H.R.Griem, Spectral Line Broadening by Plasmas (Academic, New York, 1974)
5. M.S.Dimitrijević, and N.Konjević, J. Quant. Spectrosc. Radiat. Transfer 24, 451 (1980).

INFLUENCE OF ION-DYNAMICS ON THE SHIFT OF C I 5052.17-Å SPECTRAL LINE IN PLASMA

Z. Mijatović, N. Konjević*, R. Kobilarov and S. Djurović

Institute of Physics, Trg Dositeja Obradovića 4, 21000 Novi Sad, Yugoslavia

*Institute of Physics, P.O. Box 68, 11080 Belgrade, Yugoslavia

I. INTRODUCTION

The experimental studies of ion-dynamics influence to the width and shift of isolated non-hydrogenic atom lines were concentrated to the lines of neutral helium.[1,2]. In this paper we extend these investigations to the shift of C I 5052-Å line.

II. EXPERIMENTAL PROCEDURES AND RESULTS

For the plasma source an argon atmospheric pressure wall stabilized electric arc is used. The gas mixture of $H_2 : Ar : CO_2$ (4 : 64 : 32) is introduced in the middle part of the arc. The current, ranging from 19 to 30 A, was supplied to the arc by a current-stabilized power supply with stability of 0.3%.

The plasma observations were performed end-on with a 1-m monochromator and photomultiplier tube. The low pressure lamp with microwave excitation is used as a source of unshifted spectral lines. The neon lamp contained certain amount of carbon compounds impurities so C I 5052.17-Å is detected and used for the shift measurements of the same line emitted from the arc plasma. For the shifts measurements, the light from the arc plasma and from the reference source is focused onto the slit of the monochromator. Using the chopper, the light could be obtained from the arc or from the reference source alternatively. Signals from the photomultiplier were led to the digitizing oscilloscope working in the averaging mode. The stepping motor of monochromator, chopper and oscilloscope are controlled by the personal computer. The same computer is used for data aquisition.

The spectral line profiles recorded from the reference source were fitted by least square method to the Gaussian. The spectral line profiles from the arc plasma were also treated by the computer program developed for deconvolution of Gaussian and $j_{A,R}(x)$.[3] Shift of plasma broadened lines is always measured at the half width. The test for the self absorption is carried out and all the investigated spectral lines are found optically thin.

An electron density in the range $(1.42\text{-}2.90)\ 10^{16}\ cm^{-3}$ is determined from the width of Balmer H_β line.[4] The electron temperatures are deduced from the plasma composition data evaluated as described in Ref. 5

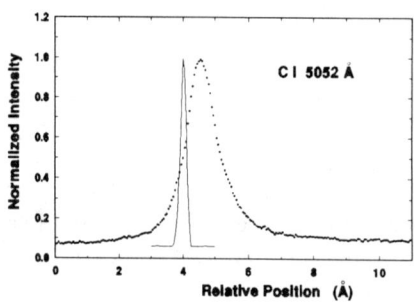

Fig. 1 An example of recorded profiles

An example of measured reference and plasma shifted profile is given in Fig.1. Experimental results d_m over theoretical values d_t for the shifts are presented in Fig. 2. D_j is ion-dynamics shift parameter, which is measure by ion-dynamics influence on the shifts. When $D_j = 1$ the static and dynamic approximations are equal. The theoretical values are calculated by using the quasistatic (open symbols) and dynamic (closed symbols) treatment of ions. Data for w_e and A are taken from Ref. 6.

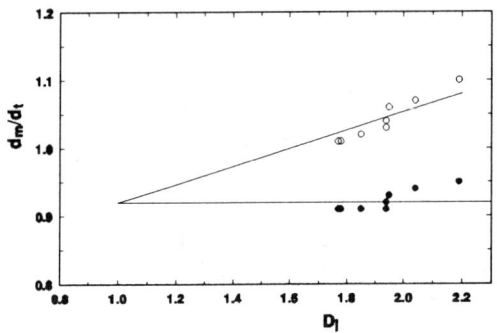

Fig. 2 d_m/d_{th} vs ion-dynamic parameter D_j

REFERENCES

1. D. E. Kelleher, J. Quant. Spectrosc. Radiat. Transfer **25**, 121 (1981).

2. R. Kobilarov, N. Konjević and M. V. Popović, Phys. Rev. A **40**, 3871 (1989).

3. Z. Mijatović, R. Kobilarov, B. T. Vujičić, D. Nikolić and N. Konjević, J. Quant. Spectrosc. Radiat. Transfer **50**, 339 (1993).

4. C. R. Vidal, J. Cooper and E. W. Smith, Astrophys. J. Suppl. Ser. **25**, 37 (1973).

5. W. B. White, S. M. Jonson and G. B. Dantzig, J. Chem. Phys. **28**, 751 (1958).

6. H. R. Griem, *Spectral Line Broadening by Plasmas* (Academic, New York, 1974).

STARK PARAMETERS REGULARITIES WITHIN SPECTRAL SERIES OF SEVERAL MULTICHARGED IONS

J. Purić and M. Ćuk

Faculty of Physics, University of Belgrade, P. O. Box 368, 11001 Belgrade, Yugoslavia

ABSTRACT

Stark width and shift dependencies on the upper level ionization potential within several N V, S VI and O VI spectral series have been found and discussed. After being well established using various theoretical calculations the dependencies have been used to predict additional Stark broadening and shift data for several N V, S VI and O VI higher members of the investigated spectral lines.

INTRODUCTION

Recently published comprehensive set of Stark broadening data of N V, S VI and O VI spectral lines,[1-3] has been used here to demonstrate the existence of Stark widths and shifts data regularities within several N V, S VI and O VI spectral series.Namely, Stark parameters dependence on the upper level ionization potential of particular line was found within 2s - np, 2p - ns, 3s - np and 3p - ns spectral series. Different kinds of regularities within Stark parameters of a given spectra can be explained on the bases of their dependence on the upper level ionization potential.[4-6] A general form of that dependence is

$$w, d = A\chi^{-b} \qquad (1)$$

where w and d are the line width and shift, χ is the corresponding upper level ionization potential and A and b depend on temperature and electron density but are independent on χ.

RESULTS AND DISCUSSION

It has been verified for the first time that the Eq.(1) is appropriate not only for the electron-impact width and shift but, also, for proton- and helium-impact parameters for the investigated ion spectral series. As the example in Fig. 1 are given: a) electron-impact width (w_e) and shift (d_e); b) proton-impact width (w_p) and shift (d_p); and helium-impact width (w_{He}) and shift (d_{He}) as the functions of the inverse value of the upper level ionization potential χ for the electron density $N=10^{17}$ cm^{-3} and different electron temperatures (o-100 000 K; Δ-200 000 K; $\square$-500 000 K and $\otimes$-800 000 K). By a compasrison of the regularities found here and those presented elsewhere (Figures 1 - 7)[3] one can conclude that the method used here differs in the choice of the variable conveying atomic structure information. Prior work[3] was based on the hydrogenic model. Consequently, it used integer principal quantum numbers instead of the upper level ionization potential. Although both parameters take into account the density of states perturbing the emitting state, the advantages of the present method are: (i) χ - based trend analyses achieve better fits; (ii) in χ values the lowering of the ionization potential[7] is taken into account, predicting merging with continuum when the

plasma environment causes a line' s upper state ionization potential to approach zero;and (iii)the χ dependence of w and d are theoretically expected.[4-6]
Using the existing Stark parameters data for the investigated lines from N V, S VI and O VI spectral series the corresponding coefficients A and b from Eq. (1) are found. The corresponding correlation's factors were

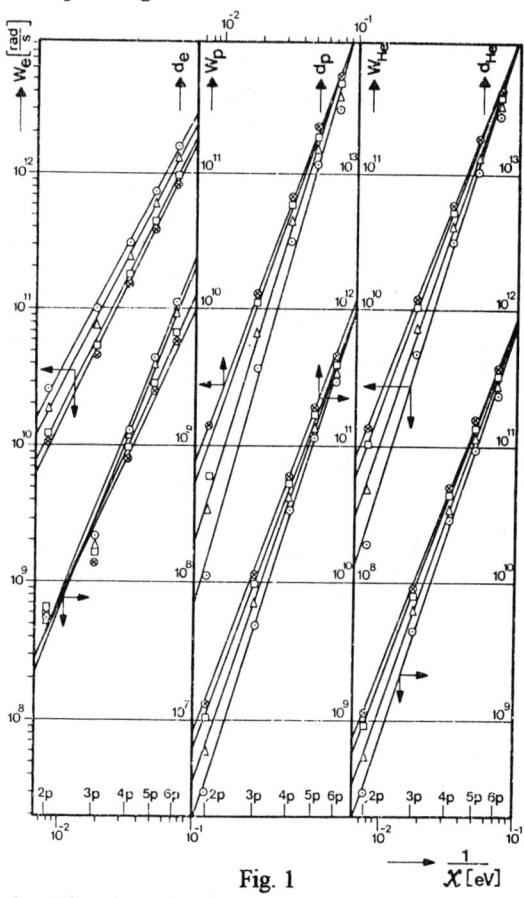

Fig. 1

almost equal to unity. Therefore, the Eq. (1) can be used to calculate Stark parameters of the higher members of the spectral series not calculated so far.

REFERENCES
1. M. S. Dimitrijević and S. Sahal- Brechot, A&AS, 95, 109 (1992)
2. M. S. Dimitrijević and S. Sahal- Brechot, A&AS,100, 81 (1993)
3. M. S. Dimitrijević and S. Sahal- Brechot, A&AS,93, 359 (1992)
4. J. Purić, M. Ćuk, M. S. Dimitrijević, A. Lesage, ApJ, 382, 353 (1991)
5. J. Purić. M. H. Miller. A. Lesage, ApJ, 416, 825 (1993)
6. J. Purić, XXI ICPIG, Invited Papers, ed. G. Ecker, U. Arendt,, J. Boseler, Arbeitgemeinschaft Plasmaphysik, Ruhr-Universitat, Bochum,1993, p.34
7. D. R. Inglis, E. Teller, ApJ, 90, 439 (1939)

"EXACT" ANALYTIC FORMULAS FOR THE STARK BROADENING OF ISOLATED ION LINES

S.Alexiou and Y.Maron
Dept. of Physics
Weizmann Institute of Science
Rehovot 76100, Israel

Starting with the Poquerusse[1] expressions for the (electron) impact-parameter integrated width function for isolated ion lines, we have integrated analytically over a Maxwellian distribution to obtain the exact (to within numerical approximations to integrals that are not doable in closed form and that give a total error of about 10% or less), closed form, semiclassical collision operator factor for a channel involving an energy transfer of $\Delta\omega$ as a function of $\Delta\omega$, density, Temperature and minimum and maximum impact parameters, assuming constant maximum and minimum impact parameters[2]. Collisions are separated into "hydrogenic", "inelastic" (meaning a substantial energy transfer, but with small velocities) and "adiabatic" and the contribution of adiabatic collisions is neglected.

The minimum impact parameter is assumed to be velocity-independent in the integration. It may be checked whether the default minimum impact parameter equal to the wavefunction extent satisfies unitarity[3]. If it does not, one can still use these formulas with the minimum impact parameter that corresponds to the average velocity. The associated program[4] has an option for doing this automatically if the wavefunction extent cutoff fails to preserve unitarity.

There are two cases, corresponding to whether for a given energy spacing $\Delta\omega$ and maximum/minimum impact parameter ρ an adiabatic regime exists or not. It is seen that no adiabatic regime exists as long as

$$\Delta\omega \leq 0.444380185\sqrt{\frac{(Z-1)e^2}{4\pi\epsilon_0 m\rho^3}} \quad (1)$$

with Z the spectroscopic charge number. In that case there is a high velocity hydrogenic regime and a low velocity "inelastic" regime. Otherwise, in addition to these two regimes, an adiabatic regime, corresponding to intermediate velocities, also exists. It is only for small $\xi \leq 0.2 - 0.3$ that one apply the straight line path adiabaticity criterion to cut off the interaction at a maximum impact parameter of $v/\Delta\omega$.

Fig. 1 illustrates the different possibilities: If η is defined as $\eta = \sqrt{\epsilon^2 - 1}$ with ϵ the eccentricity, $G = \rho(\frac{4\pi\epsilon_0 m\Delta\omega^2}{(Z-1)e^2})^{1/3}$, $B = 0.15G^{-3/2}$ and $x_0 = 0.2^{-2/3} \approx 2.92$, the first case corresponds to a large B that does not intersect the curve shown. In that case $\eta > Gx_0$ is the hydrogenic and $\eta < Gx_0$ the inelastic regime. If B intersects the curve, then the region between the two points of intersection corresponds to the adiabatic regime, the region $\eta > \eta_2$ is hydrogenic, the small η region is inelastic and if $G < 0.6217$ there is a further hydrogenic region (Gx_0, η_1).

Comparisons with fully numerical calculations[4,5] confirm a better than 10% agreement provided the minimum impact parameter is such that unitarity is conserved at the average velocity. Fully numerical calculations[6,7] can, of course, include other effects, such as quadrupole terms. These formulas should be useful for the fast and accurate (within the semiclassical dipole approximation) calculation of the weak collision contribution to Stark widths of isolated ion lines and even of the diagonal part of the partially overlapping line collision operator.

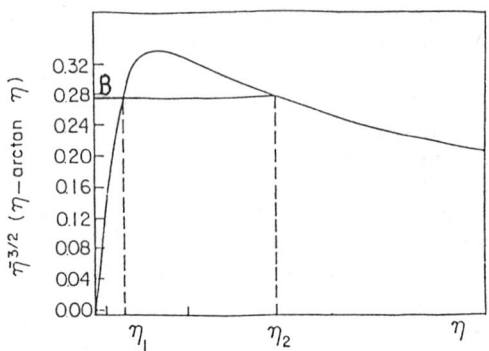

Fig.1. Hydrogenic, inelastic and adiabatic regimes.

REFERENCES

1. A.Poquerusse, Phys. Lett.**59A**, 438 (1977).
2. S.Alexiou and Y.Maron, Submitted to JQSRT.
3. S.Alexiou, JQSRT **51**, 849(1994).
4. S.Alexiou, "Manual for the Program for the quick Calculation of the Stark widths of isolated ion lines", WIS-94/1/JAN.- PH, Weizmann Institute of Science (unpublished).
5. S.Alexiou, Phys.Rev.A.**49**, 106 (1994).
6. S.Sahal-Brechot, Astron.Astrophys.**1**, 91(1969).
7. S.Alexiou and Yu.Ralchenko, Phys.Rev.A.**49**, 3086(1994).

FERMI METHOD OF EQUIVALENT PHOTONS FOR STIMULATED EMISSION OF POLARIZATION-INDUCED MULTIPHOTON RADIATION BY MULTIPLY CHARGED IONS

V.A.Astapenko[1], A.B.Kukushkin[2]

[1]Moscow Institute for Physics & Technology, Moscow, Russia
[2]RRC "Kurchatov Institute" Moscow, Russia

ABSTRACT

Polarization-induced emission[1] of radiation by an atom[2,3] in plasmas including multiphoton stimulated emission is of growing interest in recent years[1]. The mechanism of this process is based on the transformation of a part of the proper electric field of incident particle (electron) into polarization (virtual excitation) of the target (atom, ion with the core), which results in the emission of the photon(s) by the compound system {target atom + incident particle}. Most of the results for this problem pertain to calculable case of Born approximation for the motion of incident particle[1]. Application of the Fermi method of equivalent photons (EPHs) allows - for the case of multiply charged ions (MCI), within the frame of dipole approximation for the interaction between incident particle and radiating atomic electron - to extend the theory to the case of arbitrary $Ze^2/\hbar v$ (including case of quasiclassical motion of incident electron of velocity v in the field of the ion of charge Z) and arbitrary type of radiation transition (besides free-free, the free-bound[4] and even bound-bound[5] transitions, see also Ref.[6]).

In present note we present an extension of the EPHs method to the case of multiphoton stimulated emission in electron-MCI collisions under condition of negligible recoil of radiating electron. Here, the EPHs method reduces the problem of inelastic electron-MCI collision in a laser field to the scattering of the EPHs, produced by a fixed quantum/classical electric current, by the target atom, with the scattering process being influenced by stimulated multiphoton emission effects. The extension is essentially based on the fact that the probability W_n of stimulated emission of n photons in laser field with photon occupation number n_{ke}^{las} may be represented, both for the Born- approximation[7] and quasiclassical[8,9] cases for incident electrons motion, in a universal form which, for a strong laser field, $n_{ke}^{las} \gg 1$, takes the form

$$W_n = W_n^{stat} \equiv J_n^2\left[\left(n_{ke}^{las} n_{ke}^{stat}\right)^{\frac{1}{2}}\right] \qquad (1)$$

where n_{ke}^{stat} is the occupation number for the photons spontaneously emitted by the incident electron in the (static) Coulomb field of the MCI in the regime of conventional one-photon static Bremsstrahlung. Equation (1) enables us to arrive at the following result for the probability of multiphoton stimulated emission of n photons by polarization mechanism:

$$W_n^{pol} \equiv J_n^2 \left[\left(n_{ke}^{las} n_{ke}^{pol} \right)^{\frac{1}{2}} \right] \qquad (2)$$

where the occupation number n_{ke}^{pol} for the photons with wave number $\vec{k}$ and polarization $\vec{e}$, is given for the electron-MCI collision by the universal relationship[4] (r_e is classical electron radius):

$$\frac{n_{ke}^{pol}}{n_{ke}^{stat}} = (Zr_e)^{-2} \frac{3}{8\pi} \sigma_{scat}(\omega) = \left(\frac{m_e \omega^2 \alpha(\omega)}{Ze^2} \right)^2 \qquad (3)$$

Here, σ is the cross-section for the scattering of the EPHs by the MCI, which allows for the presence of laser field and is expressed in terms of polarizability $\alpha(\omega)$, for polarization-induced emission (i.e. elastic scattering of the EPHs), and scattering tensor, for the inelastic case[10] (excitation/deexcitation of the core), respectively.

Polarization-induced stimulated emission of multiphoton radiation by the MCI is to be taken into account in kinetics of radiation processes in hot plasmas with increasing laser field intensity.

REFERENCES

1. Polarization Bremsstrahlung, Eds. V.N.Tsytovich & I.M. Oiringel (Plenum, N.Y. 1992).
2. P.A.Golovinskii, Soviet Phys. JETP 94, 87 (1988).
3. P.A.Golovinskii, M.A.Dolgopolov, V.G.Khlebostroev, Fizika Plazmy (Sov.J.Plasma Phys.) 20, No.6 (1994) (to be published).
4. A.B.Kukushkin, V.S.Lisitsa, In Ref.[1], Chapt.11, p.261.
5. A.B.Kukushkin, V.S.Lisitsa, Phys. Lett.A, 159, 184 (1991).
6. V.I.Kogan, A.B.Kukushkin, V.S.Lisitsa, Phys.Reports, 213, 1 (1992).
7. F.V.Bunkin, M.V.Fedorov, Zh.Exp.Teor.Fiz. (Soviet Phys. JETP), 49, 1215 (1965).
8. I.Ya. Berson, Soviet Phys. JETP, 53, 891 (1981); 56, 731 (1982).
9. V.S.Lisitsa, Yu.A.Saveliev, Soviet Phys. JETP, 65, 273 (1987).
10. V.A.Astapenko, A.B.Kukushkin, V.S.Lisitsa, J.Phys.B: At.Mol.Opt.Phys., 25, 1985 (1992).

Lasing on a Weak Intercombination Transition $(4p^4 S_{3/2} \to 4s^2 P_{3/2})$ in ArII Plasma

Shapiro D. A., Kablukov S. I., Babin S. A., Khorev S. V.
Institute of Automation & Electrometry
Novosibirsk, 630090, Russia

In Ref. 1, it was reported that a strong generation had been obtained on a violet argon line around 438 nm. The objective of the presented work was to identify this line, optimize discharge parameters for output power, and to acquire its spectroscopic characteristics.

Experiments were carried out with a 7 mm-bore discharge tube, 0.5 m long, at current 140 A. Measures have been taken to maintain high discharge uniformity. Generation occurred under multimode multi-frequency operation.

A detailed experimental study revealed a number of oscillating lines in the vicinity of 438 nm: 438.4, 437.6, 437.1, and 436.2 nm, see Fig. 1. CW generation on the lines 437.1 and 437.6 nm has been demonstrated earlier, whereas it has been observed on the lines 438.4 and 436.2 nm for the first time in this work. The highest output power (1.2 W) was achieved on the 438.4 nm line $(4p^4 S_{3/2} \to 4s^2 P_{3/2})^2$ at the optimum coupling mirror transmittance ($\simeq$ 0.25 %). Unlike other registered lines which belong to the doublet-doublet system of transitions, the mentioned line corresponds to a quartet-doublet system. This is characterized by a considerable shift of optimal filling pressures to higher values as compared with doublet-doublet lines (Fig. 2), similar to the corresponding shift of the optimum for the long-known quartet-doublet line 514.5 nm. Branching coefficient of the investigated transition proved to be around $\simeq 0.01$, and its Einstein coefficient was $\sim 10^6$ s^{-1}, which is by 1–2 orders lower than those of doublet-doublet transitions[3]. Owing to long lifetimes of the lower level of the adjacent allowed transitions, for instance, $4p^4 S_{3/2} \to 4s^4 P_{5/2}$ with the Einstein coefficient $A = 7 \times 10^7$ s^{-1}, population inversion on them is absent[3]. On the basis of dependence of intracavity power on the coupling mirror transmittance (Fig. 3), the saturation parameter and unsaturated gain were calculated to be $\sim 10^4$ W/cm^2 and ~ 0.8 %/m correspondingly.

Considering the analogous data on the lines 514.5 and 488.0 nm, it was derived that the pumping rate of the higher laser level of the 514.5 nm line is approximately the same as that of the 438.4 nm line and is by a factor of 3 higher than for the 488.0 nm line.

The performed investigation, thus, has demonstrated the possibility of high power generation on a forbidden transition with branching coefficient 0.01, that is achieved due to high pumping rate of $4p^4 S_{3/2}$ level, the population inversion being absent on adjacent allowed transitions.

86 Lasing on a Weak Intercombination

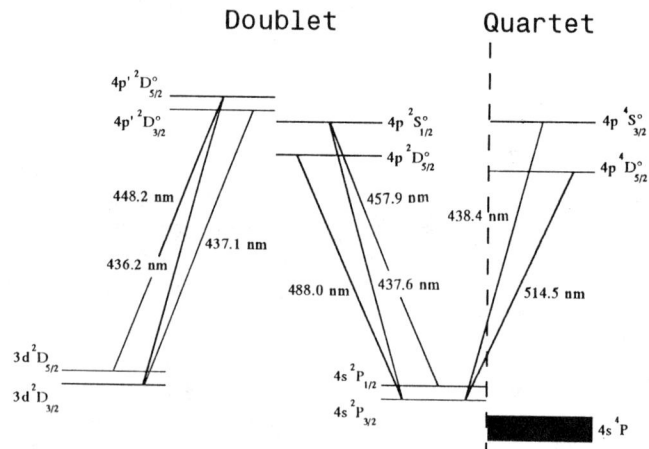

Fig 1. Partial ArII level diagram

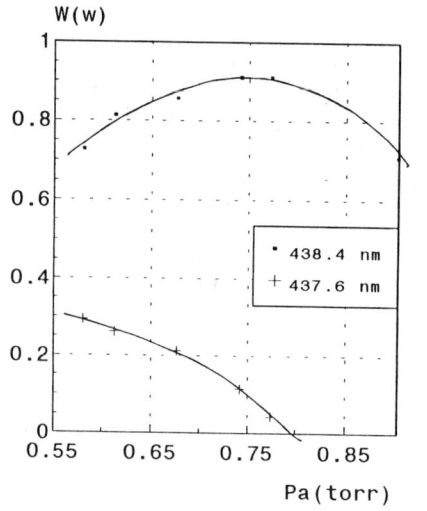

Fig 2. Output on 438.4 and 437.6 nm vs pressure

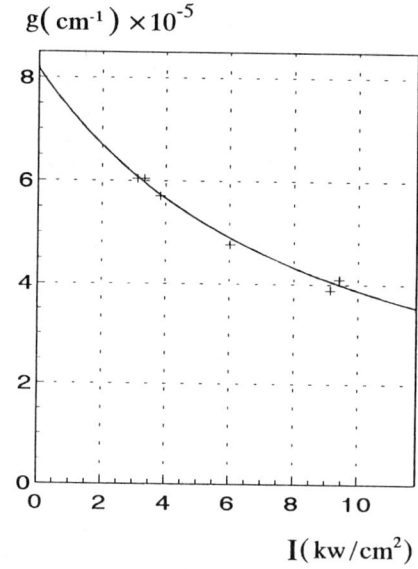

Fig 3. Gain saturation.

1. Babin S. A., Khorev S. V. Yeremenko T. Yu. "Laser Optics '93", the Book of Abstracts, St.-Petersburgh, 1993, p165.

2. Odintsova G. A., Striganov A. R. Tablitsi spektralnykh linii atomov i ionov [Tables of Ionic and Atomic Spectral Lines, in Russian], "Energoizdat", 1982.

3. B. F. Luyken// *Prog. Quant. Electr.* 1976, v.60 pp.447–454.

INFLUENCE OF THE OSCILLATOR STRENGTHS ON THE STARK BROADENING OF Rb I LINES

M.S.Dimitrijević
Astronomical Observatory, Volgina 7, 11050 Beograd, Yugoslavia

S.Sahal-Bréchot
Observatoire de Paris-Meudon, 92190 Meudon, France

INTRODUCTION

Neutral Rubidium Stark-broadening parameters are of significance for laboratory plasma research [1] as well as for Solar and stellar spectroscopy, since Rb I lines have been observed in solar and stellar spectra[2]. By using the semiclassical-perturbation formalism[3,4], we have calculated electron-, proton-, and ionized argon-impact line widths and shifts for 24 Rb I multiplets. The obtained results for Stark broadening parameters will be published elsewhere[5]. Here, we will discuss the results for Rb I, along with a comparison with experimental data and other theoretical results. We will discuss moreover, the influence of the oscillator strength (**f**) values on the obtained results.

RESULTS AND DISCUSSION

In Table I, the present results with Ar II -impact contribution included, are compared with experimental data[1]. In all cases we added to Stark broadening parameters due to electron-impacts, our results for Ar II - impact broadening. We see that the agreement between experimental and theoretical values is particularly good for shifts.

In order to see the influence of oscillator strength values on the results, calculations have been repeated with oscillator strengths calculated by using relativistic single- configuration Hartree-Fock method with allowance for core polarization, which have been taken from Table IV (values denoted as RHF+CP) in Ref. 6, and with oscillator strengths from Ref. 7, where allowance for configuration mixing and for spin-orbit interaction has been made. Different availlable results for the needed oscillator strengths are compared in Table IV of Ref. 5. One can see that the most important difference is for 5p-5d transition where with the Bates and Damgaard method the value of 0.731 have been obtained while in Refs. 6 (Table IV, values under RHF+CP) and 7 we have 0.0396 and 0.0265 respectively. In Ref. 8, several effective single-, and multiple-parameter model potentials have been compared, and for the critical 5p-5d transition the corresponding **f**-value varies between 0.0396 and 0.360. In Ref. 8

authors concluded that there is no significant improvement in computed oscillator strengths and sometimes even deterioration of accuracy is observed for more sophysticated calculations. We can see in Table I that the best agreement with experimental data is for oscillator strengths obtained within the Coulomb approximation, while in the case of more sophysticated f-values even the sign of shift is different from the experimental one. This is maybe a consequence of the fact that within the Coulomb approximation we have a summation over the complete and consistent set of atomic data. If we use better oscillator strength values for particular transition, the final result is not always better since this consistency might be disturbed if we use a mix of values from different sources.

Table I Comparison between the experimental Stark full half-widths (W) and shifts (d) of Rb I lines within the $5s^2S - 5p^2P^o$ multiplet with different calculations. The meaning of indexes is : m experimental values of Purić et al (1977)/1/; DSB-present results; fMB-present results with the oscillator strengths taken from table IV (values denoted as RHF+CP) in Ref.6; fW-present results with oscillator strengths taken from Ref.7. The electron density N is equal to $10^{17} cm^{-3}$.

λ (Å)	T (K)	Wm (Å)	WDSB (Å)	WfMB (Å)	WfW (Å)	dm (Å)	dDSB (Å)	dfMB (Å)	dfW (Å)
7800.2	15000	1.66	1.31	1.09	1.08	0.52	0.59	-0.31	-0.23
	17500	1.70	1.35	1.13	1.13	0.50	0.57	-0.32	-0.25
	20800	1.76	1.42	1.18	1.18	0.47	0.54	-0.34	-0.26
	26000	1.92	1.51	1.25	1.26	0.51	0.50	-0.35	-0.28
7947.6	15000	1.82	1.31	1.09	1.08	0.55	0.59	-0.31	-0.23
	17500	1.92	1.35	1.13	1.13	0.53	0.57	-0.32	-0.25
	20800	2.00	1.42	1.18	1.18	0.50	0.54	-0.34	-0.26
	26000	2.20	1.51	1.25	1.26	0.45	0.50	-0.35	-0.28

REFERENCES

1. J.Purić, J.Labat, Lj. Ćirković, I.Lakićević, and S Djeniže, J.Phys B 10, 2375 (1977).
2. Davis,D.N., Astrophys.J. 106, 28 (1947).
3. S.Sahal-Bréchot, Astron.Astrophys 1, 91 (1969).
4. S.Sahal-Bréchot, Astron.Astrophys 2, 322 (1969).
5. M.S.Dimitrijević and S.Sahal-Bréchot, Physica Scripta (1994) in press.
6. J.Migdalek, and W.E.Baylis, Can.J.Phys. 57, 1708 (1979).
7. B.Warner, Mon.Not.R.Astr.Soc. 139, 115 (1968).
8. J.Migdalek and E.Banasinska, JQSRT 48, 347 (1992)

EFFECT OF DISSOCIATIVE RECOMBINATION ON SPECTRAL LINE PROFILES IN NEON GLOW DISCHARGE

J.Szudy, R.Ciuryło, A.Bielski, J.Domysławska, R.S.Trawiński

Institute of Physics, Nicholas Copernicus University,
Grudziądzka 5/7, 87-100 Toruń, Poland.

Spectral line shapes analyses were performed for selfbroadened neon lines 735.5 nm ($2p^53d_5 - 2p^52p_{10}$) and 754.4 nm ($2p^53d_6 - 2p^52p_{10}$) emitted from a glow discharge at low pressures using a Fabry-Perot interferometer [1]. At pressures above 2 Torr the measured profiles were found to be well described by the Voigt profile. Contrary to that at very low pressures appreciable departures of the observed profiles from the Voigt profile have been found. Following a suggestion by Malvern et al [2] and Stacey and Thompson [3] such departures may be explained as arising from effects associated with the mechanisms by which the emitting atoms are excited. In the present study we focus our attention on the dissociative recombination of molecular ions (Ne_2^+) with electrons in the reaction

$$Ne_2^+ + e \rightleftharpoons [Ne_2^*]_{unstable} \rightleftharpoons Ne^* + Ne + kinetic\ energy,$$

which can cause incomplete thermalization of the excited Ne-atoms because of the non-Gaussian distribution of their velocities [4]. It is shown that this reaction may be regarded as the main process responsible for the distortion of the profiles of 753.5 and 754.4 nm Ne-lines emitted from a glow discharge at very low pressures. The distorted profiles were analysed as superposition of the Ballik profile [5] for thermalized atoms and the modified Ballik profile for non-thermalized atoms and the Lorentzian and Gaussian widths as well as relative densities of thermalized and non-thermalized Ne-atoms have been determined by a least-squares best fit procedure using the Marquardt algorithm. We have found that at very low pressures the Lorentzian widths differ markedly from those determined from our previous analysis [1] based on the assumption that all Ne^* - atoms are thermalized. This is seen in Fig.1 where the two sets of Lorentzian widths are plotted against the density. The widths determined in the present work when extrapolated to zero density are significantly lower than those corresponding to thermalized atoms. Moreover they are consistent with the natural width determined from the lifetime data. This provides another another convincing evidence that the well-known extrapolation anomaly is due to to excitation mechanisms [2, 3]. The excess Doppler broadening of the non-thermalized component of the line shape yielded for the dissociation energy of the molecular

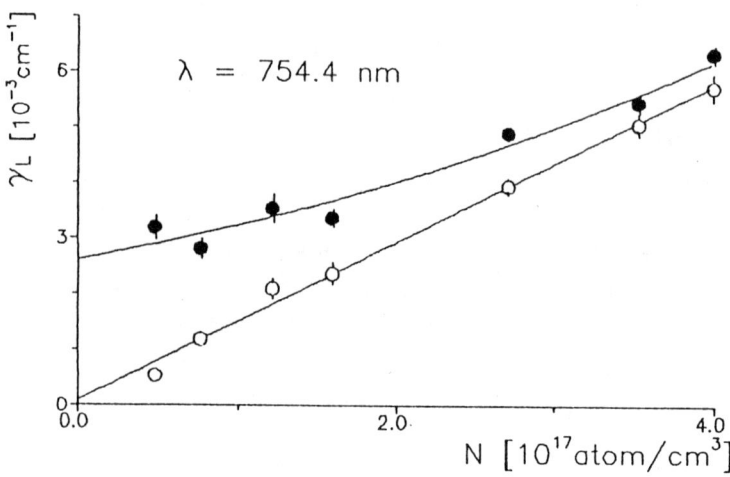

Figure 1: Comparison of Lorentzian width of the 754.4 nm Ne - line determined from best-fit procedures with (o) and without (•) inclusion of the dissociative-recombination effect.

ion Ne_2^+ a value (1.20 - 0.08) eV which is close to that determined by other method as well as to theoretical results [6, 7].

References

[1] R.S. Trawiński, A. Bielski, R. Ciuryło, J. Szudy, Ann. Physik **2**, 1 (1993).

[2] A.R. Malvern, J.L. Nicol, D.N. Stacey, J. Phys. B **7**, L518 (1974).

[3] D.N. Stacey, R.C. Thompson, J. Phys.B **16**, 537 (1983).

[4] T.R. Connor, M.A. Biondi, Phys. Rev. **140**, A778 (1965),
L. Frommhold, M.A. Biondi, Phys. Rev. **185**, 244 (1969)

[5] E.A. Ballik, App.Opt., **5**, 170 (1966)

[6] J.S. Cohen, B. Schneider, J.Chem.Phys. **61**, 3230 (1974).

[7] H.H. Michels, R.H. Hobs, L.A. Wright, J.Chem.Phys. **69**, 5151 (1978).

ION DYNAMICS AND MICROFIELD SMOOTHNESS
Spiros Alexiou
Dept. of Physics, Weizmann Institute of Science
Rehovot 76100, Israel

In the theory of spectral line broadening due to thermal plasma particle fields there are, qualitatively speaking, three regimes: Impact, quasistatic and ion dynamical (more appropriately called strong dynamical). Despite some generally successful attempts, ion dynamics is still hard to treat theoretically. The idea here is that in the ion dynamic and quasistatic regime the plasma microfield is *smooth* over the relevant time scale (memory loss time) and becomes smoother as we move towards the quasistatic regime. To illustrate, we consider the 1-state case, with a stochastic microfield V(t), which is nonzero only in some interval (0,T). Rewriting the Schrödinger equation as a Dyson integral equation in terms of the Green's function

$$G(t) = -\frac{i}{\hbar}\theta(t)U(t) \tag{1}$$

and defining

$$g(\omega) = G(\omega) - G_{as}(\omega) \tag{2}$$

with G_{as} a truncated asymptotic expansion for $G(\omega)$, such as the unperturbed Green's function $G_0(\omega)$ and

$$g_\Omega(t) = \int_{-\Omega}^{\Omega} \frac{g(\omega)}{2\pi} e^{-i\omega t} d\omega, \tag{3}$$

one makes a separable kernel approximation for

$$V(\omega - \omega') = \int_0^T dt V(t) e^{i(\omega-\omega')t} \approx \sum_k w_k V(t_k) e^{i(\omega-\omega')t_k} \tag{4}$$

based on Gaussian quadrature rules, which are efficient for a smooth V(t) and $|\omega - \omega'|$ not too large compared with 1/T. This leads to an algebraic equation

$$(\delta_{ik}\delta_{ab} + A_{ab}^{ik})g_{\Omega b}(t_k) = B_a(t_i) \tag{5}$$

where a and b refer to atomic states and i and k to Gaussian quadrature time points in (0,T). In deriving (5), an error term

$$E(t_i) = \int_{-\Omega}^{\Omega} d\omega \frac{e^{i\omega t_i}G_0(\omega)}{2\pi} \int_{\Omega}^{\infty} \frac{d\omega'}{2\pi}[V(\omega-\omega')g(\omega') + V(\omega+\omega')g(-\omega')] \tag{6}$$

has either been neglected or evaluated analytically. Such analytic evaluation is possible exactly if, for example, the plasma microfield is quasistatic over a time scale of Ω^{-1}. The net result is that one obtains for the average Green's function for $|\omega| \le \Omega$:

$$\langle G(\omega) \rangle = J(\omega) + \int_0^T dt e^{i\omega t}(\langle g_\Omega(t)\rangle - \frac{iI}{\hbar}) \tag{7}$$

with I the unit matrix and J an interaction-independent function which may be evaluated analytically in terms of sine and cosine integrals. Ω is the frequency at which the lineshape has died and can be bounded by impact considerations, as can T. Thus for a smooth microfield, its values at a few time points completely determine the time-dependent interaction V(t) and hence (via the Schrödinger equation) the U-matrix and the $g_\Omega(t)$ for the configuration in question. Therefore, G(t) is a functional of the microfield values $(V_1, V_2, \ldots V_N)$ (which may therefore be thought of as some collective coordinates) at the points $(t_1, t_2, \ldots t_N)$ and hence one can write:

$$\langle G(t) \rangle = \langle G([V_1], \ldots, [V_N], t) \rangle = \int dV_1 \ldots dV_N P_N(V_1, \ldots V_N) G(V_1, \ldots V_N, t) \tag{8}$$

where P_N is the probability density to find V_1 at t_1, V_2 at t_2 and so on with V_N at t_N. Thus, with this approach, the ion-dynamical problem reduces to the calculation of G as a function of the V's and the evaluation of P_N. An explicit expression for the first may be given (as the relevant matrix elements of the inverse of a matrix), although for large N this may be too complicated to be very useful. P_N has thus far been studied for the N=2 case in the dipole approximation[1].

Fig.1 is an example of a 60-configuration calculation for Ly–α in an Argon plasma with an electron density of 2x10^{17}e/cc and a Temperature of 16000K. The solid line is the exact solution and the dotted line is the solution according to this method using a second order expansion instead of the full quasistatic solution for the error term. Good agreement is obtained except for large times, where convergence has not yet been achieved at 60 configurations.

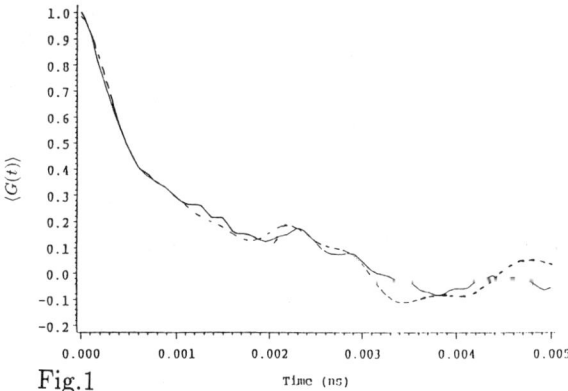

Fig.1

REFERENCES
1. For example A.Alastuey, J.L.Lebowitz and D.Levesque, Phys.Rev.A**43**, 2673(1991); J.W.Dufty and L.Zogaib, Phys.Rev.A**44**, 2612(1991); J.W.Dufty and L.Zogaib, Phys.Rev.E**47**, 2958(1993).

SPECTRA DECONVOLUTION USING BIRAUD'S METHOD

A. Lesage
DASGAL, Observatoire de Meudon, 5 place Janssen 92195 Meudon, France
M.Depiesse* , J. Richou
Laboratoire d'Optoélectronique, 83507 La Seyne sur mer, France

ABSTRACT

Biraud's nonlinear method of deconvolution[1] is applied to a plasma emitted line. Numerical experiments are performed before application to real spectra. Results show that this method gives satisfactory deconvolved results for line parameters.

BIRAUD'S DECONVOLUTION METHOD

A spectral line shape is distorted by the spectroscopic acquisition system. This is mathematically described by the convolution equation:

$$i(\lambda) = \int o(\tau) s(\lambda-\tau) d\tau + n(\lambda) \qquad (1)$$

with $i(\lambda)$ the observed spectrum, $s(\lambda)$ the apparatus function, $n(\lambda)$ the noise and $o(\lambda)$ the true spectrum. In the Fourier transform domain, the true spectrum is given by:

$$O(\omega) = I(\omega) / S(\omega) - N(\omega) / S(\omega) \qquad (2)$$

(capital letters represent Fourier transforms). Equation (2) performs a deconvolution process. The ratio $N(\omega)/S(\omega)$ becomes greater than $I(\omega)/S(\omega)$ while ω increases. $O(\omega)$ is then low-pass filtered (cut-off ω_c). The resulting function $H(\omega)$ can be expressed as the autoconvolution of a function $F(\omega)$ because $h(\lambda)$ is a positive spectral intensity distribution:

$$h(\lambda) = f^2(\lambda) \quad \text{in the Fourier transform domain: } H(\omega) = F(\omega) * F(\omega) \qquad (3)$$

where $*$ denotes convolution product. First, Biraud's method estimates the function $F(\omega)$. Then, adding new components to $F(\omega)$, this method evaluates $H(\omega)$ for $\omega > \omega_c$ insuring energy conservation ($H(0) = O(0)$). This iterative algorithm converges towards a limit given by noise variance in the signal.

RESULTS OF NUMERICAL EXPERIMENTS

Voigt shape lines are synthetized using different gaussian widths. Initial profile is a lorentzian shape. White noise is added and the deconvolution algorithm is performed. We give here the results for different ratios η (η = Apparatus function FWHM / Initial profile FWHM) and different noise levels. Relative errors on line intensity (maximum) and FWHM (Full Width at Half Maximum) between the initial and the reconstructed lines are plotted in table 1.

	0 % noise		0.3 % noise		1 % noise	
η	Intensity	FWHM	Intensity	FWHM	Intensity	FWHM
0.4	0.1 %	0.1 %	0.4 %	0.4 %	0.5 %	1.2 %
0.8	0.3 %	0.4 %	2.1 %	0.9 %	4.5 %	3.5 %
1.1	2.8 %	3.7 %	3 %	5.3 %	1 %	10.6 %
1.5	4.3 %	7.4 %	5.2 %	9.4 %	12 %	38.7 %
2.6	10 %	20 %	15.4 %	35.3 %	22.4 %	53.2 %

Table 1 - Relative errors on lines intensity and FWHM restoration

RESULTS WITH REAL SPECTRUM

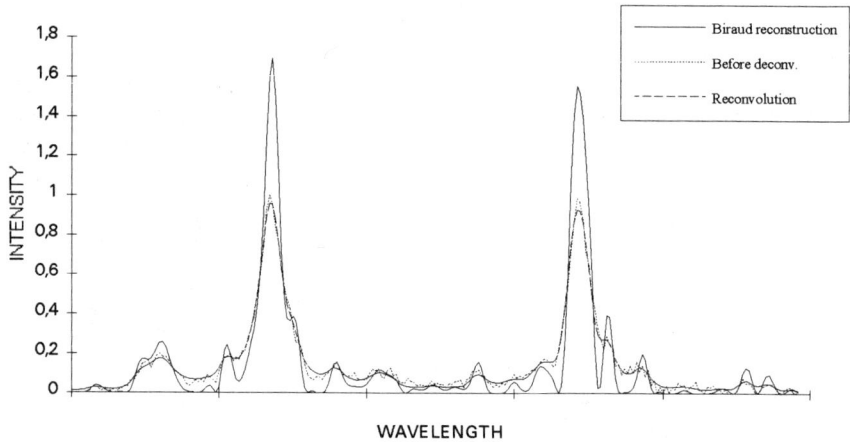

CONCLUSION

Already applied in radio astronomy[3], this reconstruction method working with any positive function seems to be well suited for plasma emitted spectra deconvolution even with 1 or 2% noise level.

REFERENCES

1. Biraud Y., Astron. Astrophys. 1, 124-127 (1969)
2. Jansson P.A., Academic Press, New York (1984)
3. Wong H.C.F, M Sc Thesis, Queen's Univ, Kingston, Ontario, Canada (1971)

* To whom all correspondence should be addressed.

STUDY OF PROFILES OF SOME CI AND NI MULTIPLETS

A. Goly and T. Wujec
Institute of Physics, Pedagogical University, Oleska 48, PL 45-052 Opole, Poland

Spectral line profiles were studied for selected CI and NI multiplets in the near infrared range of the spectrum. An argon plasma with small admixtures of H_2 (3%), CO_2 (0.4%) and traces of N_2 was produced in a cascade arc at two currents.

The radiation of the plasma was recorded using a conventional spectrometer. The studied profiles were corrected for selfabsorption of radiation as well as for contribution of the far line wings. The mostly used method of selfabsorption correction is based on the solution of the equation of radiative transfer with the Planck function as "source function". We have used another method, too. The plasma radiation, directed to the spectrometer, gives the signal S_1. A mirror behind the plasma source reflects the radiation, which passes through the plasma column - the resulting signal is S_2. The ratio of the signals (S_2/S_1) allows to calculate the absorption coefficient "k" and finally the corrected emission coefficient "ε" of the line using the formulas:

$$k = (-1/L) \cdot \ln\left((S_2/S_1 - 1)/(RD^2)\right) \quad \text{and} \quad \varepsilon = I \cdot k/(1 - \exp(-kL)),$$

where "L" is the length of the plasma column, "RD^2" is an apparatus factor, $S_1 \sim I$. Using both methods described for the center of the CI 833.5 nm profile we have obtained corrected maxima of the profile which differ less then 1%.

The electron density was evaluated from the half width of the H_β line. The other densities and the plasma temperature were determined from a set of CI, OI, AI and HI line intensities.

Table 1. Results of plasma diagnostics. O/C - density ratio of oxygen to carbon, which shows the range of the demixing effect

Experiment (N^o)	Current (A)	Temperature (K)	Electron density (10^{16} cm^{-3})		O/C
			from halfwidth	LTE assump.	
1	40	11500	4.8	4.7	4.2
2	60	12200	7.1	7.3	3.4

The very small differences between the values of electron density determined with assumption of the LTE model and from Stark broadening of the profile allow us to apply this model for diagnostic purposes without anxiety.

The evaluated relative line strengths of multiplet components were compared with other data in Table 2. Wiese's[1] values are obtained assuming LS-coupling. The excellent agreement between some results (especially for NI) indicates, that this coupling scheme may be applied for this spectrum. This is very important if overlapping of multiplet components occurs and separation procedures should be applied to obtain profiles of components.

The experimental half widths for some CI and NI lines are listed together with theoretical values[2,3] in Table 3. Voigt profiles were assumed and the "Lorentz" as well as "Gaussian" parts were determined[4]. Doppler and instrumental broadening were taken into account.

Table 2. Relative line strengths of CI and NI multiplet components

Element	Transition Array	Multiplet (N^o)	Wavelength (nm)	Line strengths (%) This work	Wiese
C	2p3s-2p($^2P^o$)3p	$^3P - {}^3P^o$ (29)	906.143	16.2	14.2
			906.247	12.4	10.9
			907.828	10.6	8.1
			908.851	9.1	10.9
			909.483	38.1	41.5
			911.180	13.7	14.4
N	2p^{2}3s-2p^2(^3P)3p	$^4P - {}^4P^o$ (22)	820.035	4.3	3.6
			821.071	8.4	7.3
			821.632	47.4	48.1
			822.312	20.0	20.7
			824.237	20.0	20.7
		$^2P - {}^2D^o$ (24)	938.681	36.7	36.0
			939.279	63.3	64.0
N	2p^{2}3s'-2p^2(^1D)3p'	$^2D - {}^2F^o$ (27)	904.588	57.0	56.5
			904.889	43.0	43.6

Table 3. Halfwidths of line profiles - mean values for multiplets

Element	Multiplet (N^o)	Wavelength (nm)	Full Halfwidth (nm) Observed	Lorentz Part	Stark(th)	Experiment (N^o)
C	31	833.52	0.120	0.118	0.126	1
			0.184	0.180	0.189	2
C	28	964.06	0.084	0.076	0.072	1
			0.137	0.132	0.110	2
C	29	908.76	0.099	0.093	0.081	1
			0.135	0.130	0.123	2
C	?	940.57	0.119	0.114	0.116	1
			0.158	0.154	0.176	2
N	22	821.18	0.070	0.063	0.045	1
			0.079	0.074	0.068	2
N	24	939.53	0.083	0.076	0.073	1
			0.132	0.127	0.111	2
N	27	904.76	0.083	0.073		1
			0.136	0.132		2

Acknowledgements: - This publication is based on the work sponsored by the Polish-American M.Curie Joint Fund II under projet MEN/NIST-94-166.

REFERENCES

1. W. L. Wiese, M. W. Smith and B. M. Glennon, NSRDS-NBS **4**, vol.I, 1966.
2. H. R. Griem, *Plasma Spectroscopy*, McGraw-Hill Book Comp., NY, 1964.
3. H. R. Griem, *Spectral Line Broadening by Plasmas*, Acad. Press NY, 1974.
4. A. Unsöld, *Physik der Sternatmosphären*, Springer-Verlag Berlin, 1968, p. 259.

DENSE ARC ELECTRODE PLASMA INSTABILITY AND LINE BROADENING

P. Serapinas
Institute of Theoretical Physics and Astronomy
2600 Vilnius, Lithuania

Stark broadening study in high density plasma and corresponding applications meet some problems till now. Due to high particle densities radiation reabsorption in plasma usually must be accounted for. Because of the strong line broadening transition from quadratic to linear Stark effect takes place for a large number of spectral lines. Besides that, as it was discussed in [1], plasma instabilities can be essential in line broadening measurements. One must be especially careful for possible presence of the small scale high frequency instabilities as detection of those in the line broadening measurements themselves is complicated.

In arc discharges nonstationary arc spot formation giving rise to electron density fluctuations is well known. Besides that in [2] aluminum arc cathode spot plasma electric conductivity and energy loss temperature dependences were regarded. In the temperature range from about 1 to 2 eV despite of the increasing role of the inelastic collisions plasma electric conductivity increases. At the same conditions plasma excitation and ionization energy is almost constant while radiation energy increases with temperature. Such arc plasma state must result in the positive feedback of the electron temperature fluctuations and plasma current filamentation at frequencies of the order of magnitude or higher than the inverse radiative lifetime. Similar plasma instabilities were observed in [3,4].

An about twofold electron temperature fluctuations, including those of the temperature dependent electron density, must result in about 30 − 70% variations of the line widths of aluminum spectral lines. Radiation noise measurement technique can be applied for experimental studies. More detailed theoretical analysis of the phenomena and results of the measurements will be presented in the report.

1. H. R. Griem, Comm. Atom. Molec. Phys. 3, 181 (1971).
2. P. Serapinas and A. V. Kupliauskienė, J. Phys. D: Appl. Phys. 27, (to be published, 1994).
3. R. J. Buteikis and P. D. Serapinas, Fizika Plazmy 17, 1495 (1991).
4. A. Anders, S. Anders, B. Jüttner and H. Pursch, J. Appl. Phys. 71, 4763 (1992).

ANOMALOUS BROADENING - ANOMALOUS ELECTRIC FIELDS?

Round table discussion

Chairman: E. Oks
Physics Dept., Auburn University, Auburn, AL 36849, USA

OKS.

There are plenty of experimental results from various groups where measured lineshapes cannot be explained by the standard theory, i.e. by the theory that accounts only for <u>individual</u> (ion and electron) microfields and treats ion broadening as quasistatic and electron broadening as impact. This is what I mean by <u>anomalous broadening</u>.

There are at least two physical reasons that might explain the discrepancy. The first one is that ions are not completely quasistatic so that their thermal motion could significantly contribute to the broadening. The second reason is that various <u>collective</u> modes (what I call <u>anomalous electric fields</u>) may develop in a plasma and drastically modify line profiles. In other words, there are situations where anomalous broadening may be explained either by anomalous electric fields, or by ion motion, or by both. Frequently both these factors bring up just an additional broadening. However there are more complicated situations where you can deduce from the same spectral line several different values of the anomalous electric field.

I'd like to offer the following example as a kick-off for the discussion. Consider the H_γ line. In the standard theory (i.e., impact electrons, quasistatic ions) this line has a central, unshifted component although it's not extremely intensive as compared to the central components of H_α and L_α. On this slide the central component has the Lorentz shape and the lateral components have quasistatic (e.g., Holtsmark) profiles, and you observe the sum, the envelope of all those curves. What if some additional quasistatic field (e.g., caused by some low-frequency turbulence) enters the game? The central component remains essentially the same but the lateral components go further into the wings. And as a lateral component broadens further to the wing, its maximum goes down – just for the sake of normalization. The resulting profile of the H_γ becomes broader, but in the very center some small bump may show up. For even greater quasistatic fields, all the lateral components will go far away from the central part. You end up with a halfwidth controlled by the electron impact width $\Delta\lambda_e$ of the central component only. So if you plot the theoretical halfwidth $\Delta\lambda^{th}_1$ versus the strength of the anomalous electric field E, it starts at some value $\Delta\lambda^{th}_1(0)$, then goes up, reaches a maximum $\Delta\lambda^{th}_{1max}$, and finally comes down to the level $\Delta\lambda^{th}_1 \approx \Delta\lambda_e < \Delta\lambda^{th}_1(0)$. Assume that your experimental halfwidth $\Delta\lambda^{exp}$ falls between $\Delta\lambda^{th}_1(0)$ and $\Delta\lambda^{th}_{1max}$. Then you get two different answers about a possible strength of the field E.

The situation becomes even more complicated if you recall that there is ion thermal motion. I will treat it very roughly. Assume for simplicity that ions, instead of being quasistatic, are now impact. (That is an extreme case of ion thermal motion.) In this case the plot of the theoretical halfwidth $\Delta\lambda^{th}_2(E)$ starts at some value $\Delta\lambda^{th}_2(0) > \Delta\lambda^{th}_1(0)$, then reaches a maximum $\Delta\lambda^{th}_{2max} > \Delta\lambda^{th}_{1max}$, and at larger fields comes down to the level $\Delta\lambda^{th} \approx \Delta\lambda_i$ ($\Delta\lambda_e < \Delta\lambda_i < \Delta\lambda^{th}_{1max}$) controlled by the ion impact width. Now for the same experimental halfwidth $\Delta\lambda^{exp}$ you deal with four different answers concerning a possible strength of the anomalous field E (they correspond to two intersections of $\Delta\lambda^{exp}$ with $\Delta\lambda^{th}_1(E)$ plus two intersections of $\Delta\lambda^{exp}$ with $\Delta\lambda^{th}_2$).

My example illustrates that an interplay between anomalous electric fields and ion motion may be more complicated than a usual assumption that both factors simply contribute to some additional broadening.

Helbig.

My point is to give a warning to experimentalists and to theorists to be careful. You can easily be fooled by some strange features in lineshapes. One example from the past – very surprising bumps on H_β that turned out to be due to molecular lines. My story: I took a capillary discharge where nice profiles of H_α were published by Ehrich. And with the same apparatus our profiles of H_α exhibited strange structures. Now I am pretty sure that I can reproduce these strange profiles – they are stable. When you try to make a theory out of experimental profiles published somewhere it's dangerous. You would better talk to the experimentalist – how much he trusts his own results.

Griem.

If you record structures that are narrower than, say electronic width, that is very suspicious. I think that for L_α, where the situation differs, the structures are real; for H_α – they are not.

Glenzer.

I'd like to focus your attention on our experimental profiles of Li-like ions obtained in a gas-liner pinch. We compared our data with profiles calculated by the Marseille' group. The comparison shows that the inclusion of ion dynamics in their calculations significantly improved an agreement with our experimental results. We also performed experiments in the gas-liner pinch trying to find some polarization effects in the H_β line of hydrogen (with Böddeker). In principle, we saw nothing. My point is that to observe spectroscopically anomalous electric fields is very difficult.

Demura.

I'll try to remind you some recent results concerning ion motion. New features here are nonlinear interference effects. They influence populations of the levels and therefore change the spectra. We initially reported some differences compared to the Marseille' group. Then we took into account ion motion and those differences disappeared. Another point is that that for comparison with molecular dynamics simulations there is a term in the equations that is very important. This term originates from the microfield fluctuations due to the motion of a radiator. Also I'd like to draw your attention to a theoretical work by Kesting presented at the previous conference. In spirit of the background paper by Seidel, he demonstrated how ion motion via an anisotropic ion broadening modifies Stark-Doppler profiles.

Konjević.

I'd like to show you a slide demonstrating an experimental problem with the P_β line. In the theoretical papers on the P_β line you find that the central part of it practically disappears. On the slide we compare our experimental profile with the standard quasistatic calculation and with a dynamic calculation by Stehle. The problem is always with the central dip of this line. What we discovered in the center are molecular lines - lines of the molecule He_2. We tried to play with plasma parameters and with the geometry of the experiment. Yet we never got rid of those molecular lines. You are always looking through a cold region - there is no way to avoid it. Our experiments were performed with a mixture of H and He in ration 1:1. Taking into account the difference of the ionization potentials, the expected ratio for perturbing ion densities should be 2:1 in favor of H.

Griem.

I'd like to remind you that in the past molecular lines were sometimes misidentified for plasma satellites of He-like forbidden lines that may be induced by electric fields. I think, only if the level of fluctuations is high enough, then we can see effects of fluctuating fields.

Voslamber.

I have a doubt concerning molecular satellites due to quasimolecules discussed this morning by Leboucher. I am not sure that, if the difference of upper and lower terms has a minimum, this is sufficient for observing a molecular satellite.

Devdariani.

You are right, it's not. The lifetime of the quasimolecule is also important. If the lifetime is not long enough, you cannot see a molecular satellite.

Oks.

You may also think about this in terms of broadening. A short lifetime of the quasimolecule translates into such a significant broadening of the molecular satellite that results into a drop of its maximum intensity below your experimental level of sensitivity.

Stamm.

Where to look for anomalous widths and anomalous broadening? Probably, you have to go to plasmas with strong gradients driving some instabilities. This plasmas you can find in fusion. Tokamaks, for instance, have very strong gradients at the edge. But the magnitude of anomalous fields there may not be high enough to observe them. If you go to laser fusion plasmas, you also have very strong gradients and there you may indeed see effects of anomalous fields. However in this case you need a comprehensive analysis like the one reported by Keane this morning.

Griem.

We tried to find anomalous fields in the MIT tokamak. We looked at the Balmer lines of deuterium and we did not find any effect of those fields.

Seidel.

For most of the experiments where anomalous fields were involved in the analysis, it was usually impossible to make independent measurements of electron density. This is the problem. If you see a structure and you attribute it to the electron plasma frequency, then you calculate an electron density out of it. But if you would have independent measurements of the density, you might see that the density is completely different. And I think the same is true even for many comparisons where ion dynamics is included.

Oks.

I'd like to comment on two points. Of course, it is very important to have independent density measurements. Excellent experiments are made by Kunze's group from Bochum in the gas-liner pinch where they measure independently the electron density by Thomson scattering. Another comment is on experiment in a capillary discharge reported this morning by Helbig. I'd like to put some numbers on the "ears" appearing in his H_α line profiles. It does not take an extremely high level of those quasimonochromatic fields. I estimated what is the thermal level for a plasma of these parameters and then what energy density of the electric field you need to explain those features. It's only five times higher than the thermal level, that's all! There are experiments where spectroscopically measured turbulent fields had an energy density a thousand times higher than the corresponding thermal level. But in relatively dense plasmas drastic modifications of lineshapes may be caused by oscillatory fields of an energy density only several times higher than the thermal level.

HIGH TO ULTRA-HIGH DENSITY PLASMAS

Application of Spectral Line Shapes to the Study of High Density ICF Plasmas

C.J. Keane, B.A. Hammel, S.H. Langer, and R.W. Lee
Lawrence Livermore National Laboratory, Livermore, CA 94550

A. Calisti, L. Godbert, R. Stamm, and B. Talin
Universite de Provence, Centre St. Jerome, Marseille, France

ABSTRACT

Spectral line broadening manifests itself in the study of high density inertial confinement fusion (ICF) plasmas in two important ways. First, comparison between measured and calculated lineshapes of individual spectral lines or groups of lines is used to diagnose plasma conditions in dense ICF plasmas, particularly in implosions. Secondly, through the emission and absorption coefficients spectral lineshapes serve as important inputs to plasma spectroscopy simulation codes which calculate simulated spectra from ICF targets. We discuss recent results from each of these areas. With regard to lineshape diagnostics, the advent of generalized line broadening codes has allowed the line profiles of complex multielectron emitters to be considered for diagnostic purposes. Particular example of this is the use of Ar He-β and its associated dielectronic satellites as a diagnostic of T_e and N_e, as well as the development of Ne-like Xe line broadening as a density diagnostic. With respect to simulation codes, the implementation of detailed lineshapes in calculations of this type is in many ways in its infancy. We present here examples of cases where effects related to spectral lineshapes such as continuum lowering and line transfer of Stark broadened lines are important so as to provide a stimulus for future work in this field.

INTRODUCTION

X-ray spectral diagnostics been extensively used as a diagnostic of density and temperature in a wide variety of plasmas relevant to inertial confinement fusion (ICF).[1-11] In the case of diagnostics of imploding capsules, trace concentrations (typically at the fraction of a percent level) of mid-Z (Z~13-20) elements such as Ar or Cl are placed in the fuel and/or ablator regions of the target. X-ray emission spectra from these dopants are used to diagnose plasma conditions in the capsule. The low concentration of these dopants results in minimal perturbation to the target hydrodynamics while at the same time yielding measurable x-ray spectra. Because such dopants may be locally placed (for example, in the gaseous fuel) this diagnostic information is intrinsically spatially resolved. Through the use of streak cameras or other techniques temporal resolution on a sub-100 ps time scale may be obtained as well. Thus, space and time resolved diagnostic information may be obtained.

X-ray spectra from these dopants typically yield a wealth of information regarding the density and temperature conditions present in the ICF target under consideration. In the case of simple, few electron emitters quantities such as ratios of individual lines, the inferred ionization balance, and individual line profiles provide diagnostic information. In the simple emitter case resonance line emission from the H- and He-like stages of the dopant (as well as associated dielectronic satellites) are typically observed.[1-8,10] For the case where more complex emitters are present ratios of groups of lines are typically used for diagnostics.[9] As has been discussed elsewhere proper interpretation of measured line ratios often implies consideration of detailed spectral simulations of these targets in order to include line transfer and other dense plasma effects.

Spectral lineshape information enters into diagnostics of this type in two ways. First, the detailed lineshape of certain lines may serve as a diagnostic of density and/or temperature. In this case a single, optically thin line is usually considered so that kinetics, opacity and other effects usually considered in spectral simulations are unimportant. An example of this is Stark broadening of H- and He-like ion dopant lines which has been extensively used as a diagnostic of N_e.[1-11]

The second area in which spectral lineshapes enter into x-ray spectral diagnostics of ICF plasmas is through spectral simulation codes.[12-17] In these calculations the coupled kinetics and transfer equations are solved to produce emissivities and opacities as a function of space and time throughout the target. These are then used to produce a simulated spectra. Lineshapes enter here through the calculation of the plasma emissivity and opacity. An example of where lineshape effects may be important in this case is Ar spectra from indirectly driven implosions.[5-9] In this case, proper calculation of the simulated spectra requires consideration of line transfer for the computed the optically thick Ar He-α and Ly-α lines. As these lines are Stark broadened the line broadening treatment used in principle can effect simulated quantities of diagnostic importance such as the ratio of Ar Ly-β to Ar He-β.

In this paper we consider recent results from these two areas. With respect to detailed lineshape calculations, we show how recent advances in the development of generalized lineshape codes have allowed us to consider more complex lineshapes for diagnostic purposes. In the area of spectral simulation, we show several cases where lineshape related effects are important. In fact, the inclusion of detailed lineshapes in spectral simulation codes is in its infancy, and a goal of this paper is to motivate further work in this area.

ADVANCES IN THE USE OF COMPLEX LINESHAPES AS TEMPERATURE AND DENSITY DIAGNOSTICS

Lineshapes of H- and He-like emitters have been used for some time as diagnostics of density in the fuel region of ICF implosions.[1-9,11] Typically, resonance lines of H- and He-like emitters are used. The most recent examples of this involve the use of Ar He-β and Ly-β as density diagnostics in indirectly driven implosions. While this is often adequate for rough estimates, in recent indirect drive work it has been necessary to increase the range of lineshapes considered for several reasons. First, satellite lines to Ar He-β have been observed to significantly affect the intensity of the Ar He-β feature.[5-9] Hence, it is desirable to calculate the line profile of Ar He-β and its satellites self consistently. Secondly, some higher performance implosions currently done on Nova have pusher opacities sufficiently high so as to attenuate Ar β line emission at 3.5 - 4 keV.[5,19] We are interested in developing line broadening diagnostics operating at higher photon energy. An obvious candidate for this is Ne-like Xe. This makes sense for two reasons. First, Xe gas is easily placed in the fuel region of Nova capsules. Secondly, Ne-like Xe 3-2 and 4-2- emission falls at roughly 4.5 and 6 keV, respectively. The 4-2 emission thus has particular promise as a density diagnostic.

Consideration of both of these problems requires consideration of lineshapes more complex than simple H- and He-like line profiles. Indeed, the tools to calculate such lineshapes have recently been written. Specifically, two codes, TOTAL[20] and MERL,[21,22] are now available. Both codes take as input an atomic data set consisting of energy levels, statistical weights, and reduced matrix elements for an arbitrary emitter. Lineshapes are then calculated by working in the "standard" quasistatic ion/impact electron limit. Ion motion effects are currently being incorporated into each. The development of these codes has been crucial to the advancement of the field of diagnostic lineshapes for ICF. Results obtained with these codes for the Ar He-β/satellite and Ne-like Xe problems have been presented elsewhere;[4-9] we briefly summarize the major results here.

We start by showing a typical spectrum from an indirectly driven ICF implosion in Fig. 1. The features seen here are common to all Ar spectra observed from indirectly driven implosions performed with either shaped or square pulses. The broadened Ar He-β and Ly-β lines

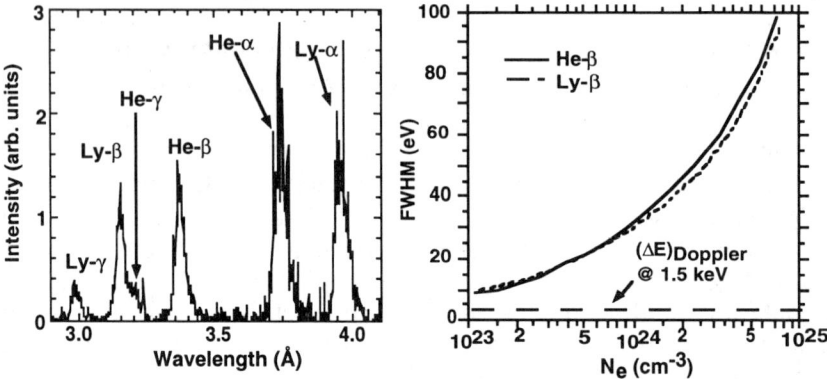

Figure 1. Typical spectrum at peak emission time for an Ar doped, indirectly driven implosion.

Figure 2. Width of H- and He-like Ar 3-1 lines vs. electron density for an electron temperature of 1 keV.

are apparent. (As discussed elsewhere, the ratio of Ly-β to He-β[5-9] serves as a measure of spatially averaged fuel electron temperature.) The lineshapes shown in Fig. 1 have widths of order 30 eV which is much greater than the Doppler width of about 1.5 eV, and are easily resolvable with current streaked spectrographs. The width of the Ar He-β and Ly-β lines thus provides a density diagnostic.

The density dependence of these linewidths as calculated using a simple K- shell quasistatic ion/impact electron treatment is shown in Fig. 2. It is clear that for conditions typical of current implosions ($N_e \sim 10^{24}$ cm^{-3}), Stark broadening dominates and the linewidth is a useful electron density diagnostic.

Things become more complicated, however, when Li-like satellites to Ar He-β are considered. These satellites are commonly seen in Nova implosion spectra[5-9] and can be seen at a fairly low level in Fig. 1.

Transitions of the type $1s^2nl - 1snl3l'$ lie just on the long wavelength side of the Ar He-β line and potentially can affect the electron density inferred from Stark broadening of Ar He-β. In order to consider this more quantitatively detailed line shape calculations including both Ar He-β and its dielectronic satellites have been carried out using both TOTAL[8-10] and MERL.[18] Some comparisons between these two codes for this case are presented elsewhere.[7] In Fig. 3 TOTAL computed lineshapes of the Ar He-β line and dielectronic satellites are shown. TOTAL takes in as input the complete energy level structure of the He-like Ar n=3 manifold as well as the associated Li-like Ar autoionizing levels. Mixing of these levels through the ion microfield as well as electron broadening occurring through collisions is considered. To date, the 2l3l' and 3l3l' Li-like autoionizing levels have been considered; efforts are underway to include the effects of 4l3l' autoionizing levels.[23] The full non-LTE populations of these excited states are used. It should be emphasized that TOTAL solves the quasi-static line broadening problem for all of the relevant levels simultaneously. As a final point, variant of MERL has been developed which includes ion motion effects. This code is currently being used to study ion motion effects on the combined profile of Ar He-β and its dielectronic satellites.[24]

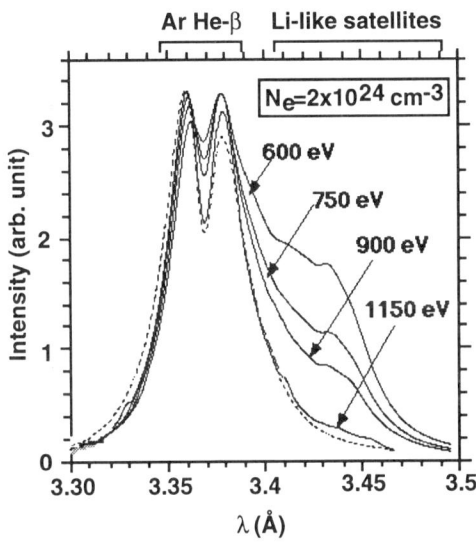

Figure 3. TOTAL computed profile of He-like Ar 3-1 emission along including associated dielectronic satellites. Results shown for electron temperature of 1150 eV. Dash: He-like Ar 3-1 profile only at 1150 eV.

Figure 3 shows that satellite emission can strongly modify the observed profile, especially at low temperatures and high densities. The result of this is that the combined He-β/satellite lineshape provides a simultaneous electron density and electron temperature diagnostic. In particular, the width of the entire feature and the intensity of the satellite contribution increase with N_e and decrease with T_e, respectively.

From Fig. 3 it is evident that fitting experimentally measured He-β/satellite line profiles to theory yields a best fit for a given value of N_e and T_e and thus a simultaneous diagnostic of these quantities. Figure 4 shows an example of this procedure for 1-ns square drive and pulse-shaped drive Nova implosions. In this figure measured He-β/satellite

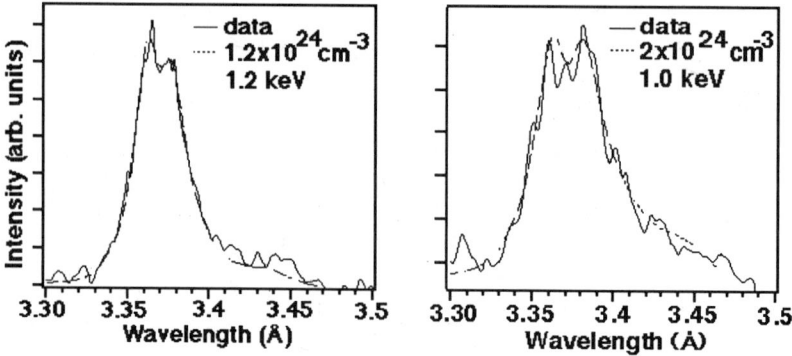

Figure 4. Measured He-like Ar 3-1 and satellite lineshapes along with best fit calculated line profile.
Left: results for 1-ns square laser pulse implosion.
Right: results for 3:1 contrast laser pulse implosion.

profiles are fit to TOTAL calculated lineshapes. The latter include contributions from the He-β line as well as the 2l3l' and 3l3l' autoionizing levels. The best fits yield inferred spatially averaged plasma conditions of (N_e, T_e) = (1.2 x10^{24} cm^{-3}, 1.2 keV) and (2.0 x10^{24} cm^{-3}, 1.0 keV) for the 1-ns square and shaped drive cases respectively. This is in accord with expectations; the shaped pulse results in a more isentropic implosion and thus a higher fuel density.

Figures 3 and 4 demonstrate the importance of generalized line broadening codes such as TOTAL and MERL in generating line profiles for complex emitters for purposes of fitting to data. TOTAL is also being used to develop line broadening of Ne-like Xe as a density diagnostic for use in high performance implosions characterized by higher fuel densities and pusher opacities compared to those typical of the targets shown in Fig. 1-4.[5,19] Initial results for Ne-like Xe lineshapes assuming static ion and impact electron broadening have been described previously.[4] More recently, the TOTAL code has been extended to include ion motion effects, as described in detail elsewhere.[25] The method used to include ion motion effects assumes that the dynamic lineshape results from a stationary Markovian process induced by the fluctuating ion microfield, which mixes together the static components.[25] This method differs from those which start with a mathematically modeled stochastic microfield.[27,28] The TOTAL method requires an estimate of the fluctuation frequency for the microfield, is required, which is usually obtained from molecular dynamics simulations.

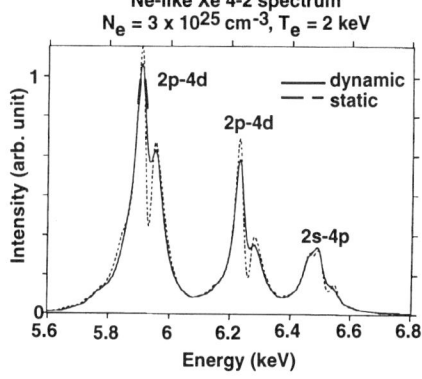

Figure 5. TOTAL calculated lineshapes for Ne-like Xe 4-2 lines showing effects of ion dynamics.

Detailed calculations of Ne-like Xe lineshapes including the effects of ion motion have been carried out and will be presented elsewhere;[26] here we show a selection of recent results. First, we recall that as discussed in Ref. 4 the 4-2 lines in Ne-like Xe show greater promise than the 3-2 lines in terms of linewidth sensitivity to density. This is because the 3-2 transitions show relatively little pressure broadening in the range 10^{24} cm^{-3}<N_e<3 x10^{25} cm^{-3}. Hence, most recent calculations assessing ion motion effects have focused on the 4-2 and 5-2 transitions, although 3-2 line profiles computed with ion dynamics are also under investigation. Figure 5 shows TOTAL computed lineshapes for the 4-2 transitions in Ne-like Xe at N_e=3 x10^{25} cm^{-3} and T_e=2 keV. Results are shown for both static and dynamic ion broadening. The inclusion of ion motion is seen to fill in "dips" near line center, as would be expected. The relatively broad width of the 4-2 lines suggests they hold promise as density diagnostics.

The 5-2 lines are also potentially useful as diagnostics as can be seen from Fig. 6, which plots the width of the 2p-5d transitions vs. N_e. Other factors must be considered when examining the 2p-5d lines as a density diagnostic, however.

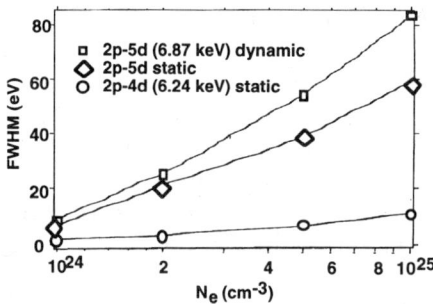

Figure 6. FWHM of Ne-like Xe 2p-4d and 2p-5d lines vs. electron density.

First, the 2p-5d lines should be somewhat weaker than the 2p-4d transitions. Secondly, the presence of satellite lines may make measurement of linewidths using all of these transitions difficult. Nearby satellites do not appear to be a problem for n=3-2 and 4-2 transitions, but it has not be considered as of yet for 5-2 lines. Finally, continuum lowering affects must be assessed. Simple estimates based on the Stewart-Pyatt formula show that the continuum is lowered to the n=5 and n=4 levels at $N_e=10^{25}$ cm^{-3} and $N_e=3\times10^{25}$ cm^{-3}, respectively. These densities thus are a rough upper limit for the region of applicability of pressure broadening of these lines for density diagnostics.

LINESHAPES IN SPECTRAL SIMULATION CODES

In terms of the number and variety of lineshape calculations required, spectral simulation codes are by far the largest "user" of line profile information in the plasma spectroscopy community. By "spectral simulation" code we mean here a code that reads in hydrodynamic information from a simulation code, experiment, or elsewhere, and solves the coupled kinetics and transfer equations to produce level populations as a function of space and possibly time. These level populations are then used to construct time and space dependent emission and absorption coefficients. A simulated spectra as seen by an observer along any given line of sight is then generated by formally solving the equation of radiative transfer along the desired ray path. Codes of this type have been heavily used by the ICF,[5-11] magnetic fusion,[29] x-ray laser,[30] and short-pulse laser-produced plasma communities,[14] among others.

Spectral lineshape information plays an important role in codes of this type because it enters directly into the calculation through the emission and absorption coefficients ε_ν and κ_ν. There are in fact two important ways in which variations in ε_ν and κ_ν arising from line profile variations can affect simulated spectra. First, variation in the line absorption profile can directly result in changes in opacity throughout the region of interest. For cases where opacity is important in driving the plasma kinetics, this can obviously lead to different answers for the computed level populations and thus the computed spectra. Secondly, assuming the kinetics is known, simply varying the lineshapes used in the formal solution of the transfer equation can changes items of diagnostic interest such as the peak to peak ratio of different lines. Given that in experimentally "noisy" situations one can often only measure peak-to peak line ratios, lineshapes can be critical to data interpretation. All of these effects are magnified many times for complex emitters where many overlapping lines are present.

In addition to the roles of spectral lineshapes themselves, there are two related theoretical problems of significance to dense plasma spectral simulation, namely the natural line width and continuum lowering.[15,31] The first of these refers to the natural linewidth in dense plasmas increasing to very large values as one proceeds to higher density and higher principal quantum numbers. Simply put, at high densities inelastic electron collisions (electron impact excitation, deexcitation, and ionization) can lead to significant natural broadening. As an example, for the H-like Ar n=3 level at $N_e=10^{24}$ cm^{-3} and $T_e=1$ keV, the time for electron impact excitation from n=3 to n=4 corresponds to a natural width of about 10 eV, which is 30% of the 3-1 static ion width! Other cases of seemingly greater importance can be found by considering more complex emitters where certain states of importance for line broadening may have levels very close by in energy to which inelastic collisions may occur very rapidly. Some of this problem may in principle by mitigated by continuum lowering. As discussed in Ref. 15, the interplay between continuum lowering and natural linewidth can be attacked by resorting to an impact approximation type calculation. This has not been carried out as of yet but it is very high priority. Other ideas for treating this problem have also been considered informally but more work is required.[32]

In an ideal world one would like detailed line-shape codes to be able to be run inline and completely self-consistently with kinetics and line transfer calculations, including the effects of continuum lowering and natural linewidth. Clearly, this goal has not been obtained as of yet. Solution of this problem is one of the major challenges facing plasma spectroscopy. Obviously we cannot present a solution of this problem here. Rather, we will present several ICF situations where lineshapes and related physics such as continuum lowering play an important role in spectral simulation. We will then discuss existing spectral simulation codes and their methods for treating physics related to spectral lineshapes.

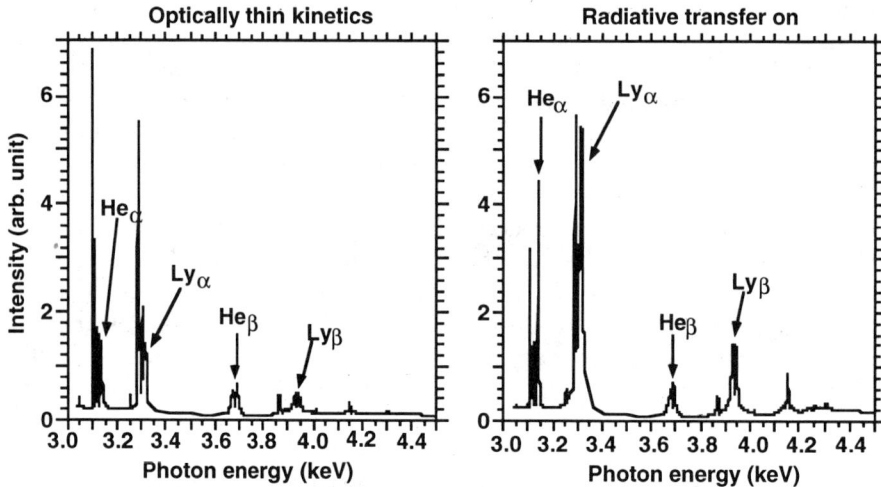

Figure 7. Simulated Ar spectra from an indirectly driven implosion calculated using the DSP postprocessing code. Left: Optically thin kinetics. Right: line photon field considered in kinetics calculation.

The first ICF situation which demonstrates the importance of lineshapes in shown in Fig. 7, which shows calculated spectra from the simulation of an Ar doped indirectly driven implosion of a DD filled plastic capsule. These spectra were calculated by postprocessing LASNEX[33] hydrodynamics output with the DSP[12] spectral postprocessing code. Spectra calculated with and without including line transfer in the kinetics calculations are shown. More specifically, in the no transfer calculation, the Ar level populations are solved using the local values of T_e, N_e, and N_i only. In the calculation including radiative transfer all lines shown in the spectrum are transferred; the line intensities and level populations are computed self consistently. Opacity is only significant for the He-α and Ly-α lines, however; these lines have an optical depth of order 10.

As discussed in detail in Ref. 9, two features regarding the effects of transfer are immediately apparent. First, the intensity of the α lines is increased. Secondly, the ratio of the optically thin He-β and Ly-β lines of Ar is varied. The latter occurs through an increase in the population of H-like Ar throughout the target when line transfer is considered.[9] This

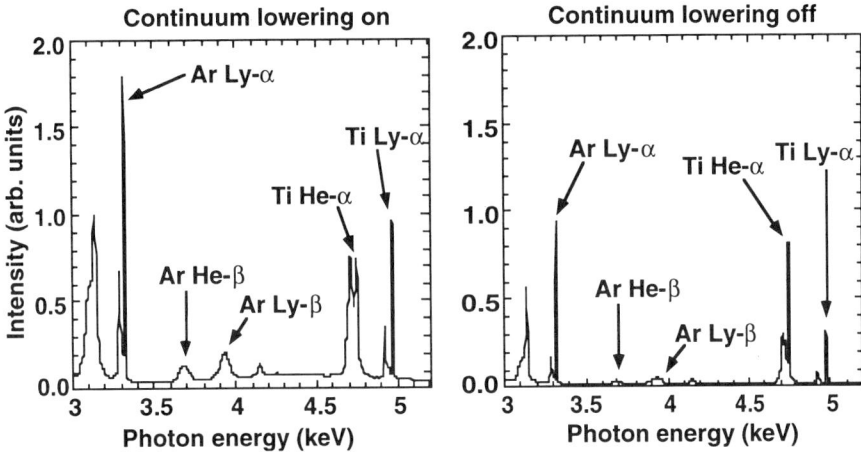

Figure 8. Computed Ar/Ti spectrum from an implosion with Ar doped into the fuel and Ti doped into the innermost 3-μm of plastic adjacent to the fuel. Left: Continuum lowering included. Right: Continuum lowering neglected.

is obviously important as the ratio of these two lines serves as a diagnostic of spatially averaged fuel electron temperature. Indeed, obtaining any kind of reasonable agreement with experiment requires that these transfer effects be considered.[8,9] Figure 7 thus represents a case where line transfer of Stark broadened lines play a crucial role in modelling experimental data. As such experiments of this type form a testbed for tests of combined lineshape/radiative transfer effects.

Figure 8 shows a second example from ICF implosions demonstrating the importance of continuum lowering in the computation of simulated spectra. (The spectra shown in Figs. 8 and 9 were generated by postprocessing LASNEX[33] hydrodynamic code output with the CRETIN[13] code.) The spectra shown in Fig. 8 is from the simulation of a capsule with dopants placed so as to diagnose fuel/pusher mix.[19,34] In this case the fuel is doped with Ar as previously. In addition, the innermost 3-μm of pusher closest to the fuel region is doped with Ti at the 0.07% atomic level. In experiments using these dopants the degree of pusher/fuel mix is varied experimentally by varying the size of surface perturbations placed on the outer capsule surface.[34] The increasing degree of mix associated with rougher capsules implies that more of the pusher material is mixed into the fuel. Thus, as the degree of pusher/fuel mix increases, one expects the ratio of Ti to Ar emission to increase.[19] In Fig. 8 the computed spectrum at peak emission for a Ar/Ti doped capsule is shown with and without continuum lowering considered. The continuum lowering model used is a modified version[31] of the Stewart/Pyatt formalism and is thus quite a simple treatment. Note the sensitivity of the ratio of Ti Ar He-α/ Ar Ly-β. This ratio has been proposed as a measure of pusher/fuel mix. The variation in this ratio as a function of continuum lowering implies that a more careful treatment of this subject in the context of spectral simulation codes is required.

The last example showing the importance of lineshapes to ICF spectral simulation codes is presented in Fig. 9. This is another "mix diagnostic" calculation of the type shown in Fig. 8, with the exception that the Ar and Ti dopants have been replaced by Xe and Cr,

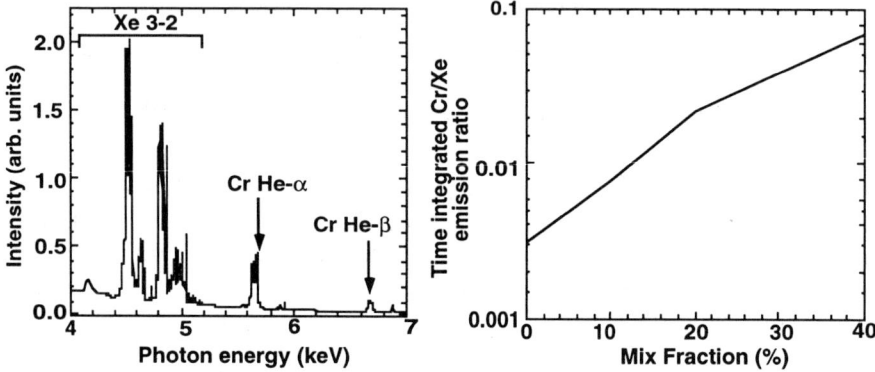

Figure 9. Computed spectrum from and indirectly driven implosion with Xe seeded into the fuel and Cr seeded into the innermost 3-μm of plastic closest to the fuel. Left: Computed spectru m at time of peak emission. Right: Time integrated ratio of He-like Cr 2-1 emission to Ne-like Xe 3-2 region emission plotted vs. mix fraction.

respectively. Figure 9 shows a calculated spectrum at time of peak emission for such a case, as well as the behavior of the ratio of Cr to Xe emission as a function of mix fraction. (The "mix fraction" is approximately the extent of the mixed region at the fuel/pusher interface expressed as a function of the fuel radius.) In this case the Cr He-α line is optically thick, as are the 3-2 Ne-like Xe resonance lines. These lines have been transferred in the calculation shown in Fig. 9, and hence the assumed lineshapes for these transitions enter into the kinetics calculation. The remainder of the spectrum consists of many hundreds of overlapping satellite lines. All lines considered in this problem are treated with simple Voigt profiles with an ad-hoc prescription for Stark broadening included in the Lorentzian portion of the lineshape. This is clearly a very crude treatment for this case with its many overlapping lines. (Simple estimates indicate that for this case electron impact broadening is greater than ion quasistatic effects and is of order or slightly greater than Doppler broadening.) Figure 9 thus represents a good example of a "benchmark" case where hundreds of detailed lineshapes are needed at a given time step.

We conclude this section by reviewing examples of available plasma simulation codes and how they treat the physics described here. Table I compares some major characteristics of existing spectral simulation codes. The "lineshapes (kinetics)" heading refers to the method by which lineshapes are incorporated into the line transfer. In a similar vein "lineshapes (spectrum)" refers to how lineshapes are used in the generation of the emission and absorption coefficients given the level populations. The "level structure" column indicates how the level structure is modified in the kinetics to account for plasma microfields. (This is not generally done in existing simulation codes.)

Table I. Spectral Simulation Code Characteristics

Code Type	Continuum Lowering	Lineshapes (kinetics)	Level Structure	Lineshapes (spectrum)
1	None	None	Zero field	Doppler
2	Stewart-Pyatt	Voigt w/ crude Stark estimate in Lorentzian	Zero field	As kinetics
2A	Stewart-Pyatt	Voigt w/ crude Stark estimate in Lorentzian	Zero field	Full inline K-shell Stark profiles
3	Stewart-Pyatt	Inline K-shell Stark	Zero field	As kinetics
4	Stewart-Pyatt	Detailed profiles split up as in levels	Zero field	As kinetics
5	Self consistent	Self consistent	Self consistent	As kinetics

The DSP code used in Fig. 7 is a code of type "2A" above. DSP[12] uses a simplified form of solution of the coupled transfer/kinetics equations. The CRETIN[13] code used in Figs. 8-9 is of type 2 above. It has a simpler lineshape treatment than DSP but includes a more sophisticated complete linearization radiative transfer package. Most codes in existence fall into categories 1-3. A code along the lines of type 4 above has been developed[16] and is being extended to handle dense plasma phenomena necessary for ICF plasma modeling.

Note that we have not considered different line transfer models as a separate category above. This is because by and large most ICF line transfer problems are not difficult, *assuming a given (usaully simple) lineshape prescription*. That is, the solution of the coupled transfer and kinetics equations is not especially difficult due to the relatively modest optical depths and minimal photon scattering present. (The latter is equivalent to stating that the probability of photon destruction is generally high due to high rates of collisional deexcitation between excited states.) The difficult part of the line transfer problem is in defining the associated kinetics and line broadening problems so as to be self consistent.[15]

Indeed, a "type 5" code as defined above represents the "ultimate" simulation capability in that line broadening, continuum lowering, and plasma microfield effects on kinetics are handled self consistently. (It is clearly a long way from step 4 to step 5.) These phenomena are strongly coupled in dense plasmas and an adequate, self consistent treatment is a major goal for plasma spectroscopy research.

The main point of Table I, then, is that the treatments of continuum lowering, Stark broadening, and plasma microfield dependent kinetics in plasma simulation codes are quite crude to date. In particular, as shown in this paper, assumptions regarding lineshapes and continuum lowering affect simulated spectra from ICF targets. There is much room for improvement of models of dense plasma processes in existing spectral simulation codes and it is the authors' hope that the results shown here will motivate further work in this area.

SUMMARY

In this article we have explored the two major ways in which spectral lineshapes play a role in the study of high density ICF plasmas. The first and most well known use of lineshapes is to provide a means of diagnosing density and temperature in high density implosions. There has been much progress in this area in recent years, primarily due to the advent of generalized line broadening codes which allow line profiles to be computed for complex emitters. As a result the behavior of the line profile of Ar He-β and its associated dielectronic satellites has been used as a diagnostic of plasma conditions in Nova implosions. In a similar vein, line broadening of 3-2, 4-2, and 5-2 transitions in Ne-like Xe has been quantified and shown to be a viable density diagnostic in future high density targets.

The second major role lineshapes play in the study of dense ICF plasmas is in dense plasma spectral simulation. The hundreds to thousands of lineshapes required by these codes must be generated relatively quickly. In addition, the problem of self consistently treating line broadening, kinetics, and general plasma microfield related effects such as continuum lowering in a simulation context remains essentially unsolved. As particular examples relevant to ICF we have shown that line transfer of Stark broadened lines as well as continuum lowering are important. We hope that the demonstration of the importance of these effects to simulation of ICF experiments motivates further work in this very interesting and rich area of physics.

ACKNOWLEDGMENTS

The authors acknowledge the support of the LLNL ICF program in carrying out this research. This work was performed under the auspices of the U.S. Department of Energy by the U.S. Department of Energy by Lawrence Livermore National Laboratory under contract No. W-7405-ENG-48.

REFERENCES

1. M.H. Key, C.L.S. Lewis, J.G. Lunney, A. Moore, J.M. Ward, and R.K. Thareja, Phys. Rev. Lett. 44, 1669 (1980).
2. C.F. Hooper, Jr., D.P. Kilcrease, R.C. Mancini, L.A. Woltz, D.K. Bradley, P.A. Jaanimagi, and M.C. Richardson, Phys. Rev. Lett. 63, 267 (1989).
3. A. Hauer, R.D. Cowan, B. Yaakobi, O. Barnouin, and R. Epstein, Phys. Rev. A 34, 411 (1986).

4. C.J. Keane, R.W. Lee, B.A. Hammel, A.L. Osterheld, L.J. Suter, A. Calisti, F. Khelfaoui, R. Stamm, and B. Talin, Rev. Sci. Inst. 61, 2780 (1990).
5. B.A. Hammel, C.J. Keane, D.R. Kania, J.D. Kilkenny, R.W. Lee, R. Pasha, R.E. Turner, and N.D. Delamater, Rev. Sci. Inst. 63, 5017 (1992).
6. B.A. Hammel, C.J. Keane, D.R. Kania, J.D. Kilkenny, R.W. Lee, R. Pasha, R.E. Turner, and N.D. Delamater, J. Quant. Spec. Radiat. Transfer 51, 113 (1994).
7. C.J. Keane, B.A. Hammel, D.R. Kania, R.W. Lee, A.L. Osterheld, L.J. Suter, R.C. Mancini, C.F. Hooper, Jr., and N.D. Delamater, J. Quant. Spec. Radiat. Transfer 51, 147 (1994).
8. B.A. Hammel, C.J. Keane, M.D. Cable, J.D. Kilkenny, R.W. Lee, and R. Pasha, Phys. Rev. Lett. 70, 1263 (1993).
9. C.J. Keane, B.A. Hammel, D.R. Kania, J.D. Kilkenny, R.W. Lee, A.L. Osterheld, L.J. Suter, R.C. Mancini, C.F. Hooper, Jr., and N.D. Delamater, Phys. Fluids B 5, 3328 (1994).
10. H.R. Griem, Phys. Fluids B 4, 2346 (1992).
11. C.J. Keane, B.A. Hammel, A.L. Osterheld, and D.R. Kania, Phys. Rev. Lett. 72, 3029 (1994).
12. C.J. Keane, R.W. Lee, and J.P. Grandy, in *Proceedings of the 4th International Workshop on the Radiative Properties of Hot Dense Matter,* edited by W. Goldstein, C. Hooper, J. Gauthier, J. Seely, and R.W. Lee (Singapore, World Scientific, 1991),p.233.
13. H.A. Scott and R.W. Mayle, Appl. Phys. B 58, 35 (1994).
14. O. Peyrusse, J. Quant. Spec. Radiat. Transfer 51, 281 (1994).
15. R.W. Lee, J.I. Castor, C.A. Iglesias, F.J. Rogers, in *Spectral Line Shapes: Vol. 7,* (R. Stamm and B. Talin, eds.), Nova Science Publishers, NY, NY (1994).
16. J.I. Castor, P. Dykema, and R.I. Klein, Ap. J. 387, 561 (1992).
17. G. Olson, J. Comly, J.K. La Gattuta, and D.P. Kilcrease, J. Quant. Spec. Radiat. Transfer 51, 255 (1994).
18. R.C. Mancini, C.F. Hooper, Jr., N.D. Delamater, A. Hauer, C.J. Keane, B.A. Hammel, and J.K. Nash, Rev. Sci. Inst. 63, 5119 (1992).
19. C.J. Keane, R.C. Cook, T.R. Dittrich, B.A. Hammel, W.K. Levedahl, O.L. Landen, S.H. Langer, D.H. Munro, and H.A. Scott, "Diagnosis of Pusher-Fuel Mix in Spherical Implosions Using X-ray Spectroscopy," to be published in Rev. Sci. Inst.
20. R. Stamm, B. Talin, E.L. Pollock, and C.A. Iglesias, Phys. Rev. A 34, 4144 (1986).
21. L.A. Woltz and C.F. Hooper, Jr., Phys. Rev. A 38, 4766 (1988).
22. R.C. Mancini, D.P. Kilcrease, L.A. Woltz, and C.F. Hooper, Jr., Comp. Phys. Comm. 63, 314 (1991).

23. N.D. Delamater, G.D. Pollak, A.A. Hauer, R.C. Mancini, C.F. Hooper, Jr., C.J. Keane, B.A. Hammel, and J.K. Nash, Bull. Am. Phys. Soc. 37, 1431 (1992).
24. D. A. Haynes, Jr., C.F. Hooper, Jr., R.C. Mancini, D.K. Bradley, J. Delettrez, R. Epstein, and P.A. Jaanimagi, "Spectroscopic Analysis of Ar-doped Laser Driven Implosions," to be published in Rev. Sci. Inst.
25. A. Calisti, L. Godbert, R. Stamm, B. Talin, C.J. Keane, and R.W. Lee, to be submitted to Phys. Rev. E.
26. A. Calisti, L. Godbert, R. Stamm, and B. Talin, J. Quant. Spec. Radiat. Transfer 51, 59 (1994).
27. D.B. Boercker, in *Spectral Line Shapes: Vol. 7*, (R. Stamm and B. Talin, eds.), Nova Science Publishers, NY, NY (1994).
28. D.B. Boercker, C.A. Iglesias, and J.W. Dufty, Phys. Rev. A 36, 2254 (1987).
29. A.E. Koniges, D.C. Eder, A.S. Wan, H.A. Scott, H.E. Dalhed, R.W. Mayle, and D.E. Post, "Role of Radiation in Vapor Shielding of First Wall During Disruption," to be published in J. Nuc. Materials.
30. R.A. London, M.D. Rosen, M.S. Maxon, D.C. Eder, and P.L. Hagelstein, J. Phys. B 22, 3363 (1989).
31. R.M. More, J. Quant. Spec. Rad. Transfer 27, 345 (1982).
32. H.R. Griem, private communication at Lawrence Livermore National Laboratory, 1994.
33. G.B. Zimmerman and W.L. Kruer, Plasma Phys. 11, 51 (1975).
34. T.R. Dittrich, B.A. Hammel, C.J. Keane, R. McEachern, R.E. Turner, S.W. Haan, and L.J. Suter, "Diagnosis of Pusher-Fuel Mix in Indirectly Driven Implosions," to be published in Phys. Rev. Lett.

SPATIALLY RESOLVED X-RAY EMISSION SPECTROSCOPY FROM DENSE PLASMAS

E. Leboucher-Dalimier, P. Angelo, H. Derfoul, P. Gauthier, A. Poquérusse

Physique Atomique dans les Plasmas Denses
LULI / Université Paris VI 75252 Paris cedex 05, France
/ Ecole Polytechnique 91128 Palaiseau cedex, France

ABSTRACT

We present instrumental spectroscopic techniques used to diagnose hot and very dense plasmas created either by irradiation of a massive target or by colliding foils.

The techniques lead to different diagnostics of plasma *macroscopic parameters* "*in situ*" such as : density, temperature and their longitudinal and transverse gradients ; the radiative properties (emission and absorption) and their gradients ; the transverse thermal conductivity. These diagnostics are useful for Initial Confinement Fusion experiments and they could be fruitful for any type of 2D inhomogeneous emitting medium.

On the other hand, for what concerns *microscopic properties* of the plasmas, the technique has led to the exhibition of effects related to the formation of transient ionic molecules in highly correlated plasmas. These effects, which were observed for the first time in a laboratory plasma, are of a great interest for Astrophysics plasmas where similar observations have been made.

In all cases the probe chosen for the analysis of either a macroscopic or a microscopic property is a space-resolved H-like line shape or its frequency integrated line intensity.

INTRODUCTION

The present article deals with X-ray emission spectroscopy in dense, correlated, inhomogeneous plasmas created either by laser irradiation of massive plane targets or by colliding foils. These plasmas are interesting for two important reasons at least : first, for the important energy transfer due to the strong axial and transverse temperature inhomogeneities ; second, for the new atomic physics involved in these highly correlated environments. For both kinds of effects, space-resolved emission spectroscopy is a powerful diagnostic tool because it allows the exhibition of specific effects.

Regarding first the energy transfers in these inhomogeneous plasmas, although the radiative transfer is much less important than the thermal conduction process, it can be used as a probe of the temperature gradients and of the associated thermal diffusion. By means of the analysis of the spatially resolved X-ray transverse emision of the plasma and of a theoretical modeling, a *direct diagnostic "in situ" of the emission coefficient* has been developed initially in the case of soft initial temperature transverse gradients.[1] In this paper we present experiments devoted to the creation of strong transverse gradients. The leading part of the *non-uniformity of the laser irradiation* is investigated. Since the interpretation of the transverse emission coefficient in terms of bundle of independent beams will be found to be inadequate, we will conclude that *the lateral thermal conduction* is important.[2,3]

Concerning now highly correlated plasma effects, X ray profiles showing *molecular emission* due to the formation of *transient ionic molecules* have been obtained for the first time in laboratory plasmas.[4,5,6] More recently, the compressed zone created in colliding foil experiments provided favourable conditions for the excitation of *forbidden transitions*. All these experimental results presented here give new motivations for theoretical computations of line wings including properly dipolar interaction[7] or multipolar perturbations at high orders,[8,9] and leading to the exhibition of strong asymmetries[10,11] and forbidden lines.[12,13] The emission spectroscopy issued from ultradense plasmas also calls for attempts to resolve the molecular contribution beside the ionic one. Such developments need an intensive analysis of the correlations between ions, electrons and transient ionic molecules, all these entities being present when the ion-ion coupling parameter Γ is larger than unity. In these conditions one has to take into account the mixing between the atomic levels of the emitter ion[12,13] and the formation of molecular potential wells[14] where an ionic molecule can remain stable at least during its emision. These microscopic aspects could be of a great interest for some macroscopic properties of correlated ionized matter such as their electrical and thermal conductivities.[15]

It is worthy of note that absorption spectroscopy is not adequate for our two studies. First of all, as this technique gives information on the spatial dependence of the ground state population, it is ineffective to exhibit temperature gradients. Concerning the analysis of ultradense effects, when a plane massive target is irradiated by a laser beam, the densest plasma which is located in a deep crater[4,5] cannot be accessed by absorption techniques. Finally, in the colliding foils experiments we will talk about here, the compressed dense zone which is very thin and strongly inhomogeneous is much better space-resolved by emisssion spectroscopy.

In section 1 we present the specific emission techniques needful for the specific dense plasma effects. In section 2 we outline the method to access the transverse emission coefficient and the transverse thermal transfer "in situ" in the plasma in the case of strong initial temperature gradients. In section 3 we report on new results concerning the excitation of molecular satellites and of forbidden atomic transitions in strongly correlated plasmas and we discuss their connection with theoretical predictions.

1 - SPECIFIC EMISSION TECHNIQUES FOR DENSE PLASMA EFFECTS

Emission spectroscopy is the only relevant technique for a discriminating analysis of dense plasma effects. Its implementation is not however straightforward, because the emission from cold dense plasmas created by irradiation of massive targets is weak, and difficult to access and to discriminate from emission issued from less dense plasma regions. It also depends drastically on the strong bidimensional gradients.

The experiments have been carried out by using the Nd-glass laser of the Laboratoire pour l'Utilisation des Lasers Intenses (LULI) with wavelength $\lambda_L = 0.263 \mu m$ and a pulse duration $\tau \approx 500 ps$. Technical improvements allowing accurate emission measurements hitherto unobtainable have been made. They aim at amplifying the dense plasma effects, at amplifying the emission and at space locating the effects.

Amplification of dense plasma effects :
Improvements in the laser beam generation such as the characterization of the focal spot (size and shape) allowed us to reach intensities greater than $4.10^{14} W.cm^{-2}$ and to control the initial transverse inhomogeneities. The choice of the detected emissive zone and the creation of new ultradense plasma conditions by colliding foils were two other means for enhancing density effects.

Amplification of the emission :
The target structure and implantation provided an extensive enhancement of the emission ; for instance, strips of emissive material sandwiched between a substrate were used : the width of these strips was taken to be narrow enough so as

to minimize and to control the reabsorption in the observation direction ; their length was larger than the whole focal size in order to maximize the emissive volume. The choice of the observation direction in such implantations is a decisive parameter, since it can be determinant for the emission issued from the densest zone : the spectrograph axis has an inclination with respect to the target plane. The creation of a hot dense plasma by irradiation of foils colliding inelastically is a new attempt to generate emissive ultradense matter. Finally, it may be remarked that it is our purpose to maintain time integration in our experimental results because it is necessary for the increasing of the emission. This integration induces a spatial smoothing over a plasma path length of the order of the ablation depth ($\approx 1.5\mu m$ for a laser intensity $2.10^{14} W.cm^{-2}$).

Choice of the spatial integration domain :

As mentioned before, space-resolved spectroscopy is required because of the existence of strong gradients. In our experimental conditions, our requirements will be a $1\mu m$ resolution in order to diagnose the lateral density and temperature gradients and a $5\mu m$ resolution in order to diagnose the transverse temperature gradients.

2 - TRANSVERSE EMISSION COEFFICIENT AND TRANSVERSE THERMAL TRANSFER

It is worth to point out that the diagnostic of these macroscopic properties are interesting for Initial Confinement Fusion and also for any type of 2D inhomogeneous plasma.

The transverse inhomogeneities of the plasma issued from extended targets are caused by the laser inhomogeneities leading to initial transverrse gradients. Bidimensional codes for plasmas produced by laser irradiation of plane targets[16,17] show that, along a transverse direction, the electron density N_e gradient is much smoother than the electron temperature T_e gradient. Thus two irreversible processes are prevailing simultaneously in the plasma : the lateral thermal conduction and the inhomogeneous radiative transfer. We demonstrate that, although the second process is less important that the first one with respect to the energy balancing, it can be used as a probe of the first process.

Direct diagnostic of transverse emission coefficient :

The problem of measuring the transverse emission coefficient can be explained as follows : since the frequency integrated line intensity is the solution of the radiative transfer equation, it couples in an emisive volume and in an observation direction the emission coefficient ε and the effective absorption coefficient K'. Both coefficients are local properties of the plasma and then intrinsically inhomogeneous. We will show in what follows that we can access directly to the transverse emission coefficient and to the transverse temperature gradient even if the intensity of the line measured is reabsorbed. This diagnostic has been made possible thanks to a realistic theoretical modeling and to the use of adapted target design and spatial resolution.

Theoretical modeling for the transverse emission and absorption coefficients:

The emission coefficient for the frequency ω can be factorized as follows

$$\varepsilon(\omega,x,y,z) = E_0(z)\varphi_1(x,a_{\text{eff}})\varphi_2(y,b_{\text{eff}})P(\omega,z) \qquad (1)$$

where $E_0(z)$ stands for the emission coefficient along the laser-target axis. The line profile $P(\omega,z)$ can be chosen transversally homogeneous because it depends weakly on T_e and because the transverse N_e gradient (i.e. along Ox and Oy) is very smooth.[16,17] The factorization (1) is justified in our experimental conditions (very weak lateral expansion, hierarchy among the temperature gradients along Ox, Oy, Oz). In (1), $2a_{eff}$ and $2b_{eff}$ represent the emission effective dimensions along the transverse directions Ox and Oy respectively.

The effective absorption coefficient $K'(\omega,z)$ can be assumed transversally constant for the same reasons as $P(\omega,z)$.

Experimental set up adapted to the diagnostic of strong transverse gradients : (Fig.1)

We created a strong transverse gradient in a preference direction by focusing the laser beam along a pseudo-elliptical spot onto a structured target. Such inhomogeneous focal spots can be obtained by rotating an angle the lens which is not, for the purposes of these experiments, corrected from geometrical aberrations (coma, astigmatism). We chose a very prolate spot (100μm x 360μm) having a strong inhomogeneity along Oy (focal dimension $2b_{foc} = 100\mu m$). A bidimensional analysis of the laser irradiance I(x,y) has been made thanks to an imaging of the focal spot with a CCD frame. After a numerical smoothing of the data, a factorization has been possible except at the periphery ($\approx 20\mu m$), i.e.,

$$I(x,y) = \frac{I_1(x,0)I_2(0,y)}{I(0,0)} \qquad (2)$$

where I(0,0) stands for the irradiance maximum, and $I_1(x,0)$ and $I_2(0,y)$ represent the laser intensities according to the major and minor axis respectively. This factorization (2) justifies to some extent the modeling chosen for the emission coefficient (1). The measured irradiance I(x,y) has been calibrated thanks to the following surface integral

$$\int_S I(x,y)dxdy = E/S \qquad (3)$$

E being the laser shot energy and S the focal area.

It is time to point out that our diagnostic, depending on this theoretical modeling (i.e.factorizations) and on a numerical smoothing of the experimental data, is ineffective to detect microscopic transverse inhomogeneities such as hot points in the focal spot or in the emission coefficient.

The target is an Al strip sandwiched between a SiO_2 substrate. This target design is well adapted to the diagnostic of strong transverse inhomogeneities ; indeed the strip, the width of which 2l is controlled by fabrication, is chosen parallel to the minor axis Oy (i.e. the strong transverse temperature gradient direction) and well centred on the laser beam. The aluminum emission is then confined by the substrate in the observation direction Ox and can extend along the direction Oy where the lateral thermal conduction can be foreseen because of the strong gradient. The spectroscopic diagnostic is set up in the direction Ox and ensures a spatial resolution along Oy (see Fig.1). The transverse magnification ratio $G_T \simeq 20$ is well adapted for the analysis of the detected $\Delta y \simeq 100\mu m$ plasma strip height. The emission is spectrally resolved by a convex PET crystal.

The theoretical resolution 5000 is drastically reduced down to 2500 due to the important longitudinal magnification ratio $G_L \simeq 0.69$ in the direction of the dispersion on the film which is also the plasma extension direction along the laser-

dispersion on the film which is also the plasma extension direction along the laser-

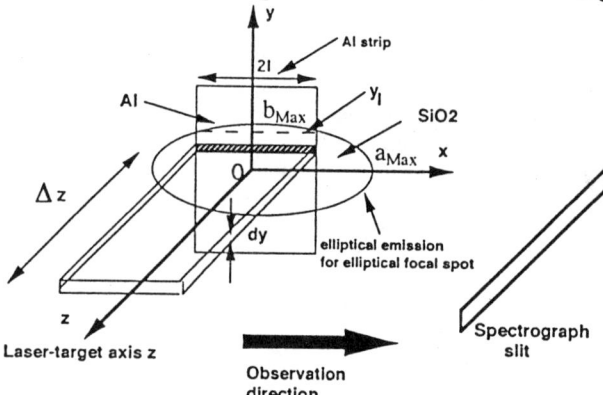

Fig.1. The experimental setup and the spatial resolution along the emissive strip height.; $2a_{Max}$ and $2b_{Max}$ are the maximum dimensions for the emission.

target axis. This reduction has however no incidence on the integrated line intensities we study. In this work we focus our intension on Al Lyß integrated line intensity from which the emission coefficient $\varphi_2(y, b_{eff})$ along the strong gradient direction can be inferred. We then examine the leading part of the laser irradiance $I_2(O,y)$ in this direction.

Direct diagnostic of transverse emission coefficient :
The frequency integrated line intensity emitted in the transverse direction Ox by a plasma slice dy located at y in front of the strip (see Fig.1) is

$$I(y,l)dy = \varphi_2(y, b_{eff})dy \int_{-x_0(y,l)}^{x_0(y,l)} dx\, \varphi_1(x, a_{eff}) \int_0^{\Delta z} dz\, E(z)\, \Phi(\tau, z) \quad (4)$$

The emissive volume involved in (4) is $dy * \Delta z * 2x_0(y,l)$ where Δz represents the emission extent in front of the target and $x_0(y,l)$ corresponds to the border of the emissive strip. It may be seen on Fig.1. that one has

$$x_0(y,l) = l \quad \text{if} \quad |y| \leq |y_l| = \frac{b_{Max}}{a_{Max}}\left[a_{Max}^2 - l^2\right]^{1/2} \quad (5)$$

i.e. in the straight part of the aluminum strip.

The quantities $2a_{Max}$ and $2b_{Max}$ represent the maximum transverse emission dimensions along Ox and Oy. The function $\Phi(\tau, z)$ introduced in (4) represents the transmission factor accounting for the reabsorption effect on the line intensity. It depends basically on the line broadening processes through $P(\omega, z)$, it is given by

$$\Phi(\tau, z) = \int_{-\infty}^{+\infty} d\omega\, P(\omega, z)\, \exp\left[-\tau P(\omega, z)/P(\omega_0, z)\right] \quad (6)$$

Here τ stands for the optical depth corresponding to the line unperturbed frequency ω_0 and to the geometrical depth $[x_0(y,l)-x]$, namely

$$\tau[x, z, x_0(y, l)] = K'(\omega, z)[x_0(y,l) - x]P(\omega_0, z)/P(\omega, z) \quad (7)$$

In the straight part of aluminum emission $I(y,l)$ can be written in the following reduced form

$$I(y,l) = \varphi_2(y, b_{eff})G_{\omega_0}(a_{eff}, l) \qquad (8)$$

with
$$G_{\omega_0}(a_{eff}, l) = \int_{-1}^{+1} dx\, \varphi_1(x, a_{eff}) \int_0^{\Delta z} dz\, E(z)\Phi(\tau, z) \qquad (9)$$

It is important to notice that G_{ω_0} depends on the selected line, on the radiative transfer and on the inhomogeneities along Ox and Oz. In the straight part of aluminum strip, i.e. for $|y| \leq |y_1|$, G_{ω_0} is independent of y as well as the optical depth τ. We now define $A(y,l)$ as

$$A(y,l) = -\left(\frac{\partial \operatorname{Ln} I(y,l)}{\partial y}\right)_l \qquad (10)$$

Thus, for $|y| \leq |y_1|$, one has

$$A(y,l) = -\left(\frac{\partial \operatorname{Ln} \varphi_2(y, b_{eff})}{\partial y}\right)_l = B(y) \qquad (11)$$

where B(y) is independent of the strip width 2l. It represents the logarithm derivative of the emission coefficient $\varphi_2(y, b_{eff})$ in the strong gradient direction. We give in Fig.2 the quantity A(y,l) versus y deduced from the Al Lyβ emission issued from varying the width srips 2l. Each curve corresponds to a laser shot and all the shots have the same energy 29J. The different curves share a common behaviour for $|y| \leq |y_1|$.

The common behaviour of the curves gives access to the transverse emission coefficient $\varphi_2(y, b_{eff})$. We have reported this coefficient on Fig.3. From the present analysis we can make the following comments : several laser shots having the same energy onto various trip targets give complementary results. Wide

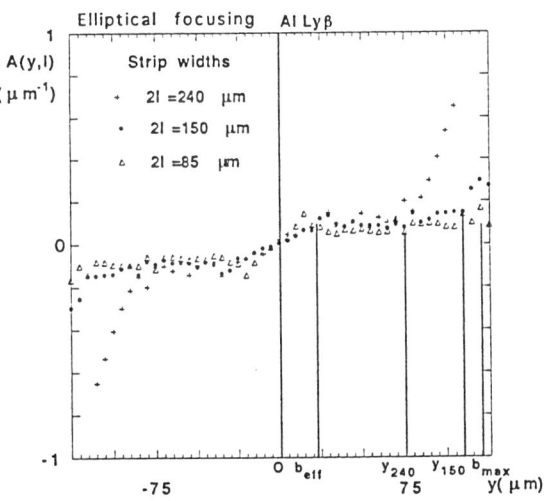

Fig.2. The quantity A(y,l) is given for Al Lyβ and various strip widths 2l. Each curve corresponds to a shot onto a given target. The laser conditions are the same for all shots (0.263μm, 500ps, 29J). y_{240} and y_{150} correspond to the limit of the straight part of the various strips.

emissive targets are efficient for an accurate knowledge of the "core" of $\varphi_2(y, b_{eff})$ target axis, while narrow strips are the only ones to give the "wings" of $\varphi_2(y, b_{eff})$.

In figure 3 one can see that, in this case of a strong transverse gradient, the ings are important : the emission extension ($2b_{Max} \cong 240 \mu m$) is wider thant the laser irradiation extension ($2b_{foc} \cong 100 \mu m$). It is reasonable to think that it is due to the lateral thermal conduction process induced by the strong temperature gradient.

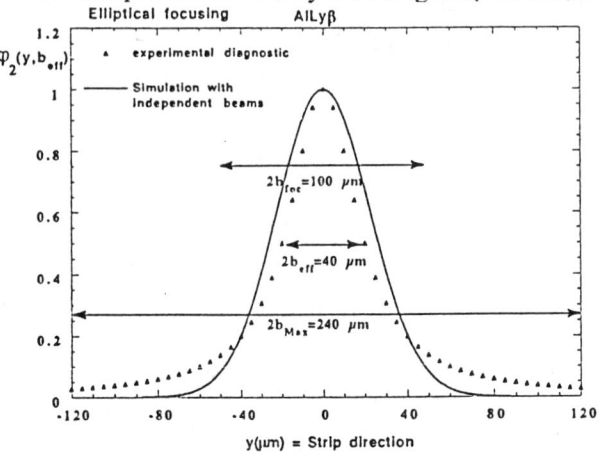

Fig.3. Transverse emission coefficient $\varphi_2(y, b_{eff})$ in the case of a strong transverse gradient (elliptical focusing). The intense line Lyβ has been chosen in this diagnostic because it gives accurate results. The triangles ▲▲▲▲▲▲ correspond to the experimental diagnostic : it has been possible to identify $\varphi_2(y, b_{eff})$ to a lorentzian function characterized by an effective width $2b_{eff}$ and an extension $2b_{Max}$. We have reported the focal dimension $2b_{foc}$. The solid curve ────── corresponds to a simulation made with an "independent beams" model.

Numerical simulation with a bundle of independent beams :

We tried to analyse the leading part of the non uniformity of the laser irradiation on the transverse emission coefficient. To do so we considered an "independent beams" model where the target would be irradiated by a bundle of tiny laser beams, these beams being parallel to the laser-target axis Oz and not interacting between each other (no transverse particle diffusion, no transverse thermal conduction).

Thus, using the monodimensional hydrogynamic code FILM,[18] we have simulated the Al Lyβ emission coefficient $\varepsilon(x = 0, I_2(0, y), z)$ along the laser target axis for each tiny beam irradiating the transverse directionOy in the plane x=0, this tiny beam having an intensity $I_2(0,y)$ measured thanks to the imaging of the focal spot. The factorization of this coefficient gives :

$$\varepsilon(x = 0, I_2(0, y), z) = E(z) \varphi_1(0) \varphi_2(y, b_{eff}) \qquad (12)$$

Here a frequency integration has been made.

Then a space integration along Oz gives access to $\varphi_2(y, b_{eff})$. This numerical simulation for $\varphi_2(y, b_{eff})$ has been reported on figure 3. There is a disagreement with the diagnosed results. The wider extension given by these latter has a plausible explanation in the lateral thermal conduction, not taken into account in the "independent beams" model. However the interpretation in terms of independent beams had been found to be correct in the case of soft gradient.[19]

3 - EXCITATION OF MOLECULAR SATELLITES AND OF FORBIDDEN ATOMIC TRANSITIONS

All the different microscopic aspects of the highly correlated emissive plasmas created by laser interaction with matter have an interest because they are obtained in laboratory conditions quite similar to astrophysical situations. In this regard, some ultradense features in the wings of hydrogen Lα have already been observed in the white and brown dwarfs[20] and interpreted in terms of the formation of molecular structure (H-H^+ and H-H^+-H^+).

When the emitter is strongly coupled with the plasma, excited forbidden transitions can be expected.[12,13,21] Moreover, for high densities ($N_e \geq 10^{23} cm^{-3}$), the average interionic spacing $<D_i>$ can be comparable to the spatial extent of the excited-state orbitals and in these conditions there has been predictions of formation of transient ionic molecules.[14,20] As pointed out before, the exhibition of the ultradense effects is difficult because of the weakness of the emissivity and also because there are specially constrained temperature conditions for the molecular transitions observation. More precisely, such transitions would need equilibrium positions R_o for the ionic molecules, these positions being different from the average interionic spacing $<D_i>$. Therefore, the equilibrium positions can only be reached if there are fluctuations around $<D_i>$, that is if the temperature is high enough. However, the temperature cannot be too high because the transient ionic molecule must remain stable during its emission. For all these reasons we need a high spatial resolution for the detection of the molecular features because they take place in a very small plasma path along the laser-target axis ($\simeq 2\mu m$) due to the existence of strong gradients.

We now present two kinds of experiments devoted to the creation of ultradense plasmas: laser beam interaction with massive targets and colliding foil experiments. In both situations, fluorine has been chosen as the emitter because of its large Stark effect and of its sensitiveness to correlation effects accordingly.

Massive LiF or CF_2 targets experiments[4,5]:

Regarding the targets we used strips of lithium fluoride (LiF), or polytetrafluorethylen (CF_2) sandwiched between a carbon substrate.

In these experiments the two following techniques allowed access to dense plasma in the crater:

- first, the dense plasma volume was increased due to its selection by an angled spectrograph slit (Fig.4); the emission from this region becomes then sufficient to explore the spectral line shapes using experimental results from only one laser shot. Thus the amplification of the path length Δz corresponding to the ultradense plasma (Fig.4) leads to an enhancement of the line wings and of the dense plasma effects accordingly.

- secondly, the high spatial resolution along laser-target axis Oz (ensured by a 40μm slit) joined to a high transverse magnification ratio 120 gives access to 1μm plasma resolution.

We present in Fig.5 F Lyβ line shapes observed in region I on the film. This region has been impressed by the signals arising from those lines-of-sight which are

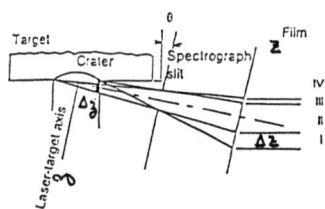

Fig.4. Spatial resolution of the spectrograph. The inclination θ of the target with respect to the spectrograph axis allows a spatial integration over a large volume of dense plasma. $\Delta Z = 120 \Delta z$ in region I on the film involves the dense part of the plasma. For a 40μm slit and a 10° inclination, $\Delta z = 7 \mu m$.

integrals over progressively more important dense parts of the plasma. The high spectral resolution (2000) of the bent concave KAP crystal allows the observation of very fine structures. The red wing exhibits two satellites (1) and (2) in cases (c) and (d) only, that is for spectra corresponding to space integrations lying within the range (3-5μm). For this interval of 2μm, the plasma volume added has the temperature required for the molecular components (101g)-(000u) and (220u)-(000g) (notations with parabolic and azimutal quantum numbers used for the separated atoms) observations. These F Lyß satellites have been identified with F^{8+}-F^{9+} ionic molecule emission, they are associated to transition energy extrema of this molecule. Such extrema have been theoretically predicted[14] for electronic densities $(1-3) \times 10^{23} cm^{-3}$ and electronic temperatures 200-300eV. These plasma parameters are consistent with the numerical estimates for the target crater.[18] Let us emphasize that all the experimental work reported here gives the first exhibition of the molecule F^{8+}-F^{9+} in laboratory plasmas.

Finally the molecular satellites (1) and (2) have been discriminated from the He-like satellites F lyß[4] observed in spectra issued from plasma path lengths involving higher temperatures and lower densities. Moreover they cannot be indentified with dielectronic satellites with an n>3 spectator.[4]

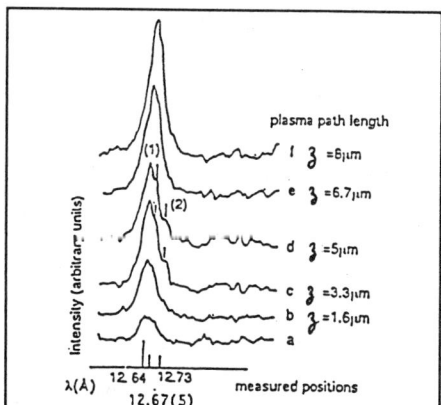

Fig.5. F LyB space integrated spectra (Target : strip 40μm, θ =10°;laser beam : 16J, 500ps, 0.263mm). Curves (a)-(f) correspond to varying dense plasma path length Δz < 8μm. Two spectra have been vertically shifted by arbitrary amounts. The red wing exhibits two satellites (1) and (2) in cases (c) and (d) only, corresponding to space-integration over the whole crater.

Al and CF_2 colliding foils experiments :

More recently we have developed colliding foils experiments with the aim of creating "hotter" ultradense plasmas. These plasmas provided favourable conditions for the excitation of molecular transitions **and** of forbidden ionic transitions. To produce such optimal conditions, we tried to realize a plane compression by colliding thin foils using two opposite laser beams. Our idea was to choose the foil thickness and initial distance so that the collision occurs with the maximum kinetic energy during the laser pulse duration. The kinetic energy is favourable for a good compression (high density) and for a strong emission (high temperature). Moreover the foils must be ablated enough to make their rear face "hot" before the collision, but not completely ablated for the efficiency of the compression. The optimum ablation depth also depends on the foil thickness and initial distance.

These experiments have been modeled by hydrodynamic simulations : [17,18] the predictions from the 1.5D FILM version[18] (LULI) and 2D LASNEX code[17] (Livermore) are quite similar. These codes indicate that 2D effects are extremely important and unfavourable for an optimum compression. They also estimate the emission duration for each line in the compressed central plasma (a few 10 picoseconds). This duration gives the order of magnitude of the temporal resolution obtained in our spectroscopic diagnostics. The experiments have been carried out on the LULI facilities, by using two beams in the same conditions as for massive target experiments $(\lambda_L = 0.263 \mu m, \tau = 500 ps)$. The available 20-30J energy shots are focused onto Al or CF_2 foils, with initial thickness $1.5 \mu m \leq e \leq 10 \mu m$ and initial distance $50 \mu m \leq d \leq 250 \mu m$ (see Fig.6). The small focal spots (Ø 90 to 100μm) allowed laser intensities as high as $4.10^{14} Wcm^{-2}$.

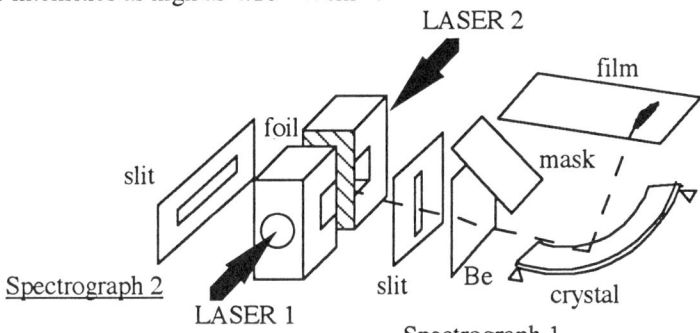

Fig.6. Implantation of the targets and of the diagnostics for colliding foils experiments. The spectrograph 1 ensures a space resolution along the collision axis. The spectrograph 2 thanks to a space resolution perpendicularly gives access to the plasma transverse dimension.

The X-ray emissison merging transversally (i.e. perpendicularly to the collision axis) from the aluminum or fluorine plasmas (inside the foils and forward the laser beams) is spectrally resolved and space-resolved in the axial (i.e. collision axis) and the transverse directions.
For the axial space-resolved diagnostic (spectrograph 1) we used the same spectrograph as for the massive target experiments (R=2000 for F Lyß, G_T=120) with an entrance slit 10μm. As the longitudinal magnification ratio is low (G_L=0.02 for Al Lyß), the plasma extension in the transverse direction does not affect the line broadening and the spectral resolution. We observed two possible collision regimes depending on the collision parameters : collision between two foils followed by a compression with creation of a dense and emissive plasma / collision between two plasmas resulting from the explosion of the foils before their collision.

We report on Fig.7 a film corresponding to the first regime, the desired one for the creation of a dense plasma. The two external symmetric zones of the spectra are due to the emission forward the beams (coronas). In the central zone, the recorded emission reveals intense and very broadened lines over a distance corresponding to a plasma spatial integration of about 10μm.

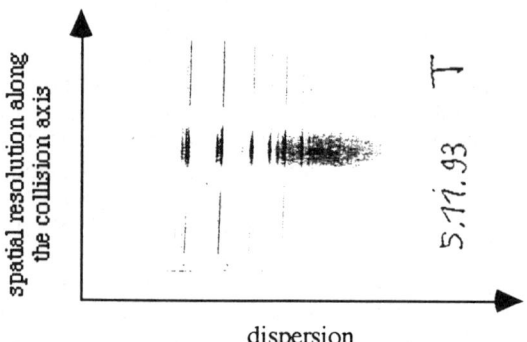

Fig.7. Spatially and spectrally resolved X-ray emission during and after a compression of aluminum foils. The spatial resolution takes place along the collision axis (e = 1.5μm, d = 100μm, $I_L = 2.10^{14} W cm^{-2}$ for each laser beam).

Concerning the transverse space-resolved diagnostic (spectrograph 2) we used the same spectrograph as for the transverse gradients analysis (R = 2500 for Al Lyß, $G_T = 6.5$) with an entrance slit 40μm. As this spectrograph has a significant longitudinal magnification ratio ($G_L = 0.85$ for Al Lyß), the plasma size has an important effect in the dispersion direction (i.e. collision axis). Thus bidimensional photographic images on each line (spectroheliograms) were obtained showing expanded coronas and a compressed zone (see Fig.8). This diagnostic giving access to the plasma transverse dimension in the compressed zone and in the coronas has been studied in collaboration with C. Back[22].

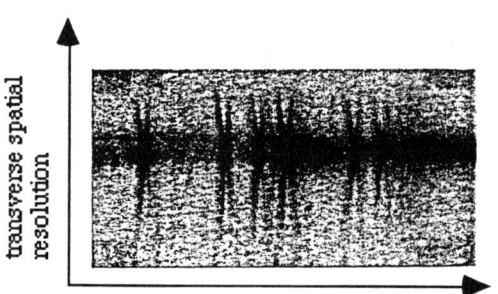

Fig.8. Spatially and spectrally resolved X-ray emission during and after a compression of aluminum foils. The spatial resolution takes place perpendicularly to the collision axis. Each line is the signature of a bidimensional image of the compressed zone and of the coronas. (e = 6μm, d = 150μm, $I_L = 2.3\ 10^{14} W cm^{-2}$ for each laser beam).

We now focus our attention on F Lyß profil diagnosed from CF_2 colliding foils experiments. We have reported on Fig.9 a profile which is spatially integrated over the central compressed zone ($\simeq$ 10μm). The red wing still exhibits the molecular satellites (1) and (2). Moreover in the blue wing a forbidden transition has been excited. This transition identified as 1s-3d had already been predicted by

Salzmann.[21] On the same figure we have reported two theoretical profiles. The computation made by R. Stamm et al.[19] includes properly both the dipolar interaction in the ionic sphere atomic model and the electron broadening. Salzmann introduced a cylindrically symmetric atomic model which accounts for the effects of the nearest neighbour on the emitter. In this model, the shape of the molecular volume is approximated by two adjacent truncated spheres, the forbidden line is excited but the electron broadening is neglected. Both theoretical profiles have been computed for an electronic density 10^{23}cm^{-3} and an electronic temperature 300eV. For a decisive comparison with the experimental profile we whould have to take into account all densities and temperatures involved in the spatial integration (10μm) during the emission of F Lyβ : the hydrodynamic codes show very strong gradients and the latter should be introduced in theoretical profiles for relevant comparisons with experimental profiles.

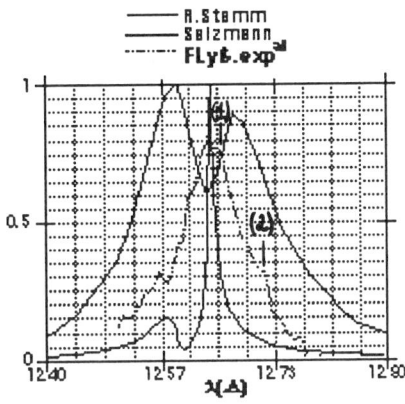

Fig.9. F Lyβ spectrum spatially integrated over the central compressed zone. The red wing exhibits two molecular satellites (1) and (2). In the blue wing the forbidden transition 3d-1s is excited (e=3μm, d=100μm, IL=3E14 W/cm2 for each laser beam). Computations made by Stamm and Salzmann for Ne=1E23 cm-3, Te=300 eV are reported.

REFERENCES

1. H. Derfoul et al., Laser and Particle Beams **12**, 485 (1994)
2. H. Derfoul et al., LULI, Report, (1993)
3. H. Derfoul, Thesis, Université Paris VI (1994)
4. E. Leboucher-Dalimier et al., Phys.Rev. E **47**, R1467 (1993)
5. E. Leboucher-Dalimier et al., J.Quant.Spectrosc.Radiative Transfer. **51**, 187 (1994)
6. E. Leboucher-Dalimier, *Pour la Science,*, n°199 (May 1993)
7. R. Stamm et al., Phys. Rev. **34**, 4144 (1986)
8. A.V. Demura et al., J.Quant.Spectrosc.Radiative Transfer. **15**, 881 (1975)
9. R. Joyce et al. Phys.Rev. A **35**, 2228 (1987)
10. L. Woltz and C. Hooper, in *Proceedings of the 2nd International Conference on Radiative Properties of Hot, Dense Matter*, edited by W. Goldstein (World Scientific, Singapore, 1987) p.476
11. E. Leboucher-Dalimier et al., in *Spectral Line Shapes 3*, edited by F. Rostas (de Gruyter, Berlin, 1985) p.251
12. J. Stein et al., Phys.Rev. A **39**, 2078 (1989)
13. D. Salzmann et al., Phys.Rev. A **44**, 1270 (1991)
14. Ph. Malnoult et al., Phys.Rev. A **40**, 1983 (1989)
15. S. Younger et al., Phys.Rev.A **40**, 5256 (1989)
16. R.I. Klein et al. J.Quant.Spectrosc.Radiative Transfer.**41**, 199 (1989)
17. T. Shepard, Private communication
18. LULI "FILM" code
19. P. Angelo et al., LULI Report, (1993)
20. J. Kielkopf, in Proceedings of the 11th ICSLS
21. D. Salzmann, Private communication
22. P. Angelo et al., in Proceedings of 12th ICSLS, Poster PB-13.

LINE BROADENING OF NONHYDROGENIC IONS IN PLASMAS

Siegfried Glenzer

Institut für Experimentalphysik V, Ruhr-Universität, 44780 Bochum, Germany

ABSTRACT

Recent measurements of the Stark broadening of isolated spectral lines along isoelectronic sequences of nonhydrogenic ions in a well-diagnosed dense plasma of the gasliner pinch show a systematic discrepancy between measured widths and the results of some theoretical approximations with increasing spectroscopic charge number Z. For $Z > 4$ various experiments give systematically larger values for the Stark widths compared with calculations in the electron-impact approximation. Since proton collisions may help to explain the observed scaling the influence of proton collisions on the linewidth of the $3s\ ^2S_{1/2} - 3p\ ^2P^o_{3/2}$ transition in Ne VIII is investigated experimentally. Although there is an improvement between theory and experiment the result suggests that further broadening mechanisms are also important. Effects on the theoretical predictions are discussed when strong (electron) collisions, ion and electron quadrupole effects, elastic and inelastic proton collisions are taken into account.

INTRODUCTION

The experimental verification of theoretical predictions of spectral line profiles emitted by dense and hot plasmas is needed for plasma diagnostic applications and radiative transport calculations[1,2]. In particular, the investigation of plasmas produced for x-ray laser or inertial confinement fusion research depends on the accurate knowledge of spectral line profiles emitted by multiply ionized atoms. On the other hand, most of the experimental data which are obtained from well-diagnosed plasmas[3] deal with isolated transitions of lowly ionized nonhydrogenic emitters ($Z \leq 5$). In these cases broadening by electron collisions is the dominant broadening mechanism, and the comparison with calculations in the electron-impact approximation indicates that modern theoretical approximations[1,4] reproduce the experimental linewidths with an accuracy of about 20 %. For more highly ionized emitters ($Z \geq 5$), however, experimental data from well-diagnosed plasmas are scarce while they are urgently needed for a test of several theoretical approximations.

For theoretical calculations of emission profiles of nonhydrogenic ions the electron-impact (dipole) approximation is usually applied giving Lorentzian line profiles with half-widths w which are determined by the product of cross sections for electron impact excitation and electron velocity averaged over an appropriate velocity distribution function and multiplied by the electron density[5]. Since the cross sections scale with the spectroscopic charge number as Z^{-2} a corresponding systematic trend is expected for the linewidths of nonhydrogenic ions, i.e. $w \sim Z^{-2}$. However, this scaling may become questionable for the linewidths of highly ionized atoms because higher multipole interactions[6] as well as strong collisions[7] may become important.

Indeed, the relative contribution of strong collisions to the linewidth estimated after Griem[1] or by Hey[8] and Alexiou[9] grows with increasing spectroscopic charge number Z. Since it is not possible to calculate them semiclassically with as high accuracy as inelastic dipole-allowed electron collisions theoretical results are becoming more uncertain. Furthermore, the contribution of ion perturbers to the linewidth is difficult to evaluate since ions are often neither static nor in conformity with the validity criteria of the impact approximation.

Experimental investigations of the scaling mentioned above by measuring Stark widths of spectral lines along isoelectronic sequences are reported in Refs. [10,11,12]. Only isoelectronic sequences consisting of three transitions in ions of relatively low charge states have been measured. There is only one study[12] involving ions with $Z > 4$ where the linewidths of the $3s - 3p$ transitions of the lithiumlike ions C IV, N V, and O VI show a Z^{-1} scaling in contradiction to Z^{-2}. However, the agreement of the experimental values of C IV and N V of Ref. [12] with other experiments[3,13] and with several theoretical approximations[1,4,7] is not satisfactory.

Therefore, we performed systematic measurements of the $3s - 3p$ and $3p - 3d$ transitions along the isoelectronic sequences of lithium[14] and boron[15] with a maximum spectroscopic charge number of $Z = 8$ in a well-diagnosed plasma of the gas-liner pinch discharge. Each plasma condition is diagnosed independently by 90° Thomson scattering and the homogeneity of the discharge in radial and axial direction is measured. Furthermore, radiative transport (self-absorption) of the investigated spectral lines is shown to be negligible. In this paper the experimental results are discussed and compared with the results of several theoretical approximations[1,4,7,9,16,17]. Furthermore, by comparing the linewidths of the Ne VIII $3s\,^2S - 3p\,^2P°$ transitions measured in plasmas with hydrogen and with helium perturbers the role of proton collisions is studied experimentally. The experimental result is compared with estimated inelastic proton collision widths after Griem[18].

Apart from the discussion of the linewidths of isolated spectral lines of nonhydrogenic ions we also discuss briefly recent measurements of $n = 4 - 5$ and $n = 3 - 4$ transitions in the lithiumlike ions C IV, N V, and O VI. For these transitions forbidden components and ion dynamical effects have to be taken into accout. A recently developed computer code[19] gives line profiles which are in excellent agreement with the experiment[20].

EXPERIMENT

A. THE GAS-LINER PINCH. A suitable source for line broadening studies is the gas-liner pinch as described at an earlier conference on spectral line shapes[21]. Figure 1 shows the experimental setup. The gas-liner pinch resembles a z–pinch characterized by a special gas inlet system. By means of a fast electromagnetic valve hydrogen or helium as driver gas is injected through an annular nozzle into the vacuum chamber. The diameter of the vacuum chamber is 18 cm and the electrode separation is 5 cm. Initially, the gas forms a hollow gas cylinder near the wall. The gas is preionized, and by discharging the main capacitor (capacitance 11.1 μF,

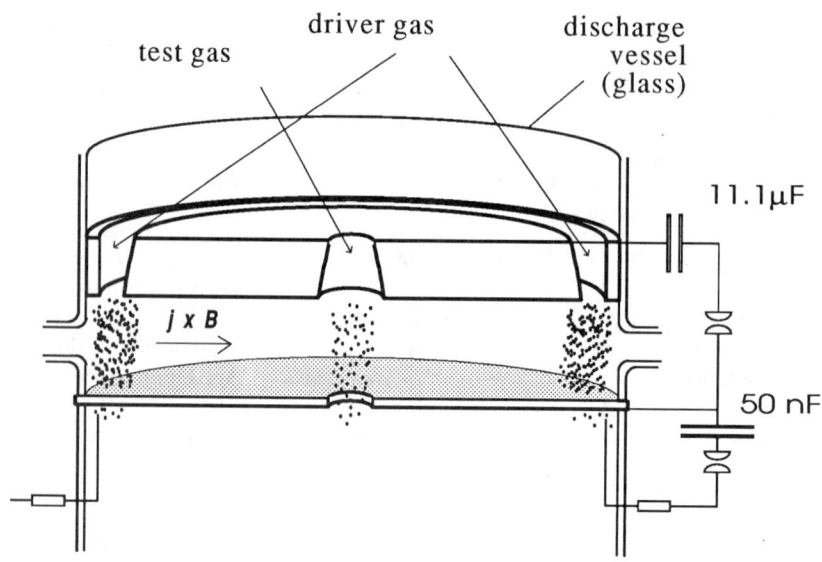

Figure 1: Schematic of the gas-liner pinch.

voltage 25-35 kV) it is compressed on the axis to a reproducible plasma column of 1-2 cm diameter and 5 cm length. Typical electron densities and temperatures reached on the axis are between $0.5 < n_e < 4 \times 10^{18} \mathrm{cm}^{-3}$ and $7.5 < k_B T_e < 50$ eV which is sufficiently hot and dense to produce multiply ionized atoms with Stark broadening as the dominant broadening mechanism of their emission lines. The compression time and the life time of the plasma depend on the discharge conditions and for the present studies they are about 2.5 μs and 0.5 μs, respectively.

The essential feature of this plasma source, however, is a second fast electromagnetic valve. It injects independently the so-called test gas along the axis of the discharge chamber through a nozzle in the center of the upper electrode. The test gas is dissociated and ionized by the imploding driver gas and by ohmic heating. We used CH_4, N_2, CO_2, a mixture of 10 % SF_6 in hydrogen or Ne as test gas in order to produce the lithium-like and boron-like ions C IV, N III, N V, O IV, O VI, F V, F VII, Ne VI, and Ne VIII. If a very small amount of test gas is used (about 1 % of the density of the driver gas) and if the injection is properly timed the test gas ions become confined in the central part of the discharge where the plasma is homogeneous in radial and axial direction. This has been verified experimentally. Furthermore, radiative transport of the emission lines is easily controlled by varying the amount of test gas because a cold boundary layer of the investigated ions is effectively absent.

The plasma is accessible through four ports in the midplane of the discharge tube. Two 1m-spectrometers for the visible and one 1m-spectrometer for the vacuum ultraviolet (vuv) spectral range equipped with optical multichannel analyzers are available. The gating time of these detectors is chosen to be 30 ns or 35 ns. This is approximately the same time resolution as that given by the ruby laser which is used for Thomson scattering.

B. PLASMA DIAGNOSTICS. A prerequisite for testing theoretical line shapes is an independent measurement of all relevant plasma parameters. Indeed, Thomson scattering yields the electron density and the temperature of the plasma with a high spatial and temporal resolution[22]. We use a ruby laser ($\lambda_o = 694.3$ nm) which is focussed into the center of the plasma and the Thomson scattering spectrum is detected at an angle of $\theta = 90°$. This gives typical values for the scattering parameter α

$$\alpha = \frac{1}{k\lambda_D} = \frac{\lambda_o}{4\pi \sin \theta/2} \sqrt{\frac{n_e e^2}{\epsilon_o k_B T_e}} > 1 \qquad (1)$$

where k is the absolute value of the scattering vector, λ_D the Debye length, k_B the Boltzmann constant, and ϵ_o is the permittivity of free space. In this regime scattering on electrons which are bunched in the Debye sphere of the ions dominates, and from the width of the scattering spectrum, the so-called ion feature, the temperature of the ions is obtained. Furthermore, the ion acoustic wave plays an important role for the shape of the scattering spectrum. The phase velocity of the ion acoustic wave depends on the ion and on the electron temperature, and from an exact measurement of the shape of the scattering spectrum the ratio T_e/T_i is obtained (T_i denotes the temperature of the driver gas ions). In order to determine the electron density from the intensity of the scattering spectrum the detection system is calibrated by Rayleigh scattering on propane. For the present investigations multiply ionized heavy atoms are added as test gas ions to a hydrogen or helium plasma. Hence, the shape of the scattering spectrum and the total scattering amplitude is modified. A central narrow peak arises with increasing concentration and charge number of the test gas ions[23] and the width of the peak is determined by the temperature of the test gas ions. Thereby, the total scattering amplitude, i.e. the frequency-integrated form factor, is increased. These effects are taken into account by the theoretical form factor of Evans[24] which is fitted to the measured scattering spectra. The procedure is described in detail in Ref. [25] and an example of a measured spectrum is shown in Fig. 2 along with the fitted form factor. In that case hydrogen as driver gas and neon as test gas have been used, and by estimating the charge number of the neon ions from spectroscopic measurements the concentration of neon ions is determined from this measurement to be 0.5 % of the electron density. In general, we found that the temperatures of the electrons, of the driver gas ions and of the test gas ions are equal within the error bars. Furthermore, the test gas ion concentration was determined to be about or less than 1 % of the electron density in all cases. Recently, we observed asymmetric Thomson scattering spectra when producing plasmas with very high test gas concentrations (~ 10 %)[26]. This effect is most probably related to a magnetohydrodynamic instability[25]. For the present investigation, however, no indication of instabilities or plasma turbulences is found from the measured Thomson scattering spectra.

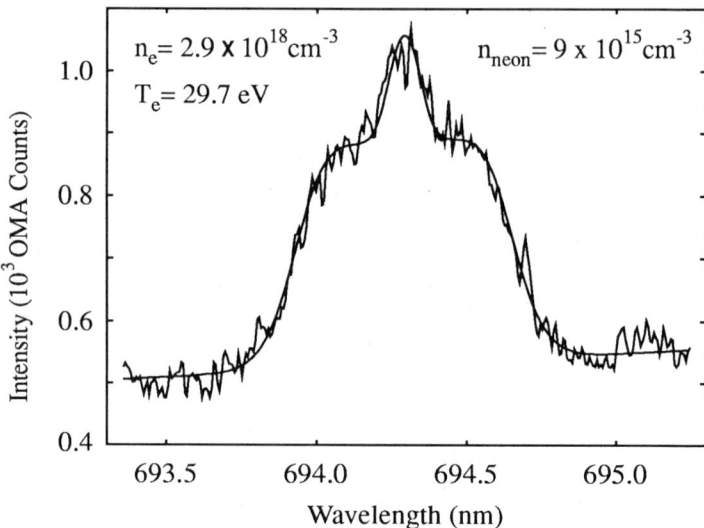

Figure 2: Example of a measured Thomson scattering spectrum along with the fit.

C. HOMOGENEITY OF THE PLASMA. In the midplane of the discharge the homogeneity in radial direction was verified by two independent methods[14,15]. The measurement of the continuum and test gas radiation in the visible spectral range of a single discharge shows after Abel inversion that the electron density changes by less than 10 % in the central part of the plasma where more than 95 % of the test gas radiation is emitted[14]. This advantageous feature prevail from maximum pinch compression up to at least 200 ns after that. More recently, we measured the electron density distribution in the midplane of a single discharge with Thomson scattering[15]. The result obtained at the time of maximum pinch compression is shown in Fig. 3 along with the emission coefficient for line radiation from test gas ions which is obtained from measured radial intensities and Abel inversion. It is obvious that the test gas ions are distributed in the central homogenous part of the plasma column over a diameter of about 1 cm. Furthermore, it is of interest to show that the plasma is also homogeneous in axial direction because magnetohydrodynamic instabilities could arise. Therefore, we measured the $2s - 2p$ doublet of Ne VIII in the vacuum ultraviolet at $\lambda = 78.03$ nm ($\frac{1}{2} - \frac{1}{2}$) and $\lambda = 77.04$ nm ($\frac{1}{2} - \frac{3}{2}$) with a charge-coupled-device (CCD) camera. About 1 cm of the 5 cm long plasma column is observed (which is, for the present investigation, the interesting part of the plasma column). The measured intensities of a single discharge are plotted in Fig. 4 as a function of the wavelength and of the height of the pinch. Even though the $2s_{1/2} - 2p_{3/2}$ transition at $\lambda = 77.04$ nm shows radiative transport effects the homogeneity along the axis of the discharge is clearly demonstrated.

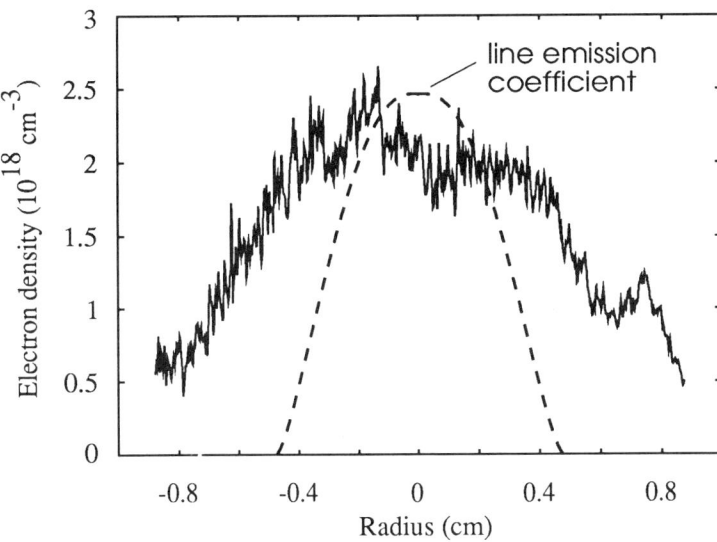

Figure 3: Electron density as a function of the radius of the discharge at maximum pinch compression. Also shown is the emission coefficient for line radiation emitted from test gas ions.

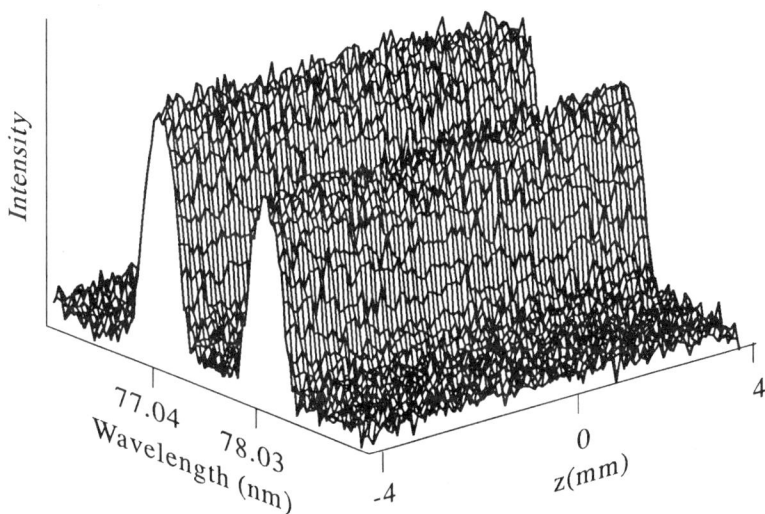

Figure 4: Example of a measured $2s - 2p$ doublet of Ne VIII as a function of the height of the pinch.

D. RADIATIVE TRANSPORT. For transitions with highly populated lower levels self-absorption is a serious problem which leads to line profile distortions. This is especially the case for resonance transitions as shown in Fig. 4. On the other hand, the lower levels of the $3s-3p$ and $3p-3d$ transitions are sufficiently lowly populated and, therfore, we expect that radiative transport effects are negligible. This is verified by measuring relative intensities of spectral lines within a multiplet for different plasma conditions and test gas ion concentrations. Since high particle densities of the plasma result in sufficiently high collision rates, level population densities within a multiplet are certainly given by the Boltzmann statistics. Hence, we compare measured line intensities with the predictions of the LS coupling approximation via

$$\frac{I}{I'} = \frac{\lambda'^4}{\lambda^4} \frac{D^2_{Line}}{D'^2_{Line}} \exp\left(\frac{E'-E}{k_B T_e}\right), \qquad (2)$$

where I, λ and I', λ' are the total intensities (integrated over the line profile) and wavelengths of two multiplet components being compared, and E and E' are the energies of the upper levels of the two components, respectively. D^2_{Line} are the so-called line factors as given in Ref. [27]. For the $3s\ ^2S - 3p\ ^2P^\circ$ transitions of the present study they are $D^2_{Line} = 2/3\ (\frac{1}{2} - \frac{1}{2})$ and $D^2_{Line} = 4/3\ (\frac{1}{2} - \frac{3}{2})$, and for the $3p\ ^2P^\circ - 3d\ ^2D$ transitions $D^2_{Line} = 10/15\ (\frac{1}{2} - \frac{3}{2})$, $D^2_{Line} = 2/15\ (\frac{3}{2} - \frac{3}{2})$, and $D^2_{Line} = 18/15\ (\frac{3}{2} - \frac{5}{2})$ (see e.g. Appendix I of Ref. [27]). For the present study we can fortunately ignore accidental cancellations of transition matrix elements in the length form because the transitions are between levels with the same principal quantum numbers, for which the relevant radial functions peak at similar distances from the nucleus.

We find for all plasma conditions and spectral lines that the measured line intensity ratios are in good agreement with the predictions of the LS coupling approximation. There is only one exception for the $3p\ ^2P^\circ - 3d\ ^2D$ transitions in Ne VI where the measured intensity ratio is 9 % larger than that predicted by the LS coupling approximation. This result can not be explained by self-absorption because for the chosen ratios self-absorption, if present, would give smaller values for the measured intensity ratios than the LS coupling approximation (see Ref. [28]). In general, the good agreement between measured and predicted ratios does not change even when increasing the concentration of test gas ions in the discharge by a factor of two. Examples of two experimental spectra of the Ne VIII $3s - 3p$ transitions at $\lambda = 286.01$ nm $(\frac{1}{2} - \frac{1}{2})$ and $\lambda = 282.07$ nm $(\frac{1}{2}\ \frac{3}{2})$ measured with hydrogen as driver gas are shown in Fig. 5. The first spectrum is measured for a neon ion concentration of 0.5 % of the electron density (which was the standard experimental condition) and a second one for a concentration of 1 %. The first experimental condition yields an intensity ratio of $I(3s_{1/2} - 3p_{3/2})/I(3s_{1/2} - 3p_{1/2}) = 2.04 \pm 6$ % and the second condition $I(3s_{1/2} - 3p_{3/2})/I(3s_{1/2} - 3p_{1/2}) = 2.05 \pm 8$ %. These results are in excellent agreement with the LS coupling approximation $I(3s_{1/2} - 3p_{3/2})/I(3s_{1/2} - 3p_{1/2}) = 2.11$. Since self-absorption would seriously affect these results we conclude that this effect is entirely negligible for our experimental conditions.

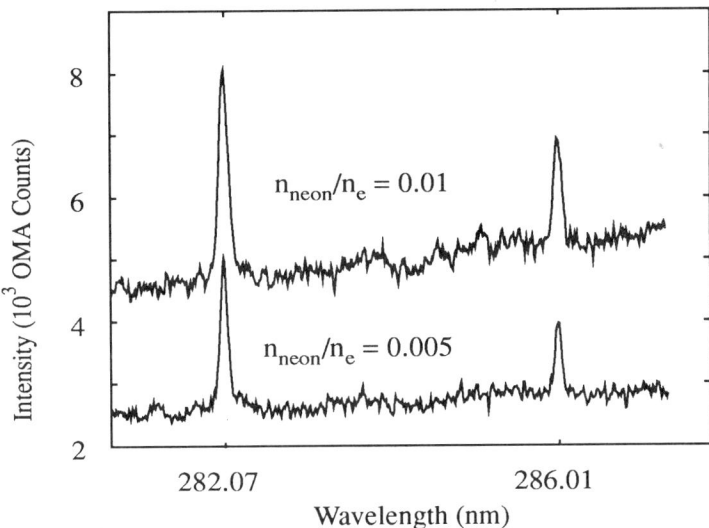

Figure 5: Examples of the measured $3s - 3p$ doublet of Ne VIII for two different emitter concentrations.

EXPERIMENTAL RESULTS AND DISCUSSION

In order to determine the Stark width of the measured spectral lines we fitted Voigt functions employing a least-squares procedure to the experimental data. The Voigt functions consist of the measured apparatus profile convoluted with a Doppler profile which was calculated according to the measured temperature of the emitters for each plasma condition. The resulting profile is convoluted with a Lorentzian profile with a variable width for the Stark broadening. The fitting procedure also takes into account a variable continuum which is checked to be linear by recording spectra without the injection of test gas. This procedure is also useful to verify that no radiation from unwanted impurities affects the measured line profiles. For all transitions measured with the detection system for the visible spectral range both the contribution of the Doppler profile and of the apparatus profile were smaller than 15 % of the overall linewidth. For transitions in the vuv the apparatus profile was up to 50 % of the linewidth.

It should be pointed out that for the measurements in the vacuum ultraviolet spectral range Thomson scattering could be performed simultaneously. However, for most of the measurements in the visible spectral range Thomson scattering was carried out immediately after the spectroscopic measurements with the same detection system. For that reason only reproducible plasma conditions with a smooth variation of electron density and temperature have been investigated. Typical error bars obtained from the rms value of 10 measurements for the same time in the discharge are 15 % for the electron density and temperature. Line profiles, however, show a much higher reproducibility, and the error in the determination of the linewidths which is also obtained from the rms value of 10 measurements is typically about 5 % and is simply due to noise.

In order to investigate the dependence of the measured linewidths on the spectroscopic charge number Z the dependence on the electron density and temperature has to be studied. In our experiment the temperature had to be increased by a factor of about 5 to obtain the desired ionisation stages of the investigated atoms. Simultaneously, there was also a small variation of the electron density.

It is a fundamental concept of the impact theory that the contribution to the linewidth owing to electronic or ionic collisions is proportional to their number density[5]. Since the dominant contribution to the linewidths of the investigated ions is due to electron collisions and criteria[7,29] for the application of the electron-impact approximation are comfortably fulfilled we accept a linear scaling with the electron density. Indeed, a linear scaling of linewidths of nonhydrogenic ions has also been proven in a number of experiments[1,3,30]. It should be mentioned that deviations from a linear scaling may be possible due to Debye shielding effects which are, however, unimportant in our studies[31]. A general temperature dependence of the linewidths, on the other hand, is not predicted theoretically. For example, the assumption of a constant Gaunt factor in the case where the energy differences between all important atomic levels are large in comparison to the temperature of the perturbing electrons results in a $1/\sqrt{k_B T_e}$ scaling of the impact widths. For higher temperatures, however, the value of the Gaunt factor increases, and the linewidth becomes less dependent on the electron temperature. Experimentally, we have investigated the dependence of the Stark width on the electron temperature for six transitions[14,15]. All data show a similar result and, therefore, only one example is discussed here, i.e. the Stark width (FWHM) of the $3s^2 S - 3p^2 P°$ transitions in O IV. Our experimental widths which are measured for a density range of 0.56×10^{18} cm^{-3} < n_e < 1.63×10^{18} cm^{-3}, the experimental result of Ref. [32] measured at $n_e = 2.18 \times 10^{17}$ cm^{-3}, and experimental data from another experiment[33] which is performed for a density range of 2.1×10^{17} cm^{-3} < n_e < 6.4×10^{17} cm^{-3} (see also Konjević et al, this book) are scaled linearly to a value of the electron density, i.e. $n_e = 1 \times 10^{18}$ cm^{-3}. The data are plotted in Fig. 6 and compared with calculated values according to several theoretical approximations[1,4,7,33,34]. For the sake of a better survey error bars are shown only for our experimental data. The vertical error bars include the uncertainty from the determination of the linewidths as well as that from the electron density measurement. The agreement between our experimental data and that of Ref. [33] is excellent. Taking into account the error bars of the experimental result of Ref. [32] it also matches our experimental data.

The comparison of the experimental data with the results of several theoretical approximations, on the other hand, shows good agreement with some of them and poor agreement with others. The semiclassical electron-impact approximation after Griem[1] is acceptable within the error bars of the experiment. Griem deduced an effective Gaunt factor from the calculation of the classical hyperbolic orbits of the perturbing electrons. In this way the contribution of inelastic electron collisions are calculated. Also included is an estimate of higher multipole interactions and of

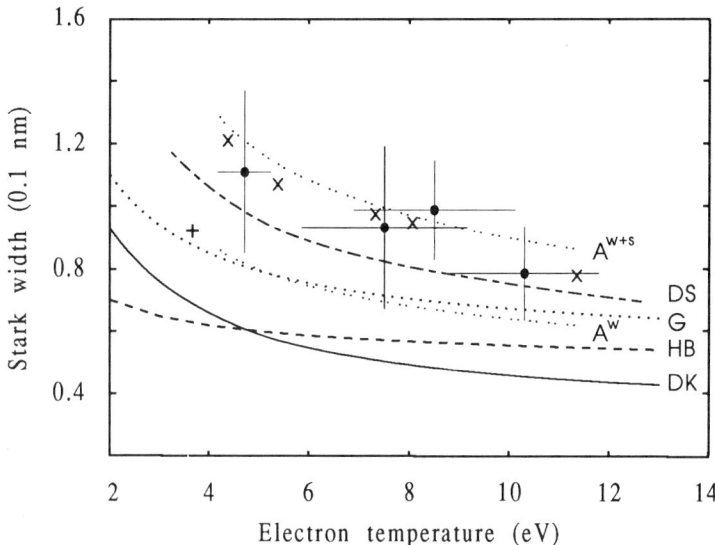

Figure 6: Comparison of Stark width data for the $3s^2S - 3p^2P^\circ$ transitions in O IV for $n_e = 1 \times 10^{18}$ cm^{-3}. Theoretical results: $\cdots$, after Griem [1] (index G); —, after Dimitrijević und Konjević [4] (index DK); - - -, by Hey [15] (index HB); $\cdots\cdots$ after Alexiou [34] (index A^w for the weak width and A^{w+s} for the weak plus strong width); ---, calculations after Dimitrijević und Sahal-Bréchot [33] (index DS); experimental values : •, Ref. [15]; × Ref. [33]; + Ref. [32].

strong collisions. The ionic contribution to the broadening, however, is difficult to estimate, because for this transition and the present plasma parameters the criterion[1] for the application of the quasi-static approximation is only close to being fulfilled. Therefore, it is not clear whether or not quasi-static ion quadrupole broadening has to be taken into account. A rough estimation after Griem[1] shows that this effect could be about 10 % of the measured linewidths. For the present calculations we did not inlude this effect, but if reliable calculations were performed for this transition and for the present plasma conditions, they should improve the agreement between the results of all theoretical approaches and the experimental values.

The calculations after the modified semiempirical approach of Dimitrijević and Konjević[4] are based on the Gaunt factor approximation for electron-impact excitation of Van Regemorter[35] and Seaton[36] and take only into account inelastic electron collisions. This theoretical approximation gives appreciably smaller results for the linewidths than the experiments.

The semiclassical calculations within the electron-impact approximation after Hey and Breger are discussed in detail in Ref. [31] for the present transitions and give also smaller results than the experiments. In comparison to the semiclassical approximation of Griem the authors choose a different procedure for the calculation of the Weisskopf radius and, in this way, their strong collision contribution to the linewidth exceeds that of Griem.

The results of Dimitrijević and Sahal-Bréchot are calculated after the semiclassical procedure described in Refs. [37]. The agreement with the experimental values is well within the experimental error bars. Strong collisions as well as the elastic contributions to the linewidth due to electronic and ionic collisions calculated within the impact approximation are taken into account. No inelastic ionic collisons are included at all.

Two different results of Alexiou[34] are shown in Fig. 6. The calculations giving the lower curve take into account only weak electron collisions, i.e. the dipole plus quadrupole contribution to the linewidth. The upper curve also includes the strong collision contribution to the broadening. Alexiou pointed out that the strong collision term denotes an error bound which gives an uncertainty in the determination of the linewidth. An exact statement of the strong collision term may be obtained only by quantum mechanical calculations. The comparison of the calculated and the experimental data favors a large strong collision contribution to the linewidths at least for the low temperature values. For higher temperatures the strong collision term becomes less important, and the calculations taking into account only weak collisions become a better approximation. Indeed, this behavior is expected for calculations within a semiclassical theory.

Although there are some discrepancies between the absolute values of the experimental data and the results of the theoretical approximations of Dimitrijević and Konjević and of Hey and Breger the dependence of the Stark width on the electron temperature is well described by all theoretical approximations. This result is also found in Refs. [14] for the linewidths of the $3s - 3p$ transitions in C IV – Ne VIII. Therefore, we scaled the experimental data which were obtained for appreciably different temperatures to a value of the electron temperature, i.e. $k_B T_e = 12.5$ eV, after the choice of the effective Gaunt factor of Griem[1,38] or of Dimitrijević and Konjević[4]. Furthermore, the data are scaled linearly to a value of the electron density, i.e. $n_e = 1.8 \times 10^{18}$ cm^{-3}.

The results for the $3s_{1/2}\,^2S - 3p_{3/2}\,^2P^\circ$ transitions in C IV – Ne VIII are shown in Fig. 7. Other experimental data, for example of Refs. [12,13], have been omitted because they are only obtained for the lowly ionized atoms of this isoelectronic sequence. The error bars in Fig. 7 include the uncertainty in the determination of the electron density, the electron temperature, and of the Stark width. Also shown is the experimental width of the $3s_{1/2}\,^2S - 3p_{3/2}\,^2P^\circ$ transition in Ne VIII measured with helium as driver gas. This value is scaled in the same way as those obtained with hydrogen as driver gas. Although both values agree within the experimental error bars we may address the discrepancy between the absolute values to the influence of inelastic proton collisions to the linewidth in the case of hydrogen as driver gas. It is already pointed out by Griem at the last conference on spectral line shapes[18] that for this transition inelastic proton collisions could account for more than 10 % of the measured linewidth. In this study the scaled width of the experiment with helium as driver gas is about 18 % smaller than the scaled width of the experiment

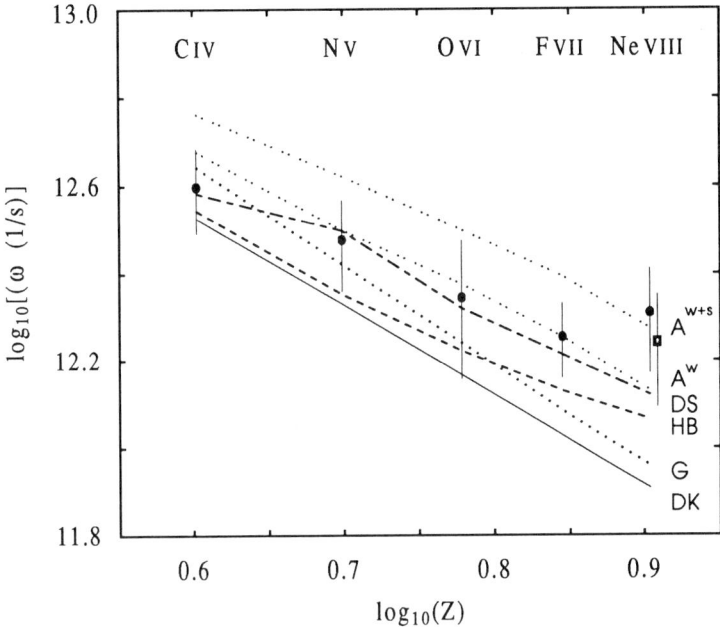

Figure 7: Experimental and theoretical Stark width in frequency units of the $3s\,^2S - 3p\,^2P^o$ transition in C IV, N V, O VI, F VII and Ne VIII as function of Z: $\cdots$, Calculations after Griem [1] (index G); —, after Dimitrijević and Konjević [4] (index DK); - - -, calculations by Hey [8] (index HB); $\cdots\cdots$, by Alexiou and Ralchenko [9,16] (index A^w and A^{w+s});---, by Dimitrijević and Sahal-Bréchot [17] (index DS), the value of F VII is interpolated; experimental values: •, data of Refs. [14] measured with hydrogen as driver gas; □, present data measured with helium as driver gas.

with hydrogen as driver gas. Inelastic collisions of helium perturbers, on the other hand, can safely be ignored for two reasons. Ion collisions depend on the mass of the perturbing ions, and in a more sensitive manner on the energy spacings between the relevant atomic states and the temperature of the perturbing ions. The temperature of the plasma of the experiment with helium as driver gas was $k_B T_i = 20$ eV. This is appreciably smaller than the temperatures of the experiments with hydrogen as driver gas: $k_B T_i = 29 - 43$ eV, and calculations according to Eq. (516) of Ref. [1] show that collisions of helium ions are indeed unimportant for the present conditions.

The ionic dipole contribution to the linewidth within the impact approximation is also investigated by Alexiou[9]. Even though the author found that in case of the $3s - 3p$ transitions in Ne VIII inelastic proton collisions are more important than for the transitions in C IV – F VII his calculated contribution to the linewidths is negligible for our experimental conditions. Also, the strong collision contribution from ions is found to be negligible[9]. More theoretical and experimental data are needed in order to verify the assumption that inelastic proton collisions are responsible for the difference between both experimental values for Ne VIII.

In Fig. 7 theoretical data calculated for $k_B T_e = 12.5$ eV and $n_e = 1.8 \times 10^{18}$cm^{-3} according to Refs. [1,4,7,9,16,17] are compared with the experiment. No ionic contribution to the calculated widths is taken into account at all. The theoretical approximation of Griem[1] gives approximately a Z^{-2} scaling of the Stark width of nonhydrogenic emitters. The comparison with the experimental data shows good agreement for the Stark widths of the lowly ionized species of the isoelectronic sequence. However, for $Z > 6$ the results after Griem[1] are appreciably smaller than the experimental data. Approximately the same scaling, i.e. $w \sim Z^{-2}$, of the linewidth of nonhydrogenic emitters is predicted by Dimitrijević and Konjević[4]. The results calculated after their theoretical approximation are inside of the error bars of the experiment for the lowly ionized species ($Z \leq 6$). Again, with increasing spectroscopic charge number the theoretical values are appreciably smaller than the experimental data. Although the theoretical widths of Hey[8] are systematically smaller than the experimental results they show a similar tendency with increasing spectroscopic charge number as the experimental data. This improved agreement for the higher ionized species of the isoelectronic sequence is due to some improvements in the calculation of the strong collision contribution (whereby both disruptive collisions and higher multipole interactions are included) to the linewidths. This contribution to the linewidth increases along the sequence from 20 – 25 % for C IV to 50 – 60 % for Ne VIII. In Ref. [4] no explicit expression for the strong collision contribution is given, but the calculations after Ref. [1] give an increasing contribution from 12 % for C IV to 14 % for Ne VIII.

For the theoretical data calculated by Dimitrijević and Sahal-Bréchot[17] no statement of the relative contribution of the strong collision term is made. Nevertheless, their theoretical results are in excellent agreement with the experimental data for the lowly ionized species of the isoelectronic sequence. Furthermore, their theoretical result for Ne VIII is inside of the error bar of the experiment with helium as driver gas. The agreement is even better when taking into account their calculated contribution to the linewidth by elastic ion collisions. In Fig. 7 this effect has been neglected because no temperature scaling appears to be reasonable. One should have in mind that this effect increases the linewidths of the calculated values by about 5 % for our experimental conditions with hydrogen as well as with helium as driver gas. A better agreement with the experiment with hydrogen as driver gas could be obtained by taking into account a contribution of inelastic proton collision in the calculations.

For the data shown in Fig. 7 Alexiou[9] gives again two calculated values for the line broadening. The lower curve includes only weak electron collisions. This calculations give results inside of the experimental error bars for all species of the sequence with the exception of Ne VIII measured with hydrogen as driver gas. On the basis of a detailed re-evaluation of the classical-path and impact approximation the author obtained a larger monopole-dipole contribution to the electron collision operator than, e.g., Dimitrijević and Konjević[4] or Hey and Breger[7]. A further improvement is achieved due to the inclusion of the electron quadrupole contribution in the weak electron collision term[16]. A more accurate calculation of this term for the linewidths of more highly ionized nonhydrogenic emitters was already suggested by Griem[6]. Indeed, this term gives a contribution to the linewidth which increases from 4 % for C IV to 17 % for Ne VIII. In this way an improved description of the trend of the linewidth along the isoelectronic sequence is obtained. In Ref. [16] it is also shown that the remaining discrepancies between the calculated weak electron collision term and the experimental widths can not be explained by ion quadrupole broadening (calculated within the impact approximation) or with resonances. Strong (electron) collisions, however, help to explain the remaining discrepancies. The upper curve of Alexiou in Fig. 7 denotes the maximum linewidth due to the inclusion of the strong collision term. In fact, the theoretical linewidth after Alexiou is between the weak width and the sum of weak and strong widths. Therefore, a satisfactory agreement with all experimental data can now be obtained after proper choice of the strong collision term within the stated error bounds. However, a more exact statement about the strong collision term would be of great help in order to predict the linewidth of highly ionized atoms with high accuracy.

In general, we investigated three more isoelectronic sequences of boron-like ions and similar results have been obtained[15]. The experimental linewidths decrease less than Z^{-2} as predicted in Refs. [1,4]. However, by taking into account a strong collision contribution to the linewidths after Hey and Breger[7] the dependence of the measured Stark widths on the spectroscopic charge number Z is described well. This results emphasize the importance to perform more calculations of the strong collision contribution to the linewidth. In this respect quantum mechanical calculations are most desirable. Moreover, those calculations should include a sufficiently large number of perturbing levels. For example, Seaton's calculations[39] of the Stark widths of the $3s - 3p$ transitions in C IV - Ne VIII neglect perturbing levels with $n > 3$ and, therefore, result in too small values for the linewidths.

So far we were concerned with isolated spectral lines of nonhydrogenic ions. Recently, we measured line profiles of non-isolated spectral lines of the lithium-like ions C IV, N V, and O VI in going to transitions of more highly exited states with close-lying perturbing levels[20]. For a comparison with theoretical line profile calculations forbidden transitions and overlapping of various $(n-1, \ell') - (n, \ell)$ transitions (for our studies we had $n = 4$ and $n = 5$) have to be taken into account. Furthermore, ion-collisional effects are of importance. The results of a suitable line shape code[19] which takes into account these effects are compared with the experimental line profiles. We find good agreement for all investigated transitions and plasma conditions (see also Stamm et al, this book). In particular, the comparison with calculated profiles in the static ion approximation and with profiles in the dynamic ion approximation[40] reveals that ion dynamical effects account for about 30 % of the total linewidth in some cases.

CONCLUSIONS

The dependence of the Stark width of isolated spectral lines of nonhydrogenic ions on the electron temperature and on the spectroscopic charge number Z is discussed on the basis of recent experimental and theoretical results. The electron temperature dependence of the experimental data is found to be in good agreement with the results of various theoretical approximations. However, the experimental Stark widths of four isoelectronic sequences show no simple scaling law, e.g. $w \sim Z^{-2}$, as suggested by some theoretical approximations. The investigation of the influence of proton collisions on the linewidth of the $3s - 3p$ transition in Ne VIII strongly indicates that proton collisions could partially explain the observed scaling. Indeed, ion collisional effects are also shown to be important to explain the broadening of non-isolated spectral lines of nonhydrogenic emitters[20]. For the linewidths of isolated spectral lines of more highly ionized nonhydrogenic emitters, however, further broadening mechanisms play a significant role. For example, by taking into account elastic contributions to the linewidth due to electronic and ionic collisions[17] or electron quadrupole broadening[16], an improved agreement between theory and experiment is obtained. The remaining discrepancies could be explained by strong collisions. However, this contribution to the linewidth is only roughly estimated within the semiclassical theories. More exact calculations are nescessary in order to verify whether or not strong collisions are an important contribution to the line broadening of nonhydrogenic ions in plasmas.

ACKNOWLEDGMENTS

The author would like to thank H.-J. Kunze for many valuable discussions. This research was supported by the Sonderforschungsbereich 191 of the DFG.

REFERENCES

1. H. R. Griem, *Spectral Line Broadening by Plasmas* (Academic, New York, 1974).
2. H. R. Griem, Phys. Fluids B4, 2346 (1992).
3. N. Konjević and W. L. Wiese, J. Phys. Chem. Ref. Data 19, 1307 (1990).
4. M. S. Dimitrijević and N. Konjević, J. Quant. Spectrosc. Radiat. Transfer 24, 451 (1980), also in *Spectral Line Shapes*, edited by B. Wende (De Gruyter, Berlin, 1981), Vol 1.
5. M. Baranger, in *Atomic and Molecular Processes*, edited by D. R. Bates (Academic, New York, 1962).
6. R. Griem, in *Spectral Line Shapes*, edited by J. Szudy (Ossolineum, Warschau, 1989), Vol 5.
7. J. D. Hey and P. Breger, S. Afr. J. Phys. 5, 111 (1982).
8. J. D. Hey, *private communication* (1991).
9. S. Alexiou, Phys. Rev. A49, 106 (1994).
10. W. L. Wiese and N. Konjević, J. Quant. Spectrosc. Radiat. Transfer 28, 185 (1982).
11. R. Kobilarov and N. Konjević, Phys. Rev. A41, 6023 (1990).
12. F. Böttcher, P. Breger, J. D. Hey, and H.-J. Kunze, Phys. Rev. A38, 2690 (1988).
13. J. Purić, A. Srećković, S. Djenize, and M. Platisa, Phys. Rev. A36, 3957

(1987).
14. S. Glenzer, N. I. Uzelac, and H.-J. Kunze, Phys. Rev. A45, 8795 (1992).
    S. Glenzer, N. I. Uzelac, and H.-J. Kunze, in *Spectral Line Shapes*, edited by R. Stamm and B. Talin (Nova Sci. Commack, New York, 1993), Vol 7.
15. S. Glenzer, J. D. Hey, and H.-J. Kunze, J. Phys. B27, 413 (1994).
16. S. Alexiou and Yu. Ralchenko, Phys. Rev. A49, 3086 (1994).
17. M. S. Dimitrijević, S. Sahal-Bréchot, and V. Bommier, Astron. Astrophys. Suppl. Series, 89, 581 (1991).
    M. S. Dimitrijević and S. Sahal-Bréchot, Astron. Astrophys. Suppl. Series, 93, 359 (1992), 95, 109 (1992).
    M. S. Dimitrijević and S. Sahal-Bréchot, Astron. Astrophys. Suppl. Series, *to be published* (1994).
18. H. R. Griem, in *Spectral Line Shapes*, edited by R. Stamm and B. Talin (Nova Sci. Commack, New York, 1993), Vol 7.
19. A. Calisti, F. Khelfaoui, R. Stamm, and B. Talin, in *Spectral Line Shapes*, edited by L. Frommhold and J. W. Keto (Am. Inst. Phys., New York, 1990), Vol 6.
    A. Calisti, L. Godbert, R. Stamm, and B. Talin, J. Quant. Spectr. Radiat. Transfer 51, 59 (1994).
    C. J. Keane, R. W. Lee, B. A. Hammel, A. L. Osterheld, L. J. Suter, A. Calisti, F. Khelfaoui, R. Stamm, and B. Talin, Rev. Sci. Instrum. 61, 2780 (1990).
20. L. Godbert, A. Calisti, R. Stamm, B. Talin, S. Glenzer, H.-J. Kunze, J. Nash, R. W. Lee, and L. Klein, Phys. Rev. E49, 5589 (1994).
    S. Glenzer, Th. Wrubel, S. Büscher, H.-J. Kunze, L. Godbert, A. Calisti, R. Stamm, B. Talin, J. Nash, R. W. Lee, and L. Klein, J. Phys. B *in print*.
21. H.-J. Kunze, in *Spectral Line Shapes*, edited by R. J. Exton (Deepak, Hampton, 1987), Vol 4.
22. H.-J. Kunze, in *Plasma Diagnostics*, edited by W. Lochte-Holtgreven (North-Holland, Amsterdam, 1968) p. 550.
23. A. W. DeSilva, T. J. Baig, I. Olivares, and H.-J. Kunze, Phys. Fluids B4, 458 (1992).
24. D. E. Evans, Plasma Phys. 12, 573 (1970).
25. Th. Wrubel, S. Glenzer, S. Büscher, and H.-J. Kunze, J. Atmos. Terr. Phys. (1994) *in print*.
26. S. Glenzer, Th. Wrubel, S. Büscher, and H.-J. Kunze, in *Proc. XXI. Int. Conf. Phen. Ionized Gases* Sept. 19-23, 1993, Bochum, eds. G. Ecker, U. Arendt, J. Böseler, APP Bochum, Vol. II, p. 367-368.
27. R. D. Cowan, *The Theory of Atomic Structure and Spectra* (University of California Press, Berkeley, 1981).
28. S. Glenzer, H.-J. Kunze, J. Musielok, Y.-K. Kim, and W. L. Wiese, Phys. Rev. A49, 221 (1994).
29. J. D. Hey and P. Breger, J. Quant. Spectrosc. Radiat. Transfer 23, 311 (1980), 24, 349 (1980), 24, 427 (1980), also in *Spectral Line Shapes*, edited by B. Wende (Walter de Gruyter, Berlin, 1981), Vol 1.

30. N. Konjević and W. L. Wiese, J. Phys. Chem. Ref. Data, $\underline{5}$, 259 (1976).
31. N. I. Uzelac, S. Glenzer, N. Konjević, J. D. Hey and H.-J. Kunze, Phys. Rev. $\underline{A47}$, 3623 (1993).
32. J. Purić, S. Djenize, A. Strećković, M. Platisa, and J. Labat, Phys. Rev. $\underline{A37}$, 498 (1988).
33. B. Blagojević, M. V. Popović, N. Konjević, and M. S, Dimitrijević, Phys. Rev. *in print*.
34. S. Alexiou, *private communication* (1994).
35. H. Van Regemorter, Astrophys. J. $\underline{136}$, 906 (1962).
36. M. J. Seaton, in *Atomic and Molecular Processes*, edited by D. R. Bates (Academic, New York, 1962).
37. S. Sahal-Bréchot, Astron. Astrophys. $\underline{1}$, 91 (1969), $\underline{2}$, 322 (1969).
38. H. R. Griem, Phys. Rev. $\underline{165}$, 258 (1968).
39. M. J. Seaton, J. Phys. $\underline{B21}$, 3033 (1988).
40. L. Godbert, A. Calisti, R. Stamm, B. Talin, R. W. Lee, and L. Klein, Phys. Rev. $\underline{E49}$, 5889 (1994).

WIDTH OF THE HYDROGEN LYMAN-α LINE WITHIN AN APPROACH BASED ON A GREEN'S FUNCTION TECHNIQUE FOR ELECTRONS AND COMPUTER SIMULATIONS FOR THE IONS

W. Olchawa, J. Halenka

Institute of Physics, Pedagogical University of Opole,
45–052 Opole, ul. Oleska 48, Poland

S. Günter

Fachbereich Physik, Universität Rostock,
Universitätsplatz 1, 18051 Rostock, Germany

INTRODUCTION

Recently, a fully quantum mechanical many particle theory of spectral line shapes based on a Green's function technique, has been developed [1, 2]. This theory has been shown to be well applicable to plasmas of electron densities above 10^{18}cm^{-3} where, as it is known, many particle effects as dynamic screening become important. Further, in difference to the line–shape theories based on the classical path approximation for electrons, a cutoff procedure for small impact parameters could be avoided taking into account quantum mechanical effects for strong collisions and going beyond a low-order perturbative treatment for the electron–atom interaction. Footing on this theory for the electrons and on a quasi-static assumption for the ions, shifts and asymmetry of hydrogen lines have been calculated [3, 4]. The theoretical results agree very well with measurements. However, the neglect of the ion dynamics leads to a considerable discrepancy between the calculated and the measured half–widths of the line profiles. As it is well-known, ion dynamical effects influence mainly the Lyman lines, Ly$_\alpha$ above all. The aim of the present paper is to make an attempt to introduce ion dynamics into the quoted theory for the electrons. Calculations have been carried out for the profiles of the Ly$_\alpha$ line for experimental plasma parameters [5].

CALCULATION AND RESULTS

With conventional assumptions and designations a Stark profile of a line from the Lyman series can be written as

$$I(\Delta\omega) = \pi^{-1}\text{Re}\int_0^\infty dt e^{i\Delta\omega t}Tr\{\mathbf{d}_{ab}\cdot\mathbf{d}_{ba'}U_{a'a}(t)\}_{av}. \quad (1)$$

Routinely, the emitter–plasma interaction is divided into V_i+V_e, where V_e and V_i are the emitter–electron and the emitter–ion interaction, respectively. In the statistic $_{ave}$ one takes into account the electron–electron and in $_{avi}$ the ion–ion as well as the ion–electron interaction. This procedure allows to perform the averaging of the time–development operator as follows:
$<U>_{av} = <U_i<U_e>_{ave}>_{avi}$. Thus, on the basis of the theory [2] for electrons, the time–development operator $<U>_{ave}$ satisfies the equation

$$i\hbar\frac{d}{dt}U_i<U_e>_{ave} = [H_0 + V_j(t) + \Sigma(\Delta\omega=0)]U_i<U_e>_{ave}, \quad (2)$$

where $\Sigma(\Delta\omega=0)$ is the self–energy operator within the impact approximation.

Under the above assumptions the Ly_α line profiles were evaluated applying the computer simulation technique as described in[6]. From results quoted in Tab I one can see that after an inclusion of ion dynamics, a good agreement between calculated and measured results is obtained.

Table I FWHM (Å) of the Ly_α line

T (K)	N_e (10^{17} cm^{-3})	$\Delta\lambda_{1/2}$(Å)		
		a	b	c
12700	1	0.054	0.214(0.164)	0.23 ± 0.02
13200	2	0.100	0.281(0.240)	0.30 ± 0.02
13200	3	0.142	0.334(0.301)	0.36 ± 0.02
14000	4	0.187	0.396(0.366)	0.42 ± 0.02

a – static ions, b – dynamic ions + Doppler (Stark only), c – experiment [5]

References

[1] S. Günter, L. Hitzschke and G. Röpke, *Phys. Rev.* A44, 6834 (1991).

[2] S. Günter *Phys. Rev.* E48, 500 (1993).

[3] St. Böddeker, S. Günter, A. Könies, L. Hitzschke and H.-J. Kunze, *Phys. Rev.* E47, 2785 (1993).

[4] A. Könies and S. Günter (to be published).

[5] K. Grützmacher and B. Wende, *Phys. Rev.* A16, 243 (1977).

[6] J. Halenka and W. Olchawa (to be published).

ANALYSIS OF LITHIUMLIKE LINE SHAPES IN A GAS-LINER PINCH

R. Stamm, A. Calisti, L. Godbert, T. Meftah, C. Mossé and B. Talin
Équipe Diagnostic dans les Gaz et les Plasmas.
URA 773 université de Provence
Centre St. Jérôme, case 232, Marseille cedex 20

S. Glenzer
Institut für Experimentalphysik V, Ruhr-Universität, 44780 Bochum, Germany

INTRODUCTION

Spectroscopic analysis is widely used for investigating various kinds of plasmas. The gas-liner pinch[1], described elsewhere (see Glenzer et al., this book), allows the creation of high density plasmas which are of interest in particular for XUV laser developments. Plasmas obtained are mainly composed of hydrogen, but include a small amount of methane, nitrogen or carbondioxide as test gas. The line shapes of lithiumlike carbon, nitrogen and oxygen have been mainly observed. Their analysis has been performed using a line shape code which takes into account the effect of ion dynamics with a frequency fluctuation model[2]. An earlier version of this code has previously been used for the density diagnostic of hot and dense plasmas. The latter plasmas have been studied using the line shapes of highly charged ions found in Inertial Confinement Fusion (ICF) experiments[3]. In this paper we focus our attention on plasmas of densities and temperatures which are intermediate between the arc plasmas and the extremely high density ICF plasmas.

FREQUENCY FLUCTUATION MODEL

We use a stochastic model of the fluctuations of the observable frequencies[2], assuming that the fluctuation of the static components represents the ion dynamic effect on the line shape. The fluctuation mechanism is assumed to obey a stationary Markov process driven by the field fluctuations. Finally, we assume that only a coarse-grained distribution of the components is necessary for the frequency fluctuation process[2].

PLASMA DIAGNOSTIC

We have focused our attention on the 4-3 and 5-4 transitions of lithiumlike carbon and nitrogen because these lines are of interest in XUV laser developments. The code has been used to analyse a coherent set of experimental profiles. We show here two examples of these comparisons between the experimental and the theoretical line shapes. The figure 1 corresponds to the 4d-3p line of the lithiumlike carbon. In this figure, the large apparatus function (approximately one third of the total width) does not hide the dynamic effect due to the ionic field. If we retain this effect in the code, the shape of the theoretical profile matches with the experimental spectrum, which is not the case if a static approximation is adopted. A similar agreement is found on figure 2 for the n=4 to n=5 line of the lithiumlike nitrogen. Other lithiumlike lines confirm that our code including ion dynamics effects allows accurate density diagnostics for gas-liner plasma conditions.

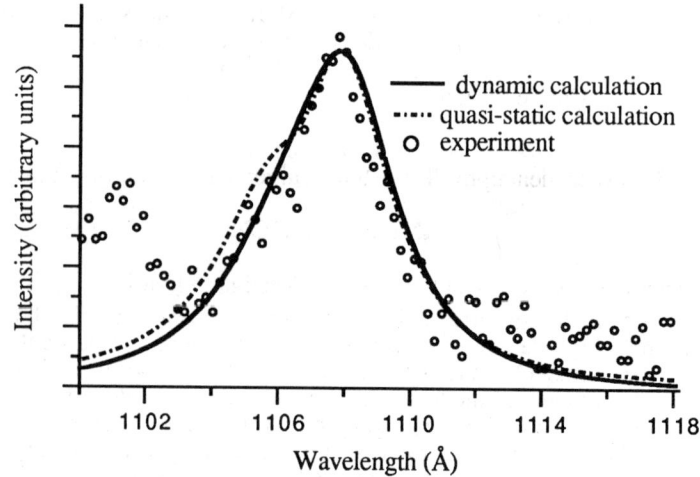

Fig. 1 Line shape of lithiumlike Carbon at $n_e = 2.5\ 10^{18}$ cm^{-3} and $T_e = 9.3$ eV

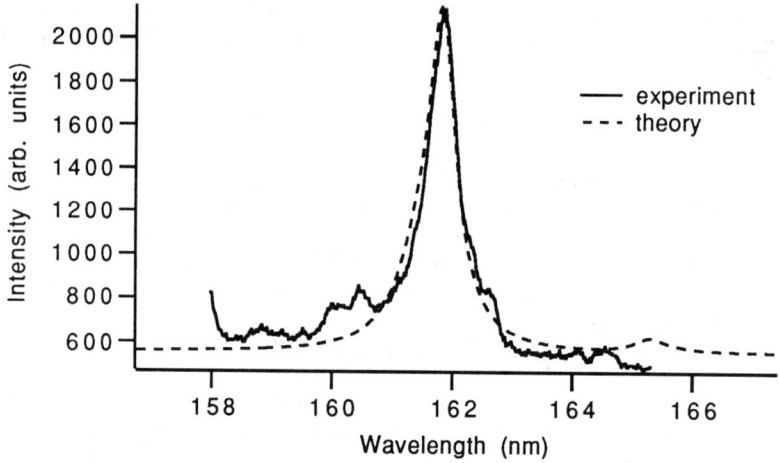

Fig. 2 Line shape of lithiumlike Nitrogen at $n_e = 1.9\ 10^{18}$ cm^{-3} and $T_e = 9.5$ eV

REFERENCES

1. A. Gawron, S. Maurmann, F. Böttcher, A. Meckler and H. J. Kunze, Phys. Rev. A38, 4737(1988)
2. B. Talin, A. Calisti, L. Godbert, R. Stamm, L. Klein and R. W. Lee, to be published
3. A. Calisti, L. Godbert, R. Stamm and B. Talin, J. Quant. Spectrosc. Radiat. Transfer, 51, 59 (1994)

Analysis of K-shell line emission spectra emitted by short-pulse laser-produced plasmas

O. Peyrusse, D. Gilles

CEA Limeil-Valenton, 94195 Villeneuve-St-Georges Cedex, France

J.C. Kieffer, C.Y. Coté, Z. Jiang

INRS Energie, 1650 Montée Ste-Julie, Varennes, Canada

ABSTRACT

The analysis of Stark-broadened K-shell emission spectra is a powerful method for extracting information from relatively low-Z dense laser-produced plasmas even in the case of short-pulse (subpicosecond) experiments where the temporal resolution is still limited. Some K-shell lines are characteristic of the period of interaction with the laser, especially the Li-like $1s2l2l'$ - $1s^22l$ satellite lines which can present a strong broadening and some features specific of short-pulse experiments. In this presentation, we will discuss the behaviour of these lines along with Stark line broadening calculations and recent experimental results obtained at INRS (Monréal) and at Ultrafast Science Laboratory (Michigan). In particular, we will show that quasi-solid density plasmas can be obtained in short-pulse experiments. Furthermore, we present some results concerning the He-like emission (He_β, He_γ lines) which is more sensitive to time-integration. In spite of this, some interesting questions are opened for interpreting the broadening of these lines in the conditions of short time-scale plasmas. What is the effect of the strong decoupling between T_i and T_e ($T_i \ll T_e$) or, why the presence of a strong red wing unexplained by standard calculations of stark line broadening.

AB INITIO NON UNIFORM MICROFIELD JOINT DISTRIBUTIONS IN PLASMAS

A.V.Demura

Hydrogen Energy and Plasma Technologies Institute
Russian Research Center "Kurchatov Institute",Moscow ,123182, Russia
and
Department of Physics,206 Allison Lab,Auburn University,Auburn, AL
36849-5311,USA

D.Gilles

CEA - Centre d'Etudes de Limeil Valenton
F-94195 Villeneuve Saint Georges, France

B.C.Huynh and C.Stehlé

DARC and UPR 176 du CNRS, Observatoire de Paris-Meudon
F-92195 Meudon Cedex, France

ABSTRACT

The microfield non-uniformity is thought to give rise to the asymmetry in line profiles. In the context of the many-body theory it was treated in details for uncorrelated plasma model[1,2] and for highly correlated plasmas with the use of the constrained average of the microfield non-uniformity tensor component dF_z/dz in the resolvent in[3,4]. On the other hand Monte-Carlo numerical simulations on the supercomputers give the independent and powerful tool for investigation of the behaviour of the multidimensional joint distributions which become necessary for the consistent treatment of the problems with the nonuniform microfield [5].

Recently the generalized many-body approach was proposed [6] based on the Baranger-Mozer cluster expansion scheme [7], that gives a statistical solution for the problem of the instant low-frequency joint microfield distribution function with the accuracy of the two-particle correlation functions. It enables to treat arbitrary mixtures of ions in plasmas with arbitrary values of the plasma parameters (density, temperature, ionic charge). Its refined current version will be presented at the conference [8].

We present here new results obtained by the extensive Monte-Carlo simulations [5] and calculations in the framework of the above mentioned formalisms for the constrained averages of the microfield nonuniformity tensor components. Our conditions cover various density conditions between low-correlated plasma (Ar^{17+}, N_e=1.5 19 cm^{-3}, T_e=800eV) to high correlated ones (Ar^{17+}, N_e=1.5 24 cm^{-3}, T_e=800eV).

We compare these results with those obtained within the two APEX's algorithms, proposed by Kilcrease et al. [4] and Demura [9], the last one being presented in more details in [8].

Results of Monte-Carlo simulations for the first time allow to obtain the surface of the truncated joint distribution of module F and dF_z/dz. This method treats correctly the many-body effects, and can thus check the validity of the previous analytical results, being often free from their limitations.

REFERENCES

1. S.Chandrasekhar, Astrophys. J.99, 25(1944).
2. A.V.Demura, G.V.Sholin, J.Q.S.R.T. 15, 881(1975)
3. R.F.Joyce, L.A.Woltz, C.H.Hooper, Phys.Rev.A 35, 2228(1987).
4. D.P.Kealcrease, R.C.Mancini, C.F.Hooper,Phys.Rev. E 48, 3901(1993).
5. A.Angelie, D.Gilles, Annales de Physique, Colloque n°3, Suppl. au n°3, 157 (1986); Spectral Line Shapes Vol.3 , Ed. F. Rostas, de Gruyter, Berlin, p. 245 (1984)
6. A.V.Demura, Preprint IAE-4632/6 (Moscow,1988) 17p.
7. M.Baranger, B.Moser, Phys.Rev. 115, 521(1959); 118,626(1960).
8. A. Demura, C. Stehlé, this issue
9. A. Demura,Abstracts of Invited Lectures and Contributed Papers, ESCAMPIG-92,EPS vol.16,ed.L.Tsendin(St.Peterburg,Russia,1992),p.63.

Line Shape Diagnostics for Solid Density Plasmas Produced by Ultra Intense Subpicosecond Laser

Z.Jiang,* J.C.Kieffer, M.Chaker
INRS-energie, 1650 montee Ste-Julie, Varennes, Canada
G.Korn, S.Coe, G.Mourou
CUOS, Univ. of Michigan, Ann Arbor, USA
O.Peyrusse, D.Gilles
CEA Limeil, Villeneuve, St.Georges, France

The recent development of a new generation of lasers has opened up new horizons in laser matter interaction.[1] The short duration of the pulse effectively eliminates the plasma hydrodynamics during the heating phase. Furthermore if there is negligible energy in the prepulse, the density gradient scale length will be much shorter than the wavelength and the laser-matter coupling will take place at near solid density plasmas. The 1D, hot, uniform solid density plasma offers a unique way to address experimentally in the laboratory some problems: i) of astrophysical interest, namely the opacity of dense matter and the physics of hot strongly correlated plasmas[2] ii) of interest for atomic physics, for instance the physics of quasi-molecular states[3] and iii) of application interests, as the fast electron ignition[4] in ICF and the generation of ultrafast x-ray source.[5,6]

Recently we have conducted several experiments to study the behaviors of solid density plasmas with an ultraclean (contrast ratio of about 10^{-10}:1) 300fs laser pulse at intensity up to a few times 10^{18}W/cm^2. Time and space integrated line profiles of He-like and Li-like emissions are used to infer the average plasma density at which various emissions take place by using Stark broadening calculations[7] (with the ions and the electrons treated in, respectively, the quasi static and impact approximation). One aspect of the line shape calculations is the use of analytic fits of the plasma microfield distribution function, deduced from a large number of Monte Carlo simulation. These analytical fits depend on the ionic correlation parameters and on the electron screening and thus the microfield will be very sensitive to the ratio of electron to ion temperature (T_e/T_i).

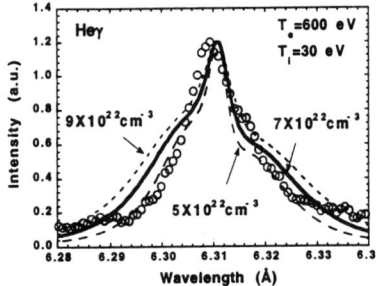

Fig.1 Al Heγ line shape (dot) and fits with different densities. $I_L=10^{18}$W/cm^2.

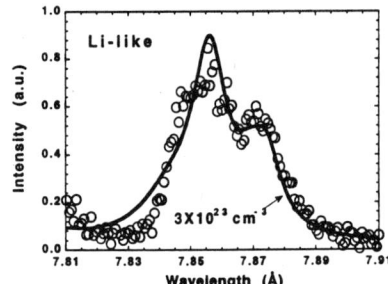

Fig.2 Al Li-like line shape (dot) and fit (solid line). $I_L=10^{18}$W/cm^2.

Fig.1 shows a Heγ line profile obtained at laser intensity of 10^{18}W/cm^2. The best fit (solid line) gives T_i=50eV and n_e=5x10^{22}cm^{-3}. We checked that this couple of parameters allows also a good fit of the Heβ line profile.The time average electron temperature is estimated as 600eV by using Lyα/Heβ line ratios calculated with time dependent collisional radiative code TRANSPEC.[8]

The density at which the Li-like emission takes place has been inferred from the 1s2l2l'-1s^{2}2l line spectra.[7] At the electron density larger than $10^{23}cm^{-3}$, the broadening is important and double excited level populations have reached the local thermodynamic equilibrium (LTE). Fig. 2 shows a Li-like spectrum taken at the laser intensity of $10^{18}W/cm^2$. The best calculation fit is also shown ($3 \times 10^{23}cm^{-3}$).

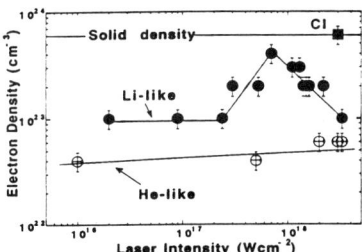

Fig.3 presents the variation of the density inferred from He-like and Li-like line shapes for Al.The density for He-like emission is constant with the laser intensity, and the inferred ion temperature is always smaller than the electron temperature. The density for Li-like emission remains constant in the intensity range $10^{16}W/cm^2$ to $3 \times 10^{17}W/cm^2$, increases up to a maximum ($4 \times 10^{23}cm^{-3}$) at intensity around $10^{18}W/cm^2$ and decreases dramatically after the peak.

Fig.3 The variation of the density inferred from He-like (open dots) and Li-like (full dots) shapes for Al. The square is the density from Cl Li-Like line shapes.

The results suggest the radiative pressure of the powerful laser pulse plays a very important role. The pressure could balance the thermal pressure and maintain a very steep gradient during the coupling thus forcing the emission to be produced at near solid density. But at the higher intensity the Al plasma is overdriven and the emission is produced during the plasma decompression on a time length longer than the laser pulse. Results obtained with a high Z target (Cl Li-like emission) indicate that when the target atomic number is matched to the electron temperature, the emission is produced at solid density (square in Fig.3) at $3 \times 10^{18}W/cm^2$. This behavior indicates that there is an optimum range set by the target material and laser intensity to produced KeV emission at solid density.

This work is supported by NSERC, FCAR and Ministere de l'education du Quebec.

Reference
* On leave from Shanghai Institute of Optics and Fine Mechanics, China
1. M.Perry and G.Mourou, Sciences 264, 917(1994)
2. F.Rogers and C.Iglesias, Science 263, 50(1994)
3. E.Leboucher-Dalimier, A.Poquerusse and P.Angelo, Phys.Rev.E47, 1467(1993)
4. M.Tabak, J.Hammer, M.Glinsky, W.Kruer, S.Wilks, J.Woodworth, E.Michael and M.Perry, Phys.Plasmas 1, 1626(1994)
5. J.C.Kieffer, M.Chaker, J.Matte, H.Pepin, C.Y.Cote, Y.Beaudoin, T.Johnston, C.Y.Chien, S.Coe, G.Mourou and O.Peyrusse, Phys.Fluids B5, 2676(1993)
6. M.M.Murnane, H.C.Kapteyn and R.W.Falcone, Phys.Rev.Lett.62, 155(1989)
7. O.Peyrusse, J.C.Kieffer, C.Y.Cote and M.Chaker, J.Phys.B:At.Mot.Opt.Phys.26, L511(1993)
8. O.Peyrusse, Phys.Fluids B4, 2007(1992)

DICKE NARROWING IN DENSE PLASMAS

D.A.Shapiro, E.V.Podivilov

Institute of Automation & Electrometry, Novosibirsk, 630090, Russia

The Dicke narrowing of an absorption line is found to be much greater for the radiation wavelength than for the ionic mean free path and in the opposite limit. Chandrasekhar's velocity dependence of the diffusion coefficients decreases the collisional narrowing by 40% in the long-wave case and increases it by 60% in the short-wave limit. For the intermediate case the simple interpolation formula is proposed, which has been tested with the aid of numerical calculation and shown to have 10% accuracy. The profile of Bennett hole induced by a laser field in an ionic distribution is calculated. It is shown that the narrowing of the hole occurs with increase in the pumping field detuning. The physical cause of the effect is the decreasing dependence of Coulomb collision frequency on the velocity of the probe ion.

1. Introduction

Observation of Doppler broadened spectral lines traditionally helps to measure a gas or plasma temperature. However, in a dense gas Doppler diagnostics may lead to incorrect results. Under elastic collisions, when there are no dephasing of the optical electron, wave trains emitted by an atom proceed to interfere. Meanwhile, Doppler shifts from an individual ion has no time to be formed when the mean free path is shorter than the wavelength of radiation, Fig. 1. An ion becomes almost fixed as in a solid state, then the Doppler broadening vanishes and the line narrows[1].

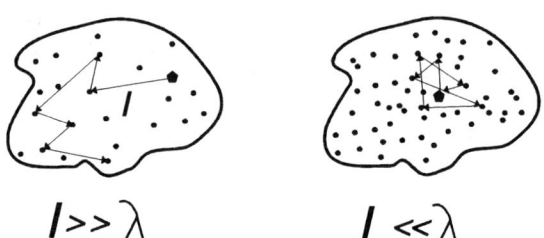

Fig. 1. *Brownian motion in media with long and short mean free path.*

The shape of spectral line $I(\Omega)$ is given by Fourier transform of correlation function $\Phi(t)$

$$I(\Omega) = \frac{1}{\pi}\Re \int_0^\infty \Phi(t)e^{i\Omega t}\,dt, \tag{1}$$

where Ω is the detuning between the field frequency ω and Bohr transition frequency ω_{mn}. The correlation function in its turn can be found by averaging over the ensemble of ions with various initial conditions

$$\Phi(t) = \left\langle e^{ik\Delta z(t)} \right\rangle. \tag{2}$$

Here z axis is directed along the wavevector $\boldsymbol{k}$. For the free motion ($\Delta z = vt$) and Maxwellian velocity distribution with thermal velocity v_T the function is

$$\Phi(t) = \exp\left(-k^2 \left\langle \Delta z^2 \right\rangle\right) = \exp\left(-k^2 v_T^2 t^2/4\right). \tag{3}$$

The line shape (1) corresponding to function (3) is the Doppler contour with the characteristic width kv_T.

When frequent elastic collisions confine the region within reach of the excited ion, function (2) decreases not as rapidly as (3). For instance, the weak (or soft) collision model studied by Rautian and Sobel'man[2], in which the velocity change is assumed to be small and collision transport frequency ν is independent of velocity $\boldsymbol{v}$, gives for coordinate variance the expression $\left\langle \Delta z^2(t) \right\rangle = v_T^2 \left(\nu t - 1 + e^{-\nu t}\right)/2\nu^2$.

Before the first collision ($\nu t \ll 1$) the motion is not too different from free motion, so that the variance depends on time quadratically, $\left\langle \Delta z^2 \right\rangle = (v_T t/2)^2$. After several collisions ($\nu t \gg 1$) the motion turns into the diffusion wandering throughout the velocity space, so that the variance grows up linearly $\left\langle \Delta z^2 \right\rangle = 2D_x t$, where $D_x = v_T^2/2\nu$ is the diffusion coefficient in the coordinate space.

Duration τ of the coherence of ionic dipole moment depends on the wavelength $\lambda = 2\pi/k$

$$\tau = \begin{cases} 1/kv_T, & \nu \ll kv_T; \\ \nu/k^2 v_T^2, & \nu \gg kv_T. \end{cases}$$

For short-wave radiation or rare collisions, while $\nu \ll kv_T$, the Doppler shift of the frequency defines coherence time τ; function $\Phi(t)$ is close to the Gaussian (3). When the wavelength exceeds the mean free path ($\nu \gg kv_T$), the diffusion controls correlation function, $\Phi(t) = \exp(-D_x k^2 t/2)$, and the spectral line width $\delta\Omega \sim 1/\tau \sim k^2 v_T^2/\nu$ becomes less than kv_T. Since $\Phi(0) = 1$, the area under the contour $I(\Omega)$ is constant, so the line width can be estimated from the intensity in the center $I(0)$. A slower fall in $\Phi(t)$ because of the collisions leads to an increase in $I(0)$ and hence to the line narrowing.

Not only collisions, but other limitations of free translational motion of the particle within the wavelength result in the line narrowing, if they follow without dephasing. Some examples are the walls spaced on distance $L < \lambda$ in hydrogen quantum standards of frequency, transverse homogeneous magnetic field in which the ionic line is spited into narrow cyclotron resonances[3] while Larmor radius is small enough $\rho_L < \lambda$ or potential well for the cooled single ion in electromagnetic trap[4].

The Dicke effect due to collisions is more difficult to detect. Nevertheless, it was observed in atomic gases on transitions between empty states of inner shells of rare-earth atoms[5]. In diatomic molecular gases the transition between vibrational levels are used to observe the effect[6].

It is interesting to observe the narrowing in plasma using spectra of ions with high charge Z, since their effective collision frequency is proportional to Z^4, whereas the dephasing is $\propto Z^2$. The hypothesis by Griem[7] about significant role of the narrowing in the dense plasma of Livermore X-ray laser has not been confirmed in the experiment on selenium lines[8]. Independently of experimental efforts the theoretical problem does exist how to calculate the ionic line width and shape in plasma. According to Chandrasekhar's formula, a charged particle collision frequency in equilibrium plasma depends strongly on the velocity v of the probe ion. For instance, the effective collision frequency of the fast ion ($v \gg v_T$) falls cubically with the velocity $\nu \propto v^3$. Then the weak collision model with the constant diffusion coefficient in the velocity space $D = \nu v_T^2/2$ could be valid only as an approximation.

The aim of the present paper is to describe how to calculate the narrowing in the ideal Maxwellian plasma. The basic equations presented in Sec. 2 are those for the density matrix with Landau collision term. The narrowing is found for limits of short (Sec. 3) and long (Sec. 4) wavelength of radiation compared to the mean free path of a probe ion[9,10]. A simple interpolating formula proposed for intermediate case is discussed in Sec. 5. Sec. 6 includes the solution of density matrix equations within the accuracy of second order in wave intensity. The Bennett hole in the velocity distribution of probe ions is shown to become narrow when the field is far from the resonance.

2. Quantum kinetic equation with Landau collision term

The quantum kinetic equation can be derived for spectroscopic density matrix $\rho(\mathbf{r}, \mathbf{v}, t)$, which is Wigner function of translational coordinates $\mathbf{r}$ and velocities $\mathbf{v}$ and the matrix over internal quantum ionic degrees of freedom. To derive it one can separate the average part of the matrix and its fluctuations[10,11]. Estimation shows that the fluctuating part varies on period of plasma oscillations, whereas the average part has undergone a change on longer time scale, time between the effective collisions. So, the rapid part follows the slow component. The kinetic equation for an ion in equilibrium plasma has Landau collision term

$$\left(\frac{\partial}{\partial t} + \mathbf{v}\frac{\partial}{\partial \mathbf{r}}\right)\hat{\rho} + i\left[\hat{V}, \hat{\rho}\right] = -\mathrm{div}_{\mathbf{v}}\mathbf{q}, \tag{4}$$

where $\hat{V} = -\mathbf{E}\hat{\mathbf{d}}/\hbar$ the Hamiltonian of interaction between the dipole moment of the probe ion and electromagnetic wave $\mathbf{E}(\mathbf{r},t)$, the commutator describes stimulated transitions. The collision integral reduces to differential operator which is the divergence of ion flux $\mathbf{q}$ through the velocity space. The flux consists of two parts: dynamic friction force $\mathbf{F}$ and diffusion flux

$$q_\alpha = \frac{F_\alpha}{m}\hat{\rho} - D_{\alpha\beta}\frac{\partial\hat{\rho}}{\partial v_\beta}, \tag{5}$$

where m is the mass of probe ion. The friction results in energy losses, the diffusion with tensor $D_{\alpha\beta}$ leads to wider distribution. The friction appears because of ion deceleration with its own static field, the diffusion arises from correlation between fluctuations of plasma microfield. In addition to the classical kinetics, the same collision term appears in the equations for off-diagonal elements of the density matrix as well as for diagonal elements. The cross-section of Coulomb interaction is independent of internal ionic state, so the collision operators coincide.

The friction and diffusion components are functionals of the velocity distribution of buffer particles. For Maxwellian distribution

$$W(\boldsymbol{v}) = \left(\sqrt{\pi} v_t\right)^{-3} \exp\left(-\boldsymbol{v}^2/v_T^2\right) \tag{6}$$

it is possible to write the explicit expressions as functions of dimensionless velocity $\boldsymbol{\xi} = \boldsymbol{v}/v_T$

$$F_\alpha = \nu m v_T \xi_\alpha \Phi_\|(\xi); \quad D_{\alpha\beta} = \frac{\nu v_T^2}{2}\left[\Phi_\| \frac{\xi_\alpha \xi_\beta}{\xi^2} + \Phi_\perp \left(\delta_{\alpha\beta} - \frac{\xi_\alpha \xi_\beta}{\xi^2}\right)\right]. \tag{7}$$

The effective transport collision frequency is given by the formula

$$\nu = \frac{16\sqrt{\pi} L N \left(Ze^2\right)^2}{3m^2 v_T^3}, \tag{8}$$

where L is Coulomb logarithm, Ze is the charge of probe ion, N is the effective density of buffer charged particles b with charges $Z_b e$ and densities N_b, $N = \sum_b Z_b^2 N_b$.

Longitudinal and transverse functions $\Phi_\|(\xi)$ and $\Phi_\perp(\xi)$ can be expressed in terms of Chandrasekhar's function[12], which can be reduced to single integrals

$$\Phi_\|(\xi) = 3\int_0^1 \lambda^2 e^{-\lambda^2 \xi^2} d\lambda, \quad \Phi_\perp(\xi) = \frac{3}{2}\int_0^1 (1-\lambda^2) e^{-\lambda^2 \xi^2} d\lambda. \tag{9}$$

The pair of functions satisfy two identities

$$\frac{1}{2\xi}\frac{d\Phi_\|}{d\xi} + \Phi_\| = \frac{\Phi_\perp - \Phi_\|}{\xi^2}; \quad \xi\frac{d\Phi_\|}{d\xi} + 3\Phi_\| = 3\exp(-\xi^2). \tag{10}$$

The diffusion tensor is anisotropic but invariant under rotation around the velocity vector of probe ion. Its components fall off at large velocities, Fig.2. At small velocity $v \ll v_T$ the spherical symmetry is restored $\Phi_\|(0) = \Phi_\perp(0) = 1$. Their expansions at small and large velocity begin from

$$\Phi_\| = \begin{cases} 1 - 3\xi^2/5, & \xi \ll 1, \\ 3\sqrt{\pi}/4\xi^3, & \xi \gg 1; \end{cases} \quad \Phi_\perp = \begin{cases} 1 - \xi^2/5, & \xi \ll 1, \\ 3\sqrt{\pi}/4\xi, & \xi \gg 1. \end{cases} \tag{11}$$

Therefore, at large velocity the friction and longitudinal diffusion become insignificant. The diffusion over angles in the velocity space occurs predominant.

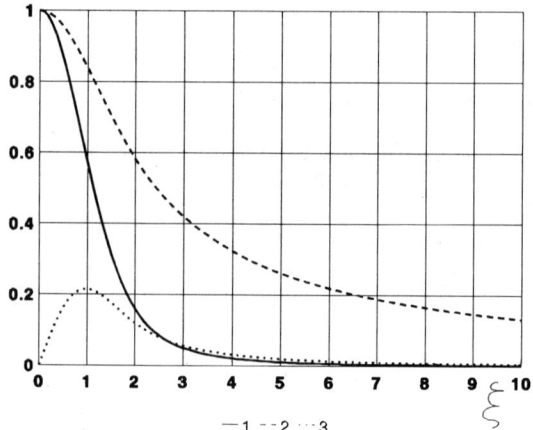

Fig. 2. *Diffusion coefficients $\Phi_\parallel$ along (curve 1) and $\Phi_\perp$ transverse (curve 2) to the velocity of probe ion, and Chandrasekhar function (curve 3) vs. dimensionless velocity ξ.*

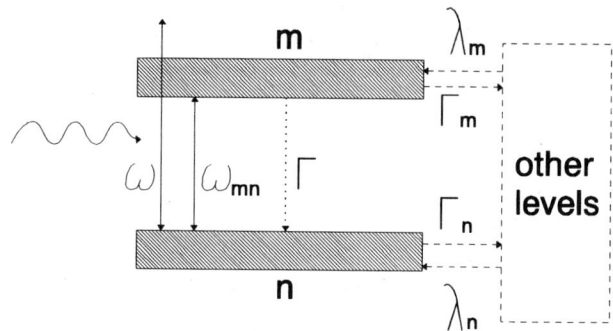

Fig. 3. *Sketch of two-level system selected by resonant field among the ionic energy levels.*

Let us describe the two-level subsystem of a probe ion the density matrix $\rho_{ij}(\boldsymbol{r}, \boldsymbol{v}, t)$, where indices $i, j = m, n$ denotes quantum ionic states. This equation allows to analyse resonant transitions between levels at the field of travelling wave $\boldsymbol{E} = \boldsymbol{E}_0 \exp(-i\omega t + i\boldsymbol{kr}) +$ c.c., Fig. 3.

The excitation of ionic levels $j = m, n$ usually occurs from the ground or metastable state, so we may assume the shape of excitation function $\lambda_j(\boldsymbol{v})$ to be Maxwellian.

$$\lambda_j(\boldsymbol{v}) = Q_j W(\boldsymbol{v}), \qquad (12)$$

where Q_j is the excitation rate of level j. Denoting the relaxation constants of level populations as Γ_j and of the coherence as Γ, we write steady-state equations for off-diagonal $\rho_{mn}(\boldsymbol{r}, \boldsymbol{v}, t) = \rho(\boldsymbol{v}) \exp(-i\Omega t + i\boldsymbol{kr})$ and diagonal $\rho_{jj} \equiv \rho_j$ elements of the density matrix

$$(\Gamma - i\Omega + i\boldsymbol{kv})\rho - \nu \hat{L}\rho = -iG\Delta N, \qquad (13)$$

$$\Gamma_j \rho_j - \nu \hat{L}\rho_j - \lambda_j(\boldsymbol{v}) = \mp 2\Re(iG^*\rho), \qquad (14)$$

where the upper/lower sign corresponds to upper/lower level. Here $\Delta N \equiv \rho_m - \rho_n$, $G = \boldsymbol{E}_0 \boldsymbol{d}_{mn}/2\hbar$, $\boldsymbol{d}_{mn}$ is the matrix element of dipole moment, collision differential operator $\hat{L}$ can be rewritten in form

$$\hat{L} = \frac{1}{2}\frac{\partial}{\partial \xi_\alpha}\Phi_{\alpha\beta}\left(\frac{\partial}{\partial \xi_\beta} + 2\xi_\beta\right), \qquad (15)$$

where $\Phi_{\alpha\beta}$ is the dimensionless diffusion coefficient: $D_{\alpha\beta} = \nu v_T^2 \Phi_{\alpha\beta}/2$.

3. Short-wave limit

The kinetic equation (13) for off-diagonal element can be solved in the case of rare collisions as an expansion in parameter $\nu/kv_T \ll 1$. For this purpose it is convenient to use spherical coordinates. The collision operator then takes the form

$$\hat{L} = \frac{1}{2}\left[\frac{\Phi_\perp}{\xi^2}\left(\frac{1}{\sin\theta}\frac{\partial}{\partial\theta}\sin\theta\frac{\partial}{\partial\theta} + \frac{1}{\sin^2\theta}\frac{\partial^2}{\partial\varphi^2}\right) + \frac{1}{\xi^2}\frac{\partial}{\partial\xi}\xi^2 \Phi_\parallel \left(\frac{\partial}{\partial\xi} + 2\xi\right)\right], \qquad (16)$$

The work done by the field of travelling wave is written as an expansion

$$\mathcal{P} = \mathcal{P}_0 \Re\left[i\sum_{n=0}^{\infty}\left(\frac{i\nu}{kv_T}\right)^n I_n(z)\right]; \mathcal{P}_0 = -2\hbar\omega|G|^2 N_{mn}\frac{\sqrt{\pi}}{kv_T},$$

$$I_n = \frac{1}{\pi^2}\int d^3\xi \left(\frac{\hat{L}}{z - \xi\cos\theta}\right)^n \frac{\exp(-\xi^2)}{z - \xi\cos\theta}, \qquad (17)$$

where $z = (\Omega + i\Gamma)/kv_T$, $N_{mn} = Q_m/\Gamma_m - Q_n/\Gamma_n$ is the unperturbed population difference.

While collisions are rare, consider only terms with $n = 0, 1$. In zero order obtain Voigt contour $I_0 = -iw(z)$ (w is the probability integral of complex argument[13]). At the line center ($\Omega = 0$) substitute $z = iy$ to I_1 and calculate the integral in the Doppler limit $y = \Gamma_{mn}/kv_T \to 0$. The work increases slightly

$$\mathcal{P} = \mathcal{P}_0(1 + b\frac{\nu}{kv_T}), \quad b = \frac{2}{\sqrt{\pi}}\left(\sqrt{2} - \ln(\sqrt{2}+1)\right) \cong 0.601. \tag{18}$$

A gain at the line maximum is proportional to frequency ν. The line, therefore, narrows, since the area conservation. The coefficient b in (18) differs from that in the weak collision model ($2/3\sqrt{\pi} \cong 0.376$) by 60%.

A reasonable question appears how to interpret the increase in coefficient b from 0.376 to 0.601. The dropping velocity dependence of the diffusion tensor seemingly should decrease, but not increase the coefficient. The matter is that one must compare the deceleration forces affecting the probe ion for correct comparison between models. Let us turn out to formula (5) for the particle flux in the velocity space. The first term corresponds to the deceleration of a single probe particle. The second term, namely the diffusion flux, is nonzero if the distribution function of probe particles has a nonzero gradient. For the sake of simplicity consider one-dimensional case. In the weak collision model with constant diffusion coefficient both diffusion fluxes at $v = u$ (shown in Fig. 4) are exactly cancelled.

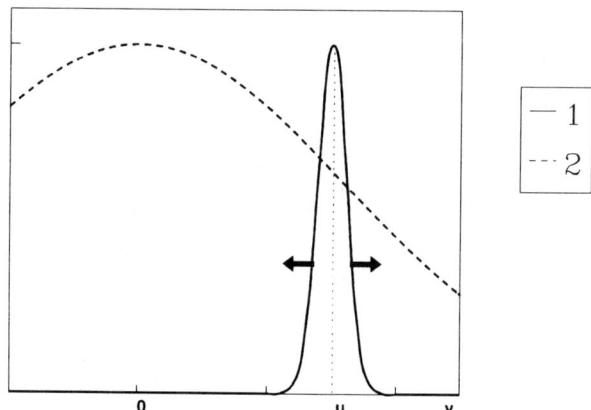

Fig. 4. *1-D distribution function of probe ions in the velocity space (curve 1) and the diffusion coefficient $D(v)$ (curve 2) with a dropping velocity dependence. Arrows indicate the directions of diffusion fluxes.*

If the diffusion coefficient $D(v)$ does decrease with the value of velocity, the flux toward small velocity exceeds the flux toward high velocity. As a result, the deceleration force increases. If the diffusion were monotonically increase ($dD/dv > 0$), then the total friction would be less.

Returning to 3-D case and using identities (10) we have $\partial D_{\alpha\beta}/\partial v_\beta = F_\alpha/m$, that is the force affecting the probe ions is doubled due to the dropping velocity dependence. The resultant friction force is really less than $2F_\alpha(0)$ owing to integration over the distribution. That is why we obtain that coefficient b is 60% more than in the model with constant collision frequency.

4. Long-wave limit

At the long-wave limit $kv_T/\nu \ll 1$ it is suitable to decompose the distribution function into a series over spherical harmonics with $m = 0$, i.e., Legendre polynomials

$$\rho(\boldsymbol{\xi}) = \frac{-iGN_{mn}}{\nu v_T^3} \sum_{l=0}^{\infty} \left(\frac{ikv_T}{\nu}\right)^l Y_{l0}(\theta) R^l(\xi) \exp(-\frac{\xi^2}{2}), \qquad (19)$$

$$Y_{l0}(\theta) = \left(\frac{2l+1}{4\pi}\right)^{1/2} P_l(\cos\theta). \qquad (20)$$

Then Eq.(13) reduces to the system of ordinary differential equations

$$\mathcal{H}^l R^l(\xi) + \frac{\xi}{\sqrt{2l+1}} \left(\frac{l R^{l-1}(\xi)}{\sqrt{2l-1}} - \left(\frac{kv_T}{\nu}\right)^2 \frac{(l+1)R^{l+1}(\xi)}{\sqrt{2l+3}} \right) = \frac{2}{\pi} \delta_{l0} e^{-\xi^2/2}$$

where

$$\mathcal{H}^l = \frac{\Gamma - i\Omega}{\nu} - \frac{1}{2\xi^2} \left[\left(\frac{d}{d\xi} - \xi\right) \xi^2 \Phi_\| \left(\frac{d}{d\xi} + \xi\right) - l(l+1)\Phi_\perp \right], \qquad l = 0, 1, 2, \ldots;$$

$$R^l(\infty) = 0, \quad R^l(\xi)\big|_{\xi \to 0} \sim \xi^l.$$

In this system equations with different l are tied by terms that are small in parameter kv_T/ν. In particular, the contribution of R^{l+1} to the equation for R^l has a smallness $(kv_T/\nu)^2$. In zero order only one term remains

$$R_0(\xi) = \frac{2}{\pi} \frac{\nu}{\Gamma - i\Omega} e^{-\xi^2/2}.$$

The distribution function is Maxwellian. In the first order two terms are necessary with $l = 0, 1$ in decomposition (19). System of two differential equations is left.

The ordinary differential equations in its turn can be reduced to an infinite system of algebraic equations, when we write $R^l(\xi)$ as a linear combination of orthogonal polynomials. It is convenient to choose Laguerre functions, eigenfunctions of the problem with constant diffusion coefficient at $\boldsymbol{k} = 0$

$$R^l(\xi) = \frac{1}{\pi^{3/4}} \sum_{n=0}^{\infty} a_n^l R_n^l(\xi), \quad R_n^l(\xi) = A_n^l \xi^l \exp(-\frac{\xi^2}{2}) L_n^{l+1/2}(\xi^2),$$

$$A_n^l = [n!2/\Gamma(n+l+3/2)]^{1/2}, \quad L_n^{l+1/2}(x) = \frac{e^x}{n! x^{l+1/2}} \frac{d^n}{dx^n} \left[x^{n+l+1/2} e^{-x} \right].$$

These functions satisfy boundary conditions. At equal l they are orthogonal on the real half-axis

$$\int_0^\infty \xi^2 d\xi\, R_n^l(\xi) R_{n'}^l(\xi) = \delta_{nn'}.$$

Multiplying equation number l by $R_n^l(\xi)$ and integrating over ξ we obtain

$$\sum_{n'=0}^{\infty} H_{nn'}^l a_{n'}^l - \left(\frac{l+1}{\eta^2} \frac{\sqrt{n+l+3/2}\, a_n^{l+1} + \sqrt{n}\, a_{n-1}^{l+1}}{\sqrt{(2l+1)(2l+3)}}\right. \tag{21}$$

$$\left. - l \frac{\sqrt{n+l+1/2}\, a_n^{l-1} + \sqrt{n+1}\, a_{n+1}^{l-1}}{\sqrt{(2l+1)(2l-1)}}\right) = \delta_{n0}\delta_{l0},$$

where $\eta = \nu/kv_T$. The matrix element

$$H_{nn'}^l = \int_0^\infty \xi^2 d\xi\, R_n^l(\xi) \mathcal{H}^l R_{n'}^l(\xi) \tag{22}$$

can be expressed in terms of gamma-functions. The power absorbed depends on coefficient a_0^0 only

$$\mathcal{P}(\Omega) = \mathcal{P}_0 \frac{kv_T}{\nu\sqrt{\pi}} \Re a_0^0. \tag{23}$$

The following formula for a_0^0 has been derived[10]

$$a_0^0 = \frac{\nu}{\Gamma - i\Omega + \frac{(kv_T)^2}{2\lambda}}, \quad \lambda = \nu \lim_{M\to\infty} \frac{\det \mathcal{H}_M^1}{(H_{00}^1)_M^*}. \tag{24}$$

Here M is the number of terms held in (21), $\mathcal{H}_M^1$ is the matrix given by (22), $H_{00}^1)_M^*$ is the cofactor to element H_{00}^1 of the matrix $\mathcal{H}_M^1$. Elements H_{mn}^1 decrease fast with $|m-n|$, then for calculation of λ we may restrict ourselves by small M.

Particularly, if $M = 0$, then

$$\mathcal{P}(\Omega) = \mathcal{P}_0 \frac{kv_T}{\sqrt{\pi}} \Re \left(\Gamma - i\Omega + \frac{(kv_T)^2}{2\left(\frac{\nu}{\sqrt{2}} + \Gamma - i\Omega\right)} \right)^{-1}. \tag{25}$$

At the line center $|\Gamma - i\Omega| \ll \nu$ parameter λ tends to $\bar{\nu} = \nu/\sqrt{2}$. It is the effective collision frequency. The value $\nu/\sqrt{2}$ has been obtained by renormalization method[9], so that the renormalization is equivalent to account only the highest order in decomposition over Laguerre polynomials.

At $M = 1$ the absorption spectrum is

$$\mathcal{P}(\Omega) = \mathcal{P}_0 \frac{kv_T}{\sqrt{\pi}} \Re \left[\Gamma - i\Omega + \frac{(kv_T)^2/2}{\frac{\nu}{\sqrt{2}} + \Gamma - i\Omega - \frac{9\nu^2}{59\sqrt{2}\nu + 80(\Gamma - i\Omega)}} \right]^{-1} \quad (26)$$

At the line center $\lambda \to \bar{\nu} = 0.599\nu$. For $M \geq 2$ the effective collision frequency is almost constant.

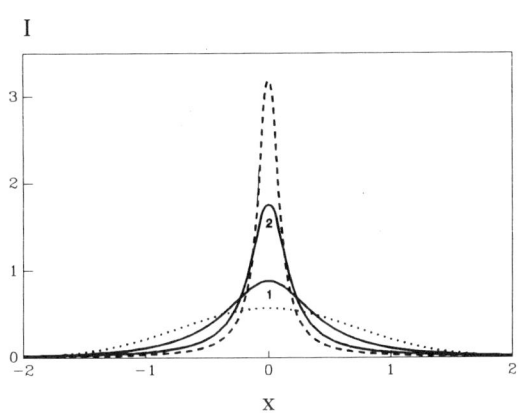

Fig. 5. *Spectral line contour $I(\Omega)$ at the long-wave limit. The detuning normalized by the Doppler width is denoted as $x = \Omega/kv_T$. Dotted line shows wide Doppler contour; dashed line corresponds to Lorentzian profile at $kv_T = 10\Gamma$. Solid curves show compressed contour at $\nu = 3kv_T$ (curve 1) and $\nu = 10kv_T$ (curve 2)*

M	0	1	2	3	4	5	6	7
$\bar{\nu}/\nu$	0.7071	0.5992	0.5942	0.5942	0.5942	0.5942	0.5942	0.5942

From the Table one can see that at $M = 1$ the value of $\bar{\nu}$ differs from its limit by 1% only. Thus, formula (26), or even (25), gives the line shape with a reasonable accuracy. The contour is shown in Fig. 5 for $kv_T/\nu = 3$ and 10 together with Doppler and Lorentzian contours, corresponding to $kv_T/\nu = 0$ and $kv_T/\nu = \infty$, respectively. The narrowing in long-wave limit is a strong effect. Asymptotics of absorption in the line center at $\Gamma \ll \nu$ is

$$\mathcal{P} = \mathcal{P}_0 \frac{kv_T}{\sqrt{\pi}} \frac{1}{\Gamma + (kv_T)^2/2\bar{\nu}}, \quad \bar{\nu} \cong 0.594\nu. \tag{27}$$

5. Intermediate case

Interaction with electromagnetic wave breaks the isotropy of ionic velocity distribution. Particles having nonzero projection of the velocity onto the wavevector direction acquire the extra phase shift $\Delta\phi = \boldsymbol{kv}t$. Collisions vary the velocity randomly, so, restore in part the isotropy. At $\nu \sim kv_T$ anisotropic part of the distribution in the domain of slow particles $|v| < v_T$ is of the order of the isotropic part. Hopefully, the decomposition over Legendre polynomials should converge rapidly. As to fast particles with $|v| \gg v_T$, for which the collision frequency is small, while the phase shift is high, their contribution into absorbed power can be ignored as a consequence of exponential decrease of the distribution function with v^2.

The dependence of intensity in the line center on parameter ν/kv_T has been determined numerically[10] by Gauss method. Coefficient a_0^0 was accepted as correct if it had a limit at $n_{max}, l_{max} \to \infty$. At $\nu/kv_T \geq 0.2$ the coefficient was occurred constant for $n_{max} = 4$ $l_{max} = 6$. The results of computation agreed well with analytically studied limiting cases.

Analytical expression for the intensity at the line center has the same dependence on values ν, kv_T, Γ as in the weak collision model, but differs in coefficients. It agrees with result by Alekseev and Malyugin[14], who showed that the line at $\nu \gg kv_T$ had the Lorentzian shape of the width $(kv_T)^2\mu^{-1}$, where μ^{-1} was the sum of inverse eigenvalues of the collision integral operator, independently of the operator specific form.

To visualize the similarity of the numerical results and those of the weak collision model the latter are also shown in Fig. 6. As it is clear from the curve, the dropping velocity dependance of collision frequency decreases the effect by 40% in the long-wave limit at $\nu \gg kv_T \gg \sqrt{\Gamma\nu}$. We can construct a simple interpolating formula

$$\mathcal{P} \propto \frac{kv_T}{\sqrt{\pi}} \frac{1}{\Gamma + \dfrac{(kv_T)^2}{2\bar{\nu} + \sqrt{\pi}kv_T}}; \quad \bar{\nu} = 0.59\nu, \tag{28}$$

which reaches the exact asymptotics at $\nu \gg kv_T$. In the opposite limit $\nu \ll kv_T$ it differs from the analytical expression only in the coefficient b at small term ν/kv_T: $\bar{\nu}/\nu = 0.59$ instead of 0.53. The relative error of formula (28) does not exceed 10% even at intermediate case $\nu \sim kv_T$. Furthermore, one can use the weak collision model with the renormalized collision frequency $\bar{\nu} = 0.59\nu$ in plasma. It has approximately the same accuracy as (28).

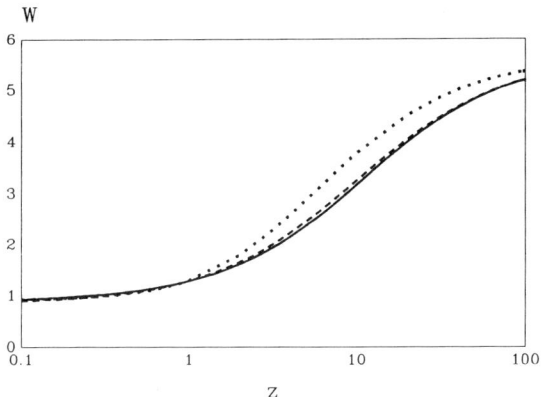

Fig. 6. *The normalized absorbed power $W = \mathcal{P}/\mathcal{P}_0$ in the line center ($\Omega = 0$) as a function of the collision frequency $z = \nu/kv_T$ at $kv_T = 10\Gamma$: numerical (solid curve); the weak collision model (dotted curve); interpolating formula (dashed curve).*

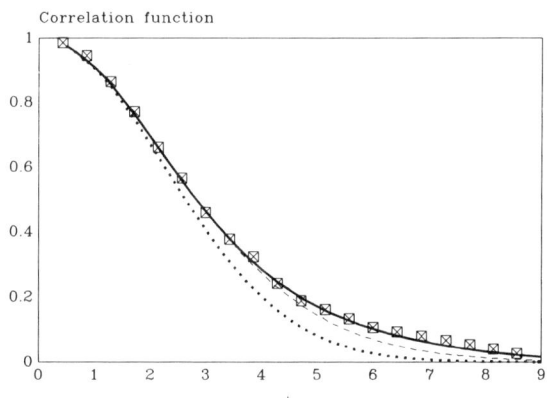

Fig. 7. *Correlation function $\Phi(t)$ obtained by molecular dynamic simulation (points) and three models: free motion (dotted curve), weak collision (dashed curve) and present theory including the velocity dependence (solid curve).*

The molecular dynamic simulation by Pollock and London[15] allows to find the correlation function for intermediate and strongly coupled plasma. The coupling parameter $\mathbf{\Gamma} = Z^2 e^2/Ta$, where $a = (4\pi N/3)^{-1/3}$ is the interparticle distance, was $\mathbf{\Gamma} = 1, 5, 10$ in the simulation. There is no theory of collisional narrowing in non-ideal plasma with $\mathbf{\Gamma} = 5, 10$, nevertheless, we can compare results for $\mathbf{\Gamma} = 1$ on the verge of applicability of the theory. Fig. 7 shows the points from simulation together with the solid line calculated in the weak collision model. The agreement is better than with free-particle and pure diffusion models. The collisional effects in this case is slight, then the theoretical curve goes closely with that of weak collision model.

6. Nonlinear effects

The velocity distribution of populations ρ_j of both levels $j = m, n$ is to be found. Let us analyze the Bennett hole in the velocity distribution induced by the electromagnetic wave[16]. To simplify the treatment we suppose electromagnetic field to be weak $|G|^2 \ll \Gamma\Gamma_j$ and neglect the effects of strong saturation. The right sides of Eqs. (13), (14) can be considered, therefore, as a perturbation. At the highest order of the perturbation theory the distributions $\rho_j(\bm{v})$ take Maxwellian shape.

$$\rho_j^{(0)} = \frac{Q_j}{\Gamma_j} W(\bm{v}), \quad \rho^{(0)} = 0.$$

Further one can substitute the unperturbed expression for the population difference $\Delta N = N_{mn} W(\bm{v})$ into the right side of the off-diagonal equation (13). The formal solution of (13) in terms of operator exponents is

$$\rho^{(1)}(\bm{v}) = -iG \int_0^\infty dt \exp(-\hat{A}t + \nu \hat{V} t) \Delta N(\bm{v}), \tag{29}$$

$$\rho_j^{(1)}(\bm{v}) = \mp \int_0^\infty dt \exp(-\Gamma_j t + \nu \hat{V} t) 2\Re \left(iG^* \rho^{(1)}(\bm{v}) \right), \tag{30}$$

where $\hat{A} = \Gamma - i\Omega + i\bm{k}\bm{v}$.

It is convenient to factor out the Maxwellian distribution $W(\bm{v})$ in Eqs. (29), (30) distinctly

$$\rho_j^{(1)}(\bm{v}) = \mp 2|G|^2 N_{mn} W(\bm{v}) R_j(\bm{v}). \tag{31}$$

Therewith only the expression for collisional operator is changed. Operator $\hat{V}$ in (29), (30) should be replaced by the new operator $\hat{B}$, where

$$\hat{V} \exp\left(-\frac{v^2}{v_T^2}\right) = \exp\left(-\frac{v^2}{v_T^2}\right) \hat{B}, \quad \hat{B} = \frac{1}{2} \Phi_{\alpha\beta} \left(\frac{\partial}{\partial \xi_\beta} - 4\xi_\beta \right) \frac{\partial}{\partial \xi_\alpha}. \tag{32}$$

Here we use identities (10) to express partial derivatives of the diffusion tensor in terms of $\Phi_{\alpha\beta}$. Thereafter one can rewrite Eq. (30) in the form

$$R_j(\boldsymbol{v}) = \Re \int_0^\infty dt \int_0^\infty d\tau \exp(-\Gamma_j\tau)\exp(\nu\hat{B}\tau)\exp(-\hat{A}t + \nu\hat{B}t). \tag{33}$$

The collisional frequency in high-current gas-discharge plasma[11] is about $\nu \sim 10^5 - 10^7 s^{-1}$. That is by several orders of magnitude less than the Doppler width $kv_T \sim 10^{10} s^{-1}$ and as a rule less than relaxation rates $\Gamma \sim 10^9 - 10^{10} s^{-1}$, $\Gamma_j \sim 10^7 - 10^{10} s^{-1}$. That is why we may use small parameters

$$\nu \ll \Gamma_j, \quad \Gamma \ll kv_T \tag{34}$$

and it is enough to hold only linear in ν terms in Eq. (33). The operator equalities given in Appendix allow us to rewrite formal solution (33) in explicit form. Using operator equalities step-by-step obtain within the first order in ν

$$R_j(\boldsymbol{v}) = \Re \int_0^\infty dt \int_0^\infty d\tau \exp(-\Gamma_j\tau - (\Gamma - i\Omega + ikv_z)t) \times$$

$$\times \exp\left[\frac{\nu}{2}\left(-\Phi_{zz}(\xi)k^2v_T^2(t^2\tau + t^3/3) + 2ikv_z\Phi_{\parallel}(\xi)(2t\tau + t^2)\right) + \mathcal{O}(\nu^2)\right], \tag{35}$$

where z axis is chosen along the wavevector $\boldsymbol{k}$ again. Function

$$\Phi_{zz}(\boldsymbol{\xi}) = \left(\Phi_{\parallel}(\xi) - \Phi_{\perp}(\xi)\right)\frac{\xi_z^2}{\xi^2} + \Phi_{\perp}(\xi)$$

is the zz - component of the diffusion tensor. This function depends on two variables $\xi = |\boldsymbol{\xi}|$ and ξ_z. The integrand in (35), obtained by the expansion, is not valid at long times t, τ. Consequently, for application of this expansion it is necessary to have the main contribution to integral (35) at the domain, where the quadratic terms in ν remain small. Since $|\partial\Phi_{\alpha\beta}/\partial\xi_\gamma| < |\Phi_{\alpha\beta}|$, the estimation of quadratic terms gives two conditions (34) of applicability of expression (35). The latter is the Doppler limit, the former coincides with the condition of smallness of the diffusion width compared to the thermal velocity.

Taking an integral over τ obtain the correction to the distribution function as a single integral over t

$$R_j(\boldsymbol{v}) = \Re \int_0^\infty dt \frac{\exp\left[-(\Gamma - i\Omega + ikv_z)t - \frac{1}{6}\Phi_{zz}(\boldsymbol{\xi})\nu k^2 v_T^2 t^3 + i\nu kv_z\Phi_{\parallel}(\xi)t^2\right]}{\Gamma_j + \frac{1}{2}\Phi_{zz}(\boldsymbol{\xi})\nu k^2 v_T^2 t^2 - 2i\nu kv_z\Phi_{\parallel}(\xi)t}. \tag{36}$$

If one were neglect velocity dependence of the diffusion tensor and take $\Phi_{\parallel} = \Phi_{\perp} = 1$ here, then the expression would differ from corresponding expression[17] for constant diffusion coefficient by factor 2 in terms proportional to $\Phi_\parallel$. The similar difference, that consists in the doubling of the friction force, was discussed and explained in Sec. 3.

The Bennett holes burnt out by the laser field in the velocity distribution are more sensitive to the velocity dependence than the Dicke narrowing. The narrowing of the Bennett hole occurs with increase of the pumping field detuning compared to weak collision model[18]. Physical cause of the effect is also the dropping velocity dependence of the collision frequency.

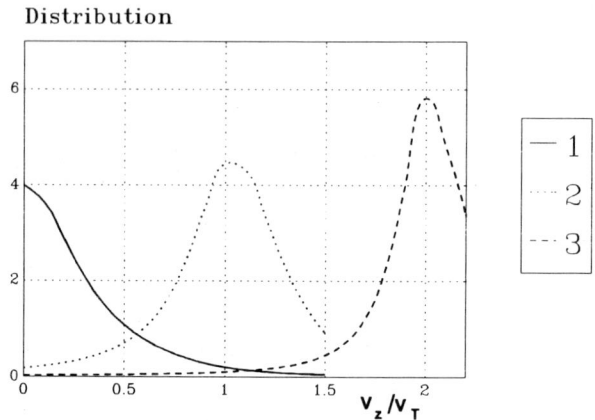

Fig. 8. *The Bennett hole profiles $R_j(v)$ in the velocity distribution of the probe ion calculated at $\nu/kv_T = 0.001$, $\Gamma_j/kv_T = 0.01$, $\Gamma/kv_T = 0.1$ and different detunings: $\Omega = 0$ (curve 1), $\Omega = kv_T$ (curve 2), $\Omega = 2kv_T$ (curve 3).*

While the detuning is less than the Doppler width ($\Omega \ll kv_T$), we may neglect the velocity dependence of diffusion coefficients and put Φ_{zz} and $\Phi_\parallel$ equal to unity in (36). In this case we may also ignore the terms proportional to kv_z. Then the integral may be simplified to the form which describes the diffusion broadening with the constant collision frequency[11]. The width of the Bennett hole, if level j is narrow ($\Gamma_j \ll \Gamma$), is determined by the maximum among the homogeneous width Γ/k and diffusion width $(\nu/\Gamma_j)^{1/2}v_T$.

Let the diffusion width exceeds the homogeneous one as in high-current ion-laser plasma. As the detuning increases, the fast particles form the hole, while function $\Phi_{zz}(\xi)$ falls. So, the diffusion width decreases. The hole narrows down to the homogeneous width far from the resonance. The profile of Bennett hole, plotted in Fig. 8 at three different detunings, undergoes the narrowing is by nearly 10% at $\Omega = kv_T$.

7. Conclusions

The exact Chandrasekhar's formulae for the diffusion tensor in the velocity space corrects the numerical value of the collisional narrowing in plasma by tens percent. Nevertheless, one can use the weak collision model with the constant, but corrected collision frequency. The interpolating formula obtained gives the value of renormalized collision frequency vs. the ratio of the mean free path and the radiation wavelength.

It seems interesting to look for the collisional narrowing in dense plasma ($n_e \sim 10^{20}$ cm^{-3}), e.g., X-ray laser plasma. The estimations by Pollock and London shows the effect to be insignificant in lasing on SeXXV lines, meanwhile their simulation proved the narrowing becomes important in strongly coupled plasma. In order to emphasize the effect, one must use lines of multi-electron spectrum of multiple ions and long-wave transitions to diminish Γ_{mn} and increase ν/kv_T. The plasma should be comparatively dense and cold. High ion density and low temperature lead to increase in $\nu \propto N_i v_T^{-3}$. However, at excessively low ionic temperature or high charged particle density the Doppler width can become less than homogeneous one.

The effect of velocity dependence appears more distinctly in nonlinear spectroscopy of plasma. The Bennett hole narrows with increase in the detuning even in gas discharge plasma at low charged particle density ($n_e \sim 10^{14}$ cm^{-3})

8. Acknowledgments

We thank S.A.Babin and S.G.Rautian for stimulating discussions. The research described in this publication was made possible in part by Grant No. RCN000 from the International Science Foundation.

Appendix

Let us consider operator $\hat{B}$ from (32) and $\hat{A} = \Gamma - i\Omega + i\boldsymbol{kv}$. Denoting a small parameter as δ and some constant as N we obtain a chain of equalities

$$[\hat{B}, \hat{A}] \equiv \hat{C} = \Phi_{\alpha\beta}\left(\frac{\partial}{\partial \xi_\alpha} - 2\xi_\alpha\right)ik_\beta v_T, \quad [\hat{C}, \hat{A}] = -\Phi_{\alpha\beta}k_\alpha k_\beta v_T^2,$$

$$\left[[\hat{C}, \hat{A}], \hat{A}\right] = 0.$$

The third commutator equals zero.

$$\hat{B}N = 0, \quad \exp(\delta\hat{B})N = N, \quad \exp(-\delta\hat{C}t)N = \exp\left\{2i\delta\boldsymbol{kv}\Phi_\|(\xi)t + \mathcal{O}(\delta^2)\right\}N.$$

The action of the operator exponent to a constant can be written, then we can derive the closed expression of the exponent with the accuracy of δ^3.

$$\exp(-\hat{A}t + \delta\hat{B}) = e^{-\hat{A}t}\exp\left\{\delta\left(\hat{B} - \frac{1}{2}[\hat{B}, \hat{A}t] + \frac{1}{6}[\hat{C}t, \hat{A}t]\right)\right\} \times$$

$$\times \exp\left\{\delta^2\left(-\frac{[\hat{C}t, \hat{B}]}{6} + \frac{\left[[\hat{C}t, \hat{A}t], \hat{B}\right]}{12} - \frac{\left[[\hat{C}t, \hat{A}t], \hat{C}t\right]}{60}\right) + \mathcal{O}(\delta^3)\right\};$$

$$e^{\delta \hat{B}} e^{-\hat{A}t} = e^{-\hat{A}t} \exp\left\{\delta\left(\hat{B} - \hat{C}t + \frac{1}{2}[\hat{C}t, \hat{A}t]\right)\right\}.$$

---

1. R.H.Dicke, Phys. Rev. **89**, 472 (1953).

2. S.G. Rautian, I.I. Sobel'man, Usp. Fiz. Nauk **90**, 118 (1966) [Sov. Phys. Usp. **9**, 701 (1967)].

3. M.I.D'yakonov, Zh. Eksp. Teor. Fiz. **51**, 612 (1966) [Sov. Phys. JETP **24** 412 (1966)].

4. J.C.Bergquist, W.M.Itano, D.J.Wineland, Phys. Rev. **A 36**, 428 (1987).

5. F.Ganikhanov, I.Konovalov, V.Kulyasov et al. Abstracts of ICONO'92 (St.Petersburg, September 24–27, 1991). Part 3, p.4.

6. J.W.Forsman, P.M.Sinclair, A.D.May et al., J. Chem. Phys. **97**, 5355 (1992).

7. H.R. Griem, Phys. Rev. **A 33**, 3580 (1986).

8. J.A. Koch, B.J. MacGovan, L.B. Da Silva, D.L. Matthews, J.H. Underwood, P.J. Batson, S. Mrowka, Phys. Rev. Lett. **68**, 3291 (1992).

9. E.V.Podivilov, D.A.Shapiro, JETP Letters **56**, 449 (1992).

10. E.V.Podivilov, A.I.Chernykh, D.A.Shapiro, Zh. Eksp. Teor. Fiz. **105**, 1214 (1994).

11. S.A.Babin, D.A.Shapiro, Phys. Rep. **241**, 119 (1994).

12. S.Chandrasekhar, Rev.Mod.Phys., **15**, 1 (1943).

13. Handbook of mathematical functions, Ed. by M.Abramowitz, I.A.Stegun, Appl. Math. Series, **55** (Government Printing Office, Washington, 1964).

14. V.A.Alekseev, A.V.Malyugin, Zh. Exp. Teor. Fiz., **80**, 897 (1981).

15. E.L.Pollock, A.London, Phys. Fluids **B 5**, 4495 (1993).

16. E.V.Podivilov, M.G.Stepanov, D.A.Shapiro (to be published).

17. S.G.Rautian, Zh. Eksp. Teor. Fiz. **51**, 1176 (1966) [Sov. Phys. JETP **24**, 788 (1967)].

18. L.A.Gel'medova, D.A.Shapiro, J. Mod. Opt. **38**, 573 (1991).

EFFECTS OF MICROFIELD NONUNIFORMITY IN DENSE PLASMAS.

A.V.Demura
Hydrogen Energy and Plasma Technologies Institute
Russian Research Center "Kurchatov Institute", Moscow, 123182, Russia
and
Department of Physics, 206 Allison Lab, Auburn University, Auburn, AL 36849-5311,USA

C.Stehlé
DARC et UPR 176 du CNRS, Observatoire de Paris-Meudon
5 Place Jules Janssen, Meudon, France

"We see so far because we are standing on giants's shoulders"
-Sir Isaac Newton.

ABSTRACT

The brief review of the current status of the investigations on the microfield nonuniformity effects in the Stark broadening in dense plasmas is provided. The generalized theoretical approach, which enables to obtain the solution for plasmas with an arbitrary composition and arbitrary coupling parameter values is presented. The relationship with the known experimental data is discussed.

INTRODUCTION

Effects of the microfield nonuniformity and the microfield time dependence in plasmas trouble since the classic work of Holtsmark[1], that made Pauli to bring specially the attention to these issues [2]. Much understanding especially of the basic problem of the hydrogen atom interaction with the nonuniform field of the single charged particle was achieved [3-8]. The main importance of the most consistent works of Kudrin and Sholin[4] was the construction of the correct perturbation approach to the secular quantum problem and the invention of the correction terms to the eigenfunctions that are linear in the perturbation parameter $n^2 a_o/R$ (n-the principal level quantum number, a_0-Bohr radius, R-the mean interparticle distance). These corrections are caused by the quadrupole terms in the multipolar expansion of the hydrogen atom interaction with the ion. In the most logically correct and complete form this was scaned in detail in the work of Sholin[6]. On the other hand Sobelman and Griem pointed out in their works the cancellation of the contributions due to the quadrupole terms from ions and electrons in the profile[7-8]. That was based on attempts to consider the radiator-pertuber interaction for the both values of the perturber charge sign on one foot[6] specifically in the frames of the quasistatic[8] approximation or the impact[7] one. But now it is understood, that due to the difference of the masses and thus in the characteristic time scales of the deviation of the electric fields from ions and electrons the range of plasma parameters, where there is no obvious cancellation of these contributions, really exist- at any rate for some wide region of detunings from the line center. It is thought, that in this region ions are almost quasistatic, while electrons are impact[6]. The theoretical works in this field, saying in words of Margenau and Lewis[9] *were inspired by a vision*, that is not so large but quite distinct asymmetry features of hydrogen lines, observed in series of bright experiments with the very high precision[10-17]. Griem[18] was the first, who tried to explain the first observations from the point of the quadratic Stark effect corrections. It may be traced through experimental results, that as a rule near the line center in one -photon emission profiles the blue side of the line is more intense. As the detuning increases the difference increases, reachs extremum and after decreases, then changes the sign and increases again [14,15,17] as one goes towards the wings. So as a rule hydrogen lines have the more intense red wing of the line than the blue one. This fact was put to ground the objections against transition of electrons to the quasistatic regime of line broadening in the far wings[6], that is intrinsically connected with the understanding of the cancellation mechanisms of quadrupole terms from ions and electrons. For much more far wings one can foresee again the change of the asymmetry due to the decreasing of the intensity of the red wing with respect to the blue one, caused by the ionisation of the atom in the microfield[19-20], and its relative increasing coming from Boltzman factors[18,21-22], which definitely appear in the quantum approach[22]. Although there is no clear understanding what is happening really, because the ionised electron may be more probably recaptured by the bare nearest perturber, providing the electric field in the far wing is created by the nearest neighbour. Moreover the bumps in the far wings are known and observed, which are due to the extrema of the quantum terms[16]. Thus it may be stated that as the quadrupole effects from ions and electrons in the quasistatic (and thus semi-classical) approximation cancel each other, perhaps, one should apply the quantum theory for the correct treatment of the line asymmetry in the far wings. There is also additional direct consequencies of the line asymmetry - line shifts[13-14,24-26], the question that needs a much caution in the experiment and the theory. The binary approach had its summit in works of Bacon[27], who calculated profiles Ly-α and Ly-β, taking the maximum precaution to the consistency of the quantum theory results, that demand to include all the terms of the same order of the perturbation parameter. Meanwhile the problem of the nonuniformity of the Coulomb field was treated by Chandrasekhar and von Neumann[28] in the case when one must consider the vector summary field of many perturber particles interacting with the test one. Unfortunately these results did not become well known and the main part of them was rederived again later. Partly it was due to the calculation in[28] the finite differences of the microfields in two points, seperated from each other by the finite distance, determined through the radius-vector $\delta \vec{r}$. Independently Milliyanchuk[29] and Müller[30] obtained distributions for single components of the Coulomb field nonuniformity tensor, rederiving some general results from[28]. As it could be possible to classify later, their results are valid when the number of particles in the Weiskopff sphere[31] of the quadrupole interaction is much more than unity. Moreover the first author also considered negative charges to be quasistatic from the beginning. Thus in their initial formulation results of these works both may be applied only for very rare conditions. In the case of ideal plasmas, the many-body problem of the nonuniform electric microfield interaction with the hydrogen atom was solved by Demura and Sholin[32]. These authors introduced the universal functions connected with the constraint moments of the nonuniformity microfield tensor at the fixed vectorial values of the electric microfield strength vector in the frames of the generalized many-body quantum perturbation approach [32]. The consideration was based on the assumption that ions are quasistatic, while electrons are impact. This enables to develop the formalism of the instant joint distribution functions of the electric microfield strength vector(of the summary ion electric field) and the independent components of the nonuniformity microfield tensor (the number of which is equal to five in the pure Coulomb case). In this work for the first time the fundamental importance of the constraint moments of the nonuniformity microfield tensor as the characteristics of the spatial and temporal microfield fluctuations was understood. Also there the formula for the shift of the line gravity center was firstly derived and the expression for the asymmetry in the line wings was rederived with account of the impact electrons. In fact the results obtained and the methods developed in this work now have been applied and widely used in many works of other authors on this subject. Several attempts were done after to incorporate in this formalism the fact of the plasma coupling parameter finiteness[24,27,33-35], while the first correct one to the accuracy of the pair radiator-perturber correlations for the ion lines was firstly published by Joyce, Woltz and Hooper[36], not talking about the approximations need to be involved to obtain the final results.

STATEMENT OF PROBLEM

In this work we present the upgrade state-of-the art in the theory of the microfield nonuniformity effects in plasmas[36-46] devoted mostly to the self-consistent and correct inclusion in the theory effects of plasma nonideality and point out some puzzles in the theoretical treatment and the relationship with the experimental data. The generalized theoretical approach[37-39] is based on the Baranger-Moser cluster expansion scheme[48-54]. It gives the statistical solution for the problem of the definition of the instant low-frequency joint microfield distribution function of the electric microfield strength vector and the nonuniformity microfield tensor with the accuracy to the two-particle correlation functions in the case of the neutral radiator and to the three-particle correlations in the case of the charged one[37-39,44]. The accuracy of the results depends on the accuracy of the correlation functions involved[47-55]. This approach enables to treat arbitrary mixtures of ions in plasmas with arbitrary values of coupling parameters and admits the usage of any type of the electric microfield definition and thus incorporates properly defined the nonlinear electron screening in the frames of the DFT-theory[52-54]. On the other hand it is shown how these results may be approximately reexpressed[42,43,45-46] in terms of the appropriate APEX version[56-57] -Adjustable Parameter Exponential Approximation method, which exploits less information about the correlation functions and became well-known due to the good coincidence with the Monte-Carlo simulation results for microfield distributions in strongly coupled plasmas. The general formulas for the first constraint moments of the microfield nonuniformity tensor, while the electric microfield strength vector being fixed, are derived for the first time for any arbitrary type of the electron screening and correlation functions and any composition of ion charges in plasmas and thus for any value of plasma coupling parameters. The interaction of the radiator with nonuniform plasma microfield is reconsidered in order to take into account the so called polarization effects[38-39,44], which appear due to the finiteness of the plasma coupling parameters and the ions "metamorphosis" into quasiparticles, bearing the cloud of screening them electrons[37-39]. Thus due to the self-consistent theoretical treatment the polarization terms are contained in the expressions for the first constraint moments of the microfield nonuniformity tensor inherently. The results of the previously elabourated perturbation approach for the theory of the hydrogen-like radiators spectral line asymmetry in plasmas[32] are generalized, and the new universal functions[38-39,44], depending on plasma coupling parameters and the plasma composition, are presented, which in fact determine the deformation of the profile due to nonuniformity effects. These functions include the polarization terms as well. The connection of the theoretical predictions with the known experimental results as well as the old inconsistencies and the new trends in the theoretical ideas are discussed.

GENERAL RESULTS FOR JOINT DISTRIBUTION FUNCTIONS

Let the radiator with the net charge Z_o is placed into the environment of field ions and electrons in plasma. The Hamiltonian of the whole system may be written in the form

$$\hat{H} = \hat{H}_o + \hat{H}_e + \hat{H}_i + \hat{H}_r + \hat{V}_{oe} + \hat{V}_{oi} + \hat{V}_{ei} + \hat{V}_{or} + \hat{V}_{ir} + \hat{V}_{er}. \tag{1}$$

Here $\hat{H}_o$-is the unperturbed Hamiltonian of the radiator; $\hat{H}_{e,i}$ -are Hamiltonians of free plasma electrons and ions accordingly; $\hat{H}_r$ - is the Hamiltonian of the radiation field; the symbol $\hat{V}$ designates operators of the pair interaction between subsystems. We count here, that the radiation field is small enough to ignore all interactions with it from the beginning. It is supposed, that the level populations of the radiator are rapidly settled, while plasmas is quasistationary and optically thin. As it stated above let the conditions of the quasistatic approximation for the interaction of plasma ions with the radiator are fulfilled[31]. The interaction with electrons may be considered either in the impact approximation[31,37] or in terms of the relaxation theory[22], or in the one-electron approximation[23]. At this point of the consideration the traditional question of the initial correlations arises[22,52−53]. Here it is assumed, that is possible to introduce the effective potentials of the ion-radiator interaction, including screening effects of plasma electrons[52−54,58]. There are several possible ways of the dividing this system on the almost uncoupled subsystems[22,26,52−54]. The only condition is presumed that the plasma microfield approach is valid[59−61]. This means in particular the neglection of the ion-perturber configurations penetrating inside the area of the bound electron orbits. Although the interaction with such configurations may be treated in the many-body approach or in the binary one depending on the physical situation it could not be expressed in terms of the microfield, while their contribution usually should be suppressed due to the Coulomb repulsion between ions, that hit this area of bound electron orbits. Thus we suppose that $\hat{V}_{oi}$ may be written in the form

$$\hat{V}_{oi} = -\hat{\vec{d}}\vec{F} + \frac{1}{6}\hat{Q}_{\alpha\beta}\frac{\partial F_\alpha}{\partial x_\beta} - \frac{e^2}{6}\hat{r}^2 div\vec{F}, \tag{2}$$

$$\hat{r}^2 = \sum_l \hat{x}_l^2 + \hat{y}_l^2 + \hat{z}_l^2, \tag{3}$$

where $\hat{\vec{d}}, \hat{Q}_{\alpha\beta}$ - are the dipole and quadrupole moment operators of the radiator accordingly with the sign of the electron charge included in their definition; $\hat{x}_l, \hat{y}_l, \hat{z}_l$ - are operators of Cartesian coordinates of the radiator "optical" electrons", numerated by the label l; $\vec{F}$ is the ion electric microfield strength vector, defined at the origin of the coordinate system as well as the derivatives of its components. The last term in (2) arises due to the plasma polarization. Let there is an arbitrary set $\{s\}$ of species of ion perturbers in plasma. Then due to the condition of the quasineutrality

$$N = N_e = \sum_s Z_s N_s, \tag{4}$$

where N_e-is the electron plasma density at the infinity, N_s - is the partial density of the ion perturbers of the sort s with the effective net charge Z_s. After these definitions we can start with general expressions for the instant joint distribution function of the individual (but nonbinary) component of the electric ion microfield $\vec{F}$ and the microfield nonuniformity tensor set $\left\{\frac{\partial F_\alpha}{\partial x_\beta}\right\}$ of six independent components at the origin of the arbitrary laboratory system of coordinates

$$W(\vec{F};\left\{\frac{\partial F_\alpha}{\partial x_\beta}\right\}) = \frac{1}{(2\pi)^9}\int d^3\vec{\rho}\prod_{m=1}^{6}\int_{-\infty}^{+\infty}d\sigma_m \cdot$$
$$exp[-i\vec{\rho}\vec{F} - i\sum_{m=1}^{6}\sigma_m\left(\frac{\partial F_\alpha}{\partial x_\beta}\right)_m] \cdot A(\vec{\rho};\{\sigma_m\}). \qquad (5)$$

In (5) $\{\sigma_m\}$ designates the set of the six Fourier variables answering to the corresponding components of the microfield nonuniformity tensor, while $A(\vec{\rho};\{\sigma_m\})$ stands for the characteristic function of the joint distribution. Equation (5) is derived assuming the additivity of the electric field and their spatial derivatives from the separate particles. Thus forming the summary many-body electric field and its tensor of the nonuniformity one obtains

$$\vec{F} = \sum_j \vec{F}_j, \qquad \frac{\partial F_\alpha}{\partial x_\beta} = \sum_j \frac{\partial (\vec{F}_j)_\alpha}{\partial x_\beta}. \qquad (6)$$

The characteristic function from (5) could be expressed in the form

$$A(\vec{\rho};\{\sigma_m\}) = exp[-N \cdot C(\vec{\rho};\{\sigma_m\})]. \qquad (7)$$

Using the generalization of the Baranger-Mozer cluster expansion approach[37-39] one can obtain

$$C(\vec{\rho};\{\sigma_m\}) = C^{(o)}(\vec{\rho};\{\sigma_m\}) - \frac{N}{2!} \cdot C^{(1)}(\vec{\rho};\{\sigma_m\}), \qquad (8)$$

$$C^{(o)}(\vec{\rho};\{\sigma_m\}) = \sum_s C_s \int d^3\vec{r} \cdot g_{sr}(\vec{r}) \cdot f_s(\vec{\rho};\vec{r};\{\sigma_m\}), \qquad (9)$$

$$C^{(1)}(\vec{\rho};\{\sigma_m\}) = \sum_{s,s'} C_s C_{s'} \int d^3\vec{r_1}\int d^3\vec{r_2} \cdot f_s(\vec{\rho};\vec{r_1};\{\sigma\}) \cdot f_{s'}(\vec{\rho};\vec{r_2};\{\sigma_m\}) \cdot$$
$$[g_{ss'}(\vec{r_1};\vec{r_2}) - g_{sr}(\vec{r_1}) \cdot g_{s'r}(\vec{r_2})], \qquad (10)$$

$$f_s(\vec{\rho};\vec{r};\{\sigma_m\}) = 1 - exp[i\varphi_s(\vec{\rho};\vec{r};\{\sigma_m\})], \qquad (11)$$

$$\varphi_s(\vec{\rho};\vec{r};\{\sigma_m\}) = \vec{\rho}\vec{E}_s(\vec{r}) + \sum_{m=1}^{6}\sigma_m\left(\frac{\partial(\vec{E}_s)_\alpha}{\partial r_\beta}\right)_m. \qquad (12)$$

Here $C_s \equiv N_s/N$, $g_{sr}(\vec{r})$ is the pair-correlation function between the perturber with the charge Z_s and the radiator with the net charge Z_o, placed at the origin of the coordinate system, $g_{ss'}(\vec{r_1}; \vec{r_2})$ is the pair-correlation function between perturbers with the charges Z_s and $Z_{s'}$ *in the field of the radiator with the charge Z_o*, $\vec{E}_s(\vec{r})$ - is the elementary electric field, provided by any single ion (quasiparticle) of the sort "s" in the origin of the coordinate system. This field is defined by the effective interaction potential for this sort of particles in plasmas and can be expressed through the following equations

$$\vec{E}_s(\vec{r}) = -eZ_s \frac{\vec{r}}{r^3} \cdot [1 - \kappa_s(r)], \qquad (13)$$

$$div \vec{E}_s(\vec{r}) = \frac{eZ_s}{r^2} \cdot \frac{\partial \kappa_s(r)}{\partial r} - 4\pi e Z_s \delta(\vec{r}),$$

$$\oint_{V \to \infty} d^3 \vec{r} \cdot div \vec{E}_s(\vec{r}) = 0. \qquad (14)$$

The result on the last line is true due to the features of the screening function $\kappa_s(r)$, that come from its definition : $\kappa_s(0) = 0, \kappa_s(\infty) = 1$. Thus the excess free electron density in the pileup near the ion Z_s is

$$\delta n_e^{(s)}(r) = \frac{1}{4\pi} \frac{Z_s}{r^2} \cdot \frac{\partial \kappa_s(r)}{\partial r}. \qquad (15)$$

The elementary ion field nonuniformity tensor components are defined then as following

$$\Phi_{ki}^{(s)}(\vec{r}) \equiv \frac{\partial (\vec{E}_s)_k}{\partial x_i} = \frac{eZ_s}{r^5} \cdot [3x_i x_k - \delta_{ik} r^2] \cdot \left[1 - \kappa_s(r) + \frac{r}{3} \frac{\partial \kappa_s(r)}{\partial r}\right] +$$
$$+ \frac{\delta_{ik}}{3} \frac{eZ_s}{r^2} \frac{\partial \kappa_s(r)}{\partial r}, \qquad (16)$$

As it follows from (13)-(16) the screening function $\kappa_s(r) \geq 0$ and could be determined, for example, in the frames of one of the recent developments of DFT- approach[58], which has the more direct connection with the spectral features, that might be really observed, while $\Phi_{ki}^{(s)}(\vec{r}) \equiv \Phi_{ik}^{(s)}(\vec{r})$. Here we will not detalize the equations depicting the connection with the distribution of the bound electron density, presuming that perturbers are the purely bare ions. Also we assume that quantum effects in the microfield distribution could be neglected(s.[52-54]). The joint distribution (5)-(16), obtained above, gives the "instant" distribution function of the low-frequency (ion) individual component of the plasma microfield and its spatial(or time) derivatives, which in fact are determined on the time scale τ of the order $\omega_{pe}^{-1} \ll \tau \ll (v_i N_i^{1/3})^{-1}$, where ω_{pe}-is the electron plasma frequency, v_i-is the relative ion "thermal" velocity with respect to the radiator, N_i-is the ion density. The basic ideas that support this derivation were put forward by M.Baranger and B.Mozer and are unchanged since that, inspite of some differences of the later approaches[41,47-57], because they are inherent for the plasma microfield formalism. It should once more underlined, that due to (2), (14)-(16) the effects of the plasma polarization (or in other words the appearance of the nonuniformity in the distribution of plasma electron density) are included properly in this treatment. The convolution of (5) over $\vec{F}$ or $\frac{\partial F_i}{\partial x_k}$ components leads to the separate distributions of the field or its nonuniformity tensor and after the appropriate appproximations recovers the previous results on the subject.

One of the most interesting properties of the joint distributions comes from the consideration of the constraint moments $<\frac{\partial F_i}{\partial x_k}>_{\vec{F}}$ that are the averages of $\frac{\partial F_i}{\partial x_k}$ over the joint distribution (5) at the fixed vectorial value $\vec{F}$

$$W(\vec{F})\left\langle\frac{\partial F_i}{\partial x_k}\right\rangle_{\vec{F}} = \frac{N}{(2\pi)^3}\int d^3\vec{\rho}\cdot exp\left[-i\vec{\rho}\vec{F}\right]\cdot A(\vec{\rho})\cdot\langle\Phi_{ik}(\vec{\rho})\rangle, \quad (17)$$

$$\langle\Phi_{ik}(\vec{\rho})\rangle = \left\langle\Phi_{ik}^{(o)}(\vec{\rho})\right\rangle - \left\langle\Phi_{ik}^{(1)}(\vec{\rho})\right\rangle, \quad (18)$$

$$\left\langle\Phi_{ik}^{(o)}(\vec{\rho})\right\rangle = \sum_s C_s \left\langle\Phi_{ik}^{(s)}(\vec{\rho})\right\rangle, \left\langle\Phi_{ik}^{(1)}(\vec{\rho})\right\rangle = \frac{N}{2}\sum_{s,s'} C_s C_{s'} \left\langle\Phi_{ik}^{(ss')}(\vec{\rho})\right\rangle, \quad (19)$$

$$\left\langle\Phi_{ik}^{(s)}(\vec{\rho})\right\rangle = \int d^3\vec{r}\cdot g_{sr}(\vec{r})\cdot exp\left[i\varphi_s(\vec{\rho};\vec{r})\right]\cdot\Phi_{ik}^{(s)}(\vec{r}), \quad (20)$$

$$\left\langle\Phi_{ik}^{(ss')}(\vec{\rho})\right\rangle = \int d^3\vec{r_1}\int d^3\vec{r_2}\left[g_{ss'}(\vec{r_1};\vec{r_2}) - g_{sr}(\vec{r_1})\cdot g_{s'r}(\vec{r_2})\right]\cdot\{\Phi_{ik}^{(s)}(\vec{r_1})\cdot$$
$$exp\left[i\varphi_s(\vec{\rho};\vec{r_1})\right]\cdot(1 - exp\left[i\varphi_{s'}(\vec{\rho};\vec{r_2})\right]) + \Phi_{ik}^{(s')}(\vec{r_2})\cdot exp\left[i\varphi_{s'}(\vec{\rho};\vec{r_2})\right]\cdot$$
$$(1 - exp\left[i\varphi_s(\vec{\rho};\vec{r_1})\right])\}. \quad (21)$$

The expressions for the first moments may be rewritten from (17)-(21) in terms of the microfield distribution function, that may be useful for the general analysis

$$W(\vec{F})\left\langle\frac{\partial F_i}{\partial x_k}\right\rangle_{\vec{F}} = N[\sum_s C_s \int d^3\vec{r}\cdot g_{sr}(\vec{r})\cdot\Phi_{ik}^{(s)}(\vec{r})\cdot W\left(\vec{F} - \vec{E}_s(\vec{r})\right) +$$
$$-\frac{N}{2}\sum_{ss'} C_s C_{s'}\int d^3\vec{r_1}\int d^3\vec{r_2}\cdot[g_{ss'}(\vec{r_1};\vec{r_2}) - g_{sr}(\vec{r_1})\cdot g_{s'r}(\vec{r_2})]\cdot$$
$$\{\Phi_{ik}^{(s)}(\vec{r_1})\cdot\left[W\left(\vec{F} - \vec{E}_s(\vec{r_1})\right) - W\left(\vec{F} - \vec{E}_s(\vec{r_1}) - \vec{E}_{s'}(\vec{r_2})\right)\right] +$$
$$\Phi_{ik}^{(s')}(\vec{r_2})\cdot[W\left(\vec{F} - \vec{E}_{s'}(\vec{r_2})\right) - W\left(\vec{F} - \vec{E}_s(\vec{r_1}) - \vec{E}_{s'}(\vec{r_2})\right)]\}. \quad (22)$$

The results (1)-(22) have the most general form. In order to obtain the more detailed expressions one ought to make reasonable approximations concerning the correlation functions. We here assume that the pair correlation functions depend only on the module of the difference radius vector in their arguments, and shall use the Kirkwood approximation for the disentangling of the triple correlations, that is[62]

$$g_{sr}(\vec{r}) \equiv g_{sr}(r), \quad g_{ss'}(\vec{r_1}; \vec{r_2}) \simeq g_{ss'}(|\vec{r_1} - \vec{r_2}|) \cdot g_{sr}(r_1) \cdot g_{s'r}(r_2), \quad (23)$$

$$h_{ss'}(|\vec{r_1} - \vec{r_2}|) \equiv g_{ss'}(|\vec{r_1} - \vec{r_2}|) - 1. \quad (24)$$

Then one can find the following general expressions for (24)

$$h_{ss'}(|\vec{r_1} - \vec{r_2}|) = \sum_{n=0}^{\infty}(2n+1) \cdot P_n(\cos[\widehat{\vec{r_1}\vec{r_2}}]) \cdot h_{ss'}(n; r_1; r_2), \quad (25)$$

$$h_{ss'}(n; r_1; r_2) = \int_0^{\infty} dk \cdot k^2 \cdot j_n(kr_1) \cdot j_n(kr_2) \cdot h_{ss'}(k), \quad (26)$$

$$h_{ss'}(k) = \frac{1}{(2\pi)^3} \int d^3\vec{r} \cdot \exp(i\vec{k}\vec{r}) \cdot h_{ss'}(r). \quad (27)$$

Here $P_n(z)$-are the Legendre polynomials with the cosine of the angle between $\vec{r_1}$ and $\vec{r_2}$ as the argument, while $j_n(y)$-are the spherical Bessel functions. Eq.(26)-(27) enable to simplify general results (1)-(22) and obtain, for example, for the microfield distribution of the reduced microfield value $\beta \equiv F/F_o$, F_o being the normal field[31], the more general formulas than were known before

$$W(\vec{F}) = 4\pi F^2 \cdot W(F), \quad A(\vec{\rho}) = A(\rho), \quad F_o = \Lambda e N^{2/3}, \quad \Lambda \equiv 2\pi(4/15)^{2/3}, \quad (28)$$

$$W(\beta) = \frac{2\beta}{\pi} \int_0^{\infty} dk \cdot k \cdot \sin k\beta \cdot A(k), \quad A(k) = exp\{-[\Psi_o(k) - \Psi_1(k)]\}, \quad (29)$$

$$\Psi_o(k) = \frac{4\pi}{\Lambda^{3/2}} \sum_s C_s I_s(k), \quad \Psi_1(k) = \frac{8\pi^2}{\Lambda^3} \sum_{ss'} C_s C_{s'} I_{ss'}(k), \quad r_o \equiv (e/F_o)^{1/2}, \quad (30)$$

$$I_s(k) = \int_0^{\infty} dx \cdot x^2 \cdot g_{sr}(r_o x) \cdot \left\{1 - \frac{\sin k\epsilon_s(x)}{k\epsilon_s(x)}\right\}, \quad \epsilon_s(x) = \frac{Z_s}{x^2}[1 - \kappa_s(r_o x)], \quad (31)$$

$$I_{ss'}(k) = \int_0^{\infty} dx_1 \cdot x_1^2 \int_0^{x_1} dx_2 \cdot x_2^2 \cdot g_{sr}(r_o x_1) \cdot g_{s'r}(r_o x_2) \cdot \sum_{n=0}^{\infty}(-1)^n(2n+1)$$
$$\cdot \{j_n[\epsilon_s(x_1)] - \delta_{on}\}\{j_n[\epsilon_{s'}(x_2)] - \delta_{on}\} h_{ss'}(n; r_o x_1; r_o x_2). \quad (32)$$

Now we can derive the expressions for the constraint moments

$$\left\langle \frac{\partial F_X}{\partial X} \right\rangle_{\vec{F}} = -\frac{2\pi Ne}{3}\left\{B_D(\beta)\left[P_2(\cos\theta) - P_2^{|2|}(\cos\theta)\frac{\cos 2\phi}{2}\right] - 2B_{D0}(\beta)\right\}, \quad (33)$$

$$\left\langle \frac{\partial F_Y}{\partial Y} \right\rangle_{\vec{F}} = -\frac{2\pi Ne}{3}\left\{B_D(\beta)\left[P_2(\cos\theta) + P_2^{|2|}(\cos\theta)\frac{\cos 2\phi}{2}\right] - 2B_{D0}(\beta)\right\}, \quad (34)$$

$$\left\langle \frac{\partial F_Z}{\partial Z} \right\rangle_{\vec{F}} = \frac{4\pi Ne}{3}\{B_D(\beta) \cdot P_2(\cos\theta) + B_{D0}(\beta)\}, \quad (35)$$

$$\left\langle \frac{\partial F_Y}{\partial X} \right\rangle_{\vec{F}} = \frac{\pi Ne}{3} \cdot B_D(\beta) \cdot P_2^{|2|}(\cos\theta) \cdot \frac{\sin 2\phi}{2}, \qquad (36)$$

$$\left\langle \frac{\partial F_Z}{\partial X} \right\rangle_{\vec{F}} = \frac{2\pi Ne}{3} \cdot B_D(\beta) \cdot P_2^{|1|}(\cos\theta) \cdot \cos\phi, \qquad (37)$$

$$\left\langle \frac{\partial F_Z}{\partial Y} \right\rangle_{\vec{F}} = \frac{2\pi Ne}{3} \cdot B_D(\beta) \cdot P_2^{|1|}(\cos\theta) \cdot \sin\phi, \qquad (38)$$

where θ and ϕ are the polar and azimuthal angles of $\vec{F}$ in the laboratory system XYZ, $P_n^{|m|}(x)$-are the generalized Legendre polynomials, while the *universal functions* $B_D(\beta)$ and $B_{DO}(\beta)$ are defined through

$$B_D(\beta) = B_D^{(o)}(\beta) - B_D^{(1)}(\beta), \quad B_{DO}(\beta) = B_{DO}^{(o)}(\beta) - B_{DO}^{(1)}(\beta), \qquad (39)$$

$$B_D^{(o)}(\beta) = \frac{12}{\pi} \frac{\beta^2}{W(\beta)} \sum_s C_s Z_s \, b_s(\beta), \quad B_D^{(1)}(\beta) = \frac{12}{\pi} \frac{\beta^2}{W(\beta)} \sum_{ss'} C_s C_{s'} \, b_{ss'}(\beta), \qquad (40)$$

$$B_{DO}^{(o)}(\beta) = \frac{2}{\pi} \frac{\beta^2}{W(\beta)} \sum_s C_s Z_s \, b_s^{(o)}(\beta), \quad B_{DO}^{(1)}(\beta) = \frac{2}{\pi} \frac{\beta^2}{W(\beta)} \sum_{ss'} C_s C_{s'} \, b_{ss'}^{(o)}(\beta), \qquad (41)$$

$$b_s(\beta) = \int_0^\infty dk \cdot k^2 \cdot A(k) \cdot j_2(k\beta) \cdot \Phi_s(k), \qquad (42)$$

$$b_s^{(o)}(\beta) = \int_0^\infty dk \cdot k^2 \cdot A(k) \cdot j_0(k\beta) \cdot \Phi_s^{(o)}(k), \qquad (43)$$

$$\Phi_s(k) = \int_0^\infty dx \cdot x^2 \cdot g_{sr}(r_o x) \cdot j_2[k\epsilon_s(x)] \cdot \Phi_s(x), \qquad (44)$$

$$\Phi_s^{(o)}(k) = 4\pi \int_0^\infty dx \cdot x^2 \cdot g_{sr}(r_o x) \cdot j_0[k\epsilon_s(x)] \cdot \{r_o^3 \delta n_e^{(s)}(r_o x)/Z_s\}, \qquad (45)$$

$$\Phi_s(x) \equiv \frac{1}{x^3}\left[1 - \kappa_s(r_o x) + \frac{x}{3}\frac{\partial \kappa_s(r_o x)}{\partial x}\right], \qquad (46)$$

$$b_{ss'}(\beta) = \int_0^\infty dk \cdot k^2 \cdot A(k) \cdot j_2(k\beta) \cdot b_{ss'}(k), \qquad (47)$$

$$b_{ss'}^{(o)}(\beta) = \int_0^\infty dk \cdot k^2 \cdot A(k) \cdot j_0(k\beta) \cdot b_{ss'}^{(o)}(k), \qquad (48)$$

$$b_{ss'}(k) = \int_0^\infty dx_1 \cdot x_1^2 \int_0^{x_1} dx_2 \cdot x_2^2 \cdot g_{sr}(r_o x_1) \cdot g_{s'r}(r_o x_2) \cdot b_{ss'}(k; x_1; x_2), \qquad (49)$$

$$b_{ss'}(k; x_1; x_2) = Z_s \cdot \Phi_s(x_1) \cdot \{ j_2[k\epsilon_s(x_1)] \cdot h_{ss'}(o; r_o x_1; r_o x_2) -$$
$$-\sum_{n=0}^\infty (-1)^n (2n+1)\left[\left(\frac{3n(n-1)}{2k^2\epsilon_s^2(x_1)} - 1\right) j_n[k\epsilon_s(x_1)] + \frac{3}{k\epsilon_s(x_1)} j_{n+1}[k\epsilon_s(x_1)]\right]$$
$$\cdot j_n[k\epsilon_{s'}(x_2)] \cdot h_{ss'}(n; r_o x_1; r_o x_2) \}, \qquad (50)$$

$$b_{ss'}^{(o)}(k) = \int_0^\infty dx_1 \cdot x_1^2 \int_0^{x_1} dx_2 \cdot x_2^2 \cdot g_{sr}(r_o x_1) \cdot g_{s'r}(r_o x_2) \cdot b_{ss'}^{(o)}(k; x_1; x_2), \qquad (51)$$

$$b_{ss'}^{(o)}(k; x_1; x_2) = 4\pi \cdot r_o^3 \delta n_e(r_o x_1) \{ j_0[k\epsilon_s(x_1)] \cdot h_{ss'}(o; r_o x_1; r_o x_2) -$$
$$-\sum_{n=0}^\infty (-1)^n (2n+1) \cdot j_n[k\epsilon_s(x_1)] \cdot j_n[k\epsilon_{s'}(x_2)] \cdot h_{ss'}(n; r_o x_1; r_o x_2) \}. \qquad (52)$$

In the end of this section it is instructive to show how the previous results for the Debye-Hückel approximation with the linearization of the perturber-perturber correlation function[37,38,39] could be obtained from the general expressions of this section. This may be achieved by the following substitutions

$$\kappa_s(r_o x) \to \kappa_s^D(x) \equiv 1 - \exp[-ax] \cdot (1 + ax), \quad a \equiv \frac{r_o}{r_D}, \quad r_D \equiv \sqrt{\frac{T_e}{4\pi e^2 N_e}}, \quad (53)$$

$$g_{sr}(r_o x) \to exp[-Z_o Z_S \cdot \Theta \cdot a^2 \cdot \frac{\Lambda^{3/2}}{4\pi} \cdot \frac{\exp[-ax]}{x}], \quad \Theta \equiv \frac{T_e}{T_i}, \quad (54)$$

$$h(n; x_1; x_2) \to -\Theta \cdot a^3 \cdot \frac{\Lambda^{3/2}}{4\pi} \cdot f_n^>(ax_1) \cdot f_n^<(ax_2), \quad (55)$$

$$f_n^>(z) \equiv (-1)^n \cdot z^n \cdot \left(\frac{d}{zdz}\right)^n \frac{e^{-z}}{z}, \quad f_n^<(z) \equiv z^n \cdot \left(\frac{d}{zdz}\right)^n \frac{\sinh(z)}{z}, \quad (56)$$

where in (56) we used the traditional designations from[48], and T_i, T_e-are the ion and the electron temperatures in plasmas accordingly.

JOINT DISTRIBUTIONS IN APEX

To our knowledge the idea of the using APEX to treat the microfield nonuniformity in strongly coupled plasmas was firstly published in[42] and secondly later, but independently in[43,45]. The detailed version of the work [42] now also is available[46]. We prefer to follow here the more general approach from the papers[43,45] and naturally original works on the APEX approach[56-57]. The APEX field distribution for the microfield is the distribution with the specified constraints on the second moment of F with the modified Debye screening of the elementary electric field. In the present case under the construction of the joint distribution function *it is necessary to use the expression for the elementary nonuniformity tensor, that does not related directly to the APEX field expression, but in principle should be obtained from the initial field expression, which comes from the physical consideration-(13) and fulfill the equation (15)*[43,45]. In other words in order to obtain the truly joint distribution one should have a freedom to convolve over the part of the independent variables in order to obtain the independent disribution over the other part of also the independent variables entering in the definition of the joint distribution. Thus the generalization of the APEX in our notations is obtained by the following substitutions in formulas of the preceding section

$$C^{(1)}(\vec{\rho}) \equiv 0, \quad k_s^{APEX}(r) = 1 - exp[-a\alpha_s r](1 + a\alpha_s r), \quad (57)$$

$$g_{sr}^{APEX}(r) = g_{sr}(r) \cdot \frac{E_s(r)}{E_s^{APEX}(r)}, \quad (58)$$

$$\varphi_s^{APEX}(\vec{\rho}; \vec{r}; \{\sigma_m\}) = \vec{\rho} \vec{E}_s^{APEX}(\vec{r}) + \sum_{m=1}^{6} \sigma_m \left(\frac{\partial (\vec{E}_s)_i}{\partial x_k}\right)_m, \quad (59)$$

where α_s is the fitting APEX parameter for the "s"-sort of perturber ions[56]. On the other hand the virtue of APEX approximation, which is formally confined in the substitution

$$g_{sr}(r) \cdot exp[i\vec{\rho}\vec{F}_s(\vec{r})] \to g_{sr}(r) \frac{E_s(r)}{E_s^{APEX}(r)} \cdot exp[i\vec{\rho}\vec{E}_s^{APEX}(r')], \quad (60)$$

inevitably renormalize the pair distributions for the other variables. This is the price to be paid for the restricting the space of the independent variables by the conditions at $Z_o \neq 0$

$$\langle F^2 \rangle = \int dF \cdot W(F) \cdot F^2 = \frac{4\pi N T_i}{Z_o} \sum_s Z_s C_s T_s, \quad (61)$$

$$T_s = \int_0^\infty dx \cdot g_{sr}(r_o x) \cdot \frac{\partial \kappa_s(r_o x)}{\partial x}, \quad a \equiv r_o/r_D, r_D \equiv \sqrt{\frac{T_e}{4\pi e^2 N_e}}, \quad (62)$$

$$\sum_s C_s \int_0^\infty dr \cdot r^2 \cdot g_{sr}(r) \cdot E_s(r) \cdot E_s^{APEX}(r) = \frac{T_i}{Z_o} \sum_s Z_s C_s T_s, \quad (63)$$

$$T_s = \frac{\Lambda^{3/2}}{4\pi} \Theta a^2 Z_o Z_s \int_0^\infty dx \cdot \frac{g_{sr}(r_o x)}{x^2} \cdot (1 - \kappa_s(r_o x)) \cdot (1 + a\alpha_s x) \cdot$$
$$\cdot exp(-a\alpha_s x), \quad (64)$$

where $\{\alpha_s\}$ is the set of parameters to be determined from (61)-(64).

It may be easily shown that the results of the preceeding section(17)-(22), (28)-(31), (33)-(38), (39)-(40), (42)-(46) may be reexpressed in terms of the modified APEX approximation, using (57)-(64). On the other hand they differ from them by the absence of the correlation terms, that stem from $C^{(1)}(\vec{\rho})$, and by the renormalized perturber-radiator pair disribution function (58),(60). Thus the results outlined in this section are quite sufficient to exploit the APEX approach together with the results of the preceding one, so we will not present here lengthy formulas that are now evident from our consideration. In the distinction with the previous works on the subject [42,45] our consideration includes the question of the ion mixtures and the construction of the joint distributions in the generalized APEX scheme. If we take the case of the single sort s_o of the ion perturbers and rewrite the expression for the first moment with account of (17)-(22) and (57)-(64) we obtain

$$W(\vec{F}) \left\langle \frac{\partial F_i}{\partial x_k} \right\rangle_{\vec{F}} = \int d^3\vec{r} \cdot \Phi_{ik}^{(s_o)}(\vec{r}) \cdot g_{s_o r}(r) \cdot \frac{E_{s_o}(r)}{E_{s_o}^{APEX}(r)} \cdot W(\vec{F} - \vec{E}_{s_o}^{APEX}(\vec{r})). \quad (65)$$

The similar result in[42,45] is the key one for these works, but in the distinction to (65) there is no renormalization of $g_{sr}(r)$! Moreover there are no results in[42,45], depicting the plasma polarization terms, and the elementary nonuniformity tensor has the artificially truncated expression, that is, in our notations - the last term in (46) is omitted by these authors. Concerning the nearest neighbour limit (NNA) from (65), it could be easily seen, that the renormalization factor goes to unity at small r, and thus the NNA limit should be recovered. On the other hand it is not so simple to make a statement what way is the more perfect in this "game" with the APEX, because from the beginning this was the *ad hoc*[56,60,61] approximation and the laws of logic may fail here. It could be only said, that the presented here theory is more correct according with the Markoff technique construction of the joint distributions[63] using consistently the APEX ideas. The case of the neutral radiator may be treated in the slightly different manner according with original works[61].

As it was stated by APEX's authors[61] its applicability was checked up to $\Gamma_i \sim 100$. On the other hand from the general considerations for more large Γ_i one should get the Gauss type function for the microfield distribution function, corresponding to the physical picture of the very low kinetic energy of particles compared to their potential energy and connected with their oscillations near the sites of the equilibrium. In this case the so simple approximation as NNA fails, and one should apply ideas of the several nearest neighbours, that are so familiar for the theory of the inter-crystal electric field. It is interesting, that when Γ_i is increasing, the plasma ion frequency is for any real value of Z_s less than ω_{pe}, but may become larger than $(v_i N_i^{1/3})^{-1}$, while $r_{Di} \ll N_i^{-1/3}$ and $r_{Di} \ll r_D \equiv r_{De}$. Thus if in the region of small Γ_i plasma ion modes are slower than the individual particle's motion, at large $\Gamma_i \gg 1$ this relation becomes vice versa.

QUANTUM PROBLEM OF LINE CONTOUR

In the outlined approximations the line contour between the upper set of sublevels $\{|\alpha><\alpha'|\}$ and the lower set of sublevels $\{|\beta><\beta'|\}$ can be written now as follows

$$I(\omega) = \frac{1}{\pi} \Re \int d^3\vec{F} \cdot \prod_{m=1}^{6} \int d\left\{\frac{\partial F_i}{\partial x_k}\right\}_m \cdot W(\vec{F}; \left\{\frac{\partial F_i}{\partial x_k}\right\}) \cdot$$
$$\rho_{\alpha'\alpha} \cdot <\alpha|\vec{d}|\beta> \cdot <\beta'|\vec{d}|\alpha''> \cdot$$
$$<<\alpha'',\beta'^\dagger|\left[i(\Delta\omega - \hat{H}_o - \hat{V}_{oi}) - \hat{\mathcal{M}}(\Delta\omega)\right]^{-1}|\alpha',\beta^\dagger>>, \qquad (66)$$

where $\rho_{\alpha'\alpha}$-is the matrix element of the density matrix between the upper sublevels, $\mathcal{M}(\Delta\omega)$ is the electron broadening operator [17,18,19,20], $\Delta\omega \equiv \omega - \omega_o$, ω_o-is the unperturbed frequency of the radiative transition $\{\alpha,\alpha'\} \to \{\beta,\beta'\}$. For the present consideration we disregard Nonlinear Interference Effects[64], and thus assume traditionally, that $\hat{\rho}$ is the unitary operator in the subspace of the upper level. Here we also would not touch the complicated effects of the line dissolution[20,34−35,46]. If one could calculate the multidimensional distribution function in an every point of the 9-dimensional phase space, and then invert the resolvent operator for the chosen values of $\Delta\omega$, the problem would be completely solved for any complete orthonormal system of wave functions. However the complete definition of (5) would need a lot of computational time and memory, not talking about the delivering the integrations. However it is possible to generate on the computer the joint distributions in the truncated phase space. The example of such a type of the joint distributions is presented by authors with Dr.D.Gilles and B.C.Huynh in this volume[65]. It should be stressed here, that the Markoff-type formalism[63] could provide the joint distribution functions that as a rule are *completely determined only in the laboratory system*[37]. On the other hand the comprehension that in the binary case the simple relation exists between the field value and its derivative over the distance between particles ($\frac{\partial F_z}{\partial z}$-component) was the ground to use the approximation, which consists in the substitution to the resolvent the constraint moments of the nonuniformity tensor[36,41,42,46]. The cost of these approximations, which become necessary due to the complicated energetic structure of the nonhydrogenlike radiators and hydrogenlike ones also, when the fine structure or the off-diagonal matrix elements of $\mathcal{M}(\Delta\omega)$ are considered, is unknown up to now. However the experience in the solution of this problem for the purely Stark pattern of the level splitting shows, that this is a questionable approach. Usually the influence of the microfield nonuniformity is bound with the Stark effect by itself, which exibits the complicated picture due to the degeneracy of the quantum states in the several regions of the F-value variation[66]. That is why the diagonalization procedure needs the special treatment in order not to loose the important contributions from the Hilbert space eigenvectors (wave functions) and to distort the result. For example, the importance of the contribution from the quadratic Stark effect instead of the numerous various calculations[27,36,67] can not be considered as clear enough till now. It is important to point here, that the contribution from the quadrupole terms in the line profile have the complicated *quantum interference character*, that is *depends on the relative phases of the matrix elements of the radiator interaction with the radiation field from one side and the microfields from the other*, thus demonstrating one of the rather unique cases in the quantum theory, when observables depend on phases of the wave functions[6,32,41]. The quantum interference effects develop themselves also in the dramatic cancellation of the quadrupole contributions from the various components of the multiplet[6,32,41]. The other important circumstance in this problem is the necessity to treat the electrons on the same grounds as ions. These means in particular the consideration of the similar terms as in (2) in $\hat{V}_{oe}$. Earlier the calculations were made for the case, when the electron velocity distribution is anisotropic and the term in the multipolar expansion, proportional to $\hat{Q}_{\alpha\beta}$, give the nonzero contribution to the electron shift[68-70]. In this work we limit ourselves to the traditionally isotropic electron velocity distributions, and find that there is no contribution to the electron shift in the first order of $\hat{V}_{oe}$ for the neutral radiators in the frames of the nonadiabatic approach to the electron impact broadening with the exclusion of the penetrating collisions in order to coincide in this point with the microfield ideas. But in the adiabatic theory the term proportional to $\hat{Q}_{\alpha\beta}$ will survive, as it was shown in earlier works[7].

GENERALIZED PERTURBATION SOLUTION FOR HYDROGENLIKE SPECTRA

Here we present the main results of the perturbation approach, generalizing to some extent the perturbation results of[6,32,37,41], the major part of which were created in[38,39,44]. The complete set of parabolic wave fuctions is used below[65]. We consider here only the terms of the order $n^2 a_o / Z_r R$, where Z_r-is the charge of the nucleus. The profile without the inclusion of the quadrupole effects is calculated with account of the off-diagonal matrix elements of $\hat{\mathcal{M}}(\Delta\omega)$, while the quadrupolar correction is obtained neglecting these matrix elements in order to be able to perform the integration over the 6-dimensional phase space[32,37]. Moreover we divide the correction terms on two parts: the contribution of the lateral Stark components (L) and the contribution of the central components (C). Thus defining this function for the lateral Stark components as $I_L{}^Q(\Delta\omega)$ we have

$$I_L{}^Q(\Delta\omega) = \frac{1}{\pi I_o} \sum_{l>0} I_l \cdot \gamma_l \cdot \int_0^\infty d\beta \cdot [I_l^{+Q}(\Delta\omega;\beta) + I_l^{-Q}(\Delta\omega;\beta)], \tag{67}$$

$$I_l^{\pm Q}(\Delta\omega;\beta) = I_{lE}^{\pm Q}(\Delta\omega;\beta) + I_{lI}^{\pm Q}(\Delta\omega;\beta) + I_{l\gamma}^{\pm Q}(\Delta\omega;\beta), \tag{68}$$

$$I_{lE}^{\pm Q}(\Delta\omega;\beta) = \pm \frac{q_l \chi_D(\beta) + q_l^o \chi_{DO}(\beta)}{(\Delta\omega \mp C_l\beta)^2 + \gamma_l^2}, \tag{69}$$

$$I_{lI}^{\pm Q}(\Delta\omega;\beta) = \pm \frac{\delta_l \Lambda_D(\beta) + \delta_l^o \Lambda_{DO}(\beta)}{(\Delta\omega \mp C_l\beta)^2 + \gamma_l^2}, \tag{70}$$

$$I_{l\gamma}^{\pm Q}(\Delta\omega;\beta) = \pm \frac{P_l\Lambda_D(\beta) + P_l^o\Lambda_{DO}(\beta)}{(\Delta\omega \mp C_l\beta)^2 + \gamma_l^2}\left[1 - \frac{2\cdot\gamma_l^2}{(\Delta\omega \mp C_l\beta)^2 + \gamma_l^2}\right], \tag{71}$$

where I_o- is the total line intensity, I_l-is the unperturbed intensity of the "l"-th lateral Stark component, and subscripts lE, lI, $l\gamma$ designate corrections in the profile induced by corrections to the frequency, the intensity and the electron width of the "l"-th Stark component. The universal functions $\Lambda_{D,DO}(\beta)$ and $\chi_{D,DO}(\beta)$ are defined as follows[38-39,44]

$$\Lambda_D(\beta) \equiv \frac{W(\beta)\cdot B_D(\beta)}{\beta}, \qquad \chi_D(\beta) \equiv -\frac{d}{d\beta}[W(\beta)\cdot B_D(\beta)], \tag{72}$$

$$\Lambda_{DO}(\beta) \equiv \frac{W(\beta)\cdot B_{DO}(\beta)}{\beta}, \qquad \chi_{DO}(\beta) \equiv -\frac{d}{d\beta}[W(\beta)\cdot B_{DO}(\beta)]. \tag{73}$$

In (67)-(71) γ_l is the electron width of the "l"-th Stark component between parabolic quantum states of the upper $\{n, n_1, n_2, m\}$ and lower $\{n', n'_1, n'_2, m'\}$ levels. In principle it depends on the microfield value[37] as well as on $\Delta\omega$ [22,23]. Here we neglect these dependences and use its simplified form in atomic units[37]

$$\gamma_l = \frac{3\sqrt{\pi}}{Z_r^2} \cdot \frac{e^2}{\hbar <v>} \cdot [Na_o^3] \cdot \mathcal{L} \cdot M_l^\gamma,$$

$$M_l^\gamma \equiv \{n^2[n^2 + (n_1 - n_2)^2 - |m|^2 - 1] +$$
$$+ n'^2[n'^2 + (n'_1 - n'_2)^2 - |m'|^2 - 1] - 4nn'(n_1 - n_2)(n'_1 - n'_2)\}, \quad (74)$$

while the "cut-off" function $\mathcal{L}$ in the straight-path approximation may be written in the the form[37]

$$\mathcal{L} = \ln \frac{\rho_D}{\rho_W} + 0.215, \quad \rho_W = \frac{\hbar n^2}{m_o <v>}, \quad <v> = (2T/m_o)^{1/2}. \quad (75)$$

where $\rho_D \equiv r_D$, ρ_W-is the Weiskopff radius[31,68], v-is thermal velocity of plasma electrons and m_o -is their mass. The other constants from (67)-(71) above could be expressed in the forms (compare with[32,41])

$$C_l = (3/2)(ea_o/\hbar)\Delta_l^d, \quad \Delta_l^d = \frac{1}{Z_r}[n(n_1 - n_2) - n'(n'_1 - n'_2)], \quad (76)$$

$$\delta_l = (2\pi/3)(Nea_o/F_o)\epsilon_l, \quad q_l = (2\pi/3)(Nea_o/F_o)\Delta_l^q/|\Delta_l^d|, \quad (77)$$

$$\delta_l^o = (2\pi/3)(Nea_o/F_o)\epsilon_l^o, \quad q_l^o = (2\pi/3)(Nea_o/F_o)\Delta_l^{oq}/|\Delta_l^d|, \quad (78)$$

$$P_l = (2\pi/3)(Nea_o/F_o)\Delta_l^\gamma, \quad P_l^o = (2\pi/3)(Nea_o/F_o)\Delta_l^{o\gamma}, \quad (79)$$

$$\Delta_l^q \equiv -\frac{1}{3Z_r^2}\{n^2[6(n_1 - n_2)^2 - n^2 + 1] - n'^2[6(n'_1 - n'_2)^2 - n'^2 + 1]\}, \quad (80)$$

$$\Delta_l^{oq} \equiv -\frac{1}{3Z_r^2}\{n^2[2n^2 + 3n + 3(n-1)(n_1 + n_2) - 6n_1n_2 + 1] -$$
$$- n'^2[n'^2 + 3n' + 3(n'-1)(n'_1 + n'_2) - 6n'_1n'_2 + 1]\}, \quad (81)$$

$$\Delta_l^\gamma \equiv \frac{1}{Z_r}\{n^3(n_1 - n_2)[(n-1)(|m|+1) + 2n_1n_2] +$$
$$+ n'^3(n'_1 - n'_2)[(n'-1)(|m'|+1) + 2n'_1n'_2]\} \cdot (M_l^\gamma/2)^{-1}, \quad (82)$$

$$\Delta_l^{o\gamma} = -\Delta_l^\gamma, \quad \epsilon_l \equiv \epsilon_l(n) + \epsilon_l(n'), \quad \epsilon_l = -\epsilon_l^o,$$

$$\epsilon_l(n) \equiv \frac{1}{2Z_r}[n[n_1(n-n_1)(n_2+1)(n-n_2-1)]^{1/2}a_{l1} -$$
$$- n[n_2(n-n_2)(n_1+1)(n-n_1-1)]^{1/2}a_{l2}], \quad (83)$$

$$a_{l1,l2} \equiv 2\frac{<n_1 \mp 1, n_2 \pm 1, m|\vec{d}|n'_1n'_2m'>(<n'_1n'_2m'|\vec{d}|n_1n_2m>)*}{|<n_1n_2m|\vec{d}|n'_1n'_2m'>|^2}, \quad (84)$$

$$a'_{l1,l2} \equiv 2\frac{<n'_1 \mp 1, n'_2 \pm 1, m'|\vec{d}|n'_1n'_2m'>(<n_1n_2m|\vec{d}|n'_1n'_2m'>)*}{|<n_1n_2m|\vec{d}|n'_1n'_2m'>|^2}. \quad (85)$$

The expression for $\epsilon_l(n')$ may be obtained from (83) by putting "'" over each symbol in (83). The case of the unshifted Stark components is more complicated, because the dipole term vanishes for these quantum states. On the other hand it is unknown how to construct the distribution function of the microfield nonuniformity tensor in the coordinate system with $\vec{OZ}||\vec{F}$ in the Markoff integral technique[63], because the two sided Fourier transform failes down due to the absence for the component $\frac{\partial F_z}{\partial z}$ the limit $-\infty$. Moreover it is impossible to convolve analytically the joint distribution function over the module of $\vec{F}$, preserving the memory of its direction that is necessary, because this is the direction of the quantization determining the existence of the unshifted Stark components by itself[37]. Therefore in this case up to now one should use the following three additional approximations. So one way is to use the function of the mean value instead of the mean of the function. This is the substitution in the resolvent the first constraint moments (33)-(38). Another way is to use the expansion over the ratio of the energy of the "quadrupole"(more correctly the two last terms in the right side of Eq.(2)) interaction $<\psi_l|\hat{U}_{oi}|\psi_l>$ to $(\Delta\omega^2+\gamma_l^2)$, because in the adiabatic theory[31] one could get the following estimate

$$\gamma_l/<n_1n_2m|\hat{U}_{oi}|n_1n_2m> \approx \frac{1}{Z_p}\frac{e^2}{\hbar v_e}, \qquad (86)$$

which is more than unity for the parameters of the dense arc plasmas with temperature around 1 eV and the charge of the ion perturbers $Z_p = 1$. Approximately this estimate also is valid for hydrogenlike ions. And the last approximation is the use of the nearest neighbour approximation[32]. We shall not concentrate on these results, but only point out, that Eq.(80),(81), multiplied by (-3) and (-3/2) give the diagonal matrix elements of Q_{zz} and r^2, that enter to the expression of the correction of the frequency of these components, while (82)-(85) are valid for the intensity correction. As it follows from (78) correction to the electron width vanishes for these states due to the equality $n_1 = n_2$, as it was known earlier[32,37]. Except for the case when $n_1 = n_2 = 0$, the correction to the *intensity* for the unshifted Stark levels is not equal to zero and we face with *the appearance of the forbidden central components induced by the quadrupole interaction, that may appear for lines, that initially have no central components at all, and thus may lead to the shift of the dip of these lines.* The appearance of the forbidden central components induced by the quadrupole interaction follows naturally from the equations (83)-(85) and was contained even in the general results of Sholin's work[6]. This fact was explicitly noticed by V.Rozhkov in the beginning of seventieth and used by Bacon[27] in Ly-β calculations.[0] Due to the absence of the zero order dipole matrix elements of the radiative transition in this case the contribution from

---

[0]The invited speaker (A.V.D.) is grateful to E.A.Oks for the pointing out the importance of this feature.

these forbidden components should be included in the next second order of the parameter $n^2 a_o/R$ simultaneously with the quadratic Stark effect, while the octupole and second order quadrupole terms do not lead to the asymmetry of the line due to the matrix elements proportionality[6] to $(n_1 - n_2)$. It also should be pointed out here, that the results above show that the terms in the interaction potential of the plasma ions with the radiator induced by the plasma polarization effects and proportional to $\hat{r}^2$ lead to the similar but different asymmetry features as compared with the "pure" quadrupole interaction[44]. As it could be seen from (67)-(85) the finiteness of the plasma coupling parameter (or $a > 0$) leads from one hand to the appearance of the new polarization corrections to the intensity, frequency and the electron impact width, and from the other one to the dependence of these corrections through the universal functions (72)-(73) on a. It is seen from (80)-(85), that polarization terms contribute to the corrections to the frequency with the same sign as the "pure" quadrupole terms and with the opposite to them sign to the corrections to the intensity and the electron impact width. It is obvious, that the expressions for the asymmetry in the line wings[32] also changes. The evolution of the universal functions $W(\beta), B_D(\beta), \Lambda_D(\beta), \chi_D(\beta)$ versus β and the parameter a for the neutral $Z = 0$ and charged points $Z = 1$, that are obtained according with (28)-(32), (39)-(52), (53)-(56), (72)-(73), is illustrated by graphs 1-8 in the end of this article. The main trends consist in the increasing of the functions values in the maxima and the shift of their positions towards the lesser values of β as the plasma coupling increases. In the NNA $\Lambda_D(\beta) \sim \chi_D(\beta) \sim \beta^{-2}$, while $\Lambda_{DO}(\beta) \sim \chi_{DO}(\beta) \sim \beta^{-3}$. This dependence is natural, because on small distances the Coulomb potential is recovered, and the polarization dissapears.

DISCUSSION OF RESULTS AND RELATIONSHIP WITH EXPERIMENTAL DATA

It appears from the various calculations[3-7,24-25,27,29-30,32-42,44,46,65,67], that line profiles become asymmetrical due to the inclusion of the microfield nonuniformity effects. As it was established in the recent atlas of hydrogen lines[41] the asymmetry of the hydrogen lines induced by the microfield nonuniformity exibits *the universal behavior versus the detuning from the line center*. The important place here is occupied by the results for the wings(s. also [71,72]), which qualitatively coincide with the results of numerous measurements of hydrogen lines profiles in plasmas. The most striking fact that is known for the first lines of Lyman-series, that the asymmetry do exist far beyond the presumed quasistatic limit for electrons (after which it should not exist, according with the theoretical ideas of the cancellation of the quadrupole effects from ions and electrons). The evolution of this asymmetry versus the detuning from the line center pretty good coincide with that, which follows from the profile calculations including the quadrupole effects basing on the assumption of the electron impact broadening. The problem, that this quasistatic limit presents by itself in the reality some region of the detunings from the line center, where the effective number of quasistatically broadening particles $R(\Delta\omega)$(see [8,68]) approaches 2. On the other hand all the theories which pretend to treat the transition of the electrons from the impact broadening regime to the quasistatic one incorporate to some extent the results of the adiabatic theory, in the frames of which the several exact solutions do exist. But again in the frames of the adiabatic theory one could get the following estimates of the Weiskopff frequency limits in the case of the pure dipole or the pure quadrupole interactions for ions (i) and electrons (e,$Z_p \equiv 1$) (in atomic units)

$$\Omega_W^d \sim \left(\frac{\hbar v_{e,i}}{e^2}\right)^2 \frac{Z_r}{n^2 Z_p}, \quad \Omega_W^q \sim \left(\frac{\hbar v_{e,i}}{e^2}\right)^{3/2} \frac{Z_r}{n^2 Z_p^{1/2}}, \quad (\Omega_W^d/\Omega_W^q) \sim \left(\frac{\hbar v_{e,i}}{e^2 Z_p}\right)^{1/2} \quad (87)$$

These estimates give the range of detunings, where the quasistatic limit is achieved separately in the dipole Ω_W^d or the quadrupole Ω_W^q interactions. For the experimental conditions[16,17] $e^2/\hbar v_e > 1$, and electrons should become quasistatic earlier in the dipole interaction, than in the quadrupole one. The values, that one can get for these limits for electrons according with experimental conditions are for Ly-α[16] $\Omega_W^d \sim 25\text{\AA}$, $\Omega_W^q \sim 48\text{\AA}$ and for Ly-β[17] $\Omega_W^d \sim 14\text{\AA}$, $\Omega_W^q \sim 27\text{\AA}$. The Ω_W^d values hit the interval of the detunings, where $R(\Delta\omega) \sim 2$, while the asymmetry still increases yet with the increasing of the frequency detuning from the line center. The values of Ω_W^q are large, and one could pay no attention to the quasistatic behaviour of electrons in the quadrupole interaction in the observed range of the frequency detunings in the experiment, although this separation may be unmeaningful, and the estimates for the dipole interaction should characterize the broadening regime of electrons. Thus one might conclude, that probably the increasing of $R(\Delta\omega)$ is too steep and early, but this contradict to the rather good coincidence of the behaviour of the mean intensity in the line wings with calculations made in the assumption of the gradual transition to the quasistatic limit(s.VCS,for example[17]), although by the way it could not be stated, that there are no other variants of the reconsideration of this problem. But on the other hand, as was shown in[27,32], the curve for the Ly-α asymmetry, which includes the change of the intensity normalization in the wings due to the impact electron contribution goes lower than the experimental data[16] and the one, which was obtained firstly by Sholin[6], disregarding the contribution from the central component and the changing of the normalization due to the electron impact broadening. Thus from this point one could state, that the electrons should go to the quasistatic regime and one could adjust this transition by the properly defined line asymmetry.

Estimating the number of the effectively broadening particles in the Weiskopff sphere $h_{e,i}$ for ions and electrons and for the dipole and quadrupole potentials separately one could get

$$h_{e,i}^d \sim \frac{n^6}{Z_r^3}\left(\frac{e^2 Z_p}{\hbar v_{e,i}}\right)^3 \cdot (N_{e,i} a_o^3), \qquad h_{e,i}^q \sim \frac{n^6}{Z_r^3}\left(\frac{e^2 Z_p}{\hbar v_{e,i}}\right)^{3/2} \cdot (N_{e,i} a_o^3). \qquad (88)$$

For the parameters of the experiments in stabilized arcs it comes from (87)-(88), that $1 < h_i^d < 10$ and $h_i^q \ll 1$ for the first lines of Lyman and Balmer series, while the value of Ω_W^q for ions lies in the line core near the center (s. also (86)). *Thus one could conclude in the accordance with the proposition in[32], that for the central Stark components the quadrupole broadening by ions is binary but predominantly quasistatic for the parameters of the known observations in the gas discharge, and one could use for these states the nearest neighbour approximation.* This facilitates the use of the higher multipoles for these states to achieve more accurate results. The h_i^q value may reach and exceed unity for the larger n and N_i, but it may occur that the perturbation approach would not be valid at such high densities as well as the ideas of the plasma microfield.

Nevertheless the main controversy *why the asymmetry does not disappear after the quasistatic limit for electrons is reached* (in the pointed above sense) still exist, and we do not know the definite answer on this question now. May be the estimates of the adiabatic theory are too crude and invalid, when one have the sum of the pole potentials, may be it is necessary to use quantum approach[73] and include higher multipoles to describe correctly the short-range interaction[74]. The problem of the improvement of the line profile calculations in the intermediate region also may reappear. The question of the investigation of the quadrupole effects in the dynamic treatment, that is without the quasistatic approximation for ions and the impact one for electrons also may be important for the understanding of the way out of this old puzzle, because h_i^d is not so large and thus the ion dynamics may noticeably affect the line contour. But it should be remembered that the estimates (88) give $h_e^d \ll 1, h_i^d > 1$ for these conditions, and thus the electrons are predominantly impact, while ions are predominantly quasistatic due to small values of $\Omega_W^{d,q}$ for them.

The nonuniform microfield could cause also the shift of spectral lines[13,14,24,32,40,41], although its contribution is considered up to now less, than that which comes from the quenching and recoil effects of the electron collisions[26,75,76]. Nevertheless the simultaneous calculations of both effects in the profile are still needed due to the probable spreading of the electron shifts intra the profile and the available now refined approach to the microfield nonuniformity presented in this paper. In this context the results summarizing the measurements of the dip position and the difference in the position and the intensity of the peaks in the profile of H_β and their connection with the quadrupole effects and the electron shift from[75] appears to be very interesting[40]. In the frames the present consideration these results, partly, may be interpreted in terms of the forbidden central compononents, induced by the quadrupole interaction. The theoretical results from[40] partially repeat general formulations from ealier works of the invited speaker[38-39] for the case of the neutral radiator, one sort of the perturber ions and the linearized Debye-Hückel approximation for the pair correlation function. But the values of the Λ and χ functions obtained in the present work and checked by the Monte-Carlo calculations by Dr.D.Gilles do not coincide with digital results presented in[40]. Moreover these results, as follows from the presented in[40] the negative sign of the quadrupole shift, contradict the detailed calculations, carried out in[41], where the sign of the quadrupole shift is positive. Perhaps, this is due to the neglection in[40] the correction of the electron impact width of the Stark components due to the quadrupole interaction[32].

The next interesting feature of the quadrupole asymmetry to be discussed is the phenomena of the changing its sign[10-15,17]. The calculations that are performed till now demonstrate the rather good coincidence with experiments on the value of detunings, where the crossing of signs occurs[27,41], although it could not be named too precise. The position of this point is found to be very sensitive to the nonuniform microfield model used in the concrete calculations[41].

Thus it may be concluded that microfield nonuniformity effects exhibit themselves in the row of very interesting physical phenomena of the fundamental importance in the profiles of the radiators in plasmas, although there is still a lot of room now for the improvement their understanding and theoretical treatment.

ACKNOWLEDGEMENTS

The support of the work of the invited speaker (A.V.D) by ISF, IUPAP, University of Toronto, Observatoire de Paris-Meudon, CEA Limeil-Valenton, Auburn University and RRC "Kurchatov Institute" is greatly acknowledged.

It is a pleasure to express our thanks to Prof. Eugene Oks for his invaluable support and discussions during the final preparations of this work at the Auburn University. We are indebted to Dr. Dominique Gilles for the fruitful cooperation.

The part of this work was stimulated by the experiments and discussions with Prof. H.-J.Kunze, whom A.V.D. express his warm gratitude. Due to the fact, that this work was carried out for a long period of time, A.V.D. patience in the pursuing this goal was kindly supported during his visit to Universitè de Provence by the interest and hospitality of Prof. Roland Stamm and Prof. Bernard Talin, whose participation thus is very much appreciated.

The invited speaker wish to thank also Panos Gavras and Andrei Derevianko the graduate students of Physics Department of the Auburn University for their assistance in the acquiring by A.V.D. the LaTex skill on the "Sun" station.

REFERENCES

1. J.Holtsmark, Ann.Physik 58,577(1919).
2. W.Pauli, Quantentheorie, Handbuch der Physik v.23, ed.H.Geiger,K.Scheil (Springer-Verlag,Berlin,1926)p.75.
3. H.Margenau,R.Meyerott, Ap.J. 121,194(1955).
4. L.P.Kudrin,G.V.Sholin, Sov.Phys.Dokl. 7,1015(1963).
5. Nguyen-Hoe,H.W.Drawin,L.Herman, J.Q.S.R.T. 4,847(1964).
6. G.V.Sholin, Opt.Spectrosc.(USSR) 26,275(1969).
7. I.I.Sobelman,L.A.Vainshtein, Doclady Academy of Sciences USSR 90,757(1953).
8. H.R.Griem, Phys.Rev. A 140,1140(1965).
9. H.Margenau,M.Lewis, Rev.Mod.Phys. 31,569(1959).
10. W.Finkelburg, Z. für Physik 70,375(1931).
11. G.Jürgens, Z. für Physik 134,21(1952).
12. G.Boldt,W.S.Cooper, Z. für Naturfosch. 19a,968(1964).
13. W.L.Wiese,D.E.Kelleher, Ap.J. 166,159(1971).
14. W.L.Wiese,D.E.Kelleher,D.R.Paquette, Phys.Rev. A 6,1132(1972).
15. R.D.Bengston,G.R.Chester, Phys.Rev. A 13,1762(1976).
16. R.C.Preston, J.Phys. B 10,523(1977).
17. K.Grützmacher,B.Wende, Phys.Rev. A 18,2140(1978).
18. H.R.Griem, Z. für Physik 137,280(1954).
19. H.Rausch,V.Traunbenberg,R.Gebauer,G.Lewin, Naturwiss. 18, 417(1930).
20. S.Stehlé,S.Jacquemot, Astron.Astrophys. 271,348(1993).
21. D.L.Huber,J.H.Van Vleck, Rev.Mod.Phys. 38,187(1966).
22. O'Brien,C.F.Hooper, J.Q.S.R.T. 14,479(1974).
23. S.Stehlé, Phys.Rev. A 34,4153(1986).
24. T.L.Pittman,V.Voigt,D.E.Kelleher, Phys.Rev.Lett. 45,723(1980).
25. J.C.Adcock,H.R.Griem, Phys.Rev.Lett. 50,1369(1983).
26. D.B.Boercker,C.A.Iglesias, Phys.Rev. A 30,2771(1984).
27. M.E.Bacon, J.Q.S.R.T. 12,519(1973);17,501(1977).
28. S.Chandrasekhar and von Neumann, Ap.J. 97,1(1943).
29. V.S.Milliyanchuk, Thesis Doctor of Sciences (L'vov State University,1956).
30. K.G.Müller, J.Q.S.R.T. 5,403(1965).
31. I.I.Sobelman, Introduction to the Theory of Atomic Spectra (Pergamon Press,Oxford,1972)609p.
32. A.V.Demura,G.V.Sholin, J.Q.S.R.T. 15,881(1975)
33. J.C.Weisheit,E.L.Pollock, SLS v.1, ed.Burkhard Wende (W.de Grüyter, Berlin,1981)p.433-445.
34. B.d'Etat,H.Nguyen, SLS v.3, ed.F.Rostas (de Grüyter,Berlin,1985)p.209.

35. E.Leboucher,M.Koenig,B.d'Etat,L.Terray,H.Nguyen, ibid,p.249.
36. R.F.Joyce,L.A.Woltz,C.H.Hooper, Phys.Rev. A 35,2228(1987).
37. A.V.Demura, Thesis (I.V. Kurchatov Institute of Atomic Energy,M.,1976), 176p.(unpublished).
38. A.V.Demura, Preprint IAE-4632/6 (Moscow,1988) 17p.[0]
39. A.V.Demura, Abstracts of Contributed Papers ICSLS-9 (Nicolas Copernicus University Press,Torun,1988) A39.
40. J.Halenka, Z.Phys. D 16,1(1990).
41. A.V.Demura,V.V.Pleshakov,G.V.Sholin, Preprint IAE-5349/6 (Moscow, 1991) 97p.
42. D.P.Kealcrease,R.C.Mancini,C.F.Hooper, Radiative Properties of Hot Dense Matter, ed.W.Goldstein,C.Hooper,J.Gaunthier, J.Seely,R.Lee (World Sientific,Singapour,1991),p.74.
43. A.V.Demura, Abstracts of Invited Lectures and Contributed Papers, ESCAMPIG-92, EPS v.16, ed.L.Tsendin (St.Peterburg,Russia,1992), p.63.
44. A.V.Demura, SLS v.7, ed.R.Stamm & B.Talin (Nova Science Publ.,1993), p.87.
45. A.V.Demura, ibid, p.89.
46. D.P.Kealcrease,R.C.Mancini,C.F.Hooper, Phys.Rev. E 48, 3901(1993).
47. M.Baranger,B.Moser, Phys.Rev. 115,521(1959); 118,626(1960).
48. H.Pfennig,E.Trefftz, Z. für Naturforsch. 21a,697(1966).
49. G.Peach, Advances in Physics 30,367(1981).
50. B.Held,C.Deutsch,M.-M.Gombert, Phys.Rev. A 29,880(1984).
51. B.Held,P.Pignolet, J.Physique 48,1951(1987).
52. F.Perrot,M.W.C.Dharma-wardana, Phys.Rev. A 33,3303(1986).
53. F.Perrot,M.W.C.Dharma-wardana, Phys.Rev. A 41,3281(1990).
54. J.Chihara, Phys.Rev. A 44,1347(1991).
55. A.Angelie,D.Gilles, Annales de Physique, Colloque $n^0 3$,supplement au $n^0 3$, 11,157(1986).
56. C.A.Iglesias,H.E.DeWitt,J.L.Lebowitz,D.McGowan,W.B.Hubbard, Phys.Rev. A 31,1698(1985).
57. J.W.Dufty,D.B.Boercker,C.A.Iglesias, Phys.Rev. A 31,1681(1985).
58. G.Massacrier, J.Q.S.R.T. 51,221(1994).
59. J.Dufty, SLS v.1, ed.B.Wende (W.de Grüyter,Berlin,1981).p.41.
60. C.F.Hooper, SLS v.4, ed.R.J.Exton (Deepak Publ.,Hampton,1987),p.161.
61. J.Dufty, Srongly Coupled Plasma Physics, ed.F.J.Rogers (Plenum Publ. Corp.,1987),p.493.
62. A.Isihara, Statistical physics (Academic Press,N.Y.-L.,1971).
63. A.A.Markoff, Warscheinlichkeitsrechnung (Leipzig,1912).
64. A.V.Anufrienko,A.L.Godunov,A.V.Demura,Yu.K.Zemtsov,V.S.Lisitsa,

---

[0]Preprints of I.V.Kurchatov Institute of Atomic Energy are protected by copyright reserved.

A.N.Starostin,M.D.Taran,V.A.Shchipakov, Sov. Phys. JETP 71,728 (1991).
65. A.V.Demura,D.Gilles,B.C.Huynh,S.Stehlé, this volume.
66. H.A.Bethe,E.E.Salpeter, Quantum Mechanics of One- and Two-Electron Systems (Handbuch der Physik,Band XXXV,ATOME I,Springer-Verlag, 1957)pp.88-436.
67. E.A.Oks,G.V.Sholin, Optics Spectroscopy(USSR) 33,217(1972).
68. H.R.Griem, Spectral Line Broadening by Plasmas (Academic Press,New York,1974).
69. A.V.Demura, unpublished manuscript(1975).
70. E.A.Oks, Plasma Spectroscopy with Quasimonochromatic Electric Fields (Energoizdat,Moscow,1990)(in Russian).
71. D.Koster,N.F.Allard, SLS v.7, ed. R.Stamm & B.Talin (Nova Science Publ., 1993),p. 493.
72. J.Kielkopf, ibid,p.271.
73. A. de Kertanguy,N.Tran Mihn,N.Feautrier, J.Phys. B 12,365 (1979).
74. N.Feautrier,N.Tran Mihn,A.R.Edmonds, J.Q.S.R.T. 23,469(1980)
75. H.R.Griem, Phys.Rev. A 38,2943(1988).
76. S.Günter, SLS v.7, ed. R.Stamm & B.Talin (Nova Science Publ.,1993) p.47.

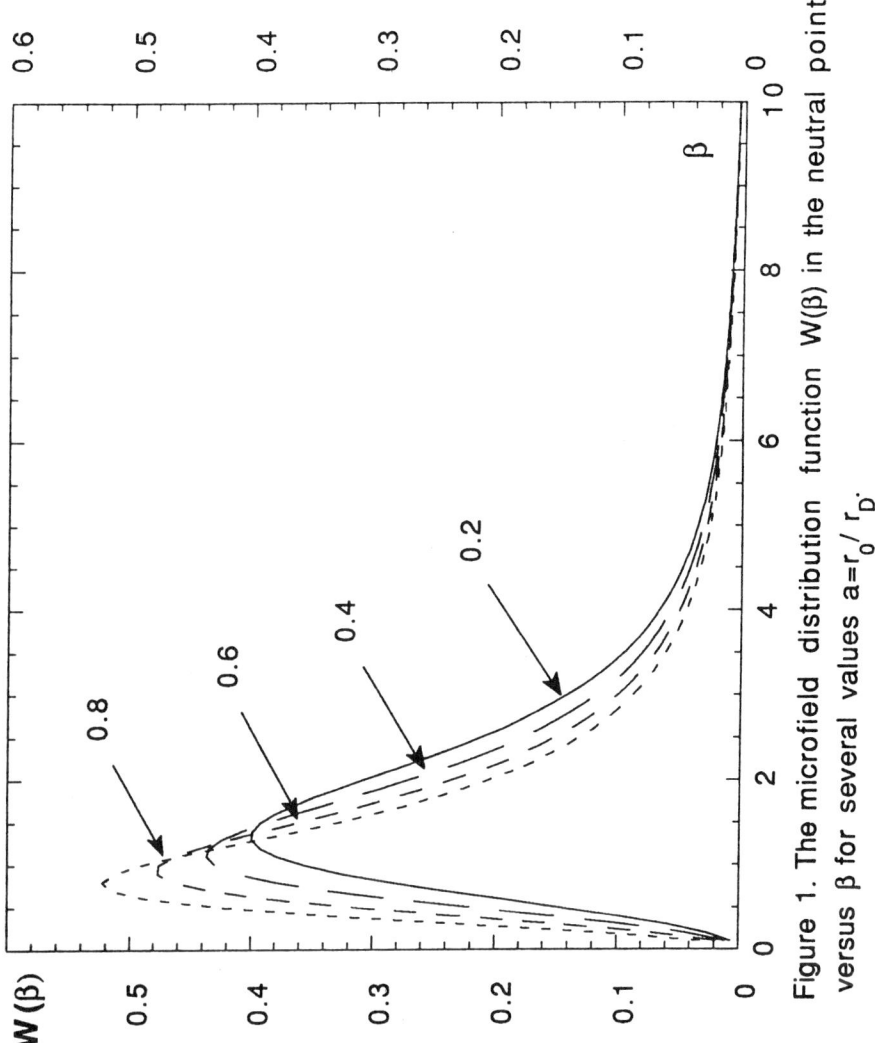

Figure 1. The microfield distribution function $W(\beta)$ in the neutral point versus β for several values $a = r_0/r_D$.

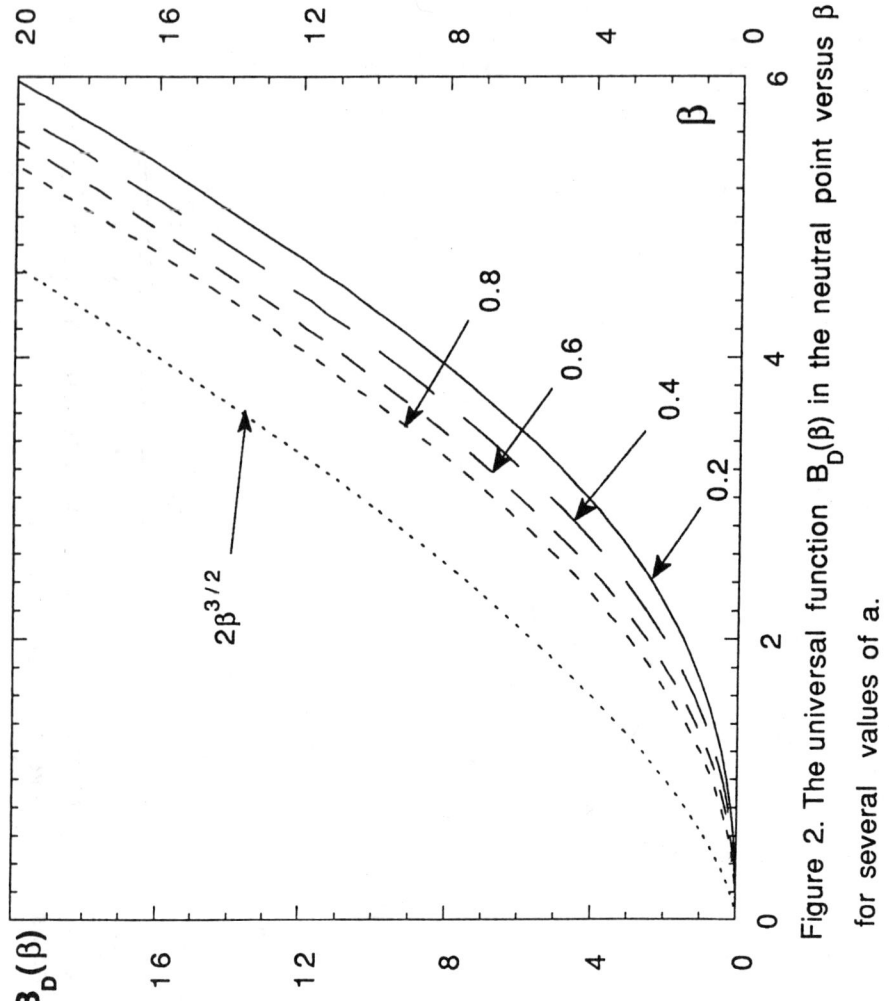

Figure 2. The universal function $B_D(\beta)$ in the neutral point versus β for several values of a.

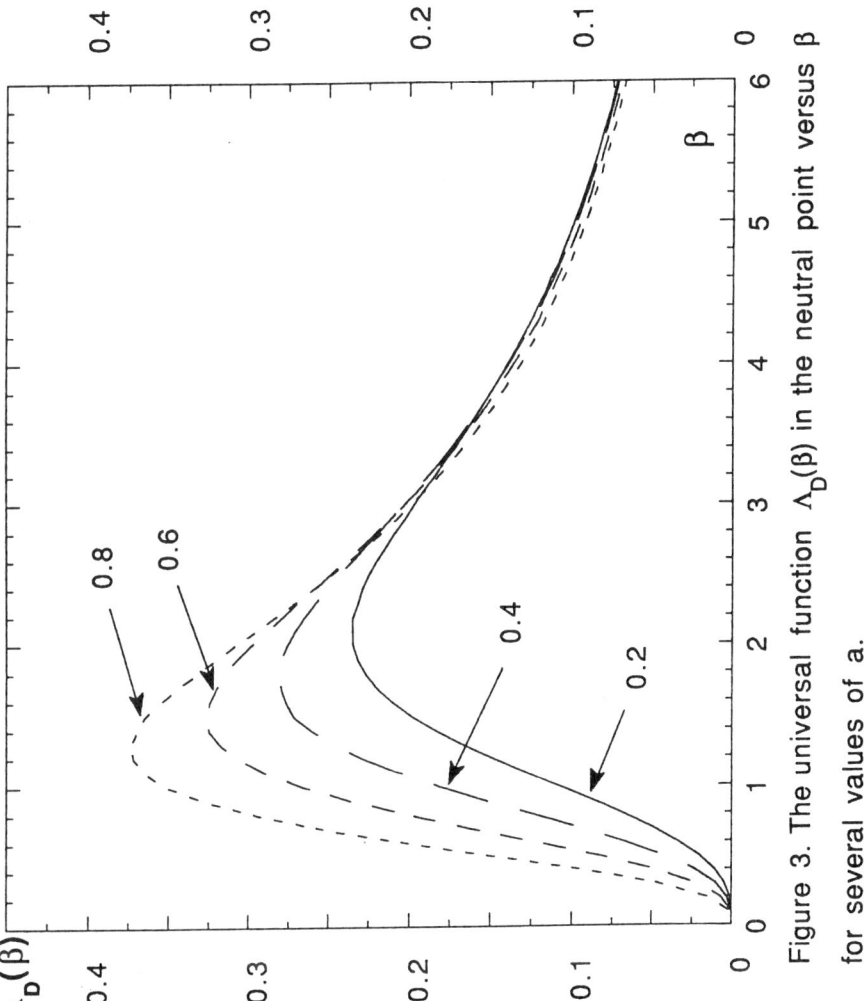

Figure 3. The universal function $\Lambda_D(\beta)$ in the neutral point versus β for several values of a.

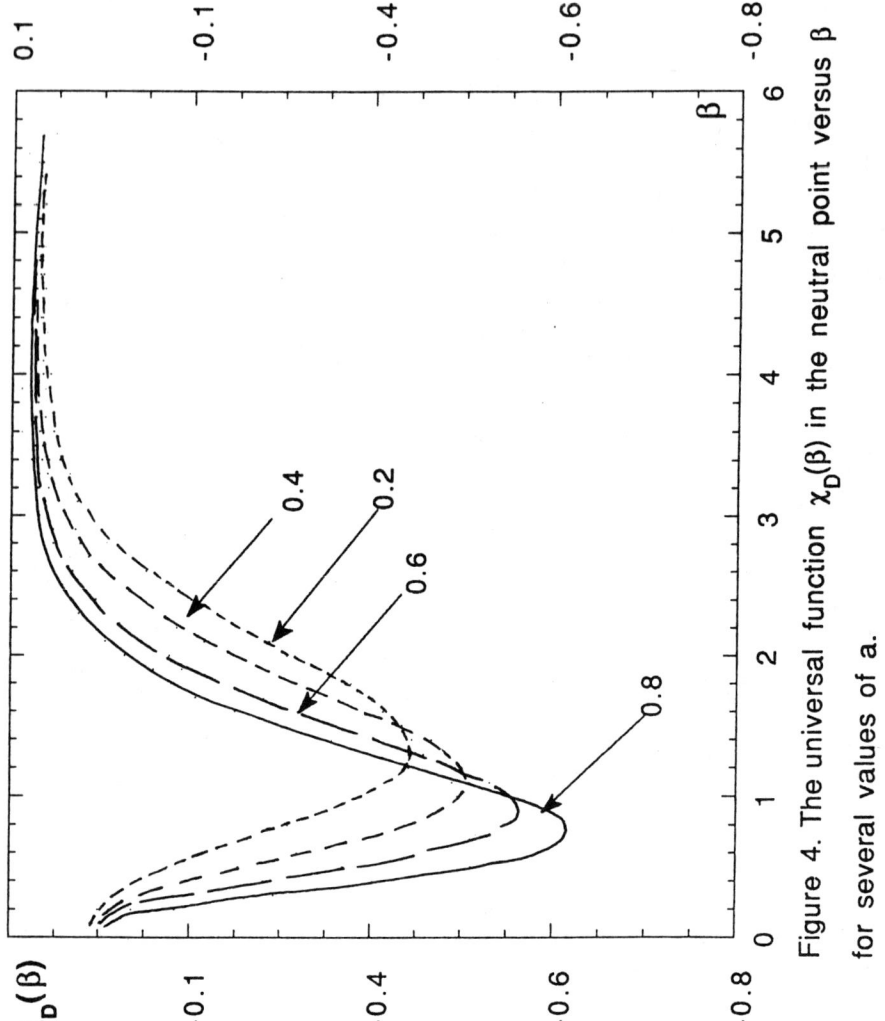

Figure 4. The universal function $\chi_D(\beta)$ in the neutral point versus β for several values of a.

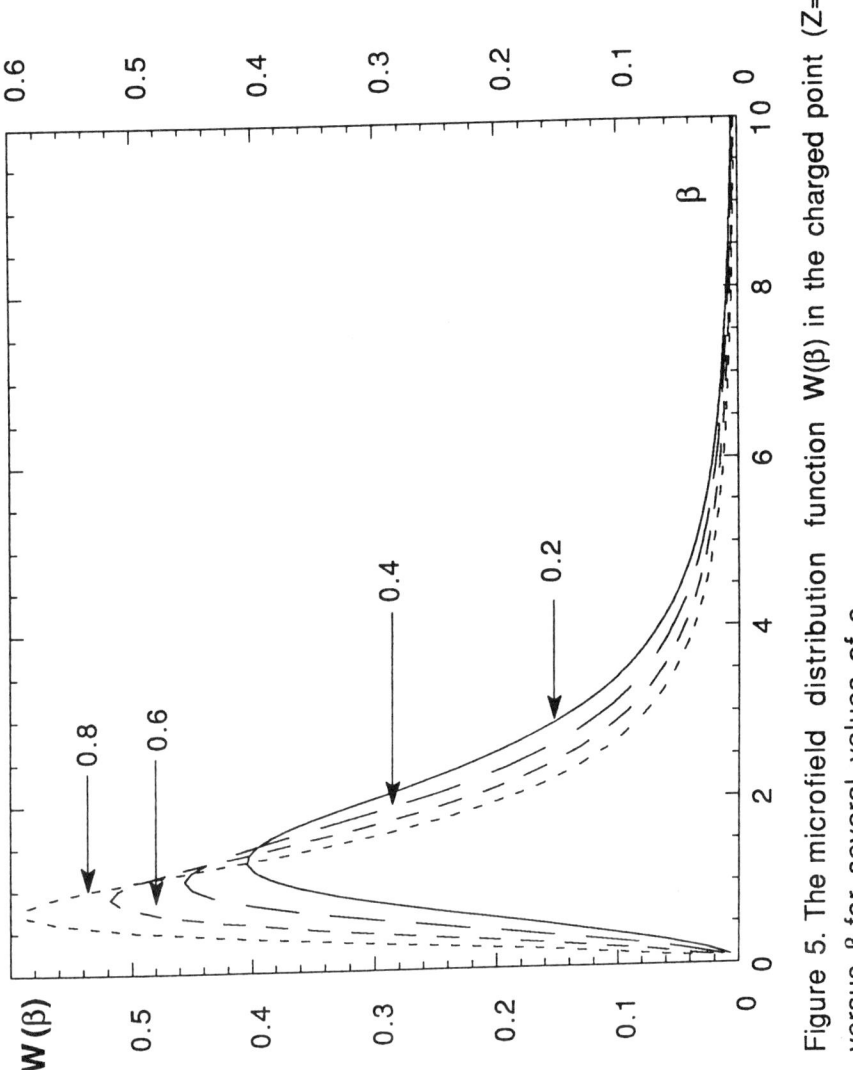

Figure 5. The microfield distribution function $W(\beta)$ in the charged point (Z=1) versus β for several values of a.

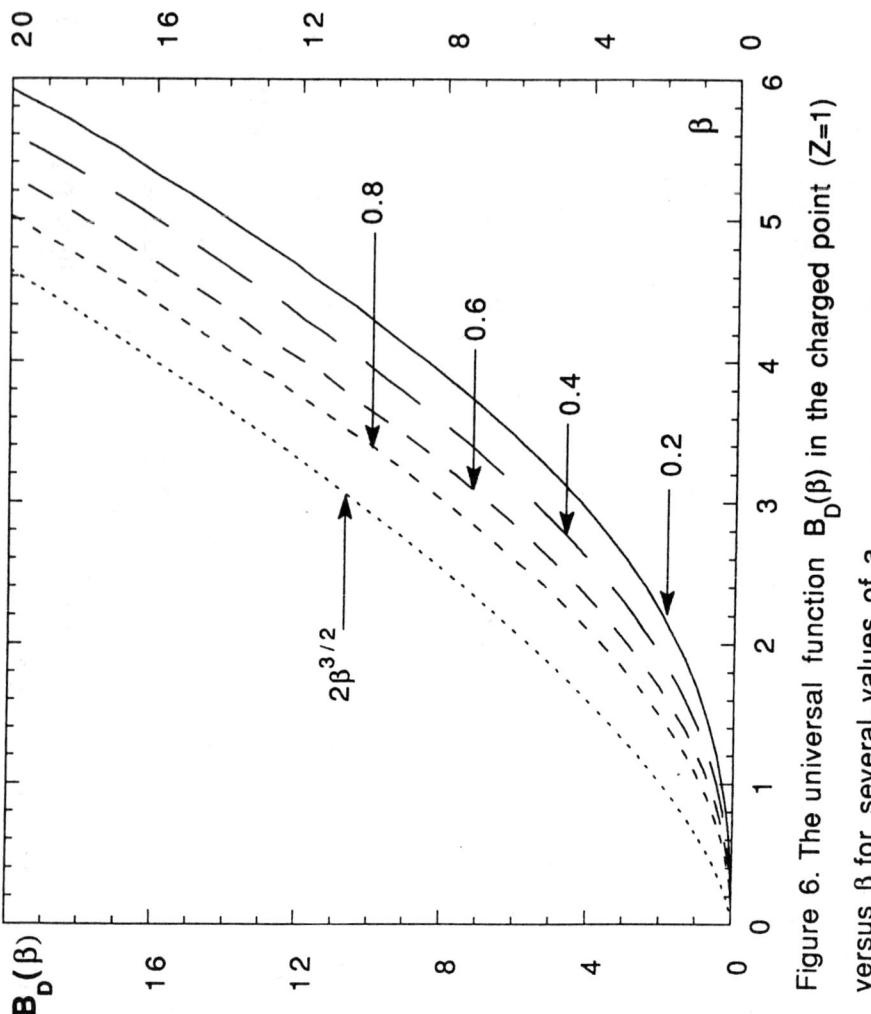

Figure 6. The universal function $B_D(\beta)$ in the charged point (Z=1) versus β for several values of a.

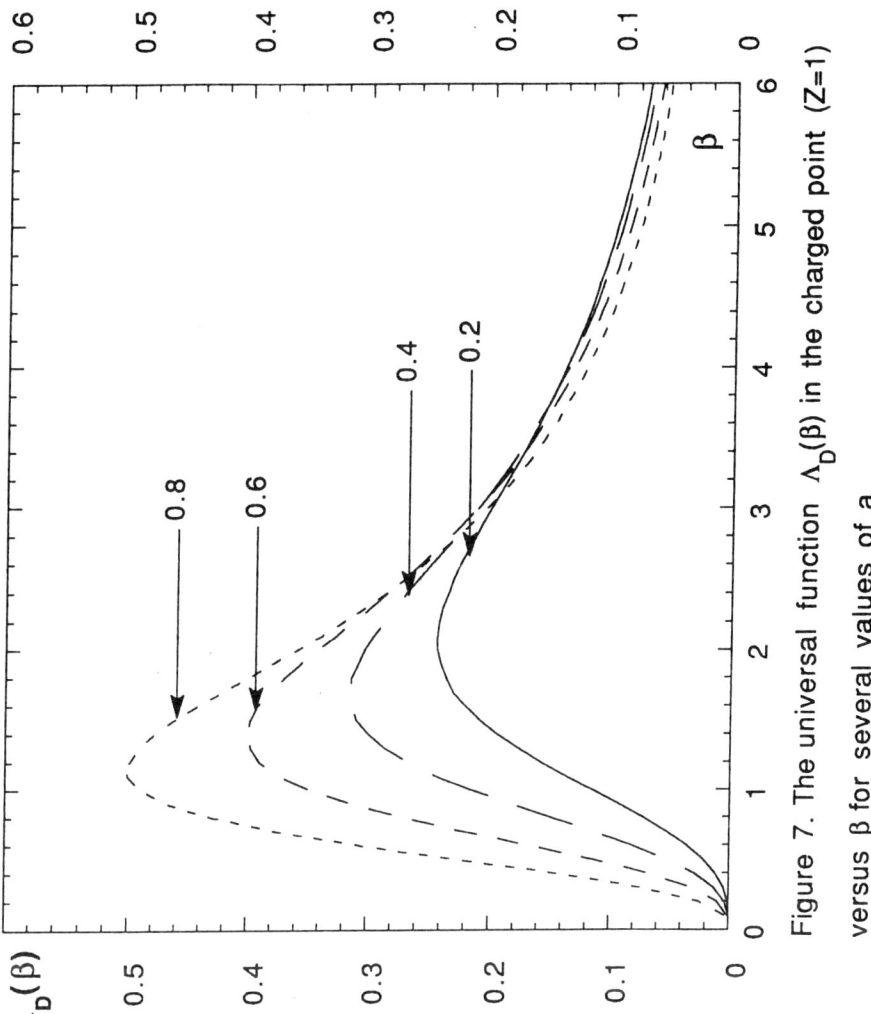

Figure 7. The universal function $\Lambda_D(\beta)$ in the charged point (Z=1) versus β for several values of a.

208 Effects of Microfield Nonuniformity

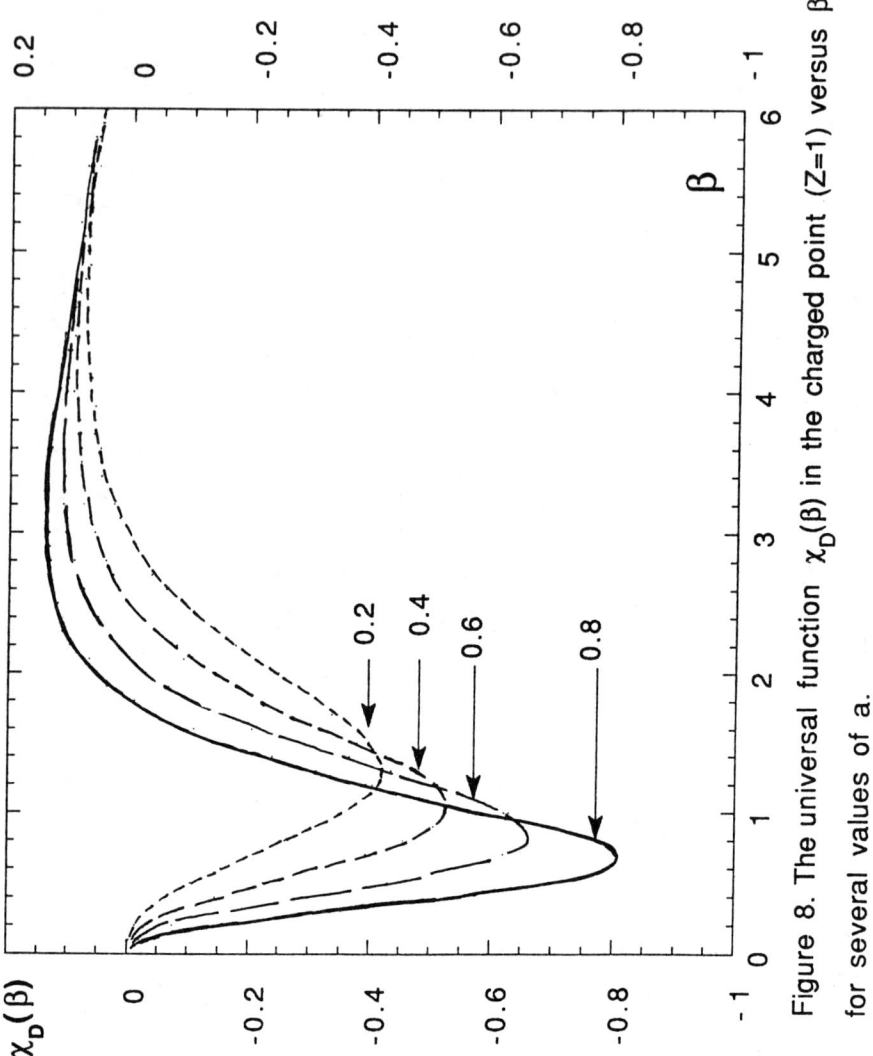

Figure 8. The universal function $\chi_D(\beta)$ in the charged point (Z=1) versus β for several values of a.

STARK WIDTHS OF HYDROGEN SPECTRAL LINES IN ONE- AND TWOFOLD IONIZED HELIUM PLASMAS

B. Grabowski, J. Halenka and W. Olchawa
Pedagogical University,
45-052 Opole, ul. Oleska 48, Poland

INTRODUCTION

Calculations of Stark profile of H-lines have conventionally been realized assuming that only single charged ions occur in plasma. In case of pure H-plasma, this assumption is self-evident. In the recent paper[1] the interesting investigations of H-α line profile, emitted from an almost entirely ionized He plasma with insignificant H-admixture, have been presented. In that case, the above quoted assumption must be revised. The aim of the present paper is to investigate the influence of twofold charged ions on FWHM of H-line profiles.

CALCULATIONS

In order to calculate the profile of a spectral line emitted from plasma, the knowledge of the electric microfield distribution functions $W_\rho(\beta)$ is necessary. For weak-coupled plasmas, like of that in the experiment[1] (mean electric energy/$kT < 1$), the Mozer-Baranger approximation is sufficiently adequate. We have calculated the functions $W_\rho(\beta)$ at a neutral point of the plasma, consisting of three components: electrons, single, and twofold charged ions.

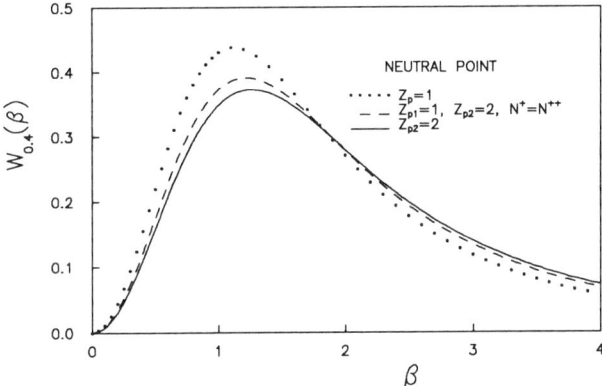

Fig. 1. The microfield distribution functions $W_\rho(\beta)$ for plasma with $\rho = 0.4$ at a neutral point; β in units of $F_0 = e/R_0^2$.

The line-profiles of Ly-α, Ly-β, H-α, and H-β we have calculated taking into account the ion dynamics. In case of Lyman lines, the computer simulation technique[2] has been applied. For Balmer lines, where the influence of the ion dynamics on FWHM is weak, the line-profiles have been calculated within the Impact Theory after its generaliziation[3]. [The starting Hamiltonian (in conventional designations) has been described as $H =$

$H_0 - \vec{d} \cdot [\vec{F}(0) + \dot{\vec{F}}(0)t] + i\Phi$. The calculations have been performed assuming that $|\dot{F}/F| < \Delta\omega_{1/2}$. So called interference term within the impact operator Φ has been modified according to the suggestions[4]. After modification of Φ, the FWHM-values of H-α line, resulting from the quasi-static approximation, $\dot{F} = 0$, differ from those obtained within Unified Theory[4] only slightly. In $\dot{F}$ Debye screening and ion–ion correlactions is included.]

Within both papers[2,3] the results of calculations agree with measurements very well.

RESULTS

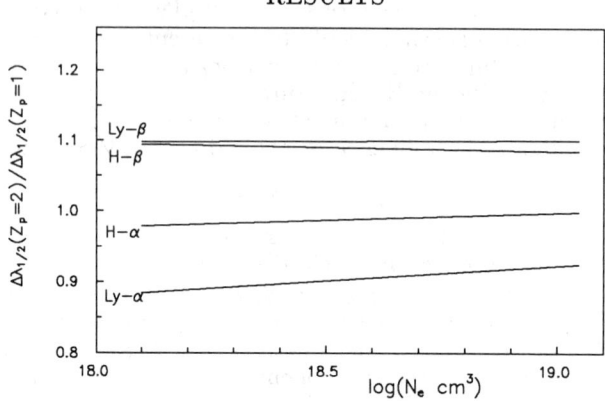

Fig. 2. Comment in text.

In Fig. 2 values of the ratio of $\Delta\lambda_{1/2}(Z_p = 2)/\Delta\lambda_{1/2}(Z_p = 1)$ for Stark profiles of Ly-α, Ly-β, H-α, and H-β are presented in the range of the electron density, N_e, from 10^{18} to 10^{19} cm^{-3}, at increasing temperature values (as in Ref. 1) from 1 to 5 eV. The $\Delta\lambda_{1/2}(Z_p = 2)$ means the FWHM of line profiles formed in plasma, in which all ions are doubly ionized, whereas the $\Delta\lambda_{1/2}(Z_p = 1)$ corresponds to a reference plasma, in which all ions are singly ionized. The calculated values of the ratio are significantly smaller compared with the value $2^{1/3}$ (the upper edge of Fig. 2), resulting from the relation $\alpha = \Delta\lambda_{1/2}(Z_p)/F_0(Z_p) = const$, often assumed and used in current literature *a priori*. In particular, for the Ly-α line this ratio reaches values smaller than unity.

REFERENCES

1. St. Böddeker, S. Günter, A. Könies, L. Hitzschke and H.-J. Kunze, Phys. Rev. E <u>47</u>, 2785 (1993).
2. J. Halenka and W. Olchawa (in press).
3. J. Halenka, W. Olchawa and B. Grabowski, Contributed Papers in *Plazma'93*, p. 141, Warszawa 1993.
4. C. R. Vidal, J. Cooper and E. W. Smith, Astrophys. J. Suppl. <u>25</u>, 37 (1973).

BV, CVI AND NVII LYMAN LINE PROFILES FROM 10 PS KrF-LASER PRODUCED PLASMAS*

Hans R. Griem, Yongzhang Leng and Julius Goldhar
University of Maryland, College Park, MD 20742

and

Richard W. Lee
Lawrence Livermore National Laboratory
Livermore, CA 94550

Plasmas with densities $\lesssim 10^{22} \text{cm}^{-3}$ and temperatures $\lesssim 200$ eV were produced near solid targets irradiated at $\sim 10^{15} \text{W/cm}^2$, with $\sim 10^{11} \text{W/cm}^2$ prepulse from ASE. Photographic spectra taken nearly side-on within $\lesssim 0.1$ mm from the target surface show strong H-like lines corresponding to $n = 2, 3$ and 4 to $n = 1$ transitions, and weaker He-like lines, with FWHM widths $\lesssim$ 4eV, see Figs. 1a, b and c.

The Lyman-α lines are all optically thick. A five-layer model with high electron density and electron temperature core layers and less dense and cooler outer layers was used for simulation. LTE is assumed for the multi-layer plasma to calculate population densities of the excited and ground levels of highly ionized ions. Central layers with high T_e contribute most of the emission while outside cooler layers cause the absorption dips of Lyman α lines. Asymmetrical self reversal is interpreted by self-absorption from radially expanding plasma. From calculated Stark profiles,[1,2] the multilayer simulation gave very convincing line profile comparisons, as shown in Figs. 2a, b and c. Contributions to the total line intensities are 42, 42, 15 and 1% from layers with electron densities of 1×10^{22}, 3×10^{21}, 1×10^{21} and $3 \times 10^{20} \text{cm}^{-3}$, respectively. A plasma shift[3] was needed for better fittings of the experimental line profiles, especially for the asymmetrical and shifted Lyman γ line of CVI. Only the largest shifts are significantly beyond the upper bound inferred from earlier measurements[4] on spherical targets, possibly because most of the emission there had been from well below critical density ($N_e = 10^{22} \text{cm}^{-3}$), say, by a factor ~ 3.

REFERENCES

1. A. Calisti, F. Khelfaoui, R. Stamm, B. Talin and R. Lee, Phys. Rev. A **42**, 5433 (1990).
2. A. Calisti, L. Godbert, R. Stamm and B. Talin, J.Q.S.R.T. **51**, 59 (1994).
3. H. Nguen, M. Koenig, D. Benredjem, M. Caby and G. Coulaud, Phys. Rev. A **33**, 1279 (1986).
4. S. Goldsmith, H. R. Griem and L. Cohen, Phys. Rev. A **30**, 2775 (1984).

*Partially supported by the U.S. Naval Research Laboratory and the National Science Foundation.

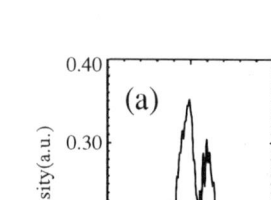
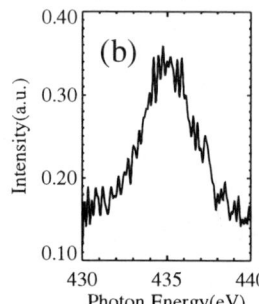
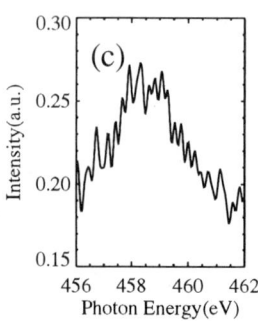

Figure 1: Measured profiles of CVI Ly-α (a), Ly-β (b) and Ly-γ (c) lines.

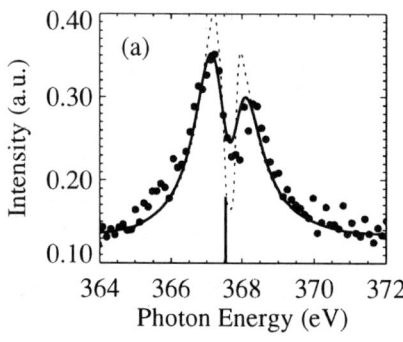
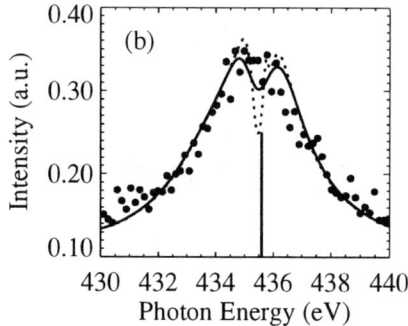
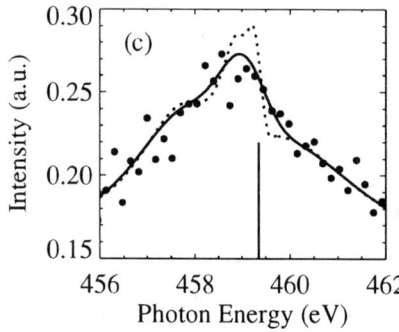

Figure 2: Comparison of measured and calculated profiles of the CVI Ly-α (a), Ly-β (b) and Ly-γ (c) lines. Vertical lines indicate unshifted line positions.

STARK BROADENING CALCULATIONS OF 3d-5f TRANSITION IN Al XI

N. Ben Nessib, Z. Ben Lakhdar
Physique Atomique et Moléculaire, Faculté des Sciences de Tunis,
1060 Campus Universitaire, Tunis, Tunisia
H. Nguyen, J. P. Arranz
D.R.P., Spectronomie des gaz et des Plasmas, Université Pierre et Marie Curie,
4 place Jussieu, 75252 Paris Cedex 05, France

In a recent work, Moreno et al. report[1] a Stark broadening measurements of 3d-nf transitions in lithiumlike ions. Stark broadening models uses the static-ion approximation and an impact approximation for the electrons. The semiclassical method[2,3] or the complete quantum formalism using the close coupling method[4,5] are generally used for the many electrons emitters. In a previous work[6], we used a fully quantum mechanical method with a Coulomb-Born-Oppenheimer and a Coulomb-Bethe approximations. Here the quantum formalism is used only for the strong collisions and the semiclassical formalism for the weak collisions.

When the perturber angular momentum l is less then l_1, the CBII approximation is used (without exchange and with unitarity of the scattering matrix). When l is between l_1 and l_D, the collision function a(z) is used :

$$\sigma^T = \sigma^Q + \sigma^{S.C} \quad (1)$$

where :

$$\sigma^Q = \frac{\pi}{k^2 [l_a]} \sum_{l,l',L} |<v,l_a;k,l;L|T|v',l'_a;k,l';L>|^2 \quad (2)$$

$$\sigma^{S.C} = \frac{8\pi}{3} |<v,l_a|r|v',l'_a>|^2 [a(z^1) - a(z^D)] \quad (3)$$

$a(z) = z [K_0(z)K_1(z)]$, $z^1 = \frac{\rho_1 \Delta E_{vv'}}{k}$, $z^D = \frac{\rho_D \Delta E_{vv'}}{k}$, $\rho_1 k^2 = l_1(l_1+1)$ and ρ_D the Debye radius.

The integration over the energy of the incident electron is obtained using a gaussian quadrature.

Using the quantum defect theory, the multipole oscillator strengths are calculated (Table I).

Table I Multipole oscillator strengths for the 3d-5f transition in Al XI

n	l	n'	l'	λ	$S_\lambda(nl,n'l')$
3	2	3	2	0	1.10
3	2	3	2	2	0.87
5	3	3	2	1	0.31
5	3	5	3	0	1.00
5	3	5	3	2	9.21

Using the CBII method, the cross-sections of the strong collision part are calculated. As an example, we present the results for the lower level (Table II).

Table II : Cross-sections for different energy k^2 of the incident electron in Rydberg.

K^2	$\sigma^Q(3d,3p)$	$\sigma^Q(3d,3d)$
1	4.51	3.61
5	0.91	0.70
15	0.27	0.22
30	0.13	0.11
50	0.07	0.06

We developed a model for the electronic broadening of a many-electron ions immersed in dense plasmas. It is valid especially when the ion charge is high enough

1. J. C. Moreno, H. R. Griem, R. W. Lee, J. F. Seely, *Phys. Rev. A* **47**, 374 (1993)
2. A. Calisti, F. Khalfaoui, S. Stamm, B. Talin et R. W. Lee, *Phys. Rev. A* **42**, 5433 (1990)
3. S. Alexiou, *Phys. Rev. A* **49**, 106 (1994)
4. V. M. Burke, *J. Phys. B* **25**, 4917 (1992)
5. T. Schöning, *J. Phys. B* **26**, 899 (1993)
6. H. Nguyen, J. P. Arranz, N. Ben Nessib et Z. Ben Lakhdar, in *Spectral Line Shapes*, Volume 7, edited by R. Stamm et B. Talin (Nova, New York, 1993)

X EMISSION FROM DENSE PLASMA CREATED BY COLLIDING FOILS

P.Angelo, P.Gauthier, E.Leboucher-Dalimier, A.Poquérusse
LULI-Université ParisVI/Ecole Polytechnique

C.Back
Lawrence Livermore National Laboratory

INTRODUCTION

Our aim is to create in a laboratory very hot and dense plasmas in order to exhibit ionic-molecule features [1,2,3]. For this, we tried to realize a plane compression by colliding thin foils using two opposite laser beams. Our idea was to choose the foil thickness and initial distance so that the collision occurs with the maximum kinetic energy during the laser pulse duration. When the foils are ablated enough to make their rear face "hot", the plasma resulting from the compression of a hot solid could be more propitious to the exhibition of transient molecules features than the plasma created in the "cold" ablated region of a massive plane target.

These experiments have been driven using hydrodynamic simulations [4].

DESIGN OF THE EXPERIMENTS

The experiments have been carried out on the LULI facilities, by using the Nd glass laser with the wavelength λ_L=0.263 μm and a pulse duration $\tau \approx$ 500 ps. The available 20-30J energy shots are focused onto Al or CF_2 foils, with initial thickness 1.5μm $\leq$ e $\leq$ 10μm and initial distance 50μm $\leq$ d $\leq$ 250μm. The X-ray emission emerging from the aluminum or fluorine plasmas (inside the foils and forward the laser beams) is spatially-resolved in the tranverse and axial directions.

FIRST DIAGNOSTIC : LONGITUDINAL SPATIAL RESOLUTION

We observed two possible collision regimes depending on the collision parameters :
• collision between two foils followed by a compression with creation of a dense and emissive plasma.
• collision between two plasmas resulting from the explosion of the foils before a possible collision.

We report on figure 1 a film corresponding to the first regime. The two external symmetric zones of the spectra are due to the emission forward the beams (coronas). In the central zone, the recorded emission reveals very broadened lines over a distance corresponding to a plasma spatial integration less than 20 μm. The first attemps to analyse the profiles show that the electronic density could be much greater than 5.10^{22} cm^{-3}.

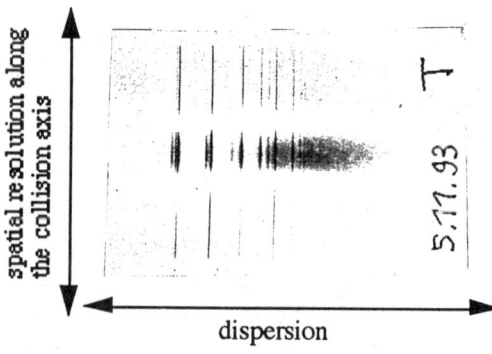

fig.1. Spatially and spectrally resolved X ray emisssion during and after a compression of aluminum foils. The spatial resolution takes place along the collision axis.
e= 1.5 μm
d= 100 μm
I_L= 2.0 E14 W.cm-2 (laser)

SECOND DIAGNOSTIC : TRANSVERSE SPATIAL RESOLUTION

The extension of the emission along the collision axis is important. This result has been obtained with a spectrograph having a consequent longitudinal magnification ratio. As a consequence the emission from the plasmas formed before and after the collision are well separeted along the dispersion direction. Figure 2 shows such a diagnostic for an aluminum foil collision.

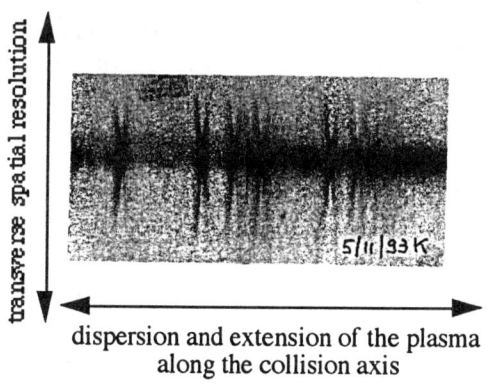

dispersion and extension of the plasma along the collision axis

fig.2. Spatially and spectrally resolved X ray emission during and after a compression of aluminum foils. The spatial resolution takes place perpendicularly to the collision axis. Every line is the signature of a bidimensional image of the plasmas formed before and after the collision of the foils.
e= 6 μm
d= 150 μm
I_L= 2.3 E14 W.cm-2 (laser)

REFERENCES

1. E.Leboucher-Dalimier, A.Poquérusse, P.Angelo,
   Phys.Rev.E 47, R1467 (1993)
2. E.Leboucher-Dalimier, A.Poquérusse, P.Angelo, I.Gharbi, H.Derfoul,
   J.Quant.Spectrosc.Radiative Transfert 51, n°1 (1994)
3. E.Leboucher-Dalimier, A.Poquérusse, P.Angelo,
   Spectral Line Shape 7 edited by R.Stamm (Nova Science, New York) 1993
4. J.C. Gauthier, J.P.Geindre, N.Grendjouan, J.Virmont
   J.Phys.D 16, 321 (1983)

SHIFT AND WIDTH OF HE II LINES

S. Günter, M. Stobbe, A. Könies,
Fachbereich Physik, Universität Rostock, 18051 Rostock, Germany,

J. Halenka,
Institute of Physics, Pedagogical University of Opole,
45-052 Opole, ul. Oleska 48, Poland

INTRODUCTION

The aim of this paper is the generalization of a former developed Green's function approach to spectral line shapes [1] for charged radiators. Calculating line profiles besides the changes in the wave functions and the energy levels one has to consider the interaction between the charged radiator and the perturbers. Further, one has to proof whether fine structure splitting has to be included. The developed theory has been applied to calculate shift and width of He II lines.

THEORY

The line profile is given by

$$I(\omega) = \frac{1}{\pi} \int_0^\infty d\beta \, W(\beta) \, \text{Im} \sum_{i',f',i'',f''} \langle i''|\vec{r}|f'\rangle \langle f''|\vec{r}|i'\rangle \times$$
$$\times \langle i'|\langle f'|[\omega - H_i^0(\beta) + H_f^0(\beta) + \text{Re}(\Sigma_f - \Sigma_i) + i\text{Im}(\Sigma_f + \Sigma_i) + i\Gamma_{if}^V]^{-1}|f''\rangle|i''\rangle,$$

where the ionic contribution containing in $H^0(\beta)$ results from the linear Stark-effect. Further, the quadrupole effect due to inhomogeneities of the ionic microfield is included. Therefore, in the coordinate system with $\vec{E}||0z$ reads [2]

$$H(\beta) = H_0 + e_0 E_0 \beta z - \frac{5}{(32\pi)^{1/2}} \frac{e_0 E_0}{2R_0} B_\rho(\beta)(3z^2 - r^2),$$

where H_0 is the Hamiltonian of the isolated emitter, $\beta = E/E_0$ denotes normalized field strength. $W_\rho(\beta)$ and $B_\rho(\beta)$ are the microfield distribution and the generalized Chandrasekhar and von Neumann functions, $\rho = R_0/D$ is the screening parameter. In the present paper the functions $B_\rho(\beta)$ have been calculated at a charged point for singly and double charged ions in Mozer-Baranger limit. Calculating the electron contributions to the self-energy, the interaction between the perturbers and the charged radiator is described by the matrix elements

$$M_{n,\alpha}(\vec{q}) = ie \int \frac{d\vec{p}}{(2\pi)^3} \Psi_n^{Z*}(\vec{p})[Z\Psi_\alpha^Z(\vec{p}) - \Psi_\alpha^Z(\vec{p}-\vec{q})], \quad (1)$$

where Z is the nuclear charge of the radiator and Ψ is the corresponding wave function.

RESULTS

Due to the small width of the first Lyman and Balmer lines, fine structure splitting and Doppler broadening become important if the electron density lies under 10^{18}cm^{-3}. This is shown in the Figure. As it has been predicted in Ref.[4], for low densities the fine structure splitting has to be considered calculating He II lines. The difference between our theoretical profile and the experimental one is due to ion dynamics which has been neglected here.

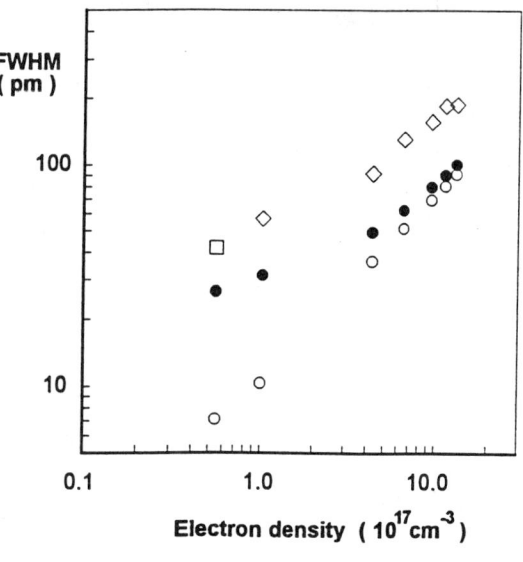

Full half-widths of the He II H_α line

theoretical results:
o without fine structure splitting and Doppler broadening
• including both effects

experimental results:
◇ [3]
□ [4]

REFERENCES

[1] S. Günter, L. Hitzschke and G. Röpke, *Phys. Rev.* A44, 6834 (1991).
[2] J. Halenka, *Z. Phys. D.: At. Mol. Cl.* 16, 1 (1990).
[3] K. Grützmacher, U. Johannsen, private communication.
[4] A. Piel, J. Slupek, *Z. Naturforsch.* 39a, 1041 (1984).

COLLISION INDUCED SPECTRA

ROTOTRANSLATIONAL COLLISION-INDUCED ABSORPTION BY H_2–H_2 PAIRS AT TEMPERATURES FROM 600 TO 7,000K

Chunguang Zheng and Aleksandra Borysow
Physics Department, Michigan Technological University

The quantum mechanical (semiclassical at high temperatures) computation of the far-infrared, rototranslational (RT) Collision Induced Absorption (CIA) spectra of H_2–H_2 pairs is presented for the first time at a temperature range from 600 to 7,000 K [1]. The work is of significant interest for those who are modeling the atmospheres of cool stars [2].

The CIA coefficient $\alpha(\omega)$ can be obtained directly from the spectral density function $g(\omega)$. Under the assumption of an *isotropic* intermolecular potential $V_0(R)$, $g(\omega)$ is known to be represented by a sum of the purely translational spectral profiles $G_{\lambda_1 \lambda_2 \Lambda L}(\omega)$. For the RT band, $G_{\lambda_1 \lambda_2 \Lambda L}(\omega)$ can be modeled by the well known 'BC' or 'K_0' model lineshapes [3]. Parameters of these model lineshapes are related to the three lowest spectral moments $M_{n;\lambda_1 \lambda_2 \Lambda L}$ ($n = 0, 1, 2$) of $G_{\lambda_1 \lambda_2 \Lambda L}(\omega)$. If $V_0(R)$ and the induced dipoles $\beta_{\lambda_1 \lambda_2 \Lambda L}(R)$ are known, $M_{n;\lambda_1 \lambda_2 \Lambda L}$ ($n = 0, 1, 2$) can be computed using the well known sum rules [4].

$V_0(R)$ and $\beta_{\lambda_1 \lambda_2 \Lambda L}(R)$ are input in our computations. The selection of $V_0(R)$ is most important for the accuracy of our high temperature computations. Several different $V_0(R)$ functions exist, they deviate mainly at short range ($R < 3$Å) which is most important at high temperatures. While functions $G_{\lambda_1 \lambda_2 \Lambda L}(\omega)$ computed from various available models of $V_0(R)$ remain almost identical in shape, their integrated spectral intensities ($M_{0;\lambda_1 \lambda_2 \Lambda L}$) vary significantly at high temperatures (up to 50% at 5,000 K). The experimental $V_0(R)$ by Ross *et al.* [5] was obtained from the shock wave measurements, where very high temperatures (up to 7,000 K) were achieved, thereby probing the potential at intermolecular separations as small as $\sim$1.5Å. Therefore we use it in our high temperature computations.

The *ab initio* induced dipoles $\beta^0_{\lambda_1 \lambda_2 \Lambda L}(R)$ (superscript '0' stands for the ground rotovibrational state of both interacting H_2) were taken from Meyer *et al.* [6]. Functions $\beta_{\lambda_1 \lambda_2 \Lambda L}(R)$ depend on the rotovibrational states (v, j) of both H_2. At long range, $\beta_{2023}(R)$ ($= -\beta_{0223}(R)$) takes the asymptotic form of a quadrupole-induced dipole ($\sqrt{3}\alpha Q/R^4$). To account for the (v, j) dependence of $\beta_{\lambda_1 \lambda_2 \Lambda L}(R)$ for the most intense terms $\lambda_1 \lambda_2 \Lambda L = 2023$ and 0223 (which contribute more than 80% to the total spectral intensity), we simply scale the entire $\beta_{\lambda_1 \lambda_2 \Lambda L}(R)$ in the same way that their asymptotic parts depend upon v and j, namely:

$$\beta_{2023}^{v_1 v_2 j_1 j_2 v'_1 v'_2 j'_1 j'_2}(R) = \left\{ \frac{\langle v_1 j_1 | Q | v'_1 j'_1 \rangle \langle v_2 j_2 | \alpha | v'_2 j'_2 \rangle}{\langle 00 | Q | 02 \rangle \langle 00 | \alpha | 00 \rangle} \right\} \beta^0_{2023}(R),$$

where $\langle v_1 j_1 | Q | v_1' j_1' \rangle$ and $\langle v_2 j_2 | \alpha | v_2' j_2' \rangle$ are matrix elements of the quadrupole moment and the polarizability of H_2 molecule. The error introduced by such a scaling procedure has been estimated to be less than 5% [1].

The influence of the *anisotropic* part of the intermolecular potential on our results has been examined. Classical RT spectral moments have been computed using the *ab initio* anisotropic potential [7], and its spherical part $V_{000}(R)$ as the isotropic part. Comparisons show that the corrections to the integrated spectral intensity due to the anisotropic part of the potential contribute less than 3% at high temperatures (>3,000 K) and even less at lower temperatures, leaving the spectral shapes almost unaffected.

Parameters of the model lineshapes are computed from $M_{n;\lambda_1 \lambda_2 \Lambda L}$ ($n = 0, 1, 2$) at temperatures from 600 to 7,000 K, thus the CIA spectra $\alpha(\omega)$ of H_2–H_2 pairs are obtained by modeling $G_{\lambda_1 \lambda_2 \Lambda L}(\omega)$. Vibrational transitions $v \to v' = v$, with $v=0, 1, 2$ and 3, are included in the computations. The RT CIA spectra of H_2–H_2 pairs computed at temperatures 1,000 K and 3,000 K are presented in Fig. 1.

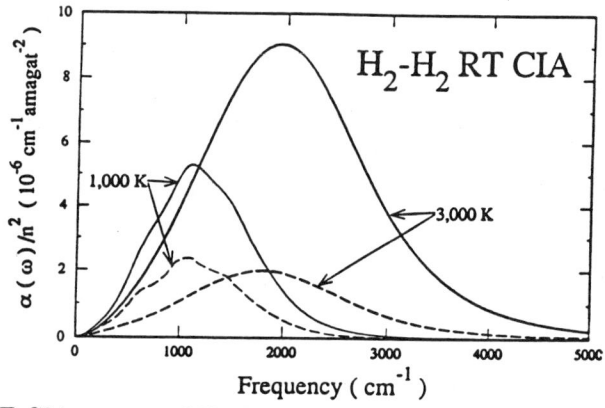

Fig. 1. RT CIA spectra of H_2–H_2 pairs. Solid lines denote the results of this work, dashed lines show previous work [8].

Acknowledgment We acknowledge the support from NASA, #NAGW - 3390.

[1] C. Zheng and A. Borysow. *Ap. J.*, 1994, in press.
[2] A. Borysow. Pressure – induced molecular absorption in stellar atmospheres. In Uffe G. Jørgensen, editor, *Molecules in the Stellar Environment*, Lecture Notes in Physics, pages 209–222. Springer-Verlag, Berlin, 1 edition, 1994.
[3] J. Borysow, L. Trafton, L. Frommhold, and G. Birnbaum. *Ap. J.*, 296:644, 1985.
[4] M. Moraldi, A. Borysow, and L. Frommhold. *Chem. Phys.*, 86:339, 1984.
[5] M. Ross, F. H. Ree, and D. A. Young. *J. Chem. Phys.*, 79(3):1487, 1983.
[6] W. Meyer, L. Frommhold, and G. Birnbaum. *Phys. Rev.*, 39A:2434, 1989.
[7] J. Schaefer and W. E. Köhler. *Z. Phys. D*, 13:217, 1989.
[8] J. L. Linsky. *Ap. J.*, 156:989, 1969.

THE EFFECT OF ROTATIONAL LEVEL MIXING IN FAR WINGS COLLISION INDUCED SPECTRA

W. Glaz, G.C. Tabisz
Department of Physics, University of Manitoba,
Winnipeg, Manitoba, R3T 2N2, Canada

Several authors have made attempts to analyse the influence of rotational level mixing effects due to anisotropic intermolecular interactions on collision induced spectra. In earlier works [1, 2, 3] approximations were made, especially regarding the collisional propagator U. In [2] only the diagonal matrix elements of the operator were considered; no propagation was taken into account.

In this work we analyse the influence of inelastic (i.e. level mixing) effects on the collision induced absorption (CIA) spectra through a different approach to the problem of the nondiagonal matrix elements of the operator U. In our calculations we use the full form of U with no reliance on a series expansion of the time dependence as discussed in [3].

We start developing our theory from the expression describing the single collision contribution to the purely collision induced spectrum [3]:

$$W(\omega) \sim \lim_{s \to 0} Re\, Tr_R \int_0^\infty \langle \mathbf{D}_e^{I\dagger}(\tau)\, U_e^I(\tau,0)\, \mathbf{D}_e^I(0)\, U_e^I(0,-\infty) \rangle_{av}\, e^{i\omega\tau - s\tau}\, d\tau \sum_B . \quad (1)$$

Here $\mathbf{D}_e$ is the induced dipole moment, $\sum_B$ is the Boltzmann distribution of states of the active molecule R. The subscript av denotes a collisional average and the superscript I the interaction representation.

The Laplace transform in this expression can be calculated for the high frequency region of the spectra by recourse to the method of stationary phase [4]. In order to apply that procedure, however, we need to calculate the matrix elements of the U operator. This is a non-trivial task if nondiagonal terms of the form, $\langle\langle J_i m_i\, J_j m_j | U_e^I | J_m m_m\, J_k m_k \rangle\rangle$, are to be considered.

In this work we show that the problem posed by such terms can be resolved by employing a two stage transformation of the wave functions to the basis in which the V operator, describing the anisotropic intermolecular interactions is diagonal. In this basis the nondiagonal elements of U can be easily found in the form:

$$\left\langle\left\langle \xi i \left| U_e^I(\tau+t_0,t_0) \right| \xi i \right\rangle\right\rangle = \left\langle\left\langle \xi i \left| \exp(\int_{t_0}^{\tau+t_0} V_a^I(s+t_0)ds) \right| \xi i \right\rangle\right\rangle$$
$$= \exp(-\frac{i}{\hbar}\int_{t_0}^{t_0+\tau} V_{\xi i}^I(s')ds') \quad (2)$$

where V_a^I is the anisotropic part of the radiator-perturber interaction and $V_{\xi i}^I \equiv \langle \xi|V^I|\xi\rangle - \langle i|V^I|i\rangle$.

We are thus able to obtain an expression describing the CIA absorption coefficient that gives the possibility of determining line shapes for a specific types of the intermolecular anisotropic interactions and molecular dynamics.

We apply our theory to a simple collision model and derive formulae describing the nondiagonal contributions to CIA spectra. The influence of the inelastic transitions is assessed by comparison of the results with those obtained when only the diagonal matrix elements of U are taken into account. Possible extension of the procedure to other collisional processes is suggested.

References

[1] Q. Ma, R.H. Tipping, and J.D. Poll, Phys. Rev. A **38**, 6185 (1988).

[2] G.C. Tabisz and J.B. Nelson, Phys. Rev. A **31**, 1160 (1985).

[3] B. Gao, G.C. Tabisz, M. Trippenbach and J. Cooper, Phys. Rev. A, **44**, 7379 (1991).

[4] M. V. Berry and M. Tabor, Proc. R. Soc. London Ser. A **349**, 101 (1976).

THE INDUCED DIPOLE MOMENT AND COLLISIONAL INTERFERENCE IN THE ROTATIONAL SPECTRUM OF HD-He AND HD-Ar

G. C. Tabisz and B. McQuarrie
University of Manitoba, Winnipeg, MB, Canada R3T 2N2

B. Gao and J. Cooper
JILA, University of Colorado, Boulder, CO, USA 80309

ABSTRACT

The contributions of various components of the induced dipole moment to interference phenomena in HD-inert gas rotational spectra are identified through calculation.

The small permanent dipole moment of HD makes it an exceptionally suitable molecule for studying interference effects between allowed- and collision-induced transitions. This paper concerns the pure rotational transitions in HD-inert gas mixtures, for which considerable experimental data have been amassed.[1]

A general theory of the phenomenon has been developed within the binary collision approximation that contains provision for allowed, induced and interference contributions.[2] Effects due to m mixing, J mixing and successive collisions are included.

The absorption coefficient $\alpha(\omega)$ for HD-X mixtures is the sum of Lorentzian and dispersion components:

$$\alpha(\omega) = \rho_{HD}\left[\frac{4\pi^2\omega}{3\hbar c}\right](1 - \exp(-\beta\hbar\omega))\, P(J)\, |\mu_{HD}|^2$$

$$\times \left[\frac{B\rho_X/\pi}{(\Delta-S\rho_X)^2 + (B\rho_X)^2}(1 + a\rho_X + b\rho_X^2) - \frac{(\Delta - S\rho_X/\pi)}{(\Delta - S\rho_X)^2 + (B\rho_X)^2}(c\rho_X + d\rho_X^2)\right]$$

Here ρ is the density, $P(J)$ is the Boltzmann distribution function for the initial rotational state J, μ_{HD} is the permanent dipole moment of HD, B and S are the broadening and shift coefficients. The parameters a, b, c and d depend on the pair induced dipole moment and describe the interference effect.

This formalism was used to perform calculations based on classical collision trajectories for the cases of HD-He and HD-Ar. The results demonstrated the crucial importance of inelastic rotational collisions and showed qualitative agreement with the experimentally determined parameters.[3] Input to these calculations included induced dipole moment functions obtained by ab initio methods.[4] Components of various origin have been calculated, of which the most important for purely collision-induced transitions are the isotropic overlap (01) and quadrupole induced (23) moments. The Table shows for HD-He the contributions of the various components to the calculation of a and c, which are involved in the leading terms describing the effect of interference on the total intensity and on the asymmetry of the profile, respectively.

In general, the (10) component which has the same symmetry as the permanent HD dipole dominates the calculation of a, determining its sign. The net effect of the other components which are all of the opposite sign is to reduce the magnitude of a, usually by more than 50%.

That the (10) component is most important is to be expected since it can produce interference effects even for an isotropic HD-X interaction. The anisotropic potential actually used in the calculations mixes in other components, whose contributions are clearly required.

Remarkably, the contribution of the (10) component to c is negligible. The profile asymmetry is thus determined by the quadrupole induced, isotropic overlap and anisotropic overlap components.

Table I: Values of interference parameters for R(1) of HD-He at 295 K

induced dipole component[4]	a (10^{-3} amagat^{-1})	c (10^{-3} amagat^{-1})
01	-0.23	-2.2
10	+6.9	-0.003
12	-0.21	-0.47
21	-1.3	+0.80
23	-0.14	-0.07
32	-0.07	+0.02
34	-0.02	+0.02
total	4.9	-1.9
expt.	5.7(9)	

REFERENCES

1. See, for example, Z. Lu, G. C. Tabisz and L. Ulivi, Phys. Rev. A47, 1159 (1993).
2. B. Gao, G. C. Tabisz, M. Trippenbach and J. Cooper, Phys. Rev. A44, 7379 (1991).
3. B. Gao, J. Cooper and G. C. Tabisz, Phys. Rev. A46, 5781 (1992).
4. A. Borysow, L. Frommhold and W. Meyer, J. Chem. Phys. 88, 4855 (1988).

The influence of the anisotropic potential on the spectral moments of collision-induced absorption of CO_2 and N_2 pairs

Marcin Gruszka and Aleksandra Borysow

Department of Physics, Michigan Technological University; Houghton, MI 49931

Abstract: Using recently developed formulas [1, 2] we computed the two lowest spectral moments of collision-induced roto-translational absorption band of two molecular systems: CO_2 and N_2. Assuming the best available anisotropic potentials we obtained [3] a good agreement with experimental values.

Introduction: In order to compute the zeroth, γ_1, and the second, α_1, spectral moments, the intermolecular potential and the induction mechanisms have to be known or else assumed. For both considered gases advanced anisotropic potentials exist and a polarizability tensor and electrostatic multipole moments are known. The measurements of far infrared CIA band are also available and therefore a quantitative comparison between the experimental and the computed spectral moments can be made.

CO_2: We used four different CO_2–CO_2 potential models: $PQ(CO_2)$ – site-site model with a point quadrupole in the center of mass [4], 3CB.II – site-site, semi-*ab initio* model [5], CS – an isotropic L-J type potential with ϵ and σ taken form the corresponding states principle [6] and CS+$\Theta\Theta$, – the CS model augmented with a point quadrupole moment. The induced dipole moment was assumed to be due to the quadrupole and hexadecapole induction, however, the second order contributions have been accounted for. In Fig.1

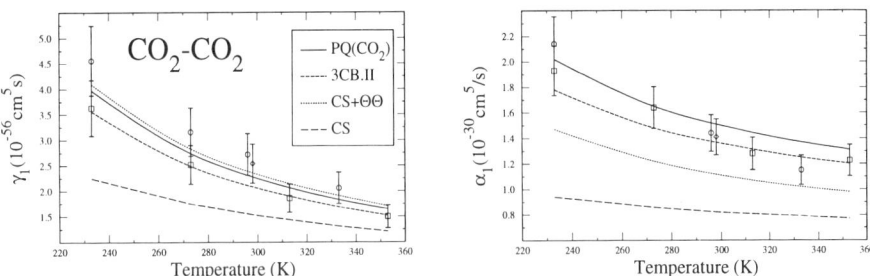

Figure 1: Computed and experimental values of γ_1 and α_1 for CO_2 pair.

we compare the values of computed spectral moments with those obtained from experiments. Both site-site anisotropic potentials $PQ(CO_2)$ and 3CB.II, gave spectral moments within an experimental error. The isotropic potential (CS) proved to be insufficient in providing a realistic representation of an intermolecular interaction. The CS+$\Theta\Theta$ model reproduced the γ_1 moment with a very good accuracy but the obtained values of α_1 were too small over the entire range of temperatures. This feature disqualifies that model as a realistic representation of CO_2–CO_2 potential.

N_2: We used two state of the art N_2–N_2 potentials [7, 8]. Since the results obtained for both models are similar, we present here only the computations done with van der Avoird *et al.* model [8], marked as AV in Fig.2. The anisotropic potential leads to α_1 which is *two times* too large, when pure electrostatic dipole induction is assumed. At the same time spherically averaged AV model (AV_{000} in Fig.2.) gives a good agreement with experiments using the same dipole components. However, a similarly good result can be achieved when the anisotropic potential is used and an overlap correction to the induction mechanism is introduced – AV(ov.cor.). We assumed the standard exponen-

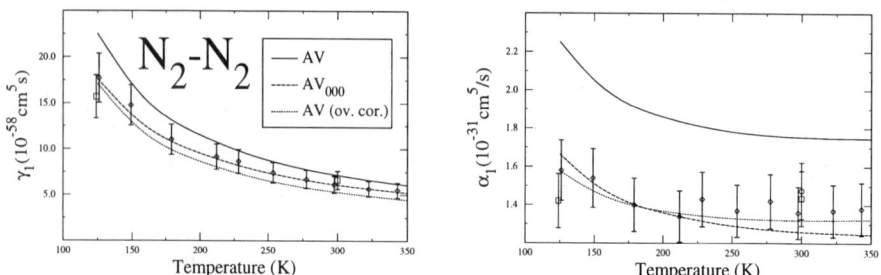

Figure 2: Computed and experimental values of γ_1 and α_1 for N_2 pair.

tial correction ($\lambda e^{-(R-\sigma)/\rho}$) to the quadrupole induced dipole terms ((2023) and (0223)) with $\lambda=1.875*10^{-3}$ au, $\sigma=7.0$ au and $\rho=0.125\sigma$.
Although it is difficult to find the exact form of the overlap based on the spectral moments only, it is evident that the overlap plays an important part in the induction process during N_2–N_2 collision.

Summary: We show that it is possible to get a good agreement between the experimental and theoretical values of two lowest spectral moments of CIA band in CO_2 and N_2 gas using a realistic anisotropic potential. Only the electrostatic induction was assumed in creation of CO_2–CO_2 supermolecular dipole but it was necessary to introduce an overlap contribution to the induction mechanism in N_2–N_2 complex.

Acknowledgments: The support by NASA, Planetary Atmospheres Program, grant # NAGW-3390 is gratefully acknowledged.

References:
1. A. Borysow and M. Moraldi. Phys. Rev. Lett., 68:3686,1992.
2. M. Moraldi. Phys. Rev. A, 1994, *to be submitted*.
3. M. Gruszka and A. Borysow. J. Chem. Phys., 1994, *submitted*.
4. C. S. Murthy, K. Singer and I. R. McDonald. Mol. Phys., 50:531,1983.
5. H. J. Bohm. Mol. Phys., 53:375, 1985.
6. G. C. Maitland, M. Rigby and E. B. Smith and W. A. Wakeham. *Intermolecular forces - their origin and determination.* Clarendon Press, 1981.
7. M. S. H. Ling and M. Rigby. Mol. Phys., 51:855,1984.
8. A. van der Avoird, P. Wormer and A. Jansen. J. Chem. Phys., 84:1629,1986.

INVESTIGATION OF COLLISION INDUCED LINE SHAPE OF A SATELLITE ASSOCIATED WITH $4s^2\ ^1S_0 - 4s3d\ ^1D_2$ TRANSITION OF CALCIUM*

M. A. Gondal, M. A. Khan and M. H. Rais
Laser Research Section, The Research Institute
King Fahd University of Petroleum & Minerals
Dhahran 31261, Saudi-Arabia

Collision induced line shapes of a satellite associated with the quadrupole transition $4s^2\ ^1S_0 -- 4s3d\ ^1D_2$ of calcium at 4575A^0 have been investigated using laser enhanced ionization technique. The blue satellite is at 48A^0 from isolated atomic calcium line. The $4s^2\ ^1S_0 - 4s3d\ ^1D_2$ transition is forbidden in dipole approximation and is normally very weak (oscillator strength ~ $6.2*10^{-7}$). However substantial oscillator strength can be induced into such a forbidden transition when absorption takes place during a collision with a perturber atom. In the presence of a perturber, such forbidden lines are often accompanied by a satellite band[1,2]. We studied the line shape of the satellite band due to the Ca-Ar complex in a single photon excitation of calcium vapor under the environment of Ar inside a thermionic diode.

Calcium vapor was generated in a stainless steel pipe whose central 80 mm length was heated to a temperature of 850 ^{0}C. A 2.5 mm diameter stainless steel rod placed at the center of this heat pipe acted as the anode while the pipe itself acting as the cathode was grounded. A small voltage (2.5 V) was applied through a current limiting resistor for detection of electrons. An excimer laser pumped dye laser using Coumarine 120 dye was used for pumping the Ca vapor. The laser enhanced ionization signal corresponding to the quadrupole transition of Ca or Ca-Ar complex was fed into a boxcar averager and a signal processor[3-5].

Fig.1 shows a typical spectrum where the laser wavelength was tuned from 4473 to 4600 A^0. The $4s^2\ ^1S_0 -- 4s3d\ ^1D_2$ transition and its satellite can be clearly identified at 4575 and 4527A^0 respectively although a number of other transitions have been observed. The Ar pressure here was 500 mbar. As seen in fig. 2, this satellite is truly sensitive to the pressure of Ar. As the Ar pressure is increased, it broadens, its intensity increases and an asymmetry appears in the shape of the profile.

The width δv of this satellite at different Ar pressure was measured and the rate co-efficient for broadening was calculated from the slope of the δv vs N plot, where N is the number density of Ar. A value of $1.14*10^{-15}$ m^3/sec was obtained.

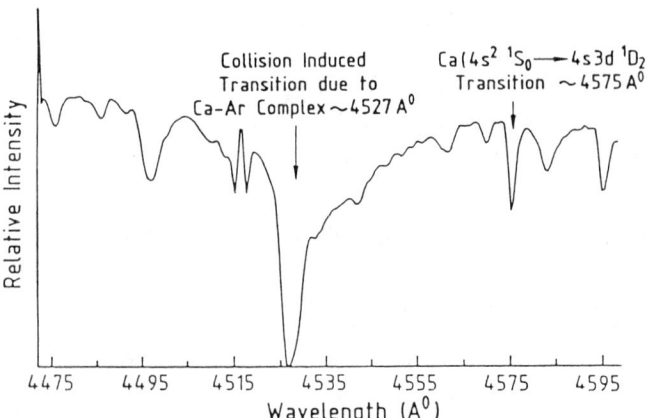

Fig.1 Typical laser enhanced ionization spectrum in the range of 4473-4600 A⁰ recorded at temperature =850 ⁰C and argon pressure=500 mbar.

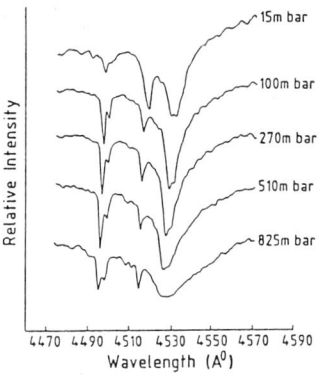

Fig.2 Laser enhanced ionization spectrum showing the broadening of the blue satellite at different argon pressure. Oven temperature=850 ⁰C.

* This work is a part of the laser research program (Project: 12043) supported by the research institute.

1. J. Coutts, S. K. Peck and J. Cooper, J. Appl. Phys. **64**, 977 (1988).
2. K. Ueda, Y. Ashizawa and K. Fukuda, J. Phys. Soc. Jpn. **50**, 632 (1981)
3. M. A. Khan, M. A. Gondal and M. H. Rais, Appl. Phys. **B57**, 123 (1993).
4. M. A. Khan, M. A. Gondal and M. H. Rais, J. Phys. B, **27** (in press).
5. M. A. Khan and M. F. Al-Kuhaily, J. Phys. B, **26**, 393(1993)

EFFECT OF THE ATOMIC POLARIZATION ON THE SPECTRAL LINE FAR WINGS INDUCED BY ANISOTROPIC COLLISIONS

A. L. Zagrebin, M. G. Lednev

Department of Optics and Spectroscopy, Institute of Physics,
St.Petersburg University, 198904, St.Petersburg, Russia

Atomic spectral line wings induced by slow anisotropic collisions of polarized atoms in crossed beams are depended on the initial polarization of the colliding atoms. This polarization effect is caused by the fact that the probabilities of the populations of the radiating quasimolecular terms in the region of the coupling type c (according to Hund) are depended on the initial polarization of the colliding atoms.

Let us consider the influence of the initial polarization of the excited $X(J)$ atoms on the shape $I(E, \Delta\omega)$ of the spectral line far wing induced by the collisions

$$X^*(J) + Y(^1S_0) \rightarrow X(^1S_0) + Y(^1S_0) + \hbar\omega$$

in crossed monokinetic beams (E is a relative kinetic energy). For an arbitrary initial polarization the spectral distribution $I(E, \Delta\omega)$ is expressed [1] in terms of the $I_{M_v}(E, \Delta\omega)$ distributions for the collisions with the fixed projection M_v of the angular momentum $\vec{J}$ on the z_v axis which is determined by the direction of the initial relative velocity $\vec{v}$. Particularly for the collisions with the fixed value of the projection M_0 of the momentum $\vec{J}$ on an arbitrary space-fixed axis z_0 we have

$$I_{M_0}(E, \Delta\omega) = \sum_{M_v = -J}^{J} \left| d^{(J)}_{M_v M_0}(\alpha) \right|^2 I_{M_v}(E, \Delta\omega),$$

where α is the angle between the z_0 and z_v axis, $\hat{d}^{(J)}(\alpha)$ is the matrix of the finite rotations with the angels $(0, \alpha, 0)$.

The spectral distributions $I_{M_v}(E, \Delta\omega)$ is given by [1]

$$I_{M_v} = \sum_{\Omega = -J}^{J} \sum_{R_c\Omega} \frac{4\pi \Gamma_\Omega(R_{c\Omega})}{|\Delta F_\Omega(R_{c\Omega})|} \int P_{\Omega M_v} \left(1 - \frac{b^2}{R_{c\Omega}^2} - \frac{U_\Omega(R_{c\Omega})}{E}\right)^{1/2} b\,db,$$

where b is an impact parameter; Ω is the projection of the total electron angular momentum on the rotating internuclear axis; U_Ω, Γ_Ω and ΔF_Ω are the interaction potential, the radiativ transition probability and the difference strength between the excited and the ground terms attributed to the Ω quasimolecular state; $R_{c\Omega}$ is Condon point; $P_{\Omega M_v}(E,b)$ is the probability of the population of the Ω - state in the region of the coupling type c during the collision with the fixed initial value of M_v in the region of the coupling type e (large R). In the approximation of sudden coupling type change $e \to c$ the $P_{\Omega M_v}(E,b)$ probabilities are expressed in terms of $|d^{(J)}_{\Omega M_v}|^2$ [1]. In the case of $J = 1$ the analytical expressions for $P_{\Omega M_v}$ are obtained without this approximation [2,3].

The spectral distributions $I_{M_v}(E, \Delta\omega)$ for the allowed $6\ ^1P_1 - 6\ ^1S_0$ (Fig. 1) and forbidden $6\ ^3P_2 - 6\ ^1S_0$ transitions induced by Ba* + He collisions are calculated on the basis of the approximations [1,2,3] for the $P_{\Omega M_v}$ probabilities. The calculated interaction potentials $U_\Omega(R)$ and the radiative transition probabilities $\Gamma_\Omega(R)$ are used for the far wing simulation.

The authors gratefully acknowledge the Russian Foudation of the fundamental researches for supporting of this investigation.

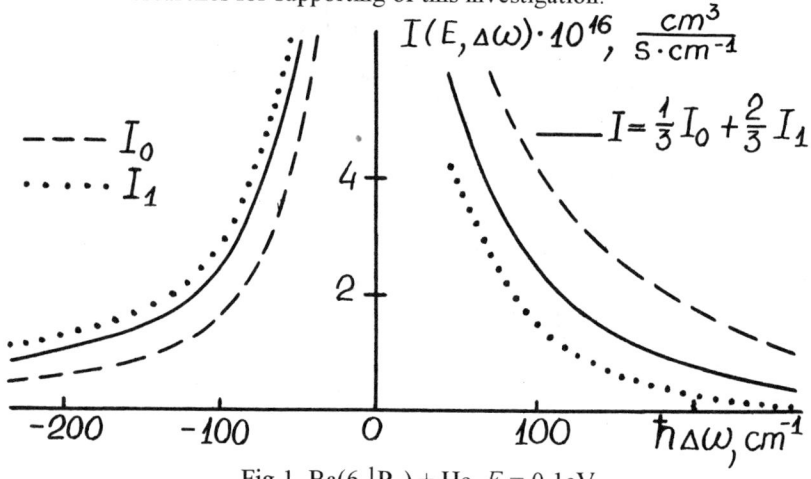

Fig. 1. Ba($6\ ^1P_1$) + He, $E = 0.1$ eV

1. A. L. Zagrebin, Sov.Phys.JETP, 70, 63 (1990).
2. A. L. Zagrebin, Opt.Spectrosk., 74, 666 (1993).
3. A. L. Zagrebin, Opt.Spectrosk., 75, 1134 (1993).

EFFECT OF COLLISIONS ON THE ATOMIC FORBIDDEN TRANSITIONS:
$He(2\ ^1S, 2\ ^3S) + Ne \rightarrow He(1\ ^1S) + Ne + \hbar\omega$

A.L.Zagrebin

Institute of Physics, St.Petersburg University, 198904, St.Petersburg, Russia

S.I.Tserkovnyi

Departement of Physics, Baltic State Thechnical University, 198005, St.Petersburg, Russia

The quasimolecular radiative decay during the collision
$$He(2\ ^1S_0) + Ne \rightarrow He(1\ ^1S_0) + Ne + \hbar\omega$$
is induced by interatomic interaction which cause the mixing of the wave functions of the resonance and metastable states. The quasimolecular radiative transition probability A (Fig.1) is determined as a function of the interatomic distance R on the basis of the multiconfiguration pseudopotential calculation of the molecular terms [1].

The spectral distributions of the photons emitted during slow collisions are calculated for the gas cell (Fig.2) and crossed beams conditions.

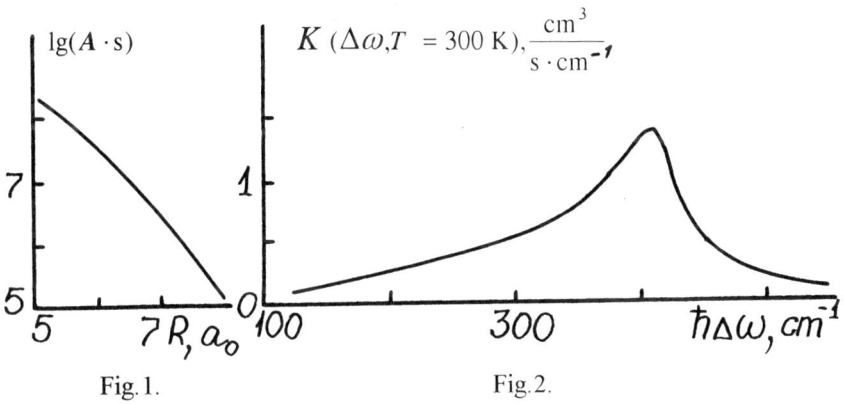

Fig.1.   Fig.2.

The radiative quenching cross sections for He($2\ {}^1S_0$) + Ne collisions are calculated as functions of relative energy in crossed beams and gas temperature. The radiative decay determine the quenching of the metastable $2\ {}^1S$ state at low energy collisions (E ≤ 0.02eV), because the excitation transfer is not essential in this case [2].

E, eV	$\sigma(E)$, 10^{-20}cm^2	T, K	$\langle\sigma(T)\rangle$, 10^{-20}cm^2
0.02	0.027	100	0.16
0.03	0.140	200	0.80
0.05	1.3	300	1.5
0.10	6.2	500	2.4
0.15	9.4	1000	3.0

The spin changing quasimolecular radiative transitions

$$\text{He}(2\ {}^3S_1) + \text{Ne} \rightarrow \text{He}(1\ {}^1S_0) + \text{Ne} + \hbar\omega$$

are induced by spin orbit and interatomic interactions together. In the classical allowed region (at E < 0.1eV) this additional calculated radiative transition probability is negligibly small compared to radiative decay probability of the He($2\ {}^3S_1$) atomic state. Therefore the radiative quenching is not essential in this case.

The experimentally investigated [3] radiative decay processes for He($2\ {}^1S_0$) + Ar, Kr, Xe collisions are discussed also.

The authors gratefully acknowledge the Internation Science Foundation for supporting of this research.

1. A.Z.Devdariani, A.L.Zagrebin, K.B.Blagoev, Ann.Phys.Fr., 14, 467 (1989).
2. A.Z.Devdariani, A.L.Zagrebin, K.Blagoev, Ann.Phys.Fr., 17, 365 (1992).
3. C.Dehnbostel, R.Feltgen, G.Hoffmann, Phys.Rev.A, 42, 5389 (1990).

COLLISION INDUCED SPECTRA:
Hg(6 $^3P_2$) + He, Ne, Ar, Kr, Xe → Hg(6 $^1S_0$) + He, Ne, Ar, Kr, Xe + $\hbar\omega$

A. Z. Devdariani, A. L. Zagrebin

Department of Optics and Spectroscopy, Institute of Physics,
St.Petersburg University, 198904, Russia

M. G. Lednev

Department of Physics, Baltic State Technical University,
St.Petersburg, 198005, Russia

The interaction between atoms during the collisions removes the prohibition for optically forbidden atomic transitions. In this case one can speak about transitions in a transient molecule or quasimolecule which exists during collisions only.

The radiative transitions in the Hg(6 $^3P_2$) + He, Ne, Ar, Kr, Xe quasimolecules are induced by spin-orbital interaction between $^1P_1$ and $^3P_1$ states in a single atom and by coupling between 1($^3P_1$) and 1($^3P_2$) quasimolecular states caused by interatomic interaction during the collisions. For the first time the probability of such transitions have been estimated on the basis of asymptotic theory in [1]. In the present study the semiempirical procedure developed in [2,3] has been used to calculate the potentials generated by the Hg(6s6p $^3P_2$) + Y($^1S_0$) configuration and the radiative transition probabilities (Y - is Ne, Ar, Kr or Xe atom). The pseudopotential calculations [4] have been used in case of Y = He.

The desired spectra have been calculated in the frame of quasistatic approach and averaged with respect to Maxwell's distribution at different temperatures in the limiting cases of low (Fig.1) and high pressure. The former corresponds to taking into account the influence of the centrifugal barrier on the movement of atoms in the 1($^3P_2$) potential. The results of the calculations are in agreement with experimental data [5,6] for the Hg* + Xe collisions.

The authors gratefully acknowledge the Internation Science Foundation for supporting of this research.

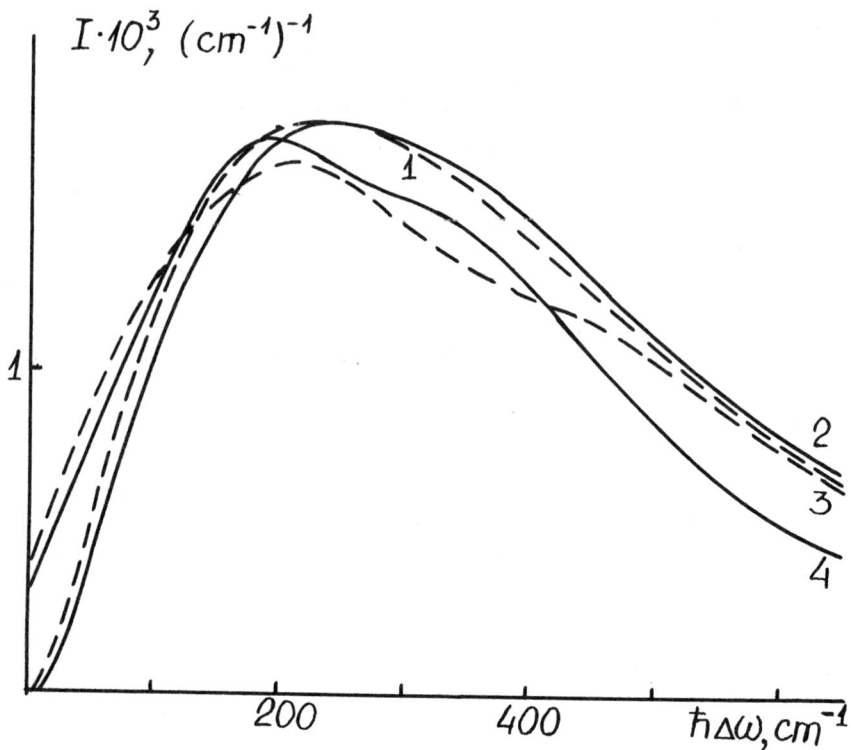

Fig.1. The Hg(6 $^3P_2$) + Y collisions induced spectra at T = 500K.
Y = Ne(1), Ar(2), Kr(3), Xe(4).

1. A.Z.Devdariani, A.L.Zagrebin, Opt. Spectrosc., 58, 752, (1985).
2. A.Z.Devdariani, A.L.Zagrebin, K.B.Blagoev, Ann.Phys.Fr., 14, 467, (1989).
3. A.L.Zagrebin, M.G.Lednev, Sov.Tech.Phys.Lett., 18, 243, (1992).
4. A.L.Zagrebin, M.G.Lednev, S.I.Tserkovnyi, Opt.Spectrosc., 74, 24, (1993).
5. A.B.Callear, K.Du, Chem.Phys., 113, 73, (1987).
6. N.A.Krjukov, N.P.Penkin, T.P.Red'ko, 66, 1235, (1989).

SPECTRAL INVARIANTS FOR THE ABSORPTION COEFFICIENT OF CO_2–AR PAIRS

Massimo Moraldi

Dipartimento di Fisica, Universita' di Firenze, 50125 Firenze, Italy

Aleksandra Borysow

Physics Department, Michigan Technological University
Houghton, Michigan 49931

We present calculations of the invariants γ_1 and α_1 of the rototranslational absorption coefficient for a mixture of CO_2 and Ar in the gaseous phase and at various temperatures. We have assumed an induced dipole moment composed of spherical components due to pure multipolar induction: the CO_2 quadrupole and hexadecapole moments inducing a dipole on Ar and this same dipole inducing a dipole on CO_2. Overlap induced dipoles are neglected which is considered a good approximation for molecules of relatively large polarizability. Six different models of the intermolecular potential have been used. These models can be found in the literature (see references cited in ref. 1) and range from an isotropic potential derived from the correspondence principle, which is able to reproduce second virial coefficient and transport properties, to more sophisticated, semiempirical potentials with anisotropic components, which reproduce also the molecular beam scattering cross sections. ¿From the comparison of our computational results[1] with the experimentally determined spectral invariants it is seen that only one[2] of the considered anisotropic potentials can reproduce the measurements at all temperatures within the experimental uncertainties. We have also examined the contributions to the absorption coefficient which result from the mixing of dipole components of different symmetry (e.g.: quadrupolar and hexadecapolar components), which are determined by the anisotropic part of the intermolecular potential. Indeed, such a contribution is ignored in most of the calculations of the rototranslational spectra. Quantum mechanical calculations of the spectral density, for example, are usually performed by neglecting the anisotropic part of the potential, because of the complexity of close coupling methods when applied to heavy molecules like CO_2. Also empirical models for the spectral density usually employ the separation of rotations and translations, in order to simplify the computational task. Our present calculations show, however, that mixing terms are not negligible, and amount to about 6% and 13% of γ_1 and α_1 respectively, when the potential of ref. 2 is used.

Acknowledgments: The grant from NASA, Planetary Atmospheres Program, NAGW-3390 is gratefully acknowledged.

1. A. Borysow, M. Moraldi, J. Chem. Phys. 99, 8424 (1993)
2. R. K. Preston, R. T. Pack, J. Chem. Phys. 66, 2480 (1977)

SELECTIVE REFLECTION
AND OTHER LINE SHAPES

ATOMIC SPECTRAL LINE SHAPES FROM SELECTIVE REFLECTION LASER SPECTROSCOPY

M. Fichet, N. Papagiorgiou, F. Schuller, D. Bloch and M. Ducloy
Laboratoire de Physique des Lasers, URA 282 du C.N.R.S.
Institut Galilée - Université Paris-Nord
F - 93430 Villetaneuse, France

ABSTRACT

Selective reflection spectroscopy at an interface with a low density resonant vapor, combined with a frequency modulation technique (FM-SR) is a high-resolution Doppler-free method for probing atoms interacting with a surface. We analyse the spectral consequences of the van der Waals - London attraction potential between the atomic dipole and its instantaneous image.
We show that FM-SR provides a way to analyse separately for each hyperfine component, collisional processus on a resonance line.
Finally, we present a new formulation for the van der Waals interaction between a dispersive dielectric surface and an atom in an arbitrary internal energy state. We show that for excited atoms with dipole couplings in emission, giant surface attraction or repulsion is predicted much larger than for perfect reflector.

I- PRINCIPLE OF FM-SR SPECTROSCOPY

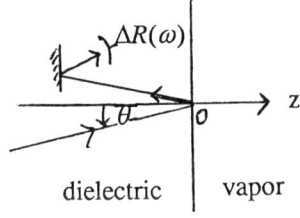

Fig.1: Principle of the selective reflection

Selective reflection consists of monitoring the spectral dependence of the reflectivity change $\Delta R(\omega)$ at a vapor-dielectric interface when the incident beam at frequency ω becomes resonant ($\omega = \omega_0$) for the vapor (fig.1).

Early works [1-4] are summarized hereafter. The incident beam induces optical dipoles in the vapor and the field reemited $\Delta E_r(\omega)$ by all induced dipoles is added to the non-resonant reflected field E_r. Under normal incident radiation ($\theta = 0$) one gets:

$$\Delta E_r(\omega) = \frac{ik}{(n+1)\varepsilon_0} \int_0^\infty p(z)e^{2ikz}dz$$

© 1995 American Institute of Physics

with p(z) the complex amplitude of the induced dipole, z the atom-wall distance, n the dielectric refractive index and $k = \omega/c$.

To evaluate the macroscopic dipole polarisation p(z) we assume that we can describe the vapor as a collection of two-level atomic systems (ground (g) and excited (e) states) whose density matrix is governed by Bloch equations. The microscopic atomic dipole is proportional to the off diagonal density matrix element σ_{eg}. The macroscopic polarisation p(z) is determined by the average over the velocity distribution of all the atomic dipole moments at position z:

$$p(z) = N\mu \int_{-\infty}^{+\infty} W(\vec{v}) \sigma_{eg}(z,\vec{v}) d^3v$$

where µ is the e-g electric dipole moment, N the atomic density and $W(\vec{v})$ the normalized velocity distribution. Atomic collisions with the surface distroy the optical excitation of arriving atoms. This implies that departing atoms ($v_z \rangle 0$) follow a transient evolution in their interaction with the beam light field, while arriving atoms ($v_z \langle 0$) are in a permanent regime. This discontinuity in the velocity integration is responsible for an extra symmetric sub-Doppler contribution in the SR response (fig.2). This singularity around $\omega = \omega_0$ may be isolated by applying frequency modulation to the incident light, and we obtain a derivative of the lineshape $dI/d\omega$ when the amplitude-modulated reflected light is detected. By this way, the large Doppler background gets negligible while the sub-Doppler contribution is turned into a pure Doppler-free dispersion lineshape which originates from atoms moving parallel to the surface ($v_z = 0$) (fig.3).

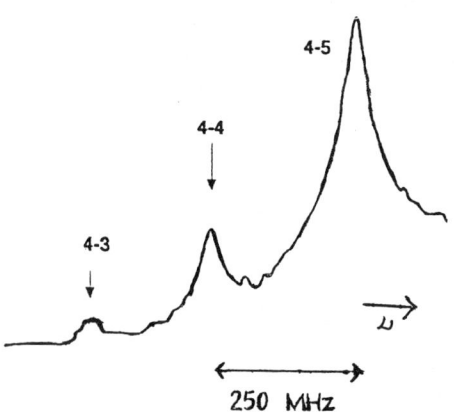

Fig.2: SR spectrum with cesium D$_2$ transition:
6S$_{1/2}$ (F=4)-6P$_{3/2}$(F'=3,4,5)
T=140°C, silica glass surface

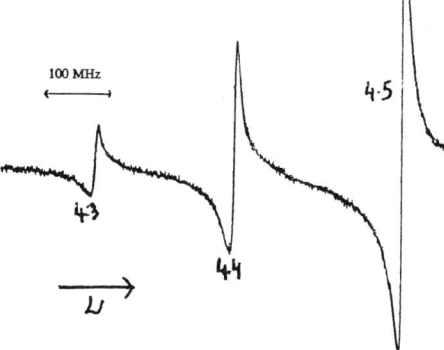

Fig.3: FM-SR spectrum with the same conditions as fig.2

II- ATOMS/SURFACE ATTRACTION EFFECTS ON FM-SR SPECTRA

Since the volume probed in the FM-SR spectroscopy is typically limited to $z \approx \lambda$ and that the dominant contribution originates from atoms almost parallel to the surface, the FM-SR technique seems a sensitive way to measure the long-range surface interaction. We have made the hypothesis [4] that the only interaction observed by this method is the London-van der Waals (vW) attraction scales in z^{-3}. This is a dipole-dipole attraction originated in the coupling between the fluctuating atomic dipole and its electric image.

We have modelled the vW attraction with the following expression for the atomic resonance frequency:

$$\omega_0(z) = \omega_0 - \frac{C_3}{z^3} \ .$$

The predicted lineshapes depend only on a single dimensionless parameter $A = \frac{2C_3 k^3}{\gamma}$ where γ is the optical linewidth. The parameter A gives the vW transition shift in $\hbar\gamma$ units at a distance $\lambda/2\pi$. Two regimes are observable: the weak vW regime (A < 1), characterized by a red shift, as expected for an attraction process, and a dispersion-like lineshape with small asymmetry. In strong vW regime (A > 1), quite anomalous lineshapes are observed including absorption-like or even inverted dispersion lineshapes (fig.4).

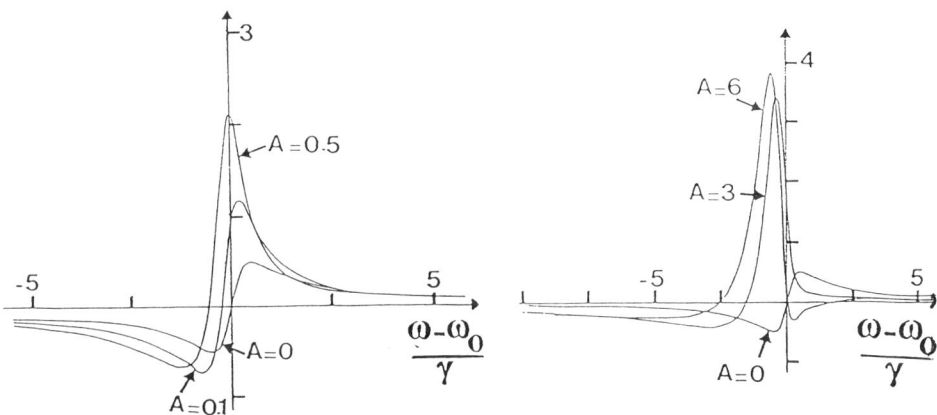

Fig.4: Theoretical FM-SR lineshapes for various strengths of the vW attraction. Vertical scales are in arbitrary units, normalized to A=0.

In Villetaneuse, we have performed [5] the first time spectral evidence of the vW attraction by the FM-SR method: red frequency shift (-3(±1)MHz) and line asymmetry with a notably broader red wing have been observed at a (silica glass)/Cs vapor interface, with the D_2 transition ($6S_{1/2} - 6P_{3/2}$, $\lambda = 852$ nm) (fig.5).

On fig.5, we present a simultaneous recording in a test cellule, filled with Cs, of the saturated absorption in volume with amplitude modulation (AM-SA). This gives a reference signal to evidence shift and asymmetry associated to surface effects. On fig.5, we give the theoretical FM-SR lineshape for A=0.2 which coincides (shift and asymmetry) with the measured spectrum (low pressure vapor (p < 2mTorr) and low power irradiation (I < 10μW/mm^2) conditions).

A strong vW regime is offered by the second transition in Cs ($6S_{1/2} - 7P_{3/2}$, $\lambda = 455$nm) related to the higher polarisability of the $7P_{3/2}$ and because the wavelength is shorter [6]. On fig.6, we present the spectrum associated to this transition with p ≈ 18mTorr. We have an excellent fit for A = 14.

Accurate fitting of the experiments has allowed us to measure the strength of the vW attraction. We have found 1.37(0.4) kHz.μm^3 for the D_2 line with a silica window and 20(4) kHz.μm^3 for the second transition with a sapphire window. In section IV, we shall show that these experimental results are in good agreement with calculated values.

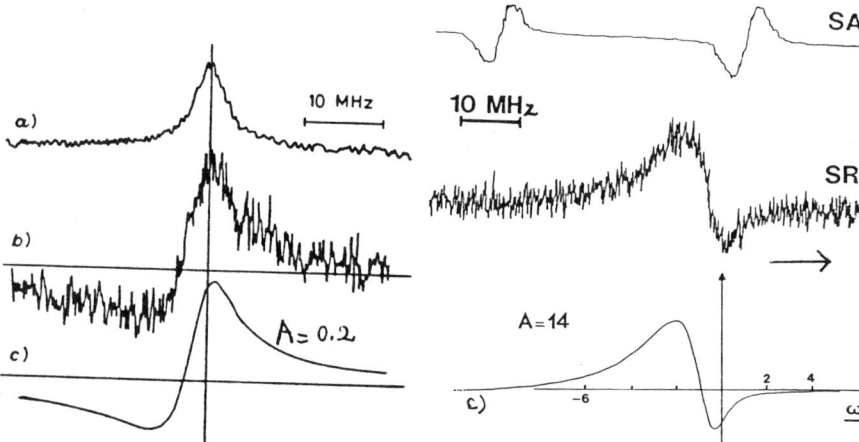

Fig.5: Simultaneous recordings of a)AM-SA and b)FM-SR spectra for one hyperfine component of the Cs-D_2 transition (F = 4, F' = 5); c) theoretical FM-SR lineshape for A = 0.2

Fig.6: Simultaneous recordings of a) FM-SA and b) FM-SR spectra for one hyperfine component of the Cs second transition (F = 4, F' = 5); c) theoretical FM-SR lineshape for A = 14

III- HIGH RESOLUTION FM-SR SPECTROSCOPIC MEASUREMENTS

We have performed series of FM-SR experiments on the D_2 line of Cs vapor possibly mixed with Kr buffer gas [7], to analyze precisely collisional effect (broadening and shift). For this purpose, we have elaborated a numerical fit which includes the distorsions induced by the vW interaction, the finite width correction for the Doppler broadening and the overlap between hyperfine components of the transition. This analysis shows that even when the collision width considerably exceeds the mean vW shift, the surface interaction remains responsible for weak but observable distorsion lineshapes, which strongly affect the collisional shift measurement. This is remarkable on fig.7 where we present the best fit obtained without vW interation (A = 0) in a case of strong collisional regime (γ_c ~ 60 MHz compared to the extrapolated value γ_n ~ 8 MHz for Cs density $\to$ 0). Notice that the distorsions caused by the vW force are quite visible, while the vW shift is only ~ 4.4 MHz for the distance $\lambda /4\pi$. The theoretical curve given by the complete fit coincides perfectly with the experimental curve and can not be distinguished on fig.7.

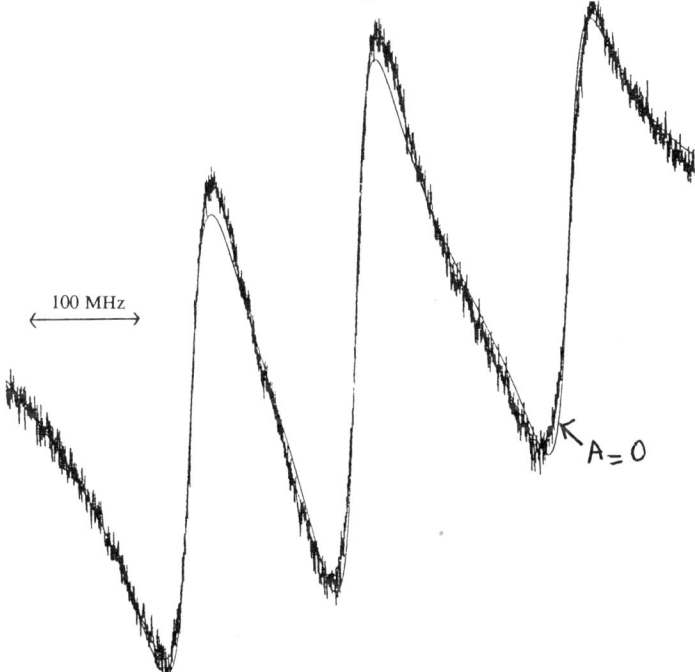

Fig.7: FM-SR spectrum for the Cs-D_2 transition (F = 3, F' = 2,3,4) in a cell with Cs and krypton (2 Torr), T ~ 150°C; we also indicate the best fit obtained with A = 0

Three different cells have been used: a pure Cs cell, a Cs cell with 0.75 Torr Kr and a Cs cell with 2 Torr Kr. We have shown that with a precise surface effects analysis that can separate the contribution of each hyperfine component and identify the distorsions caused on the lineshapes by the vW surface interaction, Doppler-free reflection spectroscopy (FM-SR) can be used for the investigation of optically dense vapor. With our method collision-induced broadenings and frequency shifts of individual hfs components of the D_2 line have been measured for Cs-Cs collisions and Cs-Kr collisions. Our results agree with previous measurements [7] and are given on tables 1 and 2. On table 1, notice that for each component the characteristic slope for the shift versus Cs density is positive except for the F = 4, F' = 5. For the self-broadening (Cs-Cs) we have found the same parameter for each hfs component of the transition: $\Delta\gamma / \Delta N = 1(\pm 0.2) \, 10^{-7}$ Hz.cm^3 (where N is the Cs density).

Transitions F-F'	4-3	4-4	4-5	3-2	3-3	3-4
shift/Cs density	8(±7)	8(±6)	-4(±4)	6(±4)	7(±6)	4(±4)

Table 1: Characteristic parameter for the Cs-Cs collisions induced frequency shift of D_2 line hfs components (shift/Cs density) in 10^{-9} Hz.cm^3

Transitions	Broadening in MHz/Torr	Frequency shift in MHz/Torr
$6S_{1/2}(F=4)-6P_{3/2}(F'=3,4,5)$	20(±2)	-4.6(±1)
$6S_{1/2}(F=3)-6P_{3/2}(F'=2,3,4)$	21(±2)	-5.4(±2)

Table 2: The Cs-Kr collision induced broadening and shift of hfs components of D_2 line

IV- vW INTERACTION BETWEEN EXCITED-STATE ATOMS AND DISPERSIVE DIELECTRIC SURFACE

We have improved [8] our model of the vW attraction for a dielectric surface, in introducing a frequency-dependent dielectric reflectivity $r(\omega)$ (equal to 1 for a perfect reflector surface). For a ground-state atom interacting with the surface we have:

$$r(\omega \rangle 0) = \frac{2}{\pi}\int_0^\infty \frac{\varepsilon(iu)-1}{\varepsilon(iu)+1} \frac{\omega \, du}{\omega^2 + u^2}$$

For excited-state atom interacting with dielectrics, we have to take account of each virtual coupling. When the virtual dipole coupling appears in absorption ($\omega \rangle 0$), transition with a superior energy level, the precedent expression is valid; but for a coupling in emission ($\omega \langle 0$), transition with a inferior energy level, we have:

$$r(\omega\langle 0) = \frac{2}{\pi}\int_0^\infty \frac{\varepsilon(iu)-1}{\varepsilon(iu)+1}\frac{\omega\,du}{\omega^2+u^2} + 2\,\mathrm{Re}\frac{\varepsilon(|\omega|)-1}{\varepsilon(|\omega|)+1}.$$

$r(\omega\rangle 0)$ is a monotonically decreasing function, always positive and less than one. On the other hand, for the case in emission, r presents sharp variations around the associated surface resonances [8]. The consequence is that giant vW attraction or repulsion ($r\langle 0$) should be observable for excited atoms whose main emission coupling should be in the resonant region.

These functions are presented on fig.8 for the birefringent sapphire. On fig.8 (b), we notice that $r(|\omega|)$ varies from 18 to -17 around $|\omega| = 0.2456\,10^{14}$ Hz ($\lambda = 12.21$ μm).

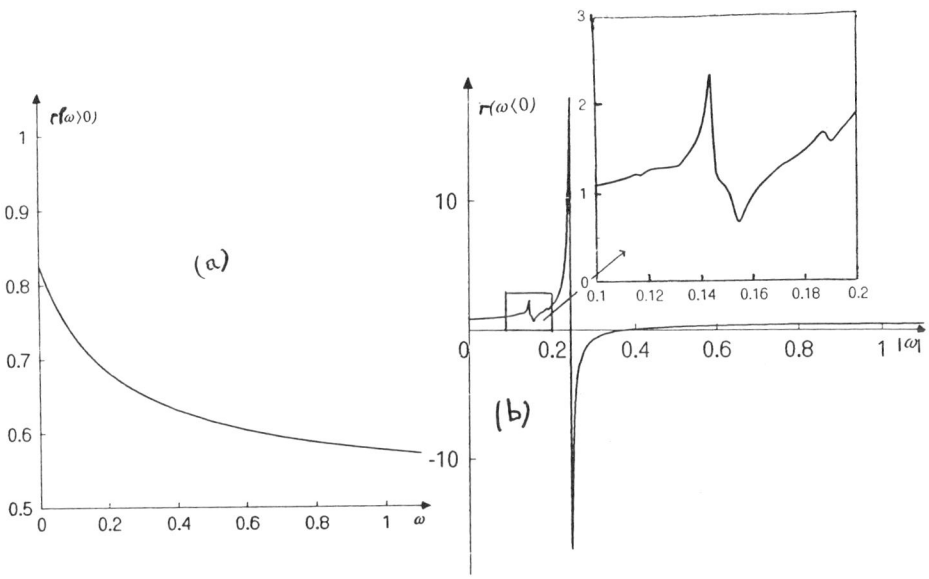

Fig.8: $r(|\omega|)$ versus ω (in units of 10^{14}Hz) for the birefringent sapphire:
(a) $\omega\rangle 0$, (b) $\omega\langle 0$

With this model the calculated value for the vW attraction for the cesium $6S_{1/2}$-$7P_{3/2}$ transition with sapphire window is of the order of 15 kHz.μm^3, in better agreement with the experimental determination (20(4) kHz.μm^3) than our previous theoretical estimation (11 kHz.μm^3) [6].

Now, in Villetaneuse, an experiment is in progress with the $6D_{3/2}$ Cs level (with sapphire surface) for which our model predicts a giant vW repulsion (-50 kHz.μm^3 compared to 19 kHz.μm^3 for a perfect reflector surface) [8].

In conclusion, it should be outlined that simultaneously to this enhanced potential, resonant surface-enhanced level broadening is also predicted and has to be included in improved model for the calculation of the resonant interaction between an excited-state atom and a dispersive dielectric surface.

REFERENCES

[1] J. P. Woerdman and M. F. H. Schurmans, Opt. Comm. 16, 248 (1975)
[2] M. F. H. Schurmans, J. Phys. (Paris) 37, 469 (1976)
[3] G. Nienhuis, F. Schuller and M. Ducloy, Phys. Rev. A 38, 5197 (1988)
[4] M. Ducloy and M. Fichet, J. Phys. II (France) 1, 1429 (1991)
[5] M. Oria et. al, Europhys. Lett. 14, 527 (1991)
[6] M. Chevrollier et. al, J. Phys. II (France) 2, 631 (1992)
[7] N. Papageorgiou et. al, "Laser Physics" 4, 392 (1994)
[8] M. Fichet et. al, to be published

BLUE SHIFT PARADOX IN SELECTIVE REFLECTION

T.A.Vartanyan
State Optical Institute, St.Petersburg 199034, Russia

D.Bloch and M.Ducloy
Laboratoire de Physique des Lasers, Université Paris-Nord, 93430
Villtaneuse, France

ABSTRACT

Blue shift of selective reflection line predicted theoretically [1] has never been seen in the experiments [2,3]. To solve this paradox we study the origin of this blue shift in more detail.

INTRODUCTION

Selective reflection spectroscopy has been shown to be a very useful tool for the study of a long range interaction between atoms and solid dielectric materials. It was utilized for the first measurements of sapphire surface van-der-Waals attraction exerted on excited ($6P_{3/2}$, 7P) cesium atoms [2]. Red shift and distortion of the line shape are in excellent agreement with our theoretical calculations performed in the limit of small atomic densities. On the other hand the nonperturbative approach to the same problem without van-der-Waals interaction [1] gives a large blue shift which has never been seen in the experiments. To solve this paradox we extended the perturbation theory to the second order in vapor density.

DOPPLER-FREE SELECTIVE REFLECTION RESONANCE

The main feature of normal incidence selective reflection line shape is that it exhibits a sub-Doppler contribution. The reason for this unexpected result is that atoms leave the surface of dielectric material in theirs ground state and it takes some time for them to adapt the external electric field. During this time they possess a transient polarization which gives rise to the contribution to the reflected field equal to that of atoms moving in the opposite direction with the same absolute value of their velocity. If N_1 and N_2 are number densities of atoms moving towards the surface and in the opposite direction, selective part of the surface reflectivity will be proportional to the following expression:

$$\frac{2\sqrt{\pi}(N_1 + N_2)D^2}{\hbar k v_T} Re \left[\int_0^\infty \frac{\exp(-y^2)dy}{y + \Omega + i\Gamma} \right] \approx$$

$$-\frac{2\sqrt{\pi}(N_1+N_2)D^2}{\hbar k v_T}\left(\frac{\gamma_E}{2}+\frac{1}{2}\ln(\Omega^2+\Gamma^2)\right), \qquad (1)$$

where $k=\omega/c$, $\Gamma=\gamma/kv_T$, $\Omega=(\omega-\omega_0)/kv_T$, ω is the frequency of the incident light, v_T is the most probable thermal velocity of atoms, ω_0, D, γ, and kv_T are the frequency, dipole moment, homogeneous and Doppler widths, correspondently, and $\gamma_E \approx 0.577$. Approximate equality (1) holds for Ω, $\Gamma \ll 1$. It is clear that in the first order perturbation theory selective reflection resonance is unshifted.

BLUE SHIFT IN THE SECOND ORDER PERTURBATION THEORY

The second order density term in the limit Ω, $\Gamma \ll 1$ is equal to

$$\left(\frac{2\sqrt{\pi}D^2}{\hbar k v_T}\right)^2 N_1 N_2 \frac{\Omega}{\Omega^2+\Gamma^2}\sqrt{8\pi}\ln(1+\sqrt{2}) \qquad (2)$$

This odd term causes the blue shift

$$\Delta\omega=\frac{N_1 N_2}{N_1+N_2}\frac{D^2}{\hbar}4\pi\sqrt{2}\ln(1+\sqrt{2}), \qquad (3)$$

First of all one should notice that the parameter of perturbation series appears to be equal to $\Delta\omega/\gamma$ but not the optical density of the vapor which is γ/kv_T times smaller. This is the reason for the shift to be rather large even if the medium is not optically dense.

The most important consequence of Eq.(3) is that the shift is proportional to the product of number densities of arriving and departing atoms. It means that in the case if arriving atoms are optically pumped, the shift will be suppressed. The nonperturbative solution of the model without arriving atoms [4] also gives zero shift.

REFERENCES

1. M. F. H. Schuurmans, J. Phys. (Paris) 37, 469 (1976).
2. M. Chevrollier, M. Fichet, M. Oria, G. Rahmat, D. Bloch and M. Ducloy, J. Phys. II (France) 2, 631 (1992).
3. A. M. Akulshin, A. A. Celikov, V. A. Sautenkov, T. A. Vartanian and V. L. Velichansky, Opt. Commun. 85, 21 (1991).
4. T. A. Vartanyan, Opt. Spectrosc. (USSR) 70, 147 (1991).

SUBNATURAL LINEWIDTH IN LARGE OPTICAL FIELDS

Lorenzo M. Narducci
Department of Physics and Atmospheric Science, Drexel University
Philadelphia, PA 19104

Christoph H. Keitel
Imperial College, Blackett Laboratory, Laser Optics and Spectroscopy
London SW7 2BZ, UK

Gian Luca Oppo
Department of Physics and Applied Physics, University of Strathclyde
Glasgow, Scotland G4 0NG, UK

Marlan O. Scully
Department of Physics, Texas A&M University
College Station, TX 77843

No excited atom lives forever, and as the excited electron decays to a lower energy level a quantum of energy, an elementary component of the fluorescence spectrum, is emitted. It is common knowledge that even the most elementary atomic transitions, for example the transitions between two non-degenerate levels, have an irreducible linewidth, and that in a collisionless environment this linewidth scales as the Einstein rate of decay of the upper state. It is also common, but incorrect, to extrapolate this observation and to conclude that the spontaneous decay rate of the upper state always controls the linewidth of the atomic fluorescence.

In fact, the fluorescence spectrum is the Fourier transform of the radiated field amplitude correlation function which, in turn, is proportional to the atomic polarization correlation function by virtue of the well known relation linking the far-field of a radiating dipole to the dipole moment itself. Thus, the fluorescence spectrum reflects the correlation and fluctuation properties of the atomic dipoles, and its width is linked to the memory of the polarization fluctuations.

With this as the premise, and with the additional observation that the atomic polarization is sensitive to the presence of possible coherent links among different atomic levels, induced for example by external coherent light sources, we claim that it is possible to "freeze" the decay time of selected polarization fluctuations with the end result that the linewidth of the associated spectral lines can become much narrower than in the case in which the atoms decay freely into vacuum.[1] As we shall see, atomic coherence is the key to the effect.

This interesting prediction has been tested experimentally by Mossberg and collaborators at the University of Oregon.[2]

1.   L.M. Narducci, M.O. Scully, G.-L. Oppo, P. Ru, and J.R. Tredicce, Phys. Rev. A 42, 1630 (1990).
2.   Y.F. Zhu, D.J. Gauthier, and T.W. Mossberg, Phys. Rev. Lett. 66, 2460 (1991).

PRECISION MEASUREMENT OF TRANSITION LINEWIDTHS

W. A. van Wijngaarden and J. Li
Department of Physics, York University
North York, Ontario, Canada M3J 1P3

Transition linewidths can be determined by exciting an atomic beam using a laser that is frequency modulated as is shown in Fig. 1. An oven produces a collimated beam of atoms that is intersected orthogonally by a laser beam. The laser frequency is scanned across the transition and fluorescence produced by the radiative decay of the atoms is detected by a photomultiplier (PMT). The electronic signal is then processed by a lock-in amplifier and sent to a computer.

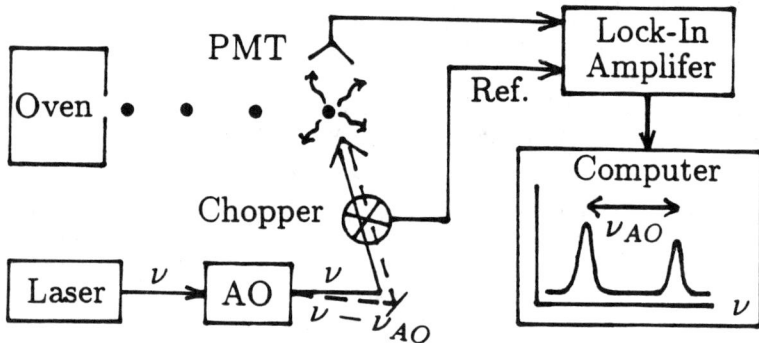

Fig. 1. Apparatus

The calibration of frequency can be accomplished by shifting part of the laser using either an electro-optic or acousto-optic (AO) modulator. Atoms are excited by both frequency shifted and unshifted laser beams, producing a pair of peaks in the fluorescence spectrum separated by the modulation frequency ν_{AO}. The latter can be generated accurate to a part in 10^6 by a frequency synthesizer (HP8647A). For alkali atoms, the frequency calibration can alternatively be accomplished using the hyperfine splitting of the two ground state levels which can be excited simultaneously during a laser scan. These splittings, used for atomic clock transitions, are known to very high accuracy.

Natural linewidths have been determined with uncertainties of a percent from lifetime measurements.[1,2] In order to directly measure a typical linewidth of 10 MHz to an accuracy of 1%, attention must be given to mechanisms that can broaden the lineshape. These include collisions and residual first order doppler shifts that can be reduced to less than 0.1 MHz using well collimated atomic beams housed in a vacuum of 1×10^{-7} torr. Magnetic fields can also perturb the lineshape by shifting the atomic energy levels. Stray fields can be reduced to a few milligauss using three pairs of Helmholtz coils. Another significant broadening mechanism is due to the laser power. Hence, the laser must be strongly attenuated and data must be extrapolated to zero power. Finally, the measured linewidth must be corrected for the laser linewidth. The manufacturer quoted linewidth of a ring dye laser is 0.5 ± 0.1 MHz while that of a Ti/sapphire laser is 0.10 ± 0.02 MHz.

This method of frequency calibration is far simpler than using a Fabry-Perot etalon to monitor laser frequency.[3] For accurate results the interferometer length must be stabilized using a HeNe laser locked to an iodine reference line. Furthermore, for small frequency shifts the cavity must have an inordinately long length to have a suitably small free spectral range. Finally, the cavity must be housed in a vacuum chamber to eliminate pressure and temperature variations. Hence, the method of using a frequency modulated laser is potentially useful, particularly for transitions having broad linewidths or correspondingly short lifetimes.

The authors wish to thank the Natural Science and Engineering Research Council of Canada and York University for financial support.

REFERENCES

1. A. Gaupp, P. Kuske and H. J. Andra, Phys. Rev. A 26, 3351 (1982).
2. C. Tanner et al Phys. Rev. Lett. 69, 2765 (1993).
3. W. A. van Wijngaarden and J. Li, submitted to JOSA, 1994.

ACCURATE PROFILES OF SOLAR INFRARED OI TRIPLET LINES

A.A.Galal, N.H.Youssef* and M.M.Behery
National Research Institute of Astronomy and Geophysics Helwan,
Cairo, Egypt
*Faculty of Science, Cairo University, Egypt

The physical conditions reproducing more accurate profiles of OI solar infrared triplet lines (7771.96-7774.18-7775.4 A°) have been defined from synthetic and empirical analyses of observed emergent line intensities at different angular distances θ from the center of the solar disk.

The synthetic computing program given by Grevesse[1] and the the empirical approach suggested by Galal and Sitnik[2] were used for this purpose. The combined analytical approaches gave a wide possibility to recheck the accuracy of preassigned atomic and atmospheric models and other physical parameters needed for theoretical computations of the line contours.

As shown in Fig.1 the LTE synthetic line profiles calculated by using various models of the solar atmosphere do not coincide with the observed contours of the lines. Full coincidence has been nearly achieved by readjusting the model parameters (see Fig.2 and Table 1) The other fitting parameters appear to be more reasonable. The abundance of OI was found to be decreasing from Log A= 9.05 at the disk center to 8.72 at the limb The micro turbulence is increasing from 0.85 kms^{-1} at the center up to 1.4kms^{-1} at the limb. The NLTE distribution of the kinetic temperature used in the profile calculation differs by about 700 K in the minimum temperature region from the temperature curve of VAL moel[3]

TABLE I. The readjusted model parameters

Logτ_{500} Parmeter	-4.809	-2.993	-1.573	-0.652	-0.357	-0.002	0.250
Te(k)	5730	4850	4990	5460	5840	6460	7220
log Pe	-1.077	-0.520	0.293	0.905	1.233	1.759	2.070
log Pg	1.847	3.467	2.244	4.755	4.907	5.059	5.170

τ_{5000} Optical Depth Te -Electronic Temperture Pe-Electronic Pressure
Pg -Gas Pressure

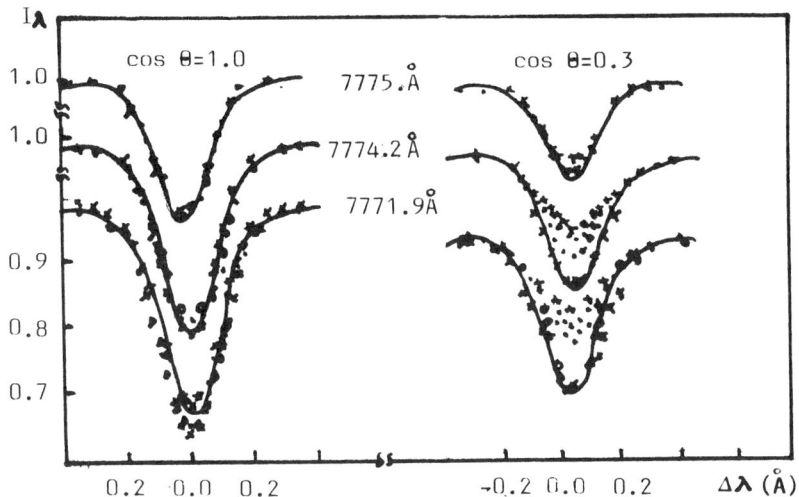

Fig.1 : Observed (Solid) and Calclated (symbols) profiles.

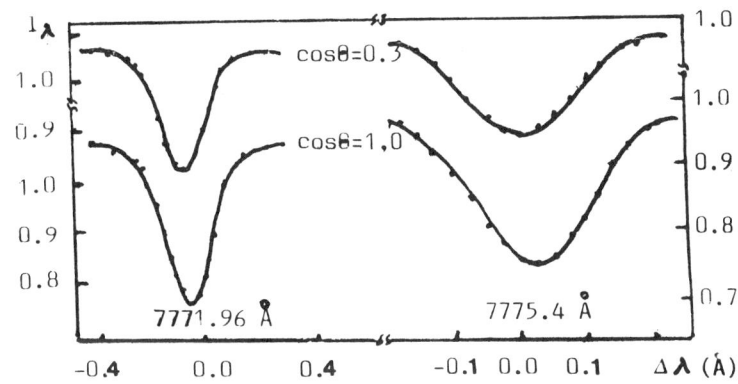

Fig.2 : Observed (solid) and readjusted (dots) profiles.

1. N.Grevesse, Private Comunication ,(1982).
2. A.A.Galal and G.F. Sitnik, Astron.Circular (A.C.) No. 963 July, (1977). (Russian).
3. J.E. Vernazza, E.H.Avrett and Losser, Astrophs. J. Suppl.45,635,(1981).

ENERGY TRANSFER AND ENERGY POOLING COLLISIONS IN Li-Cd SYSTEM

G. Pichler, D. Azinović and S. Milošević
Institute of Physics of the University, P.O.Box 304, HR-41000 Zagreb, Croatia

Collisional energy transfer is generally important elementary process in which energy is transferred between energy levels in collisions between the same or different particles.[1] Especially interesting is the energy transfer from the laser excited levels of one species into the metastable levels of another species, since the later may provide the energy reservoir for some consecutive collisional processes.[2] The latter could be collisions between two like metastable atoms leading to a very high excited levels in atoms, which are normally not accessible using a single low energy photons.

We discuss the intermetallic system of Li-Cd vapors in which Li 2p and 3d levels are laser excited. An efficient energy transfer to $Cd(^3P_J)$ levels occurs. At high $Cd(^3P_J)$ concentration the energy pooling takes place causing the appearance of $Cd(6s\ ^3S_1)$, $Cd(5d\ ^3D_J)$ and $Cd(5d\ ^1D_2)$ atoms in the mixture.

Excimer laser (XeCl at 308 nm, Model LPX 100e) pumped dye laser (Model FL 3002, Lambda Physik) tuned to 670.8 nm for Li(2p) and to 639.146 nm for two photon Li(3d) excitation were transmitted through the heat pipe oven containing lithium and cadmium vapors at temperatures between 540 and 850 °C. Helium was used as a buffer gas usually at 10-50 Torr. The laser repetition rate was 9 Hz, the energy of the single laser pulse was 1 mJ, and the duration of the laser pulse was typically of about 20 ns. The fluorescence collected at 90° was focused onto the entrance slit of the scanning monochromator for spectrally resolved measurements. Temporally resolved measurements revealed different time behavior of Li and Cd atomic spectral lines especially with respect to the time delay of the peak intensities.

Time evolution of Cd 508 nm line has a single peak at about 300 ns, after which it decays steadily, whereas Li 610 nm line has a main peak at about 500 ns time delay and in addition a weaker second peak. This second peak appears in all observed Li lines from ns (except 3s) and nd levels and its time location is from 3 to 10 µs depending on Cd and Li concentrations. We observed spectra at the first peak and at the second peak. In the first case Li and Cd spectral lines appear simultaneously, whereas in the second case only lithium lines from nd-2p (n=4-9) and ns-2p (n=4-6) transitions appear.

We observed selective wavelength excitation spectra for many Cd and Li spectral lines taken over Li resonance line at 670.8 nm at different oven temperatures and helium buffer pressures. They all show strong resonance character which is affected by absorption at exact resonance.

We studied the dependence of Cd and Li spectral line intensities on the dye laser intensity at 670.8 nm. We found a linear dependence of Li 610 nm line intensity (3d-2p transition) on the dye laser intensity, whereas Cd 508 nm line (6^3S_1-5^3P_J) exhibits cubic dependence. It is interesting that the second peak of the Li 610 nm spectral line has also a cubic dependence on the dye laser intensity.

An interesting question arises in connection with the energy transfer from Li(2p) or Li(3d) atoms to Cd(^3P_J) atoms, the latter then undergo energy pooling collisions. With laser line adjusted to lithium 2s-2p transition we may obtain the following processes in the Li-Cd vapor mixture:

$$Li(2p)+Cd(5^1S_0)+h\nu \rightarrow Li(2s)+Cd(5^3P_J) \quad (1)$$

$$Cd(5^3P_J)+Cd(5^3P_J) \rightarrow Cd(5^3D_J, 6^3S_1, 5^1D_2)+Cd(5^1S_0) \quad (2)$$

In the case of the two photon excitation to Li 3d level we have:

$$Li(3d)+Cd(5^1S_0) \rightarrow Li(2s)+Cd(5^3P_J) \quad (3)$$

followed by energy pooling[3] described by equation (2).

Due to possible saturation effects, ionization of lithium atoms, self absorption of lithium and cadmium resonance lines and complex collision processes it is difficult to clearly understand this high density intermetallic mixture. The work is in progress to reveal further experimental findings which may eventually bring more insight into the complex spectral and time behavior of the Li-Cd system.

1. M. Allegrini and S. Milosevic, Energy sharing in thermal energy collision spectroscopy, The Physics of Electronic and Atomic Collisions XVIII, T. Andersen and B. Fastrup Eds. (American Institute of Physics, 1993).
2. H. G. C. Werij, M. Harris, J. Cooper, and A. Gallagher, Phys. Rev. **43**, (1991) 2237.
3. H. Umemoto, J. Kikuma, and A. Masaki, Chem.Phys. **127** (1988) 227.—

Dynamic Measurement of Gas Temperature and Pressure using Infrared Spectroscopy

R.Berman, P.Duggan, M.P.LeFlohic, A.D.May, and J.R.Drummond
University of Toronto, Toronto, Ontario

ABSTRACT

The changing thermodynamic state of a pressure modulated gas sample near room temperature has been measured spectroscopically. Gas temperature was determined within 1K from the ratio of the integrated absorptions of two vibrorotational lines, while the integrated absorption from one line was used to measure the pressure[1].

INTRODUCTION

The purpose of these experiments was to develop a method of making non-invasive measurements of the rapidly changing temperature and pressure of a gas sample contained within a 1 cm long cell pumped by an externally driven piston. Due to the small thermal mass and rapid temporal variation, measuring the temperature of a dynamic gas sample is a challenging problem. This problem is of particular interest to remote sensing since an instrument known as a pressure modulator radiometer (PMR) uses this type of cell to make satellite based atmospheric measurements[2]. In the PMR, the cell of interest acts as a modulated spectral filter and hence, a detailed knowledge of the pressure and temperature cycle is crucial to calculating the correct spectral response.

EXPERIMENT

The absorptions of two vibrorotational lines were measured as a function of piston position and frequency. By fitting the data to the Galatry soft collision profile, both integrated absorptions were calculated. The ratio of the two integrated absorptions is equivalent to the ratio of strengths for the appropriate absorption lines. Since the line strength ratio is a function of temperature alone, by choosing lines with large differences in temperature dependence, it was possible to use the integrated absorption ratio as a sensitive probe of absolute temperature. Because the integrated absorption ratio is independent of the cell length and gas density, the experimental uncertainty was limited only by the accuracy of the absorption measurement. In addition, although the line strengths are not known to better than a few percent, the line strength ratio was precisely calculable based on the lower state populations. The use of a difference frequency spectrometer with a signal to noise ratio of 600:1, typically resulted in temperature uncertainties of 1K.

By equating the measured absorption with the product of the optical path length and known line strength, the gas density was determined. The ideal gas law was then used to calculate the gas pressure from the density and temperature. The error in the pressure measurements resulting from these calculations was dominated by the systematic 2% uncertainty in the spectroscopic line strengths.

Measurements were performed using CO under a variety of operating conditions. Figure 1 shows results for CO cycled at 10 Hz, measured through the centre of the cell and near the cell wall. In the two cases shown, the difference in the temperature cycles was due to heat transport to the walls. Measurements made with standard thermal probes and pressure transducers were consistent, within the error bars, with spectroscopic results from static cells. Additional verification with the dynamic cell, using the $C^{13}O^{16}$ isotopic lines also produced results consistent with the $C^{12}O^{16}$ results. Other results not discussed here include measurements made on water vapour. Water is of particular interest because of its atmospheric relevance and the possibility of phase changes occurring during the pressure cycle.

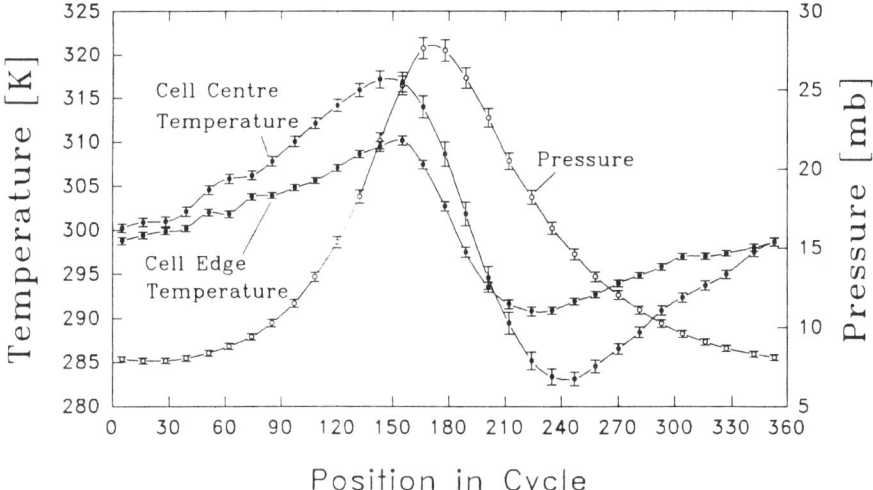

Fig. 1 The CO temperature and pressure cycle for a piston frequency of 10 Hz. The R0 and R22 lines were ratioed to measure the temperature for two different locations in the cell.

REFERENCES

1. R.Berman, P.Duggan, M.P.LeFlohic, A.D.May, and J.R.Drummond, Appl. Opt., **32**, 6280, (1993)
2. F.W. Taylor, Spectroscopic Techniques, Vol. 3, G.A. Vanesse, ed. (Academic, New York, 1983), pp.137-197.

LINESHAPE OF LIGHT ABSORPTION BY EXCITONS UNDER BAND-TO-BAND TRANSITIONS IN LOW DIMENSIONAL MOLECULAR STRUCTURE

N.I. Grigorchuk

Institute of Surface Chemistry Ukrainian Academy of Sciences, 31 Nauki Prospect, 252022, Kiev-22, UKRAINE

Usually investigations of light absorption are based on an assumption according to which the transitions to higher-lying excited states come from the ground state. Analysis of the experimental data shows that a number of molecular systems give a luminescence with quantum yield close to unity and that the radiation in such systems comes from the statistical equilibrium state. This means that during a time smaller than the radiation lifetime of excitons, their statistical distribution on sublevels of the lowest band is established, and their number may be considered as given. Therefore, the question about the photo excitation of such excitons from the lowest band to the next higher one arises.

The spectral peculiarities of the light absorption at such transitions are especially interesting for the low dimensional structures, in which the excitons may be characterized by longer lifetime, greater effective mass than for the bulk. This allows them slow coherent motion along the structure, so that their energy can be described in tight binding approximation. The damping of the excitons on the structures vibrations Γ is accounted for as parameter. Then the calculations of the lineshape of light absorption gives [1]

$$F(v) = \frac{\Omega}{I_0(\beta L_1)} Re \frac{e^{-i\frac{\alpha\beta L}{\Delta L}1}}{\sqrt{\alpha^2 + \Delta L^2}} \qquad (1)$$

for one-dimensional structure, and

$$F(v) = \frac{1}{n\Delta L} \frac{\Omega}{I_0^2(\beta L_1)} Re\left\{ \kappa\, K(\kappa) e^{-i\frac{\alpha\beta L_1}{\Delta L}} \right\}, \qquad (2)$$

for two-dimensional structure. Here $K(\kappa)$ is the complete elliptical integral of the first kind with module

$$\kappa = \frac{2\Delta L}{\sqrt{\alpha^2 + 4\Delta L^2}}, \qquad (3)$$

$$\Delta L = \hbar^{-1}(L_2 - L_1), \quad \alpha = \frac{\Gamma}{2} - i(v - \Omega), \quad \beta = (\kappa_B T)^{-1}, \quad \Omega = \hbar^{-1}(E_2 - E_1),$$

I(x) is the modified Bessel function of zeroth order.

One can see from (1) that a temperature enhancement of the system, for instance, at arbitrary given bandwidth difference ΔL only insignificantly affects the lineshape (slight asymmetry appears with temperature decrease) while as the same bandwidth difference enhancement leads to a considerable change of the shape at all possible temperatures. Lorentz curves which transform into the two maxima ones does depend on the temperature of the structure [2].

From calculations (2) one can conclude that in two-dimensional cases, the tendency for strong dependence of the lineshape of absorption on bandwidth difference as well as its weak dependence on the temperature of the system maintains. All effects came from the difference in the density of state of two considered bands, their "inequivalence". But there are some differences there. Thus, beconvecs-like curves for the great values of the ratio $2\Delta L/\Gamma$ are transformed into broad and smooth bands with approximately constant intensity. With decreasing temperature, the tendency for increase of contour asymmetry becomes brighter for all ΔL. The asymmetry sign is defined by the sign of the difference $L_2 - L_1$.

In both cases the intensity of absorption is essentially sensitive to the difference in the density of states of both bands. The intensity at the maximum of absorption falls with enhancement of difference of the state density as $[1+(2\Delta L/\Gamma)^2]^{-1/2}$ for small $\Delta L/\Gamma$ (<5/2). This downfall both for one- and two-dimensional structures is nearly the same. But for $\Delta L/\Gamma > 5/2$ there appear some distinctions due to the van Hove singularities in the density of states of two-dimensional structures [3].

REFERENCES

[1] N.I. Grigorchuk, Phys. stat. sol. (b), **153**, 633 (1989).
[2] N.I. Grigorchuk, Optika i Spektroskopija, **75**, 344 (1993).
[3] L. van Hove, Phys. Rev. **89**, 1189 (1953).

LINE MIXING AND FAR WING LINE SHAPES

Line mixing effects in the impact limit :
ECS analysis of various experiments made on some IR bands of CO_2 in Helium

C. Boulet
Laboratoire de Physique Moléculaire et Applications, UPR 136 du CNRS,
Université de Paris-Sud, Centre d'Orsay, Bât 350, 91405 ORSAY Cedex, France.

J. Boissoles
Département de Physique Atomique et Moléculaire, CNRS URA1203,
Université de Rennes I, Campus de Beaulieu, 35042 RENNES Cedex, France.

ABSTRACT

In this paper we shall describe recent experimental studies of line mixing effects observed in various IR vibrational bands of CO_2 in a bath of He over a wide range of different physical parameters. We shall show how an ECS (Energy Corrected Sudden) model derived from the IOS (Infinite Order Sudden) formalism of Green (J. Chem. Phys. **90**, 3606 (1989)) leads to theoretical predictions in rather good agreement with experiment, whatever the vibrational band considered. We shall rely this result to the ability of the formalism to predict coupling cross-sections which strongly depend on the vibrational angular momentum.

INTRODUCTION

Large deviations with respect to the superposition of Lorentzian profiles for molecular vibration-rotational spectra have been observed over a wide range of different physical parameters (perturber density, temperature, rotational structure of the band...)[1]. This will be illustrated using as examples some recent experiments made in Rennes on various vibrational bands of CO_2 in a bath of Helium. In all these experiments, it will be shown that line mixing effects are especially important.

The quantum mechanical theory of line shapes in the impact limit is now well established, and the conditions of validity of the impact approximation are now fairly well understood[1,2]. One of the criteria assumes that the duration τ_c of the efficient collisions is infinitesimally small. Therefore the relaxation matrix is expected to slightly depend on frequency for detuning $|\omega - \omega_{fi}|$ from resonance frequencies of the most intense lines smaller than τ_c^{-1}. With a light perturber like He, short range forces are dominant through head-on collisions corresponding to very short duration τ_c of about 0.05 picos ($\tau_c^{-1} \cong 100$ cm^{-1}). Therefore the domains (frequency <u>and</u> perturber density) in which the relaxation operator is a "constant" (independent of frequency), proportional to the perturber density (binary collisions) is greatly extended by considering He as a perturber.

Although the general theory for overlapping lines in the impact limit was established some time ago[3], the explicit calculation of the relaxation matrix still remains an open question. Advances in computational abilities now permit accurate quantum calculations (close coupling or coupled states methods), at least for simple systems. However the present study employs the ECS (Energy Corrected Sudden) approximation for the calculation of the relaxation matrix. Indeed, an IOS formalism for the calculation of the coupling coefficients for lines of CO_2 in a bath of He atoms has been recently proposed by S. Green[4]. Since this approach properly takes into account the angular momentum coupling : photon-rotation *and vibration*, it can be applied to every type of vibrational band, stretching as well as bending bands. Moreover its accuracy for high rotational levels may be easily improved by introducing simple corrections suggested by the ECS method[5].

Finally it must be emphasized that such an ECS calculation has been shown to be rather accurate for a "similar" system [CO-He] when compared with accurate coupled channel scattering calculations[6].

SUMMARY OF THE ECS METHOD FOR A STRETCHING BAND

We first consider a vibrational band (like the 00^03-00^00 band) in which the vibrational angular momentum is not excited. The scattering formalism follows straight forwardly from that of diatom-atom. A detailed description of that formalism exists in the literature[4,7,8] and we shall only recall the general scheme of the ECS method.

1 - The non diagonal cross-sections coupling the two lines $k \equiv j_i \rightarrow j_f$ and $l \equiv j_i' \rightarrow j_f'$ (with $j_i' < j_i$) are given by (omitting for simplicity the vibrational quantum numbers) :

$$\sigma^1(j_i' j_f' \leftarrow j_i j_f) = - \sum_L F_1(j_i' j_f' j_i j_f, L) \frac{\sqrt{\Omega_{ji} \Omega_{jf}}}{\Omega_L} (2L+1) \sigma^0(L \rightarrow 0) \quad (1)$$

where

$$F_1(j_i' j_f' j_i j_f, L) = -(2j_i'+1)$$

$$[(2j_f'+1)(2j_f+1)]^{1/2} \begin{pmatrix} j_i & j_i' & L \\ 0 & 0 & 0 \end{pmatrix} \begin{pmatrix} j_f & j_f' & L \\ 0 & 0 & 0 \end{pmatrix} \begin{Bmatrix} j_i & j_f & 1 \\ j_f' & j_i' & L \end{Bmatrix} \quad (2)$$

Upward cross sections ($j_i' > j_i$) are obtained from the detailed balance principle :

$$\rho_\ell \, \sigma^1 (k \leftarrow \ell) = \rho_k \, \sigma^1 (\ell \leftarrow k) \tag{3}$$

where ρ_k is the population of the initial level of line k. Ω_L is an adiabaticity factor [a kind of resonance function] usually defined in terms of a scaling length ℓ_c [$\tau_c = \ell_c/\bar{v}$; $\bar{v}$ mean relative velocity] by :

$$\Omega_L = \frac{1}{[1 + \frac{1}{24}(\omega_{L,L} - 2\tau_c)]^2} \tag{4}$$

2 - The diagonal elements [linewidths] have to be deduced from the sum rules (μ_k is the dipole reduced matrix element for transition k) :

$$\mu_k \, \sigma^1 (k \leftarrow k) = - \sum_{\ell \neq k} \mu_\ell \, \sigma^1(\ell \leftarrow k) \tag{5}$$

which are almost rigorous when vibrational dephasing is unimportant. Calculations which included the CO_2 vibrational dependence in a simplistic CO_2-He potential have shown a very little vibrational dependence of the vibrationally elastic cross-sections[4], a result which is substantiated experimentally since He-broadening leads to very weak lineshifts and correlatively to an unsignificant vibrational dependence of the widths[9].

3 - Finally the calculation of the relaxation matrix requires "only" through eq. (1) the knowledge of the basic cross-sections σ^0 (L → 0) which are the "ordinary" inelastic cross sections coupling all the rotational levels to the fundamental one. They can be obtained ab-initio for simple systems when an accurate intermolecular potential is available. Otherwise the alternative, which has been used in the present work, and is now quite common consists in :

i) expressing the basic rates through a reasonable analytical law characterized by some fitting parameters[7] :

$$\sigma^0 (L \to 0) = \frac{A}{[L(L+1)]^\alpha} \tag{6}$$

ii) then the adjustable parameters A, α (and ℓ_c) are deduced from a least square fit of the observed linewidths[9] by using eq (1-6).

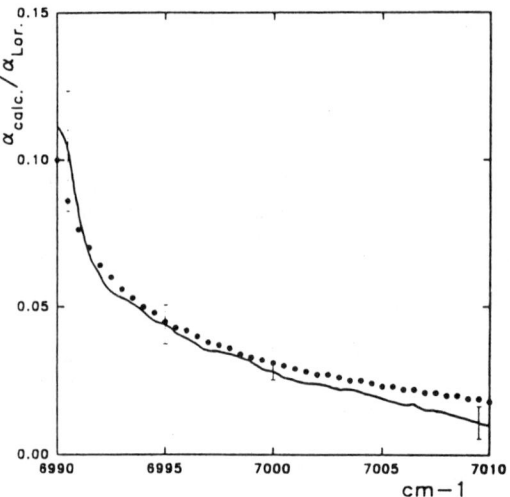

Figure 1. CO_2 [$00^03 - 00^00$ band] - He ; T = 296 K
Ratio of the absorption coefficient observed in the wing of the band (above 6990 cm^{-1}), to that expected from a sum of Lorentzian uncoupled lines.
——— : experiment
• • • : ECS predictions

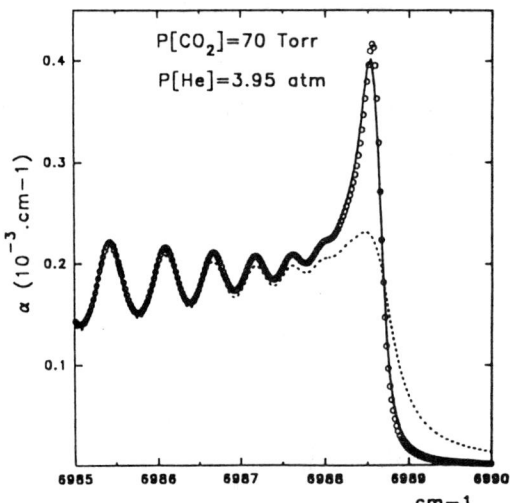

Figure 2. CO_2 [$00^03 - 00^00$ band] - He ; T = 296 K
Experimental and ECS absorption coefficients in the region just before the band head at moderate He pressures
——— : experiment
------- : Lorentzian calculation (no mixing)
O O O : ECS predictions

Once the σ^1 cross-sections are known, the absorption coefficient may be obtained from[3] :

$$\alpha(\omega) \propto \mathfrak{Im} \sum_{k,l} \mu_l \ll l \left| \frac{1}{\omega - \omega_0 - iW} \right| k \gg \mu_k \rho_k \qquad (7)$$

where ω_0 is the diagonal matrix of the line frequencies and W the relaxation matrix whose elements are defined by :

$$\ll l |W| k \gg = \frac{n_b \bar{v}}{2\pi c} \sigma^1 (l \leftarrow k) \qquad (8)$$

where n_b is the perturber density.

AN EXAMPLE OF STRETCHING BAND : 00^03-00^00 BAND

This band is of great interest since it exhibits a very sharp band head around R(40). Therefore many of the strongest lines in the R branch are crowded together within a narrow interval. Such a structure is known to promote important line mixing effects[10]. Moreover the wing beyond the band head is located close to the most intense lines of the R branch $[\sigma - \sigma_{fi} \ll \tau_c^{-1} \cong 100 \text{ cm}^{-1}]$ and the impact approximation will be sufficient to describe this "near wing" region.

As shown in Fig. 1 a strong sublorentzian behavior is observed in the wing of the band and the ECS model is in reasonable agreement with experiment.

In the region located just before the band head [corresponding roughly to R(36) → R(40] a noticeable excess in absorption with respect to a sum of Lorentzian components is observed at intermediate He pressures [up to ∼ 5 atm]. Here too, ECS results are satisfactory [cf Fig 2].

At higher pressures, where the rotational structure has disappeared, line mixing effects lead to an important enhancement of the R branch, due to its narrowing, while the P branch is only slightly affected [cf Fig 3]. Agreement is seen to be quite good with ECS predictions and finally from Fig 1 to 3 it may be reasonably concluded that ECS formalism succeds at predicting the evolution of the spectral profile over an extended range of perturber pressure, at least for this band.

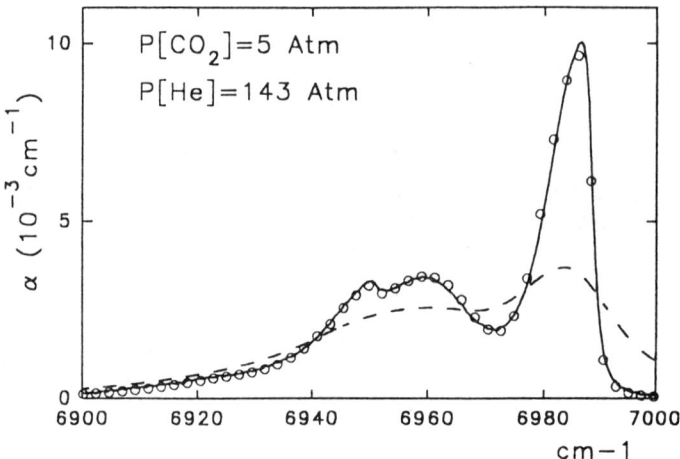

Figure 3. CO_2 [$00^03 - 00^00$ band] - He ; T = 296 K
Comparison between experimental and ECS absorption coefficients for the whole band at high He pressure
——— : experiment
------ : Lorentzian calculation (no mixing)
O O O : ECS predictions

LINE MIXING IN Q BRANCHES OF BENDING BANDS.

Line coupling effects are very important in Q branches where rotational components are closely spaced and strongly overlap and interfere, even at moderate perturber pressures.

As mentionned previously, the IOS formalism of Green[4], which can be easily improved by introducing ECS corrections, can be applied to every type of CO_2 band.

Coupling cross sections between lines in the same branch [P, Q or R] and between lines in different branches (interbranch mixing) can be written, always in terms of the same basic rates. As an example, we only give here the equation for cross-sections coupling two Q lines belonging to the $v_1 v_2 l_2 v_3 \rightarrow v_1' v_2' l_2' v_3'$ vibrational band (for the case j' < j) :

$$\sigma^l [Q(j) \rightarrow Q(j')] = (-1)^{l_2+l_2'} (2j'+1)^{3/2} (2j+1)^{1/2}$$

$$\sum_L \begin{pmatrix} j & L & j' \\ l_2 & 0 & -l_2 \end{pmatrix} \begin{pmatrix} j & L & j' \\ -l_2' & 0 & l_2' \end{pmatrix} \begin{Bmatrix} j & j & L \\ j' & j' & L \end{Bmatrix} \frac{\Omega_j}{\Omega_L} (2L+1) \sigma^0 (L \rightarrow 0) \qquad (9)$$

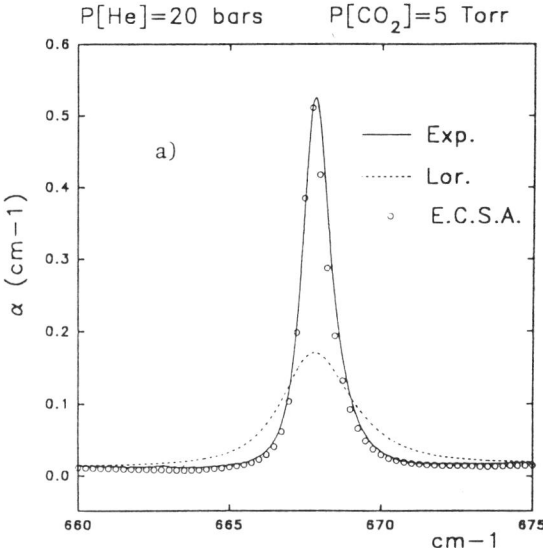

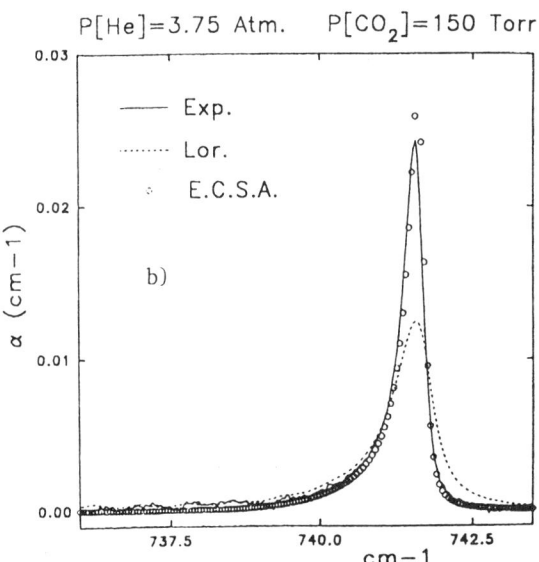

Figure 4. Comparison of ECS theoretical predictions and experimental absorption coefficients for two CO_2-Q branches (T = 296 K):
a) $01^10 - 00^00$ band : Σ ($\ell_{2i} = 0$) $\rightarrow$ π ($\ell_{2f} = 1$)
b) $11^10 - 02^20$ band : Δ ($\ell_{2i} = 2$) $\rightarrow$ π ($\ell_{2f} = 1$)

The computational procedure is simple : we keep the same basic rates (and scaling length ℓ_c) as previously deduced from fitting of linewidths. ECS equations are used for the calculation of the off diagonal elements, and the diagonal ones are obtained from a set of sum rules, similar to the following one, written for a Q component :

$$\mu [Q(j)] \sigma^1 [Q(j) \to Q(j)] = - \{ \sum_{Q(j') \neq Q(j)} \mu[Q(j')] \sigma^1 [Q(j) \to Q(j')]$$
$$+ \sum_{R(j')} \mu[R(j')] \sigma^1 [Q(j) \to R(j')] \quad (10)$$
$$+ \sum_{P(j')} \mu[P(j')] \sigma^1 [Q(j) \to P(j')] \}$$

A number of CO_2 Q branches pressurized by He and belonging to π-Σ, π-Δ and π-π bands have been investigated in Rennes[12]. Some preliminary results are given in Fig 4 from which it can be reasonably concluded that the ECS predictions are in fairly good agreement with experiment, whatever the type of band, i.e. the values of the vibrational angular momentum. The major reason of such a success may be found in the ability of the formalism to predict cross-sections -which strongly depend on the values of the vibrational angular momentum [see Fig 5]- through an accurate scheme for the coupling of the various angular momenta involved in IR absorption.

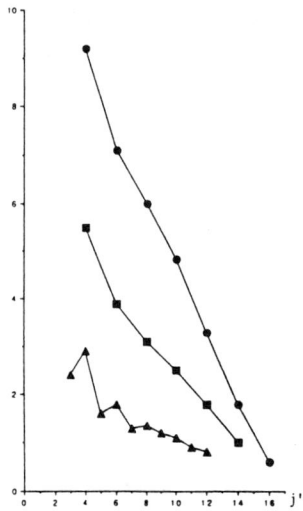

Figure 5. Vibrational dependence of Q-branch line coupling cross-sections. This figure gives the cross sections coupling the Q(2) line to the other Q(j') components for three cases (all values are in $Å^2$).
- ● isotropic Raman diffusion [$\sigma^0 [Q(j) \to Q(j')]$]
- ■ IR absorption ; $\pi \leftarrow \Sigma$ band
- ▲ IR absorption ; $\pi \leftarrow \Delta$ band

We take pleasure in making acknowledgment here of the valuable help of many friends : the Rennes' group (R. Le Doucen, F. Thibault and V. Menoux) and Prof. M. Tonkov (from St Petersburg), who have devoted much effort to experimental investigations. We are also grateful to Dr S. Green (NASA, New York) for his interest and his collaboration. This paper is also theirs.

REFERENCES

1. for recent reviews see
   A. Levy, N. Lacome and C. Chackerian, in The Spectroscopy of the Earth Atmosphere and interstellar Medium, edited by K.N. Rao and A. Weber (Academic, New York, 1992) ; S. Green, in Status and Future developments in transport properties edited by W. A. Wakeham, A. S. Dickinson, FRW Mc Court and V. Vesovic (Dordrecht : Kluwer) 1992.

2. A. Ben Reuven, Spectral lineshapes in gases in the binary collision approximation. Adv. Chem. Phys. **33**, 235 (1975) ; R. G. Breene Jr. Theories of Spectral line shape (Wiley, New-York 1981).

3. M. Baranger, Phys. Rev. **111**, 494 (1958) ; U. Fano, Phys. Rev, **131**, 259 (1963) A. Ben Reuven, Phys. Rev. **145**, 7 (1966).

4. S. Green, J. Chem. Phys. **90**, 3603 (1989)

5. A.E. De Pristo, R. Ramaswamy, S.D. Augustin and H. Rabitz, J. Chem. Phys. **71**, 850 (1979)

6. J. Boissoles, C Boulet, D. Robert and S. Green, J. Chem. Phys. **87**, 3436 (1987) ; J. Chem. Phys, **90**, 5392 (1989).

7. G. Millot, J. Chem. Phys. **93**, 8001 (1990)

8. J. Boissoles, F. Thibault, R. Le Doucen, V. Menoux and C. Boulet, J. Chem. Phys. **100**, 215 (1994)

9. F. Thibault, J. Boissoles, R. Le Doucen, J.P. Bouanich, Ph. Arcas and C. Boulet, J. Chem. Phys. **96**, 4945 (1992)

10. Y.I. Baranov, M.O. Bulanin and M.V. Tonkov, Opt. Spectrosc. **50**, 336 (1981) ; I.M. Grigorev, V.M. Tarabukkin and M.V. Tonkov, Opt. Spectrosc. **58**, 147 (1985)

11. F. Thibault, J. Boissoles, R. Le Doucen, V. Menoux and C. Boulet, J. Chem. Phys. **100**, 210 (1994)

12. M. Tonkov, F. Thibault, J. Boissoles, R. Le Doucen and C. Boulet, J. Chem. Phys. Submitted.

LINE SHAPES IN THE FAR WINGS OF ALLOWED TRANSITIONS

R. H. Tipping
Department of Physics and Astronomy, University of Alabama
Tuscaloosa, AL 35487

Q. Ma
Department of Applied Physics, Columbia University
New York, NY 10027

ABSTRACT

A far-wing line shape theory based on the binary collision and quasistatic approximations that is applicable for both the low- and high-frequency wings of the vibration-rotational bands has been developed. This theory has been applied in order to calculate the frequency and temperature dependence of the continuous absorption coefficient for frequencies up to 10 000 cm^{-1} for pure H_2O, and for H_2O-N_2 and H_2O-CO_2 mixtures. The calculations were made assuming an interaction potential consisting of an isotropic Lennard-Jones part with parameters that are consistent with values obtained from other data, and the leading long-range anisotropic part, together with the measured line strengths and transition frequencies. Current work in progress to extend the theory to multi-component anisotropic potentials, to account for the effects of molecular motion on the near wings, and to validate the theory with recent experimental data will be discussed briefly.

INTRODUCTION

The absorption of radiation by water vapor in the window regions of the Earth's atmosphere is a problem of great practical as well as theoretical interest.[1,2] In the past few years there have been a number of general reviews concentrating primarily on the experimental characterization of this absorption; for historical reasons this absorption is called the "water continuum" and the interested reader is referred to these reviews for original literature citations.[3-5] As a result of this extensive work, there is nearly unanimous agreement on the density dependence (quadratic), general agreement on the temperature dependence (strong, negative, at least in the 8-13 μm window for atmospheric temperatures), but considerable disagreement as to the magnitude and physical mechanism responsible for the absorption. Water dimers,[6] polymers,[7] collision-induced absorption,[8] and the superposition of the far wings of collisionally broadened allowed dipole lines[9] have all been suggested as possible sources of the water continuum. While the first two mechanisms are undoubtedly present and contribute to the absorption in the atmosphere (in varying amounts in the different spectral regions), there is increasing evidence to believe that they are not the primary mechanism.

Within the past decade, substantial progress for calculating the absorption far from the line centers of resonant transitions has been made. Starting from the basic formalism of Fano[10] and Davies et al.,[11] Rosenkranz[12] has developed a quasistatic theory that is applicable in the high-frequency wing of the pure rotational band (the 8-13 μm window). This theory successfully predicted both the magnitude and the strong, negative temperature dependence. This theory was extended by Ma and Tipping,[13] but because of the nature of the approximations made, was limited to the high-frequency wings of the pure rotational and/or the vibrational bands. Utilizing the same basic formalism, but proceeding in a different way mathematically, these authors also developed a theory that was applicable for the low-frequency wing of the pure rotational band, i.e., for the millimeter spectral region.[14] Again, good agreement in both magnitude and temperature dependence between theory and experiment was obtained.

More recently, a generalized far-wing line shape theory based on the binary collision and quasistatic approximations was formulated and applied to calculate the absorption coefficient, $\alpha(\omega)$, for frequencies between 300 and 1100 cm^{-1} for several temperatures.[15] The results were in excellent agreement with the extensive experimental data of Burch and co-workers.[16] In particular, by treating the resonant and anti-resonant terms separately, better agreement with the high-temperature laboratory data was obtained as compared to the theory of Rosenkranz. This new theory has been generalized to vibration-rotational bands[17] and to foreign broadening,[18] and thus constitutes a unified theoretical framework within which one can calculate the magnitude and temperature dependence of the self and foreign broadened far-wing absorption.

THEORY

The absorption coefficient per unit volume at frequency ω, $\alpha(\omega)$, of a gas in thermal equilibrium at temperature T can be written in the form

$$\alpha(\omega) = \frac{4\pi^2}{3\hbar c} n_a \, \omega \, \tanh(\hbar\omega/2kT) \, [F(\omega) + F(-\omega)], \quad (1)$$

where c, k and $\hbar$ have their usual meaning; n_a is the number density of absorbers, and $F(\omega)$, the spectral density, is given by

$$F(\omega) = \frac{1}{\pi} \text{Re Tr} \int_0^\infty [\vec{\mu}_a \cdot e^{i(\omega - L)t} \rho \vec{\mu}_a] \, dt. \tag{2}$$

In this expression $\vec{\mu}_a$ is the dipole moment operator of the absorber molecule, and L is the total Hermitian Liouville operator L which can be expressed as the sum of its components

$$L = L_0 + L_1 = L^{(a)} + L^{(b)} + L_1. \tag{3}$$

If one neglects the small non-commuting corrections (to be discussed later) and assumes

$$e^{-iLt} \approx e^{-iL_0 t} e^{-iL_1 t}, \tag{4}$$

then Eq. (2) can be approximated by

$$F(\omega) = \frac{1}{\pi} \text{Re Tr} \int_0^\infty [\vec{\mu}_a \cdot e^{i(\omega - L_0)t} e^{-iL_1 t} \rho \vec{\mu}_a] \, dt. \tag{5}$$

By further assuming that the total density operator can be approximated by the product of three parts

$$\rho = \rho^{(\text{int})} \rho^{(b)} \rho^{(a)}, \tag{6}$$

where $\rho^{(a)}$ and $\rho^{(b)}$ are the parts related to the unperturbed absorber and bath molecules, respectively, and $\rho^{(\text{int})}$ is the part related to the interaction between them, and making the binary collision approximation, one can obtain an expression for the spectral density in which formally all the variables except the internal variables of the absorber have been isolated:

$$F(\omega) = \frac{1}{\pi} \text{Im} \frac{1}{2\pi i} \int_{-\infty}^\infty \text{Tr} [\vec{\mu}_a \cdot \frac{1}{\omega - \omega' - L_a}$$

$$\times < \frac{1}{\omega' - L_1} \rho^{(\text{int})} >_b \rho^{(a)} \vec{\mu}_a] \, d\omega'. \tag{7}$$

The statistical average of the operator $<(\omega' - L_1)^{-1} \rho^{(\text{int})}>_b$ over the bath variables, denoted by the subscript b, contains all the information about the bath molecules and the interaction dynamics between the absorber and bath molecule pairs. We assume that this interaction, $V(\vec{r})$, is of the form

$$V(\vec{r}) = V_{anis}(\vec{r}) + V_{is}(r), \quad (8)$$

where $V_{anis}(\vec{r})$ is the leading term of the expansion of the anisotropic potential and $V_{is}(r)$ is a simple Lennard-Jones type model

$$V_{is}(r) = C\sigma^{-6}[(\sigma/r)^s + (\sigma/r)^6], \quad (9)$$

where C, σ and s are parameters which are consistent with values obtained from other data.

Assuming that the duration of a collision is infinite, the relative motion of the two molecules is not treated quantum mechanically, rather one uses the quasistatic approximation. In this approximation, the quantum mechanical average is replaced by an integration over r, the separation between two interacting molecules, with the statistical weight for a pair of molecules given by

$$\rho(\text{int}) = n_b\, e^{-V(\vec{r})/kT}, \quad (10)$$

where n_b is the number density of perturbers.

As a result of the approximations discussed above, the spectral density can be expressed in terms of line shape functions, $\chi_{ij}(\omega)$, which can be calculated from the interaction potential; that is

$$F(\omega) = \frac{1}{\pi} \sum_{\omega_{ij}>0} \left\{ \frac{1}{(\omega - \omega_{ij})^2} \chi_{ij}(\omega - \omega_{ij}) + e^{\hbar\omega_{ij}/kT} \frac{1}{(\omega + \omega_{ij})^2} \chi_{ji}(\omega + \omega_{ij}) \right\} \rho_i\, |\mu_{ij}|^2; \quad (11)$$

where explicit expressions for the line shape functions are given in Refs. 13 - 15. If instead of individual line shape functions, one introduces symmetrized line shape functions defined by

$$\hat{\chi}_{ij}(\omega) \equiv e^{\hbar\omega/2kT} \chi_{ij}(\omega), \quad (12)$$

then $\alpha(\omega)$ can be written in the familiar form

$$\alpha(\omega) = n_a \sum_{\omega_{ij}>0} S_{ij} \frac{\omega \sinh(\hbar\omega/2kT)}{\omega_{ij} \sinh(\hbar\omega_{ij}/2kT)}$$

$$\times \frac{1}{\pi}\{\frac{1}{(\omega-\omega_{ij})^2} \hat{\chi}_{ij}(\omega-\omega_{ij}) + \frac{1}{(\omega+\omega_{ij})^2} \hat{\chi}_{ji}(\omega+\omega_{ij})\}, \quad (13)$$

where S_{ij} are the usual line strengths.[19]

In light of the fact that the matrix elements of the dipole moment of the absorber molecule between different vibrational states are at least an order of magnitude smaller than those between the same vibrational states, and that the matrix elements of the dipole moment between the same vibrational states are only slightly dependent on the vibrational quantum number, we can assume that the lines belonging to the different vibrational bands are not coupled to each other, and that the lines within each band are coupled in the same way. Based on this approximation, the total spectral density can be expressed as the sum of its components corresponding to each band

$$F(\omega) = \sum_{v_i v_j} F_{v_i v_j}(\omega), \quad (14)$$

and the total absorption coefficient can be expressed as the superposition of the components corresponding to each band

$$\alpha(\omega) = \sum_{v_i v_j} \alpha_{v_i v_j}(\omega), \quad (15)$$

where

$$\alpha_{v_i v_j}(\omega) = n_a \sum_{\omega_{v_i i, v_j j}>0} S_{ij} \frac{\omega \sinh(\hbar\omega/2kT)}{\omega_{v_i i, v_j j} \sinh(\hbar\omega_{v_i i, v_j j}/2kT)}$$

$$\times \frac{1}{\pi}\{\frac{1}{(\omega-\omega_{v_i i, v_j j})^2} \hat{\chi}_{ij}(\omega-\omega_{v_i i, v_j j})$$

$$+ \frac{1}{(\omega+\omega_{v_i i, v_j j})^2} \hat{\chi}_{ji}(\omega+\omega_{v_i i, v_j j})\}. \quad (16)$$

To simplify the calculations, one can introduce the average positive and negative frequency resonance line shape functions, $\hat{\chi}_+(\omega)$ and $\hat{\chi}_-(\omega)$, defined by

$$\hat{\chi}_+(\omega) \equiv \sum_{\substack{\omega_{ij}>0 \\ E_i>E_j}} \hat{\chi}_{ij}(\omega) \, \rho_i \, |\mu_{ij}|^2 \Big/ \sum_{\substack{ij \\ E_i>E_j}} \rho_i \, |<i|\mu_m|j>|^2 \qquad (17)$$

and

$$\hat{\chi}_-(\omega) \equiv \sum_{\substack{\omega_{ij}>0 \\ E_i>E_j}} \hat{\chi}_{ji}(\omega) \, \rho_j \, |\mu_{ij}|^2 \Big/ \sum_{\substack{ij \\ E_i>E_j}} \rho_j \, |<j|\mu_m|i>|^2, \qquad (18)$$

which are the same for every band. In terms of these two functions, the absorption coefficient can thus be written

$$\alpha_{v_iv_j}(\omega) = n_a \sum_{\omega_{v_ii,v_jj}>0} S_{v_ii,v_jj} \frac{\omega \sinh(\hbar\omega/2kT)}{\omega_{v_ii,v_jj} \sinh(\hbar\omega_{v_ii,v_jj}/2kT)}$$

$$\times \frac{1}{\pi}\{\frac{1}{(\omega - \omega_{v_ii,v_jj})^2} \hat{\chi}_+(\omega - \omega_{v_ii,v_jj})$$

$$+ \frac{1}{(\omega + \omega_{v_ii,v_jj})^2} \hat{\chi}_-(\omega + \omega_{v_ii,v_jj})\}. \qquad (19)$$

COMPUTATIONAL RESULTS OF THE WATER CONTINUUM ABSORPTION

Using the theory outlined above together with the accurate position and intensity data of the H_2O transitions[19], and the dipole moment of H_2O, $\mu = 1.8546$ D, the quadrupole moment $Q = -1.04$ ea_o^2 of N_2, and the same parameters of the isotropic interaction used by Rosenkranz[12] [$C/k = 8.0 \times 10^5$ Å^6 K, $s = 48$, and $\sigma = 3.13$ Å for H_2O-H_2O; and $C/k = 8.2 \times 10^5$ Å^6 K, $s = 9$, and $\sigma = 2.94$ Å for H_2O-N_2], we have calculated the average positive and negative frequency resonance line shape functions for H_2O-H_2O and for H_2O-N_2 as shown in Figs. 1 and 2 for $T = 296$ K. Also shown are the corresponding results obtained from the theory of Rosenkranz.[12] These functions have the same general shape as the empirical χ functions deduced from laboratory and atmospheric data.[20] Similar results have been obtained for other temperatures in the range 200 K < T < 700 K, and for H_2O-CO_2 but are not shown.

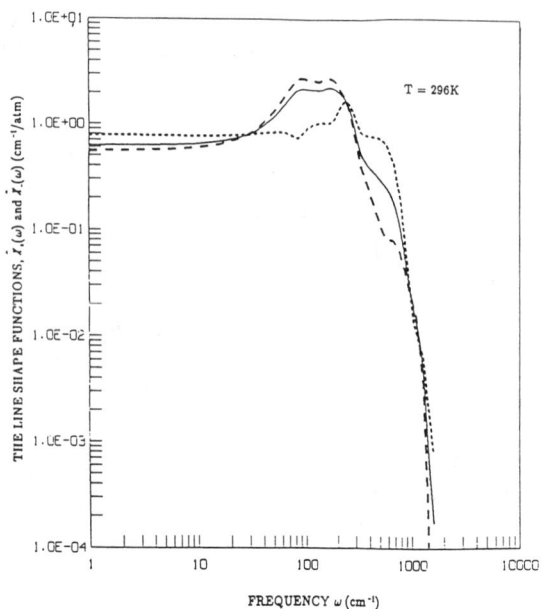

Fig. (1). The average positive and negative frequency resonance line shape functions for H_2O-H_2O broadening in cm^{-1}/atm: $\hat{\chi}_+(\omega)$ (dashed curve) and $\hat{\chi}_-(\omega)$ (dotted curve), as a function of ω calculated for $T = 296$ K. The solid curve is the intensity-weighted average line shape function.

The corresponding theoretical absorption coefficients, $\alpha(\omega)$, for frequencies to 10 000 cm^{-1} are shown in Figs. 3 and 4, respectively. We note that these results vary over 5 orders of magnitude even in this limited frequency range.

COMPARISONS BETWEEN THEORY AND EXPERIMENT AND VALIDATIONS

The theoretical absorption coefficients for self- and N_2-broadening[13-15] have been compared with the experimental results of Burch and co-workers.[16] From these and other comparisons,[17,18] we can draw the following conclusions: (i) Using the different average positive and negative resonance line shape functions gives better results than using a single average line shape function; (ii) The temperature dependence for the limited range of T is well described by the theory (in fact, the agreement is generally better for higher temperatures where the measured absorption is greater due to the increased vapor pressure and where the corresponding experimental uncertainties are lower); (iii) The agreement in the 300 < ω < 1100 cm^{-1} region is very good for both self- and N_2-broadening; (iv) The agreement in the 2400 < ω < 2700 cm^{-1} region is not as good, but the absolute magnitude of the continua are smaller than those in the 300 < ω < 1100 cm^{-1} region by approximately a factor of 100 (see Figs. 3 and 4), and the corresponding experimental uncertainties are much larger; (v) The agreement for measurements made near the centers of vibration-rotational bands (1200 - 2200 cm^{-1} and 3000 - 4200 cm^{-1}) show greater scatter than measurements made between the bands; (vi) Most of the theoretical values are below the corresponding experimental measurements.

In order to illustrate the temperature dependence in more detail over a wider range of T, we compare in Fig. 5 the theoretical calculations of the self-absorption coefficient for frequencies near 1203 cm^{-1} with the experimental data obtained by a number of researchers.[16,21-23] Also shown are some data from Varanasi[24] for a frequency near 1000 cm^{-1}. It is clear that the theoretical results reproduce the general

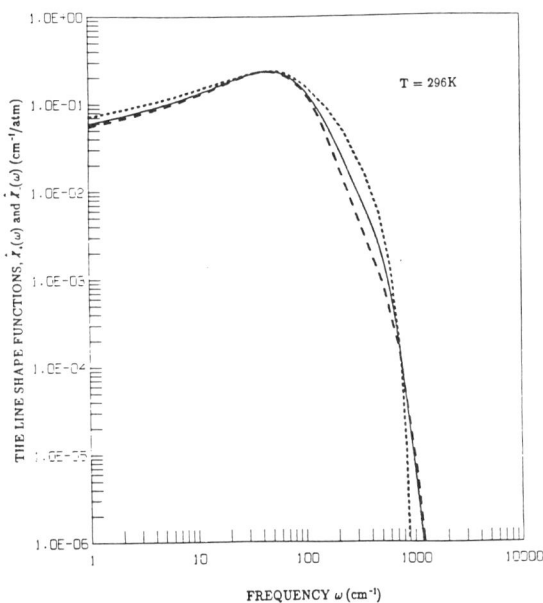

Fig. (2). The average positive and negative frequency resonance line shape functions for H_2O-N_2 broadening in cm^{-1}/atm: $\hat{\chi}_+(\omega)$ (dashed curve) and $\hat{\chi}_-(\omega)$ (dotted curve), as function of ω calculated for T = 296 K. The solid curve is the intensity-weighted average line shape function.

282   Line Shapes in the Far Wings

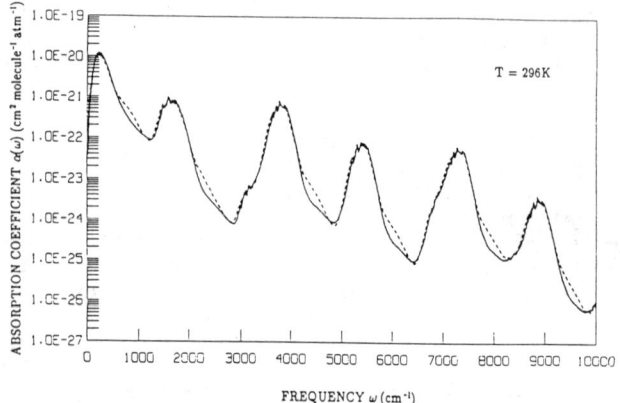

Fig. (3). The H_2O-H_2O absorption coefficient $\alpha(\omega)$ in cm^2 molecule^{-1} atm^{-1} as a function of ω in cm^{-1} calculated for T = 296 K. The results obtained from the two averaged line shape functions are the solid curve, and the results from the intensity-weighted average line shape function are the dotted curve.

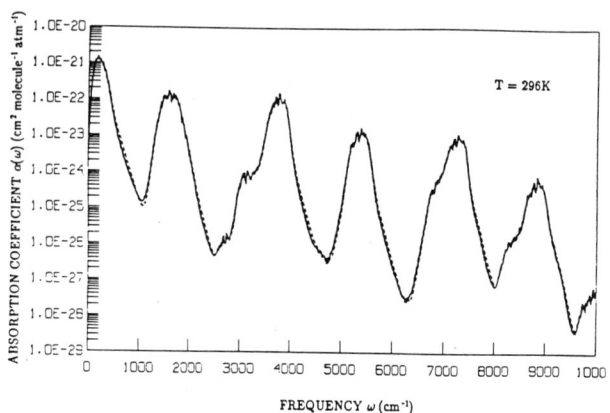

Fig. (4). The H_2O-N_2 absorption coefficient $\alpha(\omega)$ in cm^2 molecule^{-1} atm^{-1} as a function of ω in cm^{-1} calculated for T = 296 K. The results obtained from the two averaged line shape functions are the solid curve, and the results from the intensity-weighted average line shape function are the dotted curve.

trend very well, but underestimate the magnitude by 30 % or so. We note that in contrast to the empirical model of Roberts et al.[23] or predictions made assuming the dimer model, the absorption coefficient shows a positive slope for temperatures above 400 K that is corroborated by the present theoretical calculations. This change of slope with temperature depends sensitively on the frequency considered. This fact is very important, for instance, when one wants to calculate the far-wing absorption at high temperature.

We now discuss some preliminary validations of the theoretically calculated continua using atmospheric measurements. In Fig. 6 we show the radiance of the Earth as measured by the Nimbus-4 IRIS spectrometer in the 400 < ω < 1600 cm^{-1} region. Also shown is the predicted synthetic spectrum obtained using a line-by-line calculation without including the effects of the water continuum. For clarity, the difference between these is plotted in the lower part of the figure along with the IRIS standard deviation. As apparent from the figure, large differences are observed. In Fig. 7, we show a similar comparison except that the synthetic spectrum was calculated including the theoretical

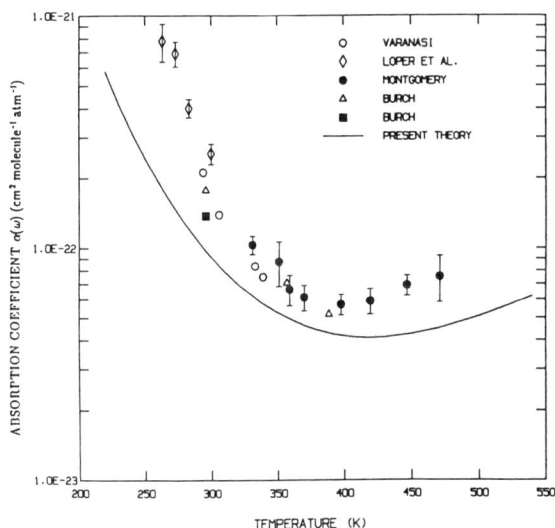

Fig. (5). Comparison between the theoretical temperature dependence and various laboratory data: ◊ Loper et al.;[21] ● Momtgomery;[22] △ Burch et al.[16]; ■ Burch data, cited by Roberts et al.;[23] all for $\omega \cong 1203$ cm^{-1}; o Varanasi[24] for $\omega \cong 1000$ cm^{-1}.

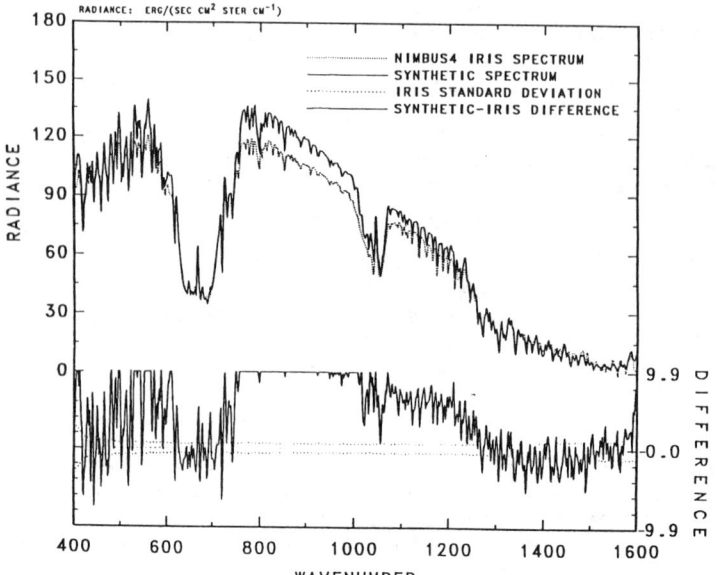

Fig. (6). Comparison between the Nimbus-4 IRIS satellite radiance spectrum in the 400 cm^{-1} < ω < 1600 cm^{-1} spectral region and the line-by-line synthetic spectrum calculated without the inclusion of the far-wing water absorption.

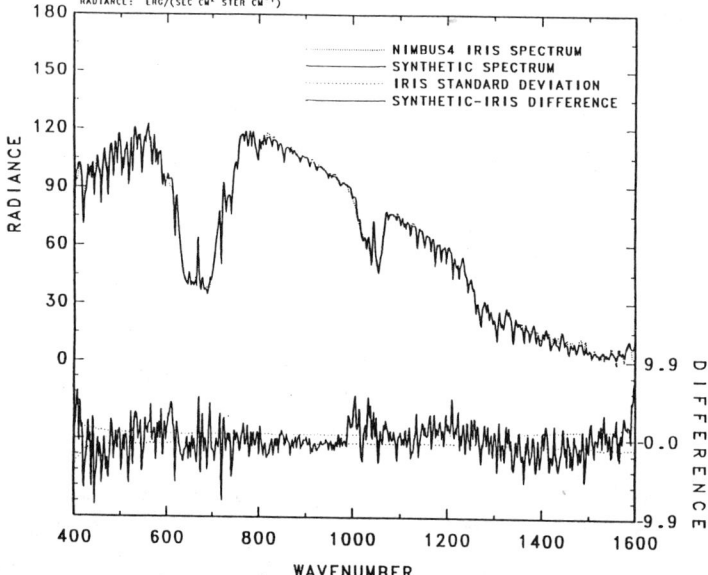

Fig. (7). The same as Fig. (6) except that the far-wing water absorption was included.

H_2O-H_2O continuum. The differences are dramatically reduced, especially in the important 700 - 1200 cm^{-1} region. We have also compared our theoretical calculations with the recent laboratory measurements of Kulp[25] with and without the continuum correction and have obtained similar agreement.[26]

THEORETICAL REFINEMENTS

The theory described above and applied to the far-wing absorption of water lines is predicated on several approximations, the most restrictive of which is the inclusion of only the leading term of the long-range anisotropic interaction. Recently we have been able to generalize the theory to multi-component anisotropic potentials.[27] The application to the high-frequency wing of the 4.3 μ band of CO_2 broadened by argon gave results that are in excellent agreement with the experimental data. Similar calculations for water have not yet been carried out. Modifications of the theory to incorporate corrections due to the time-dependence of the collisional dynamics neglected in the quasistatic approximation have also been investigated.[28] These lead to increased absorption in the near wings of the lines and thus to additional absorption closer to the band centers; this refinement slightly improves the agreement with experiment for the case of CO_2-Ar, but is less important than the inclusion of an accurate anisotropic potential. Another approximation that was made (see Eq. (4)) was to neglect the effects of the non-commutation of the operators L_0 and L_1. In the case of self-broadened water, these lead to an increase in the line shape between 10 and 200 cm^{-1}, and this increase is more important for lower temperatures than for higher ones. This generally improves the agreement between theory and experiment.[29] More recently, we have been able to show that by symmetrizing the theory with respect to the Liouville operators in Eq.(4), that is, by approximating

$$e^{-iLt} \cong 1/2 \{ e^{-iL_0 t} e^{-iL_1 t} + e^{-iL_1 t} e^{-iL_0 t} \}, \qquad (20)$$

we are able to obtain the same results[30] that were introduced in an ad hoc fashion by Boulet and co-workers[31,32] in their "quasistatic resonant" formulation.

In conclusion, a theoretical framework, based on the binary collision and quasistatic approximations that allows one to calculate the far-wing absorption using an accurate multi-component anisotropic interaction potential has been developed and applied to several different molecular systems. Although, in general, good agreement with experimental data has been obtained for several different molecular systems without the introduction of arbitrary parameters, much additional work is necessary both for the application to other systems and to incorporate further refinements.

ACKNOWLEDGMENTS

The authors would like to acknowledge support for this research from the Department of Energy Interagency Agreement under the Atmospheric Radiation Measurement Program and from the NASA Venus Analysis Program.

REFERENCES

1. A. Deepak, T. D. Wilkerson, and L. H. Ruhnke, editors, *Atmospheric Water Vapor*, Academic Press, New York, 1980.
2. R. M. Goody and Y. J. Yung, *Atmospheric Radiation*, Oxford University Press, New York, 1989.
3. J. Hinderling, M. W. Sigrist, and F. K. Kneubuhl, Infrared Phys. **27**, 63, (1987).
4. W. B. Grant, Appl. Opts. **29**, 451 (1990).
5. M. E. Thomas, Infrared Phys. **30**, 161 (1990).
6. S. S. Penner and P. Varanasi, J. Quant. Spectrosc. Radiat. Trans. **7**, 687 (1967); P. Varanasi, SPIE Proc. **928**, 213 (1988).
7. H. Carlon, J. Appl. Phys. **52**, 3111 (1981).
8. G. Birnbaum, editor, *Phenomena Induced by Intermolecular Interactions*, Plenum, New York, 1985.
9. W. M. Elsasser, Ap. J. **87**, 497 (1938).
10. U. Fano, Phys. Rev. **131**, 259 (1963).
11. R. W. Davies, R. H. Tipping and S. A. Clough, Phys. Rev. A **26**, 3378 (1982).
12. P. W. Rosenkranz, J. Chem. Phys. **83**, 6139 (1985); J. Chem. Phys. **87**, 163 (1987).
13. Q. Ma and R. H. Tipping, J. Chem. Phys. **93**, 7066 (1990).
14. Q. Ma and R. H. Tipping, J. Chem. Phys. **93**, 6127 (1990).
15. Q. Ma and R. H. Tipping, J. Chem. Phys. **95**, 6290 (1991).
16. D. E. Burch, D. A. Gryvnak, and J. D. and Pembrook, AFCRL-TR-71-0124 (1971); D. E. Burch, D. A. Gryvnak, and F. J. Gates, AFCRL-TR-74-0337 (1974); D. E. Burch, D. A. Gryvnak, and J. D. Pembrook, AFCRL-TR-75-0420 (1975); D. E. Burch, AFGL-TR-81-0300 (1982); D. E. Burch and R. L. Alt, AFGL-TR-84-0128 (1984); D. E. Burch, AFGL-TR-85-0036 (1985).
17. Q. Ma and R. H. Tipping, J. Chem. Phys. **96**, 8655 (1992).
18. Q. Ma and R. H. Tipping, J. Chem. Phys. **97**, 818 (1992).
19. L. S. Rothman, R. R. Gamache, R. H. Tipping, C. P. Rinsland, M. A. H. Smith, D. C. Benner. V. M. Devi, J.-M. Flaud, C. Camy-Peyret, A. Perrin, A. Goldman, S. T. Massie, and L. R. Brown, J. Quant. Spectrosc. Radiat. Transfer **48**, 469 (1992).
20. S. A. Clough, F. X. Kneizys, and R. W. Davies, Atmos. Res. **23**, 229 (1989).

21. G. L. Loper, M. A. O'Neill, and J. A. Gelbawachs, Appl. Opt. 23, 3701 (1983).
22. G. P. Montgomery, Appl. Opt. 17, 2299 (1978).
23. R. E. Roberts, J. E. A. Selby, and L. M. Biberman, Appl. Opt. 15, 2085 (1976).
24. P. Varanasi, J. Quant. Spectrosc. Radiat. Trans. 40, 169 (1988).
25. T. Kulp, private communication.
26. R. H. Tipping and Q. Ma, in *Induced Spectroscopy: Advances and Applications*, NATO Advanced Research Workshop, edited by G. C. Tabisz, to be published, 1994.
27. Q. Ma and R. H. Tipping, J. Chem. Phys. 100, 8720 (1994).
28. Q. Ma and R. H. Tipping, J. Chem. Phys. 100, 5567 (1994).
29. Q. Ma and R. H. Tipping, J. Chem. Phys. 100, 2537 (1994).
30. Q. Ma, R. H. Tipping, J.-M. Hartmann, and C. Boulet, J. Chem. Phys., to be published.
31. J. Boissoles, V. Menoux, R. Le Doucin, C. Boulet, and D. Robert, J. Chem. Phys. 91, 2163 (1989).
32. J.-M. Hartmann and C. Boulet, J. Chem. Phys. 94, 6406 (1991).

THE WINGS OF PRESSURE BROADENED MOLECULAR BANDS

George Birnbaum
National Institute of Standards and Technology
Gaithersburg, MD 20899

ABSTRACT

This paper reviews some aspects of the theory of the high frequency wing of infrared bands of molecular gases. Particular attention is paid to the ν_3 band of CO_2, broadened by a number of perturbers, in a small frequency region beyond the band head where the experimental absorption coefficient $\alpha(\omega)$ obeys the relation $\alpha(\omega)\omega^4$ = const. The theory giving this result is presented, and values of the mean squared torque and rotational relaxation time obtained from this plateau region of the spectrum are discussed.

INTRODUCTION

The continuum absorption of the far wings of allowed (pressure broadened) and collision induced bands in compressed gases and gaseous mixtures exhibit similar shapes and density dependences. The shapes are usually exponential-like, and at not too high densities the absorption coefficients vary as the product of the densities of the components of the mixture. However, the line shaping mechanisms differ: in the allowed spectra of dipolar molecules, line broadening and line mixing due to the effect of the anisotropic potential play dominant roles, whereas in collision induced spectra it is the dipole moments induced during collisions that are significant. Except for the case of HD and HD in gaseous mixtures, where both pressure broadening and collision induced absorption and their interference are important in shaping the far infrared and infrared spectra,[1] the effect of the anisotropic potential is generally not very significant in collision induced spectra. Whereas the entire collision-induced spectrum originates from the finite collision duration, this effect (non-Markovian behavior) is significant only in the far wings of allowed lines.

For some time, calculations of the shape of collision induced bands have produced results in close agreement with experiment.[2] On the contrary, the situation is not yet as satisfactory in treating the shape of pressure broadened bands, particularly the far wings, although much progress has been made in this direction. In general, two kinds of theories have been advanced to deal with this problem; one based on a quasistatic approximation in which the intermolecular potential is taken to be independent of the time,[3-5] and another in which the effect of molecular torques, due to a time dependent potential, appears in the theory.[6,7] This paper reviews some recent advances in the latter approach. In particular, we emphasize the high-frequency wing of the ν_3 band of CO_2, where following a very sharp decrease in absorption, $\alpha(\omega)$, due to line interference, a plateau in the high frequency wing of the $\alpha(\omega)\omega^4$ spectrum is observed.

Since the pioneering work of Winters et al.[8], the ν_3 band of gaseous CO_2 has been prominent in studies of the continuum absorption in high frequency wings. The reasons for this are clear: the rotational lines converge to a band head beyond which there is no obscuring rotational structure, and, moreover, there are no significant collision-induced bands in this region. The absorption coefficient for this band broadened by a number of perturbers is shown in Fig. 1 plotted as $\Delta_0^4 \alpha(\Delta_0)$, where Δ_0 is measured from the band center at approximately 2349 cm^{-1}. This plot makes apparent a plateau in a small region from 2450 to 2470 or 2480 cm^{-1} for CO_2-Xe or CO_2-Ar, respectively, but remains rather independent of Δ_0 over a larger region for the perturbers N_2, Ne and He. The sharp drop in absorption in the vicinity of 2440 cm^{-1} is due to line interference and has been theoretically described by Bulanin et al.[10] and Cousin et al.[11].

THEORY

We summarize the theory developed previously[7]; for related work see Filippov[12] and Bulanin et al.[6] Our approach emphasizes the use of spectral moments and certain integrals over time to develop simple model shapes appropriate for line wings. The shape of an infrared band is described by the spectral density

$$g(\omega) = \frac{1}{\pi} \operatorname{Re} \int_0^\infty e^{-i\omega t} \Phi(t) dt \qquad (1)$$

The correlation function (CF) is given by

$$\Phi(t) = \sum_{j,k} \rho_k e^{i\omega_k t} d_k^* d_j F_{jk}(t) \qquad (2)$$

where initial orientational correlations, which are significant only in the far wings, are neglected. Here ω_k is the rotational-vibrational frequency, ρ_k is the population of the active molecule, d_k^* and d_j are reduced dipole matrix elements and $F_{jk}(t)$ is a reduced CF. The subscripts j and k represent, respectively, $v_i v_f J_i' J_f'$ and $v_i v_f J_i J_f$ where v and J are, respectively, vibrational and rotational quantum numbers. $F_{jk}(t)$ may be described by the kinetic equation

$$\frac{dF_{jk}(t)}{dt} = - \sum_l \int_0^t dt' F_{jl}(t-t') B_{lk}(t') \qquad (3)$$

where $B_{lk}(t)$ is a memory function, and $B_{lk}(t)$ and $F_{jl}(t)$ are taken as real, even functions of time. In modeling $B_{lk}(t)$, t^2 may be replaced by $y^2 = t^2 - i\beta\hbar t$ to otain a spectrum which obeys

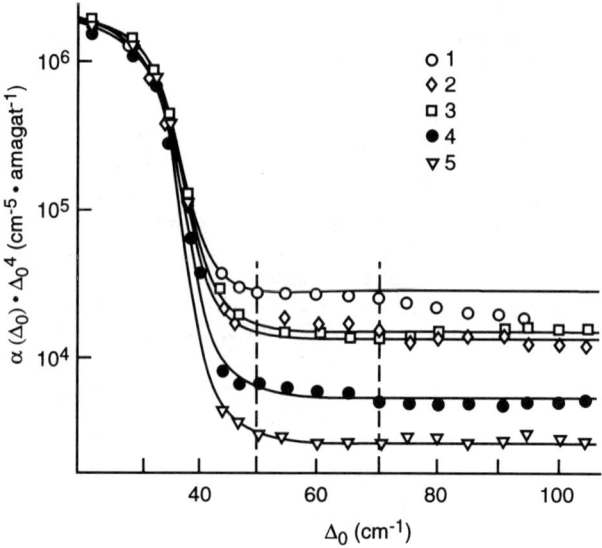

Figure 1. Shape of the wing of the ν_3 band of CO_2. Experimental points are CO_2 broadened with 1-Xe, 2-Ar, 3-N_2, 4-Ne, 5-He. The solid lines were calculated on the hypothesis of empirically adjusting the number of interfering lines. The vertical lines delineate the region in determining $\Delta\nu_J$.[9]

detailed balance, to a good approximation; however, this is only important in the far wings. Then taking the Laplace transform $\int_0^\infty dt\, F_{jk} \exp(-i\Delta_k t)$ where $\Delta_k = \omega - \omega_k$, we find that in the wings where $\Delta_k \gg \hat{B}'_{kk}(\Delta_k)$ and frequency shifts are unimportant

$$g(\omega) = \frac{1}{\pi} \sum_{j,k} \rho_k d_k^* d_j \hat{B}_{jk}(\Delta_k)/\Delta_k\Delta_j \tag{4}$$

The function

$$\hat{B}'_{jk}(\Delta_k) = \int_0^\infty dt\, B_{jk}(t)\cos\Delta_k t \tag{5}$$

is a frequency dependent line width (j = k) or line coupling parameter (j≠k), (usually negative). For $\Delta_k = 0$ and for frequencies not too far from the line center, one has

$$\hat{B}_{jk}(0) = B_{jk}(0)\tau_{jk} = \gamma_{jk} \tag{6}$$

γ_{jk} is the impact result and $\tau_{jk} = \int_0^\infty B_{jk}(t)dt/B_{jk}(0)$.

Useful sum rules have been presented by Filippov[12] and Tonkov and Filippov[13]

$$\sum_j d_j B_{jk}(t) = 0, \qquad \sum_k \rho_k d_k^* B_{jk}(t) = 0 \tag{7}$$

and similar relations by replacing $B_{jk}(t)$ by $\hat{B}_{jk}(\Delta_k)$. Thus, $g(\omega)$ approaches zero as $\Delta_j \approx \Delta_k$. However, Ma[14] suggests that Eqs. (7) are valid only approximately.

Equation (4) has been applied in the impact approximation to describe the very rapid decrease of absorption[10,11] shown in Fig. 1. Using Eq. (7), it is easy to show that Eq. (4) in the impact limit becomes

$$g(\Delta_0) = \frac{1}{\pi} \Delta_0^{-4} \sum_{j,k} \rho_k d_k^* d_j \gamma_{jk} \omega_j' \omega_k' \tag{8}$$

where $\Delta_0 = \omega - \omega_v$, $\omega_j' = \omega_j - \omega_v$, $\omega_k' = \omega_k - \omega_v$, ω_v is the solvent shifted band center, and $\omega_k'/\Delta_0 \ll 1$ and $\omega_j'/\Delta_0 \ll 1$. Thus $g(\Delta_0)\Delta_0^4$ is independent of Δ_0. The absorption coefficient is given by

$$\alpha(\omega) = \frac{4\pi^2 n}{3\hbar c} \omega (1 - e^{-\beta\hbar\omega}) g(\Delta_0) \tag{9}$$

where n is the density of active molecules. The function $\alpha(\omega)\Delta_0^4$ is also practically constant over the frequency region of interest since ω varies but little over an infrared band, and the exponential function is practically zero.

The short-time expansion of the CF, Eq. 2, in even powers of time is

$$\Phi(t) = \sum_{j,k} \rho_k d_k^* d_j \left\{ \delta_{jk} - \frac{t^2}{2} \left[(\omega_k')^2 \delta_{jk} + B_{jk} \right] \right.$$

$$\left. + \frac{t^4}{24} \left[(\omega_k')^4 \delta_{jk} + 6(\omega_k')^2 B_{jk} - \ddot{B}_{jk} + \sum B_{kl} B_{lj} \right] \right\} - \ldots \tag{10}$$

This expression is to be compared with Gordon's result

$$\Phi(t)/\Phi(0) = 1 - \frac{t^2}{2}\left(\frac{2kT}{I}\right) + \frac{t^4}{24}\left[2\left(\frac{2kT}{I}\right)^2 + \frac{\langle N^2 \rangle}{I^2}\right] - \ldots \tag{11}$$

where I is the moment of inertia for a linear molecule and $\langle N^2 \rangle$ is the mean-squared torque. The terms containing $(\omega_k')^2$ and $(\omega_k')^4$ yield, respectively, $(2kT/I)$ and $2(2kT/I)^2$; the term involving $B_{kl}B_{lk}$ is neglected because it varies as n_p^2 since B_{kl} is proportional to n_p, the density of perturbers.[7] The term involving B_{jk} sums to zero because of the sum rules, Eq.(7). The term involving $\ddot{B}_{jk}$ should give $\langle N^2 \rangle/I^2$, but the function $B_{jk}(t)$ is not known. However, $B_{jk}(t)$ may be represented by $(\gamma_{jk}/\tau_c)H(t)$, where $H(t)$ is a model function that rapidly decays to zero in a time of the order of τ_c, a collision duration parameter. Thus if $\hat{H}(\Delta_k) = \tau_c \hat{h}(\Delta_k)$, we have $\hat{B}_{jk}(\Delta_k) = \gamma_{jk}\hat{h}(\Delta_k)$. Moreover, all time derivatives of the model $B_{jk}(t)$ sum to zero by virtue of Eq. (7). The memory function that gives $\langle N^2 \rangle/I^2$ is obtained in the next section.

According to Eq. (11), the anisotropic potential, V, does not appear until the t^4 term where $\langle N^2 \rangle$ is a function of V. Apart from $\langle N^2\rangle/I^2$, the terms in Eq. (11) are those for freely rotating linear molecules. By contrast, V appears in the t^2 term in the quantum mechanical representation, Eq. (10), where $B_{jk}(t) = \hbar^{-2}\langle V_{j1}(0)V_{1k}(t)\rangle$. However, according to Eqs.(7), the sum of these terms disappear. This cancellation is related to line mixing and the loss of absorption at high frequencies.

EFFECT OF THE TORQUE

To deal with the effect of molecular torques on the spectral shape, we consider

$$\frac{\partial^4 F(t)}{\partial t^4} = -\int_0^t \frac{\partial^3 F(t-t')}{\partial t^3} B(t')dt' - \frac{\partial^2 F(0)}{\partial t^2} B(t) - F(0)\frac{\partial^2 B(t)}{\partial t^2} \quad (12)$$

obtained by differentiating a kinetic equation for $F(t)$. Here $F(t) = F_{jk}(t) + F_N(t)$ and $B(t) = B_{jk}(t) + B_N(t)$, where now $F_{jk}(t)$ and $B_{jk}(t)$ represent that part of the CF and memory function, respectively, which do not contain the effect of molecular torques, and $F_N(t)$ and $B_N(t)$ are the contributions due to the torques. Then since $F_N(0) = 0$, $\ddot{F}_N(0) = 0$, and $B_N = (0)$, we find

$$\frac{\partial^4 F_N(t)}{\partial t^4} = -\frac{\partial^2 B_N(t)}{\partial t^2} \quad (13)$$

provided the time is restricted to values so small that the integral in Eq. (12) may be neglected. We write for the spectral density, by successively integrating $\int_0^\infty \exp(-i\Delta_0 t)F(t)$ by parts

$$\hat{F}_N(\Delta_0) = (\pi\Delta_0)^{-4}\text{Re}\int_0^\infty dt e^{-i\Delta_0 t}\frac{\partial^4 F_N(t)}{\partial t^4} \quad (14)$$

where at sufficiently high frequencies the band may be regarded as originating from a single transition at the vibrational frequency, ω_v, and $\Delta_0 \neq 0$.

The function $\ddot{B}_N(t)$ is defined by

$$\frac{\partial^2 B_N(t)}{\partial t^2} = -\frac{\langle \vec{N}(0)\cdot\vec{N}(t)\rangle}{I^2} = -\frac{\langle \vec{J}(0)\cdot\vec{J}(t)\rangle^\dagger}{\langle I^2\rangle} \quad (15)$$

where $\vec{N} = \dot{\vec{J}}$, and $\vec{J}$ is the angular momentum. From Eqs. (14) and (15), we obtain

$$\pi\Delta_0^4 \hat{F}(\Delta_0) = \frac{\langle j^2\rangle}{I^2}\text{Re}\int_0^\infty dt e^{-i\Delta_0 t}\frac{\langle \vec{J}(0)\cdot\vec{J}(t)\rangle^\dagger}{\langle j^2\rangle} = \frac{\langle j^2\rangle}{I^2}\tau_N \hat{h}_N(\Delta_0) \quad (16)$$

where $\hat{h}_N(0) = 1$. The dagger at the right side of the angular brackets indicates that the quantities inside evolve in time via a projected Liouville operator, and consequently τ_N may be regarded as a projected torque correlation time.[15,16] This time is independent of density and is an average collision duration that may be estimated from

$$\tau_N \approx \bar{r}/v \tag{17}$$

where $\bar{r}$ is a parameter representing the range of the intermolecular forces producing the torque and v is a mean relative velocity. Borrowing from efforts to represent collision-induced line shapes by empirical profiles, several functions have been used with reasonable success i.e., $x^{5/2}K_{5/2}(x)$,[6] $xK_1(x)$,[7] and $x^2K_2(x)$,[7] where $x = \Delta_0/\Delta\omega_N$. In the limit $\Delta_0 = 0$, one obtains

$$\int_0^\infty dt \frac{\langle \vec{J}(0) \cdot \vec{J}(t) \rangle^\dagger}{\langle J^2 \rangle} = \frac{1}{\tau_J} \tag{18}$$

where τ_J is the angular momentum correlation time, which is inversely proportional to the density of perturbers. Then from the integral in Eq. (16) with $\Delta_0=0$ and Eq. (18), we obtain

$$(\tau_N \tau_J)^{-1} = \langle \dot{J}^2 \rangle / \langle J^2 \rangle = \langle \dot{J}^2 \rangle / I^2 \mu_2 \tag{19}$$

where $\mu_2 = \langle \omega_k'^2 \rangle = 2kT/I$. In the impact approximation where $\Delta_0 \tau_N \ll 1$, Eq. (16) becomes

$$\hat{F}_N(\Delta_0) = \frac{1}{\pi \Delta_0^4} \frac{\mu_2}{\tau_J} \tag{20}$$

Comparing this result with Eq. (8), we see that both equations decrease at high frequencies as Δ_0^{-4}, and both depend on the density of perturbers n_p. Equation (20) is the same as the result obtained by Burshtein and McConnell[17] in an approach where the fourth spectral moment, $\langle \Delta_0^4 \rangle = 2\mu_2^2 + \langle N^2 \rangle/I^2$ lacks the factor 2 before μ_2^2. We note that the condition $\Delta_0 \tau_N \ll 1$ may not be satisfied in practice since the plateau region begins at about 50 cm^{-1} and τ_N^{-1} is of the same order of magnitude.[6,7]

COMPARISON WITH EXPERIMENT

The high frequency wings of the ν_3 band of CO_2 shown in Fig. 1, plotted as $\alpha(\Delta_0)\Delta_0^4$, for a number of perturbers, exhibit a frequency independent behavior in a narrow range for Xe and Ar and over a more extended range for Ne, N_2 and He. The values of $\Delta\nu_J$ obtained by fitting Eq. (20) multiplied by the band strength $\Sigma_k \rho_k |d_k|^2$, are given in Table 1. These values are somewhat greater than the Lorentz widths $\Delta\nu_L$ except for Ne and He broadening where $\Delta\nu_L \gg \Delta\nu_J$. The data in Fig. 1 was also analyzed by Filippov and Tonkov[18] in terms of the function $K(\Delta_0)\Delta_0^2/\mu_2$. Here $K(\Delta_0)$ is Eq. (20) divided by the Lorentz wing contribution approximated by $(\tau_0 \Delta_0^2)^{-1}$, where τ_0 is an average mean-time between collisions. Thus one obtains

$$K(\Delta_0)\Delta_0^2/\mu_2 = \tau_0/\tau_J \tag{21}$$

Table 1. Experimentally determined angular momentum line widths, $\Delta\nu_J$ (cm^{-1}/amagat).

	I.R.[9]	I.R.[18]	NMR[19]	Rotat. Relax.[a]	IR[9] [b]	IR[18] [b]	NMR[19] [b]
He	0.011	0.0078		0.026	0.19	0.22	
Ne	0.021	0.014			0.36	0.41	
Ar	0.059	0.034	0.026		1	1	1
Xe	0.088	0.057	0.039		1.5	1.7	1.5
N$_2$	0.053		0.023	0.052	0.90		0.88

[a] From ref. 20, computed from the rotational relaxation rate constant for the J=19 line. However, $\Delta\nu_J$ of the various states get progressively smaller for increasing or decreasing J-values centered about J=19.[21]
[b] Normalized for Ar.

The fitting gives $\tau_0/\tau_J = (1 - G)$ where G is a parameter in the line broadening theory of Filippov and Tonkov.[18] The values of $\Delta\nu_J$, obtained from the fitted value of G and the value of τ_0 obtained in their theory of line width, which requires a knowledge of the parameter G, are given in Table 1. These $\Delta\nu_J$ are significantly less than the values given by Dokuchaev and Tonkov.[9] However, when $\Delta\nu_J$ is normalized by the value for Ar, the two determinations are in better agreement. The reason for the discrepancies in the absolute values of $\Delta\nu_J$ might be due to the evaluation of τ_0.

Nuclear spin relaxation times were measured by Jameson et al.[19] for ^{13}C in ^{13}C^{16}O$_2$ broadened by N$_2$, Ar and Xe. Since the relaxation is completely dominated by the spin rotation mechanism, values of the cross sections for rotational angular momentum transfer, σ_J, could be obtained. The corresponding values of $\Delta\nu_J$ are considerably smaller than the values of $\Delta\nu_J$ obtained from either of the infrared band wing studies.[9,18] The normalized values, however, are in better agreement. Also included in Table 1 are $\Delta\nu_J$ obtained from rotational relaxation rate constants measured for the 00°1 level in CO$_2$ by using a saturating pulse from a CO$_2$ laser at P(20) in the 10.4μ band, and measuring the subsequent repopulation.[20] The agreement with the IR result[9] is excellent for CO$_2$-Ar, but the value for CO$_2$-He is larger by a factor of nearly 3.

Table 2. Mean squared torques from the ν_3 band of CO_2.

Perturber	$\langle N^2\rangle/n_p$ [a]	$\langle N^2\rangle_X/\langle N^2\rangle_{Ar}$ [b]		
	cm^{-2}/amagat	ref. 19	ref. 9	ref. 6
Ne	320	.41	0.50	0.51
Ar	790		1	1
Xe	1,090	1.38	1.2	1.34

[a] Ref. 22   [b] Eq. (22)

Table 2 gives the values of $\langle N^2\rangle/n_p$ determined from measurements of the spectral moments of the ν_3 band of CO_2,[22] and normalized values of $\langle N^2\rangle/n_p$ computed from

$$\frac{(\langle N^2\rangle/n_p)_X}{(\langle N^2\rangle/n_p)_{Ar}} = \frac{(v\Delta\nu_J)_X}{(v\Delta\nu_J)_{Ar}} \tag{22}$$

This relation is based on Eqs. (17) and (19) with the assumption that $\bar{r}$ is similar for all perturbers. The good agreement of the normalized values for CO_2-Xe and CO_2-Ne obtained from Eq. (22) and the normalized values obtained from the spectral moments supports the view that the plateau region represents predominantly the contribution of the mean squared torque, Eq. (20), and that the rotational contribution, Eq. (8), is not significant for these broadeners.

The CO_2-Xe and CO_2-Ar spectra exhibit a decrease beyond the plateau region that is attributable to the effect of finite collision duration, i.e., the $\hat{h}_N(\Delta_0)$ function in Eq. (16), whereas no such decrease is seen for CO_2-Ne and CO_2-He. One reason for this might be attributable to the values of $\Delta\nu_N$, which are greater in CO_2-Ne and CO_2-He than in CO_2-Ar and CO_2-Xe, and hence the decrease does not appear in the frequency range shown in the figure. In the case of CO_2-He, $\langle N^2\rangle$ is so small that the plateau region could be dominated by the rotational contribution, Eq. (8). To investigate this matter, this equation was evaluated by Boulet[22] who obtained

$$A(\Delta_0) = \pi\Delta_0^4 g(\Delta_0)/\sum_k \rho_k |d_k|^2 = 3.84 \text{ cm}^{-3}\text{atm}^{-1} \tag{23}$$

To compare this result with the CO_2-He spectrum from which we get $K(\Delta_0)\Delta_0^2 = 0.125\mu_2$,[18] we divide the function $A(\Delta_0)$ by $\Delta\nu_0 = 0.060$ cm^{-1}/amagat, an average Lorentz width for CO_2-He broadening. $K(\Delta_0)$ here is the experimental deviation function obtained by dividing the measured absorption by the computed Lorentz absorption. The experimental value thus obtained is $K(\Delta_0)\Delta_0^2 = 43$ cm^{-2}, whereas Eq. (23) gives 66 cm^{-2}. In view of the uncertainties in the theoretical calculation of the γ_{jk} parameters, and the experimental uncertainty of determining the small absorption in the high frequency wing of the CO_2-He spectrum, we believe that the agreement between theory and experiment is reasonable and the absorption in the plateau region is accounted for by the rotational perturbation for CO_2-He. Further calculations along this line for other perturbers would be useful.

DISCUSSION

A plateau observed in the high frequency wing of the ν_3 band of CO_2 can give interesting information concerning molecular interactions. However, the theory predicting such a plateau is based on the impact approximation which requires that $\Delta_0\tau_N \ll 1$, which is not realized in practice because Δ_0 is the same order of magnitude as τ_N^{-1}. In fact, with data including that shown in Fig. 1 but extending to higher frequencies, fits were made to evaluate the duration of collision parameter, τ_N.[6,7]

At higher frequencies than shown in Fig. 1, the CO_2-He and CO_2-Ne spectral wings must decrease due to the effect of finite collision duration. For CO_2-Ne this may be described by the spectral shape, $\hat{h}_N(\Delta_0)$, in Eq. (16), if the plateau is mainly due to the effect of torques. Since the torque contribution for CO_2-He absorption is so small, a very high frequency decrease should arise from $\hat{B}_{jk}(\Delta_k)$ in Eq. (8). Finally, although results of practical interest have been obtained by dealing separately with the rotational (quantum) and torque (classical) contributions to the spectrum, a more satisfactory theory would deal with both in a unified way. An approach to realize this goal has not been found as yet.

The author acknowledges helpful comments from R. M. Mountain, C. Boulet, M. V. Tonkov, and S. Temkin, and the support of this work by a grant from the Planetary Atmospheres Program of NASA.

REFERENCES

1. Bo Gao, J. Cooper, and G. C. Tabisz, Phys. Rev. A<u>66</u>, 5781 (1992).
2. L. Frommhold, Collision-Induced Absorption in Gases (Cambridge University Press, Cambridge and New York, 1993).
3. P. W. Rosenkranz, J. Chem. Phys. <u>83</u>, 6139 (1985).
4. J. Boissoles, V. Menoux, R. Le Doucen, C. Boulet, and D. Robert, J. Chem. Phys. <u>91</u>, 2163 (1989).
5. Q. Ma and R. H. Tipping, J. Chem. Phys. <u>95</u>, 6290 (1991).
6. M. O. Bulanin, M. V. Tonkov, and N. N. Filippov, Can. J. Phys. <u>62</u>, 1306 (1984).

7. G. Birnbaum, Mol. Phys. 81, 519 (1994). In this paper, reference number [25] should be replaced by [14] on page 528. Reference [21] should be: Rosenkranz, P. W., J. Chem. Phys. 83, 6139. In Table 1, footnote c, $\mu_2 = (2kT/I)(2\pi c)^{-2}$.
8. B. H. Winters, S. Silverman, and W. S. Benedict, J.Q.S.R.T. 4, 527 (1964).
9. A. B. Dokuchaev and M. V. Tonkov, Opt. Spectrosc. 60, 664 (1986).
10. M. O. Bulanin, A. B. Dokuchaev, M. V. Tonkov, and N. N. Filippov, J.Q.S.R.T. 31, 521 (1984).
11. C. Cousin, R. Le Doucen, and C. Boulet, A. Henry, and D. Robert, J.Q.S.R.T. 36, 521 (1986).
12. N. N. Filippov, Sov. J. Chem. Phys. 10, 664 (1992).
13. M. V. Tonkov and N. N. Filippov, Opt. Spectrosc. 54, 475, 591 (1983).
14. Q. Ma, private communication.
15. D. Kivelson, Mol. Phys. 28, 321 (1974).
16. D. Kivelson and T. Keyes, J. Chem. Phys. 51, 4599 (1972).
17. A. I. Burshtein and J. M. McConnell, Physics 157, 933 (1989).
18. N. N. Filippov and M. V. Tonkov, J.Q.S.R.T. 50, 111 (1993).
19. C. J. Jameson, A. K. Jameson, and K. Jackowski, J. Chem.Phys. 86, 2717 (1987).
20. R. R. Jacobs, K. J. Pettipiece, and S. C. Thomas, Appl. Phys. Lett. 24, 375 (1974).
21. R. R. Jacobs, S. J. Thomas, and K. J. Pettipiece, IEEE J. Quantum Electron, QE-10, 480 (1974).
22. L. Berreby and E. Dayan, Mol. Phys. 48, 581 (1983).
23. C. Boulet, private communication (1994).

SEMICLASSICAL ANALYSIS OF THE INTERBRANCH LINE COUPLING IN THE INFRARED BAND SHAPES OF LINEAR MOLECULES

N.N.Filippov and M.V.Tonkov

Institute of Physics, St.Petersburg University, Peterhof, 198904 St.Petersburg, Russia

Abstract. The spectral evidences for interbranch line coupling have been analysed using a semiclassical approach. Interbranch line coupling causes the smooth branch broadening effect and determines the efficiency of smooth band shape transformation.

The vibration-rotation band shape $\Phi(\omega)$ in the impact approximation is given by

$$\Phi(\omega) = \frac{1}{\pi} \operatorname{Re} \left\{ \sum_{q,q',j,j'} \rho_j d_{qj} d_{q'j'} \left[\frac{1}{i(\omega - L) + \Gamma} \right]_{qj,q'j'} \right\}, \qquad (1)$$

where the pair of indices (q, j) defines the line corresponding to transition $(v_i, j) \to (v_f, j + q)$, v and j are respectively vibrational and rotational quantum numbers, ρ_j is the population of the active molecule, and d_{qj} is the reduced dipole matrix element. The branch index q may be equal to $-1, 0, 1$, which correspond to rotational branches P, Q and R. L is the line frequency matrix, Γ is the relaxation matrix. The diagonal elements of Γ-matrix determining line widths and shifts and the off-diagonal elements being responsible for non-additive effects (line coupling) when the lines overlap.

The elements $\Gamma_{qj,qj'}$ describe the coupling of lines of the same branch (intrabranch line coupling) and $\Gamma_{qj,q'j'}(q \neq q')$ describe the coupling of lines of different branches (interbranch line coupling). We find that the efficiency of interbranch line coupling of the q branch lines can be characterized by a ratio

$$\sigma_q = (A_q \Gamma_q)^{-1} \sum_{j,j'} P_j d_{qj} d_{qj'} \Gamma_{qj,qj'}, \qquad (2)$$

where Γ_q is the mean diagonal element of the q branch submatrix and A_q is the intensity of the branch q. The values of the parameter σ_q vary from 0 (isolated branch case) to 1 (intrabranch line coupling is negligible as compared with the interbranch one). They have been calculated for $\Sigma - \Sigma$ and $\Sigma - \Pi$ bands of linear molecules using a semiclassical approach (SCA) developed in Refs. 1,2:

$$\sigma_{\pm 1}^{\Sigma-\Sigma} = \tfrac{1}{2}\langle 1 - \cos\alpha\rangle, \quad \sigma_0^{\Sigma-\Pi} = \langle \sin^2\alpha\rangle, \quad \sigma_{\pm 1}^{\Sigma-\Pi} = \sigma_{\pm 1}^{\Sigma-\Sigma} + \tfrac{1}{2}\sigma_0^{\Sigma-\Pi}, \qquad (3)$$

where $\langle \ldots \rangle$ is the mean value and α is the collision induced angular momentum reorientation. It has been shown that the σ parameter characterizes the efficiency of smooth branch shape transformation with the gas density increase, the width of the smooth Q-branch, for example, is close to $2\sigma_0 \Gamma_0 + \Omega_Q$, where Ω_Q is the width of the Q-branch rotational structure.

The dependence of the σ parameters on a collision type has been analysed. In the strong collision case the σ values are

$$\sigma_{\pm 1}^{\Sigma-\Sigma} = \tfrac{1}{2}, \quad \sigma_0^{\Sigma-\Pi} = \tfrac{1}{2}, \quad \sigma_{\pm 1}^{\Sigma-\Pi} = \tfrac{3}{4},$$

which is in accordance with resultes of Ref. 3. In the weak collision case these parameters are close to zero. The σ values have been calculated for CO_2 molecules perturbed by Ar and He. For $\Sigma - \Sigma$ bands they are $\sigma_{\pm 1}(Ar)=0.34$, $\sigma_{\pm 1}(He)=0.10$, for $\Sigma - \Pi$ bands: $\sigma_0(Ar)=0.46$, $\sigma_0(He)=0.21$. The considered systems are the very convenient to analyse the line coupling effects so the averaged values of the Ar and He line broadening coefficients of CO_2 are near the same[4] but the collision types of these pairs are different. One can see in Fig. 1 that the Ar broadening efficiency of the Q-branch is about twice as much as the He one. The spectral evidences for interbranch line coupling in $\Sigma - \Sigma$ bands CO_2 perturbed by Ar and He have also been considered.

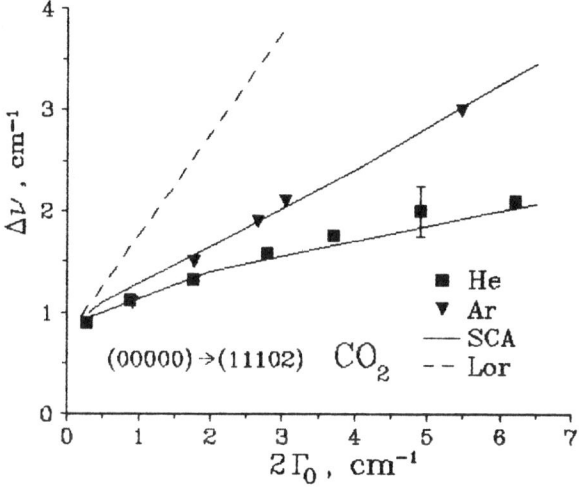

Fig. 1. Full width of Q-branch $\Delta\nu$ vs width $2\Gamma_0$ of separated vibration-rotation line for CO_2 + Ar and CO_2 + He mixtures at room temperature. The curves are SCA calculated and calculated as a sum of Lorentzians. The experimental points are taken from Ref. 2.

1. N.N.Filippov and M.V.Tonkov, JQSRT **50**, 111 (1993).
2. N.N.Filippov and M.V.Tonkov, Proc. ASA Workshop (France, Reims, 1993), p. 89.
3. S.Temkin, L.Bonamy, J.Bonamy, and D.Robert, Phys.Rev. **A47**, 1543 (1993).
4. M.O.Bulanin, A.B.Dokuchaev, M.V.Tonkov, and N.N.Filippov, JQSRT **31**, 521 (1984).

CLOSE-COUPLING CALCULATION OF LINE-MIXING PARAMETERS FOR D_2-He

R. Brezina and W.-K. Liu
Department of Physics, University of Waterloo, Ontario, Canada N2L 3G1

S. Green
NASA Goddard Space Flight Center, Institute for Space Studies, New York, New York 10025, U.S.A.

Recently, very precise lineshape measurements of the $Q(j)$ lines in D_2 using Raman gain spectroscopy[1] revealed slight asymmetry due to the mixing of the $Q(j)$ lines. From a careful fitting of the measured lineshape, line-mixing parameters can be accurately extracted, which in principle can be compared with *ab initio* scattering calculations.

While the calculation of lineshapes for pure gases is still rather difficult, the theory for a radiative active molecule infinitely diluted in a bath of inert gas atoms in the pressure-broadening regime has been well-developed. The lineshape for the isotropic $v = 0 \to 1$ Raman Q-branch can be expressed as[2]

$$I(\omega) = -\frac{1}{\pi}|\langle 1|\bar{\alpha}|0\rangle|^2 \mathrm{Im} \sum_{j',j} \mathbf{G}(\omega)_{j'j} p_j, \qquad (1)$$

where $\bar{\alpha}$ is the isotropic polarizability of the molecule, p_j is the population of the j-rotational level, and

$$\mathbf{G}(\omega)_{j'j}^{-1} = \omega \delta_{j'j} - \Phi_{j'j}, \quad \Phi_{j'j} = \omega_j \delta_{j'j} - in\bar{v}\langle \sigma(j',j)\rangle. \qquad (2)$$

In eq.(2), ω_j is the angular frequency of the $Q(j)$ line, $\bar{v} = (8k_BT/\pi\mu)^{\frac{1}{2}}$ is the average relative velocity (k_B, μ are the Boltzmann constant, and the reduced mass of the molecule-atom system, respectively),

$$\sigma(j',j) = \frac{\pi}{k^2} \sum_{J,l,l'} \left(\frac{2J+1}{2j+1}\right) [\delta_{j'j}\delta_{l'l} - S^J_{1j'l',1jl}(E_k+E_{1j})S^{J*}_{0j'l',0jl}(E_k+E_{0j})], \qquad (3)$$

and $\langle \cdots \rangle = \int_0^\infty x e^{-x}(\cdots)$ with $x = \hbar^2 k^2/k_BT = E_k/k_BT$ being the dimensionless kinetic energy denotes the thermal averaging procedure. $S^J_{vj'l',vjl}(E_k+E_{vj})$ is the S-matrix element for the collisional transition from the initial vibrational-rotational state $(vjlJ)$ to the final state $(vj'l'J)$ at a total energy (kinetic plus vibrational-rotational energies) $E_k + E_{vj}$.

Diagonalizing the matrix **G** exactly results in an expression for $I(\omega)$ consisting of a Lorentzian and a dispersion component:

$$I(\omega) = \frac{1}{\pi}|\langle 1|\bar{\alpha}|0\rangle|^2 \sum_j \frac{P_j \gamma_j + Q_j(\omega - \nu_j)}{(\omega - \nu_j)^2 + \gamma_j^2}, \tag{4}$$

where $\nu_j - i\gamma_j$ are the eigenvalues of the matrix Φ, and $P_j - iQ_j$ can be related to the eigenvectors of Φ. Physically, γ_j and $\delta_j \equiv \nu_j - \omega_j$ are the linewidth and lineshift, respectively, of the $Q(j)$ line. For well-separated lines, a first-order perturbation treatment gives $\delta_j = n\bar{v}\,\mathrm{Im}\langle\sigma(j,j)\rangle$, $\gamma_j = n\bar{v}\,\mathrm{Re}\langle\sigma(j,j)\rangle$ and $P_j = \mathrm{Re}\,A_j$, $Q_j = -\mathrm{Im}\,A_j$ where

$$A_j = p_j - in\bar{v} \sum_{j' \neq j} \frac{p_j \langle\sigma(j',j)\rangle + p_{j'}\langle\sigma(j,j')\rangle}{\nu_j - \nu_{j'}}. \tag{5}$$

The results of our close-coupling scattering calculation of the line-mixing parameters for the isotropic Raman $Q(j)$ lines of D_2 infinitely diluted in He, using the *ab initio* potential employed previously for the calculation of linewidths and lineshifts[3], are summarized below. It is clear that at 298 K and 22 amagat, the perturbation calculations are in excellent agreement with the exact results.

Line-mixing parameters at 298 K and 22 amagat.

j	exact				perturbation			
	P_j	Q_j	γ_j (GHz)	δ_j (GHz)	P_j	Q_j	γ_j (GHz)	δ_j (GHz)
0	0.27387	-0.00334	1.566	4.779	0.27389	-0.00334	1.568	4.782
2	0.57616	0.00314	1.127	5.199	0.57614	0.00313	1.126	5.196
4	0.14077	0.00020	0.882	5.398	0.14077	0.00020	0.882	5.389

REFERENCES

1. P.M. Sinclair, J.W. Forsman, J.R. Drummond, and A.D. May, Phys. Rev. **A48**, 3030 (1993).

2. A. Ben-Reuven, Phys. Rev. **145**, 7 (1966); R. Shafer and R.G. Gordon, J. Chem. Phys. **58**, 5422 (1973).

3. S. Green, R. Blackmore and L. Monchick, J. Chem. Phys. **88**, 4113 (1988); *ibid*, **91**, 53 (1989).

EXPERIMENTAL HF–Ar LINESHAPE PARAMETERS IN FAR INFRARED: BROADENING, SHIFTS, AND LINE MIXING

I.M.Grigoriev, N.N.Filippov, A.V.Rozanov, and M.V.Tonkov

Institute of Physics, St.Petersburg University, Peterhof, St.Petersburg 198904 Russia

Abstract. We studied the pure rotational lineshapes of HF in gaseous argon (30–330 Amagat) at ambient temperature. The measured absorption coefficients were fitted by a sum of Rosenkranz profiles to obtain widths, shifts, and asymmetry parameters. We compared them with the literature data and ascribed the detected line asymmetry to the line mixing effect whose influence was considered in the framework of semiclassical approach.

In recent years several research groups studied the regularities of bandshape transformations with pressure in absorption and Raman spectra of molecular gases.[1–3] In the case of slightly overlapped lines their interference yields some asymmetry[4], which was registered in vib-rot spectra of HF in Ar.[3] In FIR we have recently obtained the similar results for HF in He and Xe.[5,6] So, the measurements for the chosen system may provide a fruitful basis to reveal the nature of bandshape forming depending both on vibrational excitation and on perturbing agent. Besides, such data can be used for improvement of the known intermolecular potentials.[7]

For the experiments we used a FT spectrometer LFS–1000 with the spectral range 30–410 cm^{-1}. The spectral resolution of 0.3–0.8 cm^{-1} allowed to record the lineshapes without instrumental distortions due to density variation from 30 to 330 Amagat. We measured a series of spectra for each value of Ar density with different HF concentrations and then averaged them with the corresponding weights for each spectral point[5,6] in order to obtain the 'complete profile' with comparable relative inaccuracies in the microwindows and near the lineshape maxima.

To obtain the lineshape parameters, we fitted the complete profiles with the sum of Rosenkranz lineshapes (1) by several numerical methods[5,6] which provided the same parameter values within the intervals of uncertainty:

$$A(\nu) = C \frac{\nu\left(1 - e^{-hc\nu/kT}\right)}{1 + e^{-hc\nu/kT}} \sum_m \frac{A_m\, \gamma_m\, \rho\left[1 + y_m(\nu - \nu_m)\right]}{(\nu - \nu_m)^2 + (\gamma_m \rho)^2}, \quad (1)$$

where ν_m, A_m, γ_m, y_m are frequency, relative intensity, broadening coefficient, and asymmetry parameter for mth line, ρ is the perturber density, and C is the scaling factor. Taking into account the influence of the line asymmetry on lineshifts,[8] we determined the linear lineshift coefficients $\delta_m = (\nu_m^{exp} - \nu_m^{calc})/\rho - \rho \cdot \gamma_m^2 \cdot y_m/2$, where ν_m^{calc} were calculated from the known rotational costants.[9]

The obtained data are represented in Fig.1–3, being compared with the known experimental data[11,12] the data of the most recent semiclassical calculation of widths and shifts[10], and experimental asymmetry parameters for

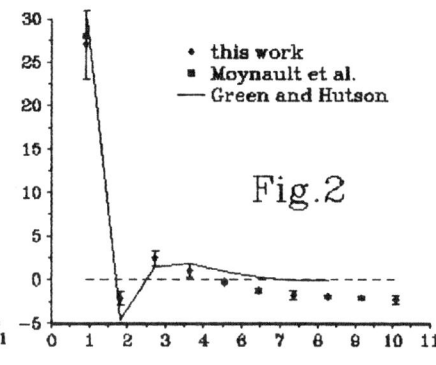
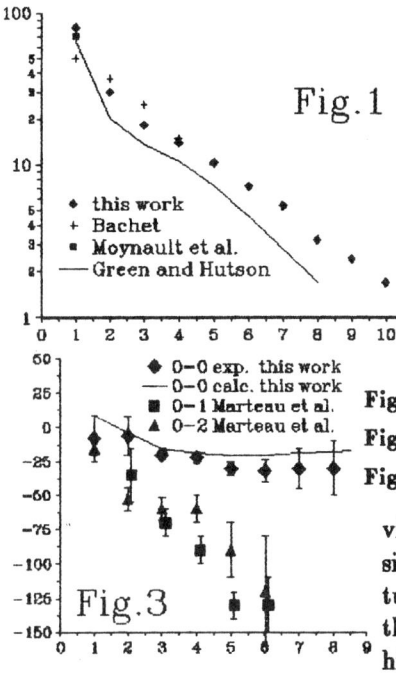

Fig.1. Broadening coefficients, 10^{-3}cm^{-1}Amagat^{-1}

Fig.2. Lineshift coefficients, 10^{-3}cm^{-1}Amagat^{-1}

Fig.3. Asymmetry parameters, 10^{-3}cm

vib-rot lineshapes from Ref.3. The semiclassical calculation represents the main features of lineshape parameter m-dependence though it underestimates the widths of higher lines far beyond the experimental inaccuracy and does not predict their 'residual' shifts which may be attributed to the higher densities influence.

In Fig. 3 it is seen the dramatic difference in asymmetry parameter m-dependence for pure rot and vib-rot lineshapes. To calculate them for our case, we used the semiclassical method[13] based on the calculation of distribution function of the probability of the collision induced changes in the angular momentum of rotational motion the adiabaticity of which was also taken into account.[5] The obtained curve is also given in Fig. 3 demonstrating the great effect of vibrational excitation on the line asymmetry.

1. G.J.Rosasco, W.Lempert, W.S.Hurst et al., Chem.Phys.Lett. **97**, 433 (1983).
2. M.O.Bulanin, A.B.Dokuchaev, M.V.Tonkov et al., JQSRT **31**(6), 521 (1984).
3. Ph.Marteau, C.Boulet, and D.Robert, J.Chem.Phys. **80**(8), 3632 (1984).
4. P.W.Rosenkranz, IEEE Trans.Ant.Propag. **23**, 498 (1975).
5. I.M.Grigoriev, M.V.Tonkov, and N.N.Filippov, Opt.Spectr. **75**(4), 753 (1993).
6. I.M.Grigoriev, M.V.Tonkov, and N.N.Filippov, Opt.Spectr., in press (1994).
7. J.Hutson, J.Chem.Phys. **96**(9), 6752 (1992).
8. F.Thibault, J.Boissoles, R.LeDoucen et al., J.Chem.Phys. **97**(7), 4633 (1992).
9. I.G.Nolt, J.V.Radostitz, G.DiLonardo et al., J.Mol.Spectr. **125**, 274 (1987).
10. S.Green and J.Hutson, J.Chem.Phys. **100**(2), 891 (1994).
11. G.Bachet, C.R.Acad.Sci. **B274**, 1319 (1972).
12. J.-M.Moynault, G.Bachet, R.Occelli et al., C.R.Acad.Sci. **B291**, 215 (1980).
13. N.N.Filippov and M.V.Tonkov, JQSRT **50**(1), 111 (1993).

Line mixing in Q-branches of Π-Δ transitions of CO_2

M.V. Tonkov*, F. Thibault**, R. Le Doucen**

*Institute of Physics, St Petersburg University, St Petersburg, 198904, RUSSIA

** Département de Physique Atomique et Moléculaire, Université de Rennes I,
35042 RENNES Cedex, FRANCE

When molecular collisions are the principal line broadening mechanism (at pressures greater than 50 Torr), line profiles in near wings can be calculated using impact approximation which leads to the Lorentz contour for isolated lines, and experiments confirm this conclusion. However, there are regions in infrared absorption spectra with very dense rotational line structure (Q-branches or band heads, for exemple) where band shape cannot be determined as the sum of Lorentz lines.

These deviations have generally been attributed to line mixing effects, also known as line interference, line coupling, etc. Since the first publications on the Q-branch shapes in IR absorption [1-3] the authors have paid most attention to the Π-Σ transitions. To our knowledge there are only remarks about Q-branch studing for the Π-Δ CO_2 band at 741.7 cm^{-1} [4,5]. In this report we present some results of Q-branch investigation for CO_2 spectra in 17 μm region in pure gas and in mixtures with He and Ar at various pressures including Π-Δ bands 11102-02201 (597.3 cm^{-1}) and 11101-02201 (741.7 cm^{-1}).

The spectra have been recorded using the Fourier-transform spectrometer BRUKER HR 120 with resolution 0.002 cm^{-1} in the region 580-800 cm^{-1}. The gas sample was inserted into the cells of stainless steel supplied with CsI and KRS-5 windows, which withstands the pressures up to 100 bars.

The transformation of Q-branches at low pressures is similar. At pressures below 100 Torr one can observe the fine structure for all Q-branches with one exception. Because of the very small difference B'-B" it is not possible to resolve in this branch the Q-component of even J for the band 597.3 cm^{-1} even at pressure 15 Torr and resolution 0.002 cm^{-1}. When the pressure rises up to 1 atm line mixing produces the deviation of the measured shape from that calculated as a sum of Lorentz lines near the band origin at low J values. A similar effect has been observed for CO_2 in pure gas [2] and in mixtures of CO_2 with He and N_2 [6]. The even Q-branch at 597.3 cm^{-1} broadens in this condition as a whole but the broadening coefficients are smaller than ones for neighboring rotational lines.

At high pressures (upper 10 atm) the Π-Σ Q-branches become symmetric with a shape close to Lorentzian profile. They continue to broaden but the broadening coefficients are much smaller than those for separate rotational lines. They are equal to about 0.016 cm^{-1}/atm for CO_2-He mixture and 0.045 cm^{-1}/atm for CO_2-Ar mixture. The similarity of these values for all transitions proves the absence of vibrational relaxation influence on the Q-branch shapes as the initial level energy changes from zero for the fundamental band to 1285.4 cm^{-1} for 791.4 band. The Π-Δ branches are asymmetrical under these conditions. We explained this fact by the superposition of two Q-branch components. To estimate their widths we divided the branch in two component of equal intensity with Lorentzian profile. The broadening coefficients found from these data coincide with those for Π-Σ transitions within the accuracy limits. It is worthnoting that the broadening coefficients for 597.3 cm^{-1} Q-branch are different at small and high pressures. This fact can be explained by the absence of coupling between lines of the two components of this Q-branch at low pressures and arising from this coupling at the pressures about 10 atm.

REFERENCES

1. N. Lacome and A. Levy, J. Molec. Structure. **80**, 257 (1982).
2. L. L. Strow and B. M. Gentry, J. Chem. Phys. **84**, 1149 (1986).
3. A. B. Dokuchaev, A. Yu. Pavlov, E. N. Stroganova, and M. V. Tonkov, Opt. & Spectrosc. **60**, 585 (1986).
4. M. Margottin-Maclou, A. Henry, and A. Valentin, J. Chem. Phys. **96**, 1715 (1992).
5. L.L. Strow, D. Tobin, and S. Hannon, ASA Workshop Proc., p. 119, Reims, 1993.
6. V. M. Tarabukhin and M. V. Tonkov, Opt. & Spectrosc. 62 199 (1987).

LINE-MIXING AND DURATION-OF-COLLISION EFFECTS IN THE ν_3 R-BRANCH BAND HEAD OF CO_2

D. Tobin, L. Strow, S. Hannon

University of Maryland Baltimore County, Baltimore, MD 21228

J.W.C. Johns

National Research Institute of Canada, Ottawa, CANADA

The line-mixing absorption coefficient, $k_{1st}(\nu)$, is computed using Rosenkranz's first-order approximation [1]. Rotational relaxation rates, K, are modelled with an empirical energy gap scaling law. The off-diagonal elements of the line-mixing relaxation matrix, W, are then given by

$$W_{j'j} = -\zeta K_{j'j} \tag{1}$$

where ζ is a single adjustable parameter representing the "strength" of line-mixing. ζ should equal 1 if the relaxations rates are modelled perfectly by the scaling law.

Birnbaum introduced a χ-function to model the effects of finite duration-of-collisions. In terms of the Lorentz line-shape, $k_L(\nu)$, a line-shape with collision duration τ_2 is given by [2,3]

$$k(\nu) = k_L(\nu)\chi_B(\nu) = k_L(\nu) A_m z K_1(z) \exp\left(\tau_2 \gamma_j + \tau_0 \Delta\nu\right) \tag{2}$$

with

$$z = \sqrt{(\gamma_j^2 + \Delta\nu^2)(\tau_0^2 + \tau_2^2)} \quad \text{and} \quad \Delta\nu = \nu - \nu_j \tag{3}$$

where $K_1(z)$ is a modified Bessel function of the second kind, $\tau_0 = \frac{0.72}{T}$, γ_j is the width of line j, and A_m is a constant representing the effect of line-mixing far from line center.

Since the line-mixing theory is valid only under the impact approximation, we include the effects of duration-of-collision by simply replacing $k_L(\nu)A_m$ in Equation (2) with $k_{1st}(\nu)$, i.e.

$$k(\nu) = k_{1st}(\nu)\chi_B(\nu). \tag{4}$$

Contributions from all lines within $\sim$100 cm^{-1} of the R-branch band head are included in the calculations. The least-squares fit for ζ and τ_2 is performed using all observed spectra simultaneously.

Almost all existing atmospheric transmission algorithms use Cousin's [4,5] empirical χ-function in this spectral region. This χ-function does not contain the proper functional form to model line-mixing and requires 25 parameters for N_2-broadened CO_2 and 14 parameters for self-broadened CO_2. Our model is more physically based and requires only 2 parameters, reproducing all observed spectra to within $\sim$0.015 error in transmission. In addition, the fitted ζ parameter is very close to its predicted value of 1 ($\sim$0.95) for N_2-broadened CO_2. For self-broadened CO_2, however, ζ is significantly lower than 1 ($\sim$0.71), implying that the scaling law used here for the rotational relaxation rates is inaccurate for CO_2-CO_2 collisions. The fitted duration-of-collision parameter, τ_2, has a self-broadened value which is greater than that for N_2-broadened, suggesting that CO_2-N_2 collisions have shorter durations than CO_2-CO_2 collisions.

1. P. W. Rosenkranz, IEEE Trans Antennas Propag. AP-23, 498 (1975).

2. G. Birnbaum, J.Q.S.R.T. v21, 597 (1979).

3. D. P. Edwards, L. L. Strow, J. Geophys. Res. v96, 20,859 (1991).

4. C. Cousin, Applied Optics, v24, 897 (1985).

5. C. Cousin, Applied Optics, v24, 3899 (1985).

LINE MIXING AND STATE-TO-STATE RATES IN D_2 DETERMINED FROM THE RAMAN Q BRANCH

A. D. May, P. M. Sinclair, J. W. Forsman, and J. R. Drummond
Department of Physics, University of Toronto, Toronto, Canada, M5S 1A7

The standard theories of pressure broadening of isolated lines predict a Lorentzian line shape. For a band of lines the profile is more complicated. Rosenkranz[1] has shown that at low densities the profile can be written:

$$I(\omega) \propto \sum_j N(j) \left\{ \frac{\rho W_{jj}^\circ}{(\omega - \omega_j)^2 + (\rho W_{jj}^\circ)^2} - \frac{(\omega - \omega_j)\rho Y_j^\circ}{(\omega - \omega_j)^2 + (\rho W_{jj}^\circ)^2} \right\} \quad (1)$$

The line mixing parameter, Y°, is given by:

$$Y_j^\circ = 2 \sum_{n \neq j} \frac{\operatorname{Re} W_{nj}^\circ}{(\omega_n - \omega_j)} \quad (2)$$

Here W_{jj}° is the real part of an off-diagonal element of the line relaxation matrix describing collisional transfer of the optical coherence from line j to line n. It is evident from this expression that each line in the band is made up of a symmetric Lorentzian describing ordinary line broadening and an asymmetric part due to line mixing.

Precise line shape measurements of the Q(0) to Q(4) lines in D_2 have been made using Raman gain spectroscopy. At the densities studied (2.5-30 amagat) the lines are well separated but slightly asymmetric due to line mixing. We have fit our individual spectral profile using Eq. 1 to extract both widths and mixing values. From plots of these values versus density we get the broadening coefficients and line mixing parameters. In Fig. 1 we show the line mixing for the ortho deuterium lines; the mixing parameters are given in Table I.

We have used the line mixing parameters along with the broadening coefficients to determine the dephasing and state-to-state rotational relaxation rates. Our values agree with those of Smyth et al.[3] determined from broadening coefficients using the empirical modified exponential gap law.

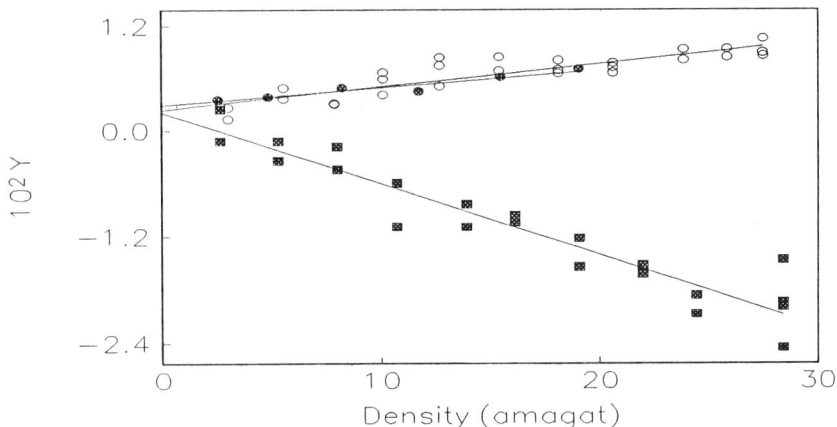

Fig. 1. Line mixing parameters in ortho deuterium. Squares J = 0, filled circles J = 2, open circles J = 4.

TABLE I. Line mixing parameters for D_2

Line	Mixing parameter Y^o (10^{-4} amagat^{-1})
Q(0)	-8.3(5)
Q(1)	-2.2(3)
Q(2)	+2.8(3)
Q(3)	+2.8(5)
Q(4)	+2.1(6)

1. P. M. Sinclair, J. W. Forsman, J. R. Drummond, and A. D. May, Phys. Rev. A. **48**, 3030 (1993).
2. P. W. Rosenkranz, IEEE Trans. Antennas Propag. **AP-23**, 498 (1975).
3. K. C. Smyth, G. J. Rosasco, and W. S. Hurst, J. Chem. Phys. **87**, 1001 (1987).

WATER VAPOR ABSORPTION IN MMW ATMOSPHERIC WINDOWS. CONTINUUMS

A. Bauer, M. Godon and J. Carlier
Laboratoire de Spectroscopie Hertzienne. Assosié au CNRS
Université de Lille I - 59655 Villeneuve d'Ascq Cedex - France

Q. Ma
Department of Applied Physics. Columbia University and Institute for Space Studies - Goddard Space Flight Center, New York, NY 10025.

The gaps between the millimeter wave spectral lines of water vapor, called atmospheric windows, are of special interest for the modeling of the transmission of radiation. In these regions, models involving simple impact or kinetic lineshapes have not been able to reproduce the available experimental data, obtained mostly from field measurements.

Various models have been proposed by several groups to describe the far wing behavior ; they are compared to our experimental data obtained in the far wings of the183 GHz water vapor line.

Absolute absorption of pure water vapor and atmospheric mixtures of $H_2O + N_2$ has been measured in the wings of the 183 GHz line. After measurements closer to the line center, we present a study at 239 GHz, at +56 GHz from the line center. Series of measurements have been performed at different temperatures between the ambiant and 356 K, with varying pressures and concentrations, using a Fabry Perot interferometer allowing a path length of 230 meters.

The experimental absorptions α have been compared to those obtained from various models using different line profiles.

Besides the usual Van Vleck-Weisskopf (VVW) and Gross, rewritten by Zhevakin-Naumov (ZN) lineshapes, profiles involving an empirical or semi empirical continuum have been used. The Clough-Kneizys-Davies (CKD) model, which is usually included in the FASCODE atmospheric transmission computer code, has been used in our calculations ; it involves a multiplying factor applied to a VVW line shape far from the center. Our absorption data have also been compared to those obtained from the Liebe (L) model made for the millimeter range ; this model involves a totally empirical continuum added to a VVW lineshape contribution.

The new theory developed for pure water vapor by Ma-Tipping (MT) in the millimeter wave range, based on the quasistatic statistical approach, yields a new continuum, that accounts for farwing behavior. Absorption calculations using this model were carried out at 239 GHz and compared to the experimental data.

$H_2O - H_2O$.

The expected quadratic pressure dependence is confirmed. The temperature dependence is usually obtained from $\alpha(T)/\alpha(T_0) = (T/T_0)^n$, where n is partially dependent on the N temperature dependence of the collisional broadening of all the lines involved. In the wings, the temperature dependence is much stronger than that predicted from the VVW and ZN shapes (-3.7). At 239 GHz, the n value is - 7.7.

$H_2O - N_2$

In our experimental conditions, where the H_2O pressure is much smaller than the N_2 pressure, a combination of $H_2O - N_2$ and $H_2O - H_2O$ collisions is expected to lead to a H_2O pressure dependence implying a negligible quadratic term added to a linear one. The quadratic component obtained experimentally is not negligible, its relative value increases further in the linewings. Separate temperature dependences could be obtained for the linear and quadratic term.

A satisfactory agreement between experiment and models can only be obtained with models involving a continuum. The best agreement with semi empirical models is obtained with the MT one for pure water vapor.

1 - A. Bauer and M. Godon, JQSRT 46, 211 (1991).
2 - M. Godon, J. Carlier and A. Bauer, JQSRT 47, 275(1992).
3 - A. Bauer, M. Godon, J. Carlier, Q. Ma and R.H. Tipping, JQSRT 50, 463(1993).
4 - S.A. Clough, F.X. Kneizys and R.W. Davies, Atmospheric Research 23, 229(1989).
5 - H.J. Liebe. Int. J. Infrared Mill. Waves 10, 631(1989).
6 - Q. Ma and R.H. Tipping, J. Chem. Phys. 93, 6127(1990).

CLOSE EXAMINATION OF THE LINE MIXING CONCEPT IN THE LINE WING ABSORPTION

S.D.Tvorogov, O.B.Rodimova, L.I.Nesmelova
Institute of Atmospheric Optics, Tomsk, 634055, Russia

The physical and mathematical reasonings in favour of the line mixing effect in the problem of the spectral line shape periphery in vibrational-rotational spectra of gases are discussed.

Today, an extensive literature (see, for example, Refs. 1-4) has evolved which claims that the line mixing effect is a decisive factor in forming the line wing absorption. Their arguments, which can be devided , by convention, into two groups, appear in the context of (i) the use of the analytical perturbation theory with respect to the non-diagonal matrix elements of the relaxation superoperator, (ii) the consequences of the sum rules for these matrix elements, (iii) the extension of the kinetic equations to the superoperators, (iv) the direct numerical estimates.

It is shown in the report that the evidences of the importance of the line mixing in the line wings should not be considered as mathematically unexeptionable and physically consistent.

Thus, one of the conditions of the applicability of the perturbation theory (the procedure of the reversion of the relaxation superoperator on the base of the perturbation theory is now well- accepted) is that the frequency detuning $\delta\omega$ should exeed the halfwidth of the line, and therefore the superoperator in the point of resonance $\delta\omega = 0$ (as it often takes place) could not be used. It should be calculated at large frequency detunings (the opposite asymptotic case) and it leads to quite different estimates (with respect to the values and to the dependence on the inherent parameters) of the non-diagonal matrix elements.

The sum rules are usually treated as the equations for the matrix elements of the relaxation operator. In fact, the sum rules are the identities which are to be fulfilled for any solution of the problem, as, for example, the unitarity conditions of the scattering matrix.

The above conclusions as well as the conclusions on some peculiarities of the construction of the kinetic equations are obtained as a result of rigorous mathematical considerations, see Ref.5.

Within the framework of the spectral line shape theory proposed by the present authors (see, for example, Ref.6)) the interference of the quantum states can, of course, take place, however, it is much less significant than the other factors defining the regularities of the line wing behavior (such as the Hibbs factor with the classical potential of the intermolecular interaction, the directed diffusion of the molecules in the process of the relaxation, etc.). The proposed theory quantitatively explains the experimental data on absorption (including that of the strong laser field) and radiation of the light both in the line wings and in the troughs between the lines inside the bands. In some cases the successful interpretation of the experimental data and the qualitative conclusions, for example, on the role of the intermolecular interaction, on the basis of the theory in hand occured before the emergence of the concept "the line mixing - the line periphery" itself.

One of the curios features of the situation is the appearance, in some cases, of similar figures and, that is most noteworthy, of similar formulas from remarkably different initial prerequisites. The way out, to the authors' opinion, consists (i) in the *ab initio* analysis of the problem and (ii) in the prolongation of the tentative assumptions of these approaches into the neighbour fields, such as the radiation of light, the nonlinear optics, etc., where their consequences would be drastically different, and the appropriate experiments could solve the problem of choice immediately. Both versions are applied for the comparative analysis of the concepts of the line mixing and of the line wing theory.

REFERENCES

1. M.O.Bulanin, A.B.Dokuchaev, M.V.Tonkov and N.N.Filippov, JQSRT **31**, 521 (1984).
2. C.Cousin, R.Le Doucen, C.Boulet, A.Henry and D.Robert, JQSRT **36**, 521 (1986).
3. C.Boulet, Spectral Line Shape, No.5, Proceedings of the 9-th Int. Conf. on Spectral Line Shapes, Poland, 1988, p.539.
4. J.M.Hartmann, M.Y.Perrin, Q.Ma and R.H.Tipping, JQSRT **49**, 675 (1993).
5. S.D.Tvorogov, O.B.Rodimova and L.I.Nesmelova, Optika Atmosphery **3**, 468 (1988).
6. L.I.Nesmelova, O.B.Rodimova and S.D.Tvorogov, *Spectral Line Shape and Intermolecular Interaction* (Nauka, Novosibirsk, 1986).

EXTENSION OF THE QUASISTATIC FAR-WING LINE SHAPE THEORY TO MULTI-COMPONENT ANISOTROPIC POTENTIALS

Q. Ma[†] and R. H. Tipping[‡]
[†]Goddard Institute for Space Studies, New York, NY 10025
[‡]Department of Physics, University of Alabama, Tuscaloosa, AL 35487

During the past decade, substantial progress has been made on the calculation of the absorption by the far wings of allowed lines using the quasistatic theory and the theoretical results for the specific cases (H_2O-H_2O, H_2O-N_2, and CO_2-Ar) are reported.[1-3] However, all these studies have been carried out using only a single-component anisotropic potential model. Because of the inability to represent well the anisotropic interaction by a single-component, and the sensitivity of the results to it, it is necessary to generalize the formalism to multi-component anisotropic potentials.

It is well known that, the spectral density $F(\omega)$ is given by

$$F(\omega) = \frac{1}{\pi} Re \int_0^\infty dt\, Tr[\mu \cdot e^{i(\omega - L_0 - L_1)t} \rho \mu]. \tag{1}$$

By assuming $\exp(-iL_0 t - iL_1 t) \approx \exp(-iL_0 t) \times \exp(-iL_1 t)$,[3] we can express $F(\omega)$ as

$$F(\omega) \approx \frac{1}{\pi} Im \frac{1}{2\pi i} \int_{-\infty}^\infty d\omega'\, Tr[\mu \cdot \frac{1}{\omega - \omega' - L_a} < \frac{1}{\omega' - L_1} >_b \rho^{(a)} \mu]_a. \tag{2}$$

In order to obtain the matrix elements of the resolvent operator $1/(\omega - L_1)$ we have to find the basis in which L_1 is diagonal. However, it is not feasible to diagonize an anisotropic interaction which consists of several components each having a specific spherical symmetry and r dependence. Consequently, only the main component is included in the calculations and all others are ignored.

In order to overcome this fundamental drawback, we assume that $L_1 = L_1^{(0)} + L_1^{(1)} + \cdots$ and make the approximation, $\exp(-iL_1 t) \approx \exp(-iL_1^{(0)} t) \times \exp(-iL_1^{(1)} t) \times \cdots$. Then, we are able to express the spectral density $F(\omega)$ as

$$F(\omega) \approx -\frac{1}{\pi} Im\, (\frac{1}{2\pi i})^2 \int_{-\infty}^\infty d\omega' \int_{-\infty}^\infty d\omega''$$

$$\times Tr[\mu \cdot \frac{1}{\omega - \omega' - \omega'' - L_a} < \frac{1}{\omega' - L_1^{(0)}} \times \frac{1}{\omega'' - L_1^{(1)}} >_b \rho^{(a)} \mu]_a. \tag{3}$$

For simplicity, only the formula corresponding to two components is given above. Thus, we obtain an expression of $F(\omega)$ in which the resolvent operator takes a product form and each term is related to one component only. Then, we are able to obtained its matrix elements by diagonalizing each component separately.

In practice, the main task is focused on the calculation of the intensity weighted band-averaged line shape function $\hat{\chi}(\omega)$ based on the knowledge of the interaction potential. In the formalism in which only one component is involved ($V_{aniso}(r) = GR_1(r)$), we denote by $|\alpha\rangle$ and G_α, the eigenvectors and eigenvalues of the operator G. Then, the explicit expression of $\hat{\chi}(\omega)$ is given by

$$\hat{\chi}(\omega) = 4\pi n_b \omega^2 \sum_{\alpha\beta}\sum_m \langle\alpha|\mu_m|\beta\rangle^* \frac{1}{|G_{\alpha\beta}R_1'(r_c)|} \qquad (4)$$

$$\times r_c^2 e^{-V_{iso}(r_c)/kT - (G_\alpha + G_\beta)R_1(r_c)/2kT} \langle\alpha|\rho^{(b)}\rho^{(a)}\mu_m|\beta\rangle / \sum_{ij} \rho_i |\mu_{ij}|^2,$$

where r_c are the positive solutions of the equation $\omega = G_{\alpha\beta}R_1(r_c)$ and $G_{\alpha\beta} = G_\alpha - G_\beta$. In general, $V_{aniso}(r) = GR_1(r) + FR_2(r) + \cdots$ and more than one term has to be included in the practical calculations. With the help of Eq. (3), we are able to derive a generalized expression for $\hat{\chi}(\omega)$,

$$\hat{\chi}(\omega) = 4\pi n_b \omega^2 \sum_{\alpha\beta}\sum_{\gamma\delta}\sum_m \langle\alpha|\mu_m|\beta\rangle^* \langle\alpha|\gamma\rangle\langle\delta|\beta\rangle \frac{1}{|G_{\alpha\beta}R_1'(r_c) + F_{\gamma\delta}R_2'(r_c)|}$$

$$\times r_c^2 e^{-V_{iso}(r_c)/kT - [(G_\alpha + G_\beta)R_1(r_c) + (F_\gamma + F_\delta)R_2(r_c)]/2kT} \langle\gamma|\rho^{(b)}\rho^{(a)}\mu_m|\delta\rangle / \sum_{ij} \rho_i |\mu_{ij}|^2,$$

(5)

where r_c are the positive solutions of $\omega = G_{\alpha\beta}R_1(r_c) + F_{\gamma\delta}R_2(r_c)$.

We consider the ν_3 band of CO_2 broadened by Ar in which both the measurement[2] and potential energy surface[4] are available; this case has previously been studied by Boissoles et al. using the one-component model.[2] With the assumption that the anisotropic interaction takes the form $V_{aniso}(r,\theta) = P_2(\cos\theta)V_2(r) + P_4(\cos\theta)V_4(r) + P_6(\cos\theta)V_6(r)$ and $V_l(r) = 4\epsilon[B_l(\sigma/r)^{12} - A_l(\sigma/r)^6]$, we obtained the parameters by fitting the published potential data.[4] After calculating the line shape function $\hat{\chi}(\omega)$ with Eq. (5), the absorption coefficient is easily calculated and good agreement with the measurement data is obtained.

1. P. W. Rosenkranz, J. Chem. Phys. 83, 6139 (1985); 87, 163 (1987).
2. C. Boulet, J. Boissoles, and D. Robert, J. Chem. Phys. 89, 625 (1988); J. Boissoles, V. Menoux, R. Le Doucen, C. Boulet, and D. Robert, J. Chem. Phys. 91, 2163 (1989).
3. Q. Ma and R. H. Tipping, J. Chem. Phys. 95, 6290 (1991); 96, 8655 (1992); 97, 818 (1992); 100, 2537 (1994); 100, 5567 (1994).
4. G. A. Parker, R. L. Snow, and R. T. Pack, J. Chem. Phys. 64, 1668 (1976).

PHOTOABSORPTION STUDIES OF QUASIMOLECULES OF ALKALINE-EARTH AND RELATED METAL ATOMS

Yukinori SATO
Research Institute for Scientific Measurements, Tohoku University
2-1-1 Katahira, Aoba-ku, Sendai 980, Japan

ABSTRACT

Far-wing absorption profiles of the resonance lines have been studied for the metal atoms (Ca, Sr, Ba, Yb, and Hg) perturbed by rare-gas atoms (He, Ne, Ar, Kr, and Xe) and some diatomic molecules by means of the classical double-beam absorption and dispersion method and a laser-pump and probe method.

Reduced absorption coefficients (RAC's) of the far-wing continuum bands are determined in absolute scale for the $nsnp\ ^1P_1 - ns^2\ ^1S_0$ resonance lines of Ca, Sr, Ba and Yb perturbed by rare gases (the foreign-gas broadening) and by the metal atoms (the self broadening) at temperatures of $950 \sim 1050$K. RAC's for the Hg $6s6p\ ^3P_1 - 6s^2\ ^1S_0$ line broadened by N_2, CO, H_2, and D_2 are determined in absolute scale at temperatures of $410 \sim 480$K. The spectral ranges covered are about $20 \sim 2000$ cm^{-1} on either side of the respective line center.

Using an inversion method based on the quasi static theory and employing the theoretical ground-state potential curves calculated by the pseudopotential method, the RAC's for the Ca-He, Ca-Ne, and Ca-Ar systems are transformed to potential curves of the $A\ ^1\Pi$ and $B\ ^1\Sigma$ states.

Blue-wing oscillations observed for the rare-gas broadened Ba and Yb resonance lines are analyzed in detail based on the uniform-semiclassical approximation. These oscillations are due to the existence of a phase-difference which is stationary against the change in the orbital angular momentum of the relative nuclear motion. Two main sources of such stationary phase-differences are pointed out and relative importance of the two sources is discussed.

A far-wing excitation and probe method is applied to observe the partial line-shapes of the Hg $6s6p\ ^3P_1 - 6s^2\ ^1S_0$ line broadened due to the fine-structure transitions: $Hg(^3P_1)+N_2, CO \to Hg(^3P_0)+N_2, CO$, and the chemical reaction: $Hg(^3P_1) + H_2 \to HgH(X,v,j) + H$. The partial line-shapes broadened due to these specified processes are compared with the total line-shapes obtained by the double-beam absorption method. The red-wing excitation is much more effective than the blue-wing excitation in causing the fine-structure transition and the chemical reaction.

INTRODUCTION

Much experimental work has been devoted to study rare-gas-broadened far-wing spectra of the resonance lines of metal atoms such as alkali[1-7], alkaline earth[8-18], Tl[19,20] and Hg.[21,22] It seems to be now well understood that the far-wing continuum is due to a molecular transition which occurs during a binary collision of radiator and perturber atoms. The quasistatic theory[23] (QST) based on the classical Franck-Condon principle is practically important to derive information about interatomic potentials from the far-wing spectra. Semiclassical treatments, developed by Sando and Wormhoudt,[24] Szudy and Baylis,[25] and Bieniek,[26] are much more realistic for analysing the satellite structures than the classical one in which the satellite appears as a singularity.

Broadening of the resonance line by collisions with the same-gas atoms (self broadening) has also been extensively investigated both experimentally[27-31] and theoretically[32,33] and now seems to be well characterized by the first-order dipole-dipole (resonance) interaction.

Far-wing line-shapes have also been studied for much more complicated systems of atom-diatom collisions.[34-37] In these cases, a variety of competing processes are involved, including energy transfer (electronic to vibrational, rotational, or translational) and chemical reaction.

We have started a series of studies on the far-wing spectra of the resonance lines for the alkaline earth and related metal atoms broadened by atomic and diatomic perturbers. A part of the studies has been briefly reviewed before.[38]

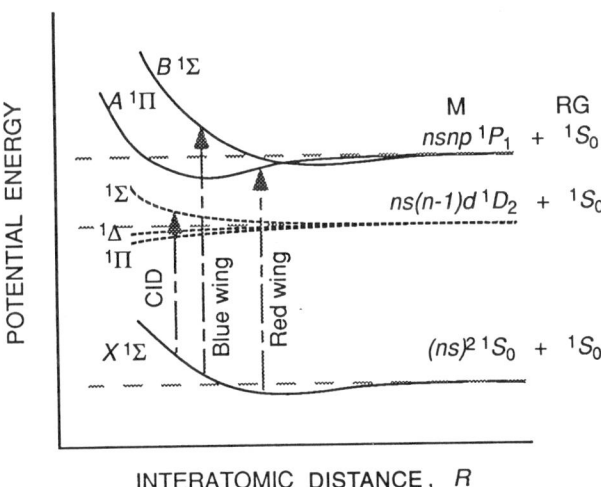

Fig. 1. Qualitative sketch of the interatomic potentials for alkaline-earth–rare-gas systems.

Figure 1 gives a sketch of the potential energy curves of the molecular states associated with the far-wing absorption for a system of an alkaline-earth atom (M) and a rare-gas atom (RG). The potential curves for the Sr-Ar system has been discussed in detail by Julienne and Mies.[42] The molecular $A\ ^1\Pi$–$X\ ^1\Sigma$ transition is expected to give a red-wing spectrum and the $B\ ^1\Sigma$–$X\ ^1\Sigma$ transition to give a blue-wing spectrum. The satellite absorption features observed by Harima et al.[9–12] at near red-wing regions are attributed to the shallow well of the B state potential. The collision-induced-dipole (CID) absorption bands observed by Ueda et al.[43,44] may be related to the molecular state shown in Fig. 1, which correlates adiabatically to the $^1D_2$ state of the metal atom.

We will summarize in this article the absolute measurements of the reduced absorption coefficients (RAC's) for the far-wings of the $nsnp\ ^1P_1 - ns^2\ ^1S_0$ resonance lines of Ca, Sr, Ba[39] and Yb[40] broadened by rare-gas atoms.

An inversion procedure of the RAC's into potential curves will be presented. This allows us to estimate the potential curves of the $A\ ^1\Pi$ and $B\ ^1\Sigma$ states of Ca-He, Ca-Ne, and Ca-Ar systems from the present absolute RAC's with the aid of theoretical ground-state potentials calculated by Shimakura.[45].

The $6s6p\ ^1P_1 - 6s^2\ ^1S_0$ resonance lines of Ba and Yb perturbed by rare gases show prominent satellite structures accompanied by shallow oscillations on their blue wings.[38,39] These blue-wing oscillations will be analyzed in detail in terms of the uniform-semiclassical approximation to know the sources of the oscillations.

We will summarize quite briefly our results on the absolute measurements of RAC's for the far wings of Hg $6s6p\ ^3P_1 - 6s^2\ ^1S_0$ line broadened by N_2,[41] CO, H_2, and D_2. We have performed these measurements in conjunction with the following far-wing excitation and probe (FEP) experiments.

Finally, a part of the results of our FEP experiments will be briefly reviewed. This is a pump and probe method using two tunable lasers. The pump laser is tuned to excite a far-wing band as has been done by Kleiber et al.[35] We have applied this method to the far wings of Hg resonance line, $6s6p\ ^3P_1 - 6s^2\ ^1S_0$ transition at 253.7 nm, broadened by the diatomic molecules AB (AB = N_2, CO, H_2, and D_2). In these cases, the collisional qusimolecules Hg*AB excited by the pump laser may decay into various competitive channels such as

(a) Adiabatic process to yield Hg($^3P_1$):   Hg*AB → Hg($^3P_1$) + AB,
(b) Nonadiabatic intra-multiplet transitions to yield Hg($^3P_0$):
$$\text{Hg*AB} \rightarrow \text{Hg}(^3P_0) + \text{AB*},$$
where AB* may be rotationally and/or vibrationally excited,
(c) Chemical reaction: Hg*AB → HgA(α, v, j) + B,
where (α, v, j) denotes a set of the internal (electronic, vibrational and rotational) states,

(d) E-V energy transfer to yield the ground state Hg:
$$Hg^*AB \to Hg(^1S_0) + AB^*,$$
where AB* may be in vibrationally highly excited states, and

(e) Radiative decay: $\quad Hg^*AB \to HgAB + h\nu$,
where HgAB may be in a bound state (van der Waals complex) or in a free state dissociating into Hg + AB.

We have observed far-wing spectra of the Hg resonance line broadened due to the decay channels (a) and (b) by detecting the final states $Hg(^3P_1)$ and $Hg(^3P_0)$ for N_2 or CO being employed as a perturber gas. We have observed the line-shapes broadened due to the reactive channel (c) by detecting the final states $HgH(X,0,j)$ or $HgD(X,0,j)$ for H_2 or D_2 being employed as a perturber gas. Examples of the results from these FEP experiments will be briefly reviewed.

EXPERIMENTAL

A. Double-beam absorption method

One experimental approach we have employed is a classical double-beam absorption method using a Xe short-arc lamp as a continuum light source. This method is combined with an interferometric dispersion measurement for obtaining the column density of the metal vapor. This approach allows us to determine the reduced absorption coefficient (RAC) for a far-wing continuum band in absolute scale. The experimental setup and procedures of the double-beam absorption method have been described in our previous papers.[40,44] Briefly, a heat-pipe metal-vapor cell is placed in a test-beam section of the Mach-Zender interferometer. The cell for Hg vapor is operated at temperatures of 300~480 K and the cell for the other metal vaper is operated at 800~1250 K. One of the perturber gases is admixed in the cell at a pressure in the range of 50-700 Torr. The light beam from a xenon short-arc lamp is divided into two: one passes through the absorption cell, while the other serves as a reference for the absorption. These two light beams and the interference beam which is the superposition of these two beams are alternately dispersed by a Czerny-Turner spectrometer and are detected by a 1024-channel photodiode array. From the interference spectrum measured around the metal resonance line, the number density of the metal atoms integrated over the line-of-sight (or column density) $(N_M l) = \int_{vapor-length} N_M(l) dl$ is determined. Absorption spectrum, in terms of the optical depth, is given by

$$k(\nu)l = -\ln[I_T(\nu)/I_R(\nu)],$$

where $I_T(\nu)$ and $I_R(\nu)$ are the relative spectral intensities of the test beam and the reference beam, respectively.

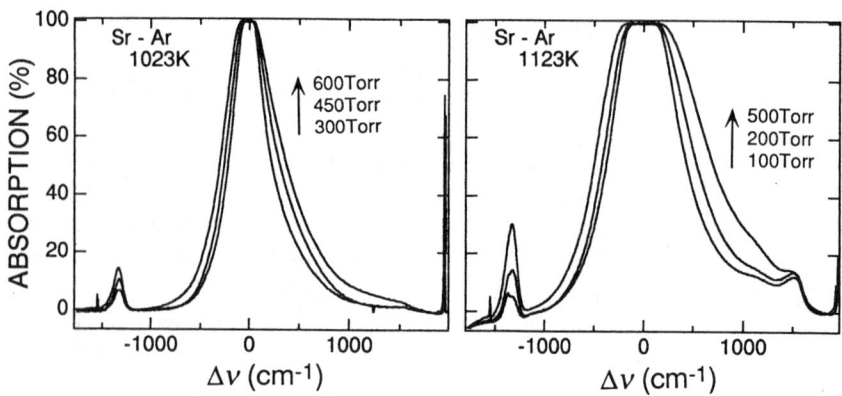

Fig. 2. Absorption spectra, in terms of the percentage absorption $[1 - I_T(\nu)/I_R(\nu)] \times 100$, in the vicinity of the Sr $5s^2\ {}^1S_0 - 5s5p\ {}^1P_1$ line at λ=460.7nm. The cell temperatures are 1023 and 1123 K for the left and right traces, respectively. the Ar pressures are 300, 450, and 600 Torr at 1023 K and 100, 200, and 500 Torr at 1123 K. The absorption increases as the Ar pressure increases.

Absorption spectra of the Sr/Ar mixtures, observed at cell temperatures of 1023 and 1123 K, are shown in Fig. 2 in terms of percentage absorption, $[1 - I_T(\nu)/I_R(\nu)] \times 100$, plotted against the wavenumber shift (detuning) $\Delta\nu$ (in cm^{-1}) from the Sr $5s5p\ {}^1P_1 - 5s^2\ {}^1S_0$ resonance line at 460.7nm. It is clearly seen that the absorption depends on the Ar pressure. It is also seen from the absorption data at 1123K that the satellite structure around $\Delta\nu \sim$ +1500 cm^{-1} depends scarcely on the Ar pressure, implying that these satellite bands may not be ascribed to the Sr-Ar collisions but to the Sr-Sr collisions (self-broadening). The CID bands associated with the Sr $5s$-$4d$ transition are seen on the absorption data in Fig. 2, which have been reported previously[44] and will not be discussed here. Assuming that the far-wing spectrum consists of a superposition of the Sr-Ar and Sr-Sr components, we have

$$k(\nu) = \gamma_{\text{SrAr}}(\nu) N_{\text{Sr}} N_{\text{Ar}} + \gamma_{\text{SrSr}}(\nu) N_{\text{Sr}}^2,$$

and, therefore,

$$\frac{k(\nu) l}{(N_{\text{Sr}} l)^2} = \gamma_{\text{SrAr}}(\nu) \frac{N_{\text{Ar}}}{N_{\text{Sr}} l} + \frac{\gamma_{\text{SrSr}}(\nu)}{l},$$

where $\gamma_{\text{SrAr}}(\nu)$ and $\gamma_{\text{SrSr}}(\nu)$ are the reduced absorption coefficients (RAC's) for the Sr-Ar and the Sr-Sr systems, respectively. Thus, we can obtain $\gamma_{\text{SrAr}}(\nu)$ and

$\gamma_{\text{SrSr}}(\nu)/l$ through the linear fit on the plot of $k(\nu)l/(N_{\text{Sr}}l)^2$ against $N_{\text{Ar}}/(N_{\text{Sr}}l)$. Aiming to obtain the RAC γ_{MM} for the self-broadening, we have proposed a method to estimate the effective absorption length l_{eff} by observing the resonance broadening of the metal resonance line.[46,47]

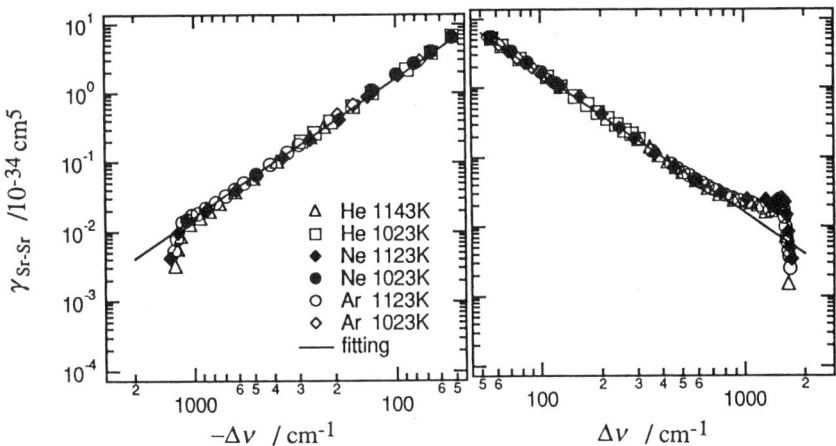

Fig. 3. Reduced absorption coefficient γ_{SrSr} for the self-broadened far wings of the Sr $5s^2\ ^1S_0 - 5s5p\ ^1P_1$ line. γ_{SrSr}, determined from the Sr-He mixture at 1023 and 1143 K and the Sr-Ne and Sr-Ar mixtures at 1023 and 1123 K, are plotted in log scale as a function of detuning $\Delta\nu$ (in cm^{-1}) from the line center.

Figure 3 gives logarithmic plots of γ_{SrSr} as a function of $\Delta\nu$, determined from the experiments on three different rare-gas mixtures. Quite a good consistency is seen in the γ_{SrSr} determined from Sr-He, Sr-Ne and Sr-Ar mixtures at different temperatures. Such a consistency gives in turn a measure of accuracy with which we determine the rare-gas broadened component γ_{SrRG}.

As seen in Fig. 3, the measured γ_{SrSr} values are proportional to $1/(\Delta\nu)^2$ in the ranges $50 \leq \Delta\nu \leq 600$ cm^{-1} and $-1100 \leq \Delta\nu \leq -50$ cm^{-1} of the blue and red wings, respectively. This $\Delta\nu$ dependence agrees with the quasistatic line shape for the resonance broadening arising from the first order dipole-dipole interaction. We have observed this typical behavior of the resonance broadening for the self-broadened lines of Ca $4s4p\ ^1P_1 - 4s^2\ ^1S_0$, Sr $5s5p\ ^1P_1 - 5s^2\ ^1S_0$, Ba and Yb $6s6p\ ^1P_1 - 6s^2\ ^1S_0$ transitions. The γ_{SrSr} values of the blue wing in Fig. 3 deviate from the resonance behavior for $\Delta\nu$ larger than 600 cm^{-1} and show a satellite peak at around $\Delta\nu = 1500$ cm^{-1}.

B. Far-wing excitation and probe (FEP) method

The second and third harmonics of a pulsed Nd:YAG laser are used to excite two dye lasers simultaneously. The output of the first dye laser operated with Coumarin 500 is frequency doubled in a BBO crystal to generate the pump laser pulse. Its wavelength covers from 251 to 259 nm in the vicinity of the Hg $6^3P_1 - 6^1S_0$ resonance line. The output of the second dye laser is frequency-converted to excite the product state: $Hg(^3P_1)$, $Hg(^3P_0)$, or HgH(or HgD)$(X^2\Sigma, v=0, j)$ for its detection by the laser-induced fluorescence (LIF) method. The $Hg(^3P_1)$ and $Hg(^3P_0)$ are detected by tuning the probe-laser wavelength to the Hg transitions $7^3S_1 - 6^3P_1$ at 436.0 nm and $7^3S_1 - 6^3P_0$ at 404.8 nm. The fluorescence of the Hg $7^3S_1 - 6^3P_2$ transition at 546.2 nm is detected with a bandwidth of ~6 nm. The probe laser pulse is optically delayed by 3 ns from the pump pulse. This delay is short enough to make the effect of secondary collisions negligible. The pump and probe pulses are directed coaxially into a heat-pipe cell which contains the Hg vapor (~20 mTorr) and one of the diatomic gases at a total pressure of about 6 Torr. The fluorescence is sampled perpendicularly to the laser axis from the cell.

We have measured the line shapes broadened due to the decay channels (a) and (b) by detecting the product states of $Hg(^3P_1)$ and $Hg(^3P_0)$ and by scanning the pump-laser wavelength in the vicinity of the Hg resonance line. The absolute ratio of the nascent product state populations, $N(^3P_0)/N(^3P_1)$, has been determined as a function of the pump-laser wavelength. A detailed description of the procedures for determining the ratio $N(^3P_0)/N(^3P_1)$ will be given elsewhere.[48]

For the perturber gas of H_2, we have measured the rotational state distribution of the nascent product $HgH(X^2\Sigma, v=0, j)$ by scanning the probe-laser wavelength with the pump-laser wavelength being fixed at some positions in the far-wing bands. The line shape broadened due to the formation of a certain rotational state of the HgH is measured by scanning the pump-laser wavelength.

RESULTS AND DISCUSSIONS

A. Absolute RAC's for far-wings of Ca-RG and Sr-RG systems

Figure 4 shows the resultant RAC, γ_{CaRG}, for the rare-gas-broadened Ca $4s4p\ ^1P_1 - 4s^2\ ^1S_0$ line at 423 nm. The γ_{CaRG}'s are plotted on a semilogarithmic scale against the wavenumber shift (detuning) $\Delta\nu$ (in cm^{-1}) from the line center. The γ_{CaRG}'s for different rare gases are shifted vertically for clarification. The γ_{CaRG} values are obtained at 1048 K for detunings of about $\pm 30 \sim \pm 160$ cm^{-1} and at 1163 K for detunings from about ± 120 to ± 900 cm^{-1}. The γ_{CaRG}'s obtained at temperatures 1048 and 1163 K coincide well at their overlapped regions, implying that no significant dependence on temperature is appreciable within this small temperature variation.

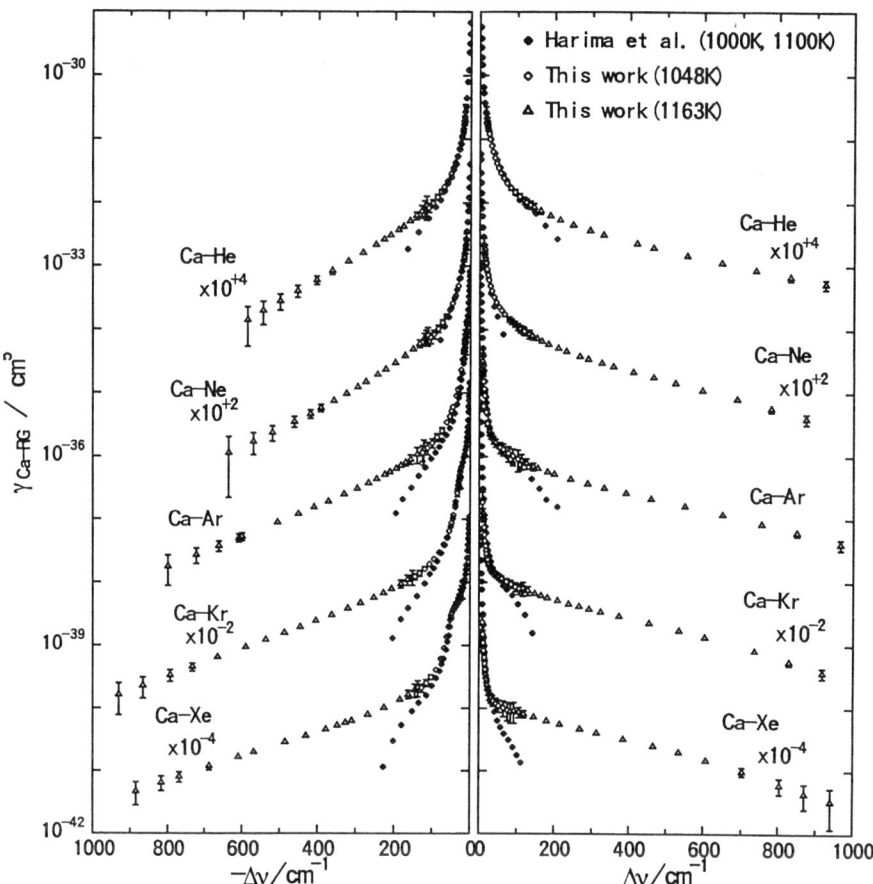

Fig. 4. $\Delta\nu$ dependence of the absolute reduced absorption coefficients γ_{CaRG} for the rare-gas-broadened far wings of the Ca $4s^2\ ^1S_0 - 4s4p\ ^1P_1$ line. The absolute values by Harima et al.(Ref. 12) are plotted for comparison.

Thus the γ_{CaRG} values give a smoothly connected curve over the whole range of data for each of the rare-gas perturbers. The present γ_{CaRG} values are compared with the absolute measurements by Harima et al.[12] in Fig. 4. Both absolute measurements are in good agreement in small detunings. The red-wing satellites (shoulders) observed by Harima et al.[12] at $\Delta\nu \sim -30\ \text{cm}^{-1}$

for Ca-Kr and $\Delta\nu \sim -50$ cm^{-1} for Ca-Xe are reproduced in the present results. However, the γ_{CaRG}'s by Harima et al. drop much faster than ours at $|\Delta\nu|$'s larger than 80 cm^{-1} on either side of the line center. The blue-wing profiles from Ar to Xe show a very rapid fall near the line core followed by a slow fall in the extended far wings. We cannot see any appreciable sign of the blue-wing shoulders suggested by Harima et al.[12]

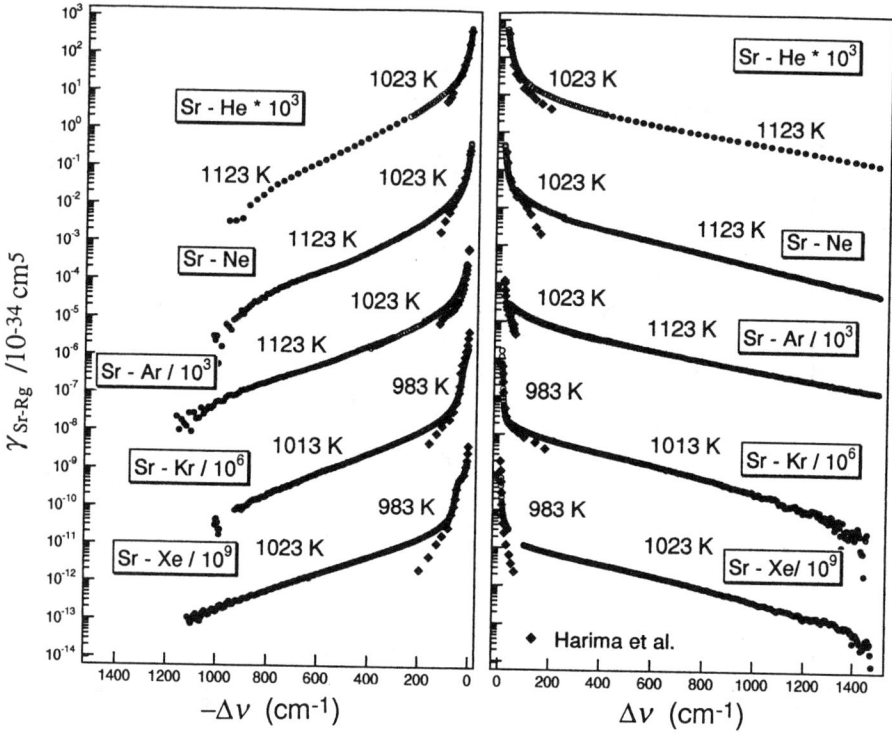

Fig. 5. $\Delta\nu$ dependence of the absolute reduced absorption coefficients γ_{SrRG} for the rare-gas-broadened far wings of the Sr $5s5p\ ^1P_1 - 5s^2\ ^1S_0$ line. The absolute values by Harima et al.(Refs. 9,11) are plotted for comparison.

Figure 5 gives the absolute values of γ_{SrRG} plotted on a semi-logarithmic scale as a function of detuning $\Delta\nu$ (in cm^{-1} from the Sr $5s5p\ ^1P_1 - 5s^2\ ^1S_0$ line at 461 nm. The γ_{SrRG}'s for Sr-He and Sr-Ne are obtaind for detunings of 20 $\sim$ 1500 cm^{-1} in the blue wing and 20 $\sim$ 950 cm^{-1} in the red wing. The γ_{SrRG}'s for Sr-Ar, Sr-Kr and Sr-Xe are obtaind for detunings of 10 $\sim$ 1500 cm^{-1} in the blue wing and 10 ~ 900 cm^{-1} in the red wing. The cell temperatures are denoted in the figure. Temperature dependence is not appreciable within the present temperature variation. The data by Harima et al.[9,11] are also plotted in Fig. 3 for comparison. Both absolute measurements are in good agreement each other in small detunings. The red-wing satellites observed by Harima et al.[9,11] for Sr-Ar ($\Delta\nu \sim -20$ cm^{-1}), Sr-Kr ($\Delta\nu \sim -35$ cm^{-1}), and for Sr-Xe ($\Delta\nu \sim -50$ cm^{-1}) are reproduced in the present results. However, the γ_{SrRG}'s by Harima et al. drop too fast at larger $|\Delta\nu|$'s than 100 cm^{-1} compared to ours. As seen in Fig. 2, the percentage absorption drops sharply in far wings and, therefore, a higher metal-vapor density is necessary to have a more reliable absorption spectrum for these larger $\Delta\nu$'s. This is the reason why we had to set higher gas temperatures in going from near- to far-wing regions. Figure 2 shows a significant increase in the percentage absorption for far-wing regions on changing the temperature from 1023 to 1123 K. We think that the absorption observed by Harima et al. are too weak in intensity at larger $|\Delta\nu|$'s than 100 cm^{-1}. On higher metal-vapor densities, however, the contribution from the metal-metal self-broadening must be properly discriminated from the total absorption. This is what we have done with our double-beam absorption-dispersion method.

B. Inversion of $\gamma_{\text{CaRG}}(\Delta\nu)$ to $\Delta V(R)$

The well-known QST expression for the reduced absorption coefficient is

$$\gamma(\Delta\nu) = \frac{8\pi^3 e^2 \hbar R_c^2 g f}{m_e c |\Delta V'(R_c)|} \exp(-\frac{V_X(R_c)}{k_B T}), \tag{1}$$

with

$$\Delta V(R_c) = V_\alpha(R_c) - V_X(R_c) = hc(\Delta\nu + \nu_0), \tag{2}$$

where g is the weight factor of the upper state ($g = 2/3$ for the $A\,^1\Pi$ and $1/3$ for the $B\,^1\Sigma$) and f is the oscillator strength of the atomic transition ($f = 1.75$ [49] for the Ca resonance transition).

The above expressions can be rewritten in an integral form as

$$\int_{\Delta\nu_\infty}^{\Delta\nu} \gamma(\Delta\nu) d\Delta\nu = \pm \frac{8\pi^3 e^2 \hbar g f}{m_e c} \int_0^R R^2 \exp(-\frac{V_X(R)}{k_B T}) dR, \tag{3}$$

where $\Delta\nu_\infty$ is a large value of $\Delta\nu$ corresponding to a vanishingly small γ and $\pm$ signs correspond to the positive (blue wing) and negative (red wing) $\Delta\nu$. If the ground-state potential $V_X(R)$ is known, Eq. (3) gives a relation between $\Delta\nu(=\Delta V)$ and R. We have inverted our $\gamma_{\text{CaRG}}(\Delta\nu)$ data into $\Delta V(R)$ using Eq. (3) with a compromised manner that the calculated $V_X(R)$ by the pseudo-potential method[45] is substituted in the r.h.s. of Eq. (3). Figure 6 gives results of the inversion for the Ca-Ar system. The potentials of the excited states ($A^1\Pi$ and $B^1\Sigma$) are evaluated by adding the difference potentials $\Delta V(R)$ determined by the inversion to the calculated ground-state potential. The $A^1\Pi$ and $B^1\Sigma$ potentials are evaluated from the $\gamma_{\text{SrAr}}(\Delta\nu)$ values (Fig. 4) for the red and blue wings respectively. The excited-state potentials thus determined are semi-experimental because the calculated ground-state potential $V_X(R)$ are used in the inversion. The upper-state potentials by the pseudo-potential methods[45] are also shown in Fig. 6. An excellent consistency is seen between the (semi)experimental and the calculated potentials for the $A^1\Pi$ state, while a discrepancy is seen for the $B^1\Sigma$ state.

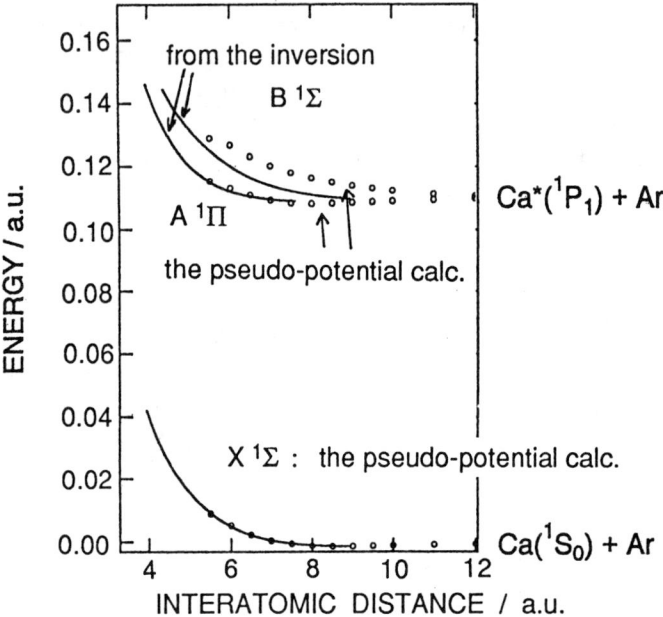

Fig. 6. Interatomic potentials for Sr-Ar systems. The ground state potential V_X is the pseudo-potential calculation by Shimakura (Ref. 45). $A^1\Pi$ and $B^1\Sigma$ potentials from the inversion of $\gamma_{\text{SrAr}}(\Delta\nu)$ are plotted by solid lines. $A^1\Pi$ and $B^1\Sigma$ potentials from the pseudo-potential calculation are plotted by open circles.

The same inversion procedures have been performed for the Ca-He and Ca-Ne systems and quite a similar result has been found for these systems as found for the Ca-Ar systems, *i.e.*, an excellent consistency is seen for the A $^1\Pi$ state between the (semi)experimental and the calculated potentials, while the calculated potentials are more repulsive than the semi-experimental one for the B $^1\Sigma$ state.

C. Blue-wing satellite of the rare-gas-broadened Yb and Ba lines

Both of the Ba and Yb first resonance lines are assigned to the $6s6p\ ^1P_1 - 6s^2\ ^1S_0$ transition. The far-wing profiles of them are much alike showing quite similar dependence on the rare-gas perturbers. Figure 7 shows the measured $\gamma_{\text{BaRG}}{}^{16}$ and $\gamma_{\text{YbRG}}{}^{12}$ for the blue-wing of the resonance line broadened by rare gases. It is clearly seen that the main satellite peak is accompanied by shallow oscillations for both of the Ba-RG and the Yb-RG systems except for the Ba-He and Yb-He cases.

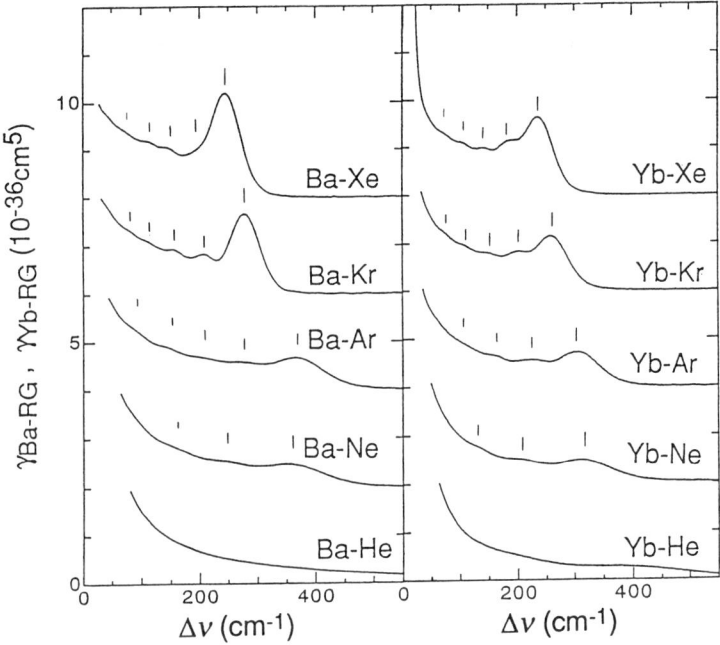

Fig. 7. Absolute reduced absorption coefficients γ_{BaRG} (Ref. 39) and γ_{YbRG} (Ref. 40) for the rare-gas-broadened blue-wings of Ba and Yb $6S - 6P$ transitions.

Similar blue-wing oscillations can be seen besides the blue-wing satellite of the self-broadened resonance line of Sr shown in Fig. 3 when it is plotted on an expanded scale. These oscillations are not probably related to vibrational structures of a bound state. The blue-wing satellites accompanied with shallow oscillations has already been observed for the Rb resonance lines broadened by Kr and Xe[3] and for the Cs($6s - 5d$) CID bands perturbed by rare gases.[50,51] One of the origins of such blue-wing oscillation may be the interference between two contributions at the same wavelength coming from different R. However, the blue-wing oscillation does not seem to be fully understood and deserves to a more detailed analysis.

We have analized the observed oscillations shown in Fig.7 using the uniform-semiclassical approximation described by Bienniek.[26]

The appearance of a satellite peak in the blue wing is known to arise from the existence of a maximun on the difference potential $\Delta V(R) = V_f(R) - V_i(R)$, where $V_f(R)$ and $V_i(R)$ are the potential energy curves of the upper and lower states of the transition. In such a case, two Condon points, R_1 and R_2, can contribute to the transition for a given absorption wavenumber.

The reduced absorption coefficient (RAC) for the blue wing at a given photon wavenumber $\nu = \nu_0 + \Delta \nu$ and temperature T may be given by

$$\gamma_T(\Delta \nu) = 8\pi^4 \alpha^2 gf \frac{hc}{Q_T} \int_{E_m}^{\infty} [\sum_L (2L+1)| <\Psi_f|\Psi_i>|^2] \exp^{-E/k_B T} dE, \quad (4)$$

where α is the hyperfine-structure constant, Q_T the translational partition function, L the orbital-angular-momentum quantum number of the relative nuclear motion, $<\Psi_f|\Psi_i>$ the overlap integral of the initial and final nuclear wavefunctions, and E the initial kinetic energy. If two Condon points, R_1 and R_2 ($R_1 < R_2$), contribute to the transition at $\Delta \nu$, the summation over L must be divided into two parts:

$$\sum_L (2L+1)|<\Psi_f|\Psi_i>|^2 = q_1 + q_2, \quad (5a)$$

where

$$q_1 = \sum_{L=0}^{L_{1M}} (2L+1)|<\Psi_f|\Psi_i>|^2, \quad (5b)$$

$$q_2 = \sum_{L_{1M}+1}^{L_{2M}} (2L+1)|<\Psi_f|\Psi_i>|^2, \quad (5c)$$

and L_{nM} (n=1 or 2) is the maximum L that allows the collision pair to approach the Condon point R_n:

$$L_{nM} + \frac{1}{2} = k_E R_n [1 - \frac{V_E(R_n)}{E}]^{1/2}, \quad (6)$$

where $k_E = \sqrt{2\mu E}/\hbar$ is the initial kinetic wavenumber. The q_1 term in Eqs. (12) includes the contributions of both Condon points while the q_2 term is contributed from the outer Condon point R_2 alone. According to the uniform-semiclassical expression by Bieniek[26], these two terms are written as,

$$q_1 = \sum_{L=0}^{L_{1M}} (2L+1)|T_1^L + T_2^L|^2, \quad (7a)$$

$$q_2 = \sum_{L_1M+1}^{L_2M} (2L+1)|T_2^L|^2. \qquad (7b)$$

T_n^L in Eqs. (7a,b), which describes the contribution from R_n, is written as

$$T_n^L = \pi \left(\frac{\hbar^2}{2\mu|\Delta V''(R_n)|k_n^2}\right)^{1/3} [\cos\beta_n Ai(-z_n) - (-1)^n \sin\beta_n Bi(-z_n)], \qquad (8)$$

where $Ai(-z)$ and $Bi(-z)$ are the regular and irregular homogeneous Airy functions respectively,

$$\beta_n = \Delta\phi_n + \frac{2}{3}|z_n|^{3/2}, \qquad z_n = \pm\left|\frac{(\Delta\phi_n'')^3}{2(\Delta\phi_n''')^2}\right|^{2/3}, \qquad (9,10)$$

$$\Delta\phi_n = \int_{R_f^t}^{R_n} k_f(R)dR - \int_{R_i^t}^{R_n} k_i(R)dR, \qquad (11)$$

R_i^t and R_f^t are the classical turning points on the initial and final states respectively, and $k_{i,f}(R)$ is the semiclassical local wavenumber:

$$k_i(R) = k_E\left[1 - \frac{V_i(R)}{E} - \left(\frac{L+\frac{1}{2}}{k_E R}\right)^2\right]^{1/2}, \qquad (12a)$$

$$k_f(R) = k_E\left[1 - \frac{V_f(R) - hc\nu}{E} - \left(\frac{L+\frac{1}{2}}{k_E R}\right)^2\right]^{1/2}, \qquad (12b)$$

$$k_n = k_i(R_n) = k_f(R_n). \qquad (12c)$$

Another uniform-semiclassical expression which includes only the second derivative of the phase difference $\Delta\phi_n$ may be possible for the q_1 term:

$$q_1 = \sum_{L=0}^{L_1M} (2L+1)(U_a^2 + U_b^2), \qquad (13)$$

$$U_a = (u_1 + u_2)\zeta^{1/4} \cos[\frac{1}{2}(\Delta\phi_1 + \Delta\phi_2)]Ai(-\zeta), \qquad (14a)$$

$$U_b = (u_1 - u_2)\zeta^{1/4} \sin[\frac{1}{2}(\Delta\phi_1 + \Delta\phi_2)]Bi(-\zeta), \qquad (14b)$$

$$u_n = \frac{\pi}{\sqrt{2}} k_n^{-1} |\Delta\phi_n''|^{-1/2}, \qquad \zeta = \pm\left|\frac{3}{4}\Delta\phi_{12}\right|^{2/3}, \qquad (15,16)$$

$$\Delta\phi_{12} = \Delta\phi_2 - \Delta\phi_1 = \int_{R_1}^{R_2} [k_f(R) - k_i(R)]dR, \qquad (17)$$

Equation (13) gives, with shorter computation time, almost the same numerical results for q_1 as Eq. (7a). We have, therefore, used Eq. (13) for the present analysis.

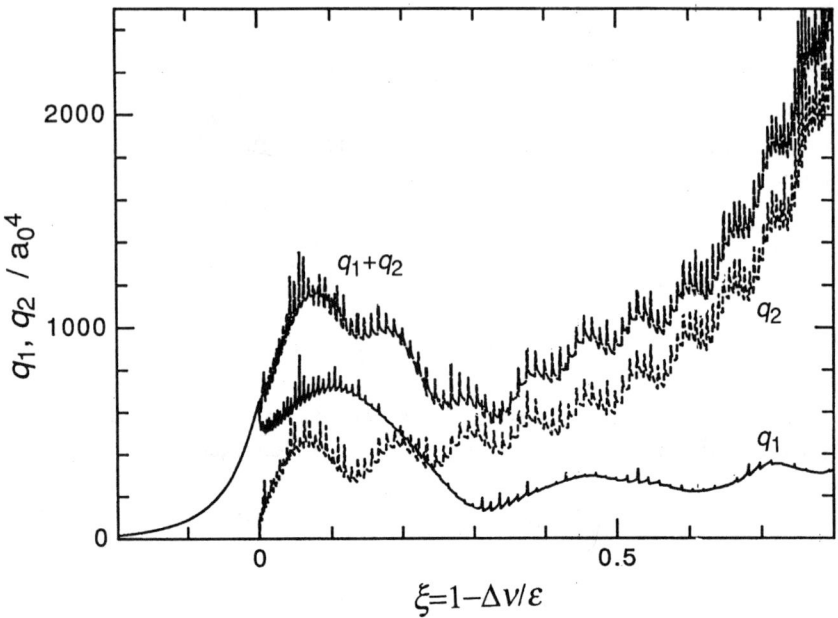

Fig. 8. Dependence of q_1 and q_2 on $\Delta\nu$.

Figure 8 shows $\Delta\nu$-dependence of the q_1 and q_2 calculated from Eqs. (13) and (8), where the q_1 and q_2 at an initial kinetic energy $E = 600$ cm^{-1} are plotted against the reduced detuning ξ ($= 1 - \Delta\nu/\epsilon$). ϵ is the classical satellite wavenumber, i.e., the maximum of the difference potential in unit of cm^{-1}. We have assumed a Morse function for the difference potential and a simple exponential function for the upper-state potential in the calculation:

$$\Delta V(R) = V_B(R) - V_X(R) = -hc\epsilon[e^{-2\alpha(R-R_0)} - 2e^{\alpha(R-R_0)}] + hc\nu_0, \quad (25)$$

$$V_B(R) = hcV_2 e^{-\alpha_2(R-R_0)}. \quad (26)$$

It is seen from Fig. 8 that oscillations occur in both of q_1 and q_2. The oscillation in q_1 comes from the fact that the phase difference $\Delta\phi_{12}$ given by Eq. (24) depends quite weakly on L when L changes from 0 to L_{1M}. The phase difference $\Delta\phi_{12}$ occurs due to an interference between the contributions from R_1 and R_2. This can be understood from Fig. 9 where the local wavenumbers,

$k_f(R)$ and $k_i(R)$ of Eqs. (19a,b), are plotted against R. $\Delta\phi_{12}$ is given by the area between R_1 and R_2 enclosed by the two $k(R)$ curves.

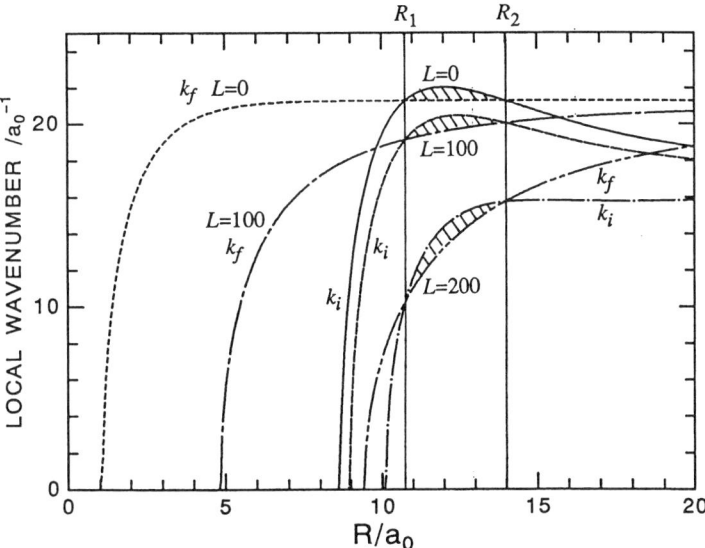

Fig. 9. Local wavenumbers $k_f^L(R)$ and $k_i^L(R)$ for $L = 0$, 100 and 200. $\Delta\nu = 240$ cm^{-1}, $\epsilon = 300$ cm^{-1}, $E = 600$ cm^{-1}. The hatched areas give $\Delta\phi_{12}^L$.

The enclosed area is seen to be steady against the change in L. This is the reason why the oscillation occurs in q_1 after the summation over L being performed. The random-phase approximation[55] in the summation over L in Eq. (13) results in a significant error when $\Delta\nu$ is near the classical satellite wavenumber. We have previously discussed[38] this type of oscillation in terms of the semiclassical S-matrix approach. This type of oscillation mechanism is very similar to the Rosenthal mechanism[52] proposed for the oscillations in the integrated cross-sections of inelastic atomic collisions.

The oscillation in q_2 comes from the fact that the phase difference $\Delta\phi_2$ in Eq. (11) is stationary against the change in L in the region $L_{1M} < L < L_{2M}$. The phase difference $\Delta\phi_2$ occurs between the two waves: one propagates along the upper-state potential and the other along the lower-state potential. The simple stationary-phase approximation for the T_2 term in Eq. (7b) is

$$U_2 = \left(\frac{\pi\hbar^2}{2\mu|\Delta V'(R_2)|k_2}\right)^{1/2} \sin[-\Delta\phi_2 + \frac{\pi}{4}]. \tag{20}$$

Equation (20) coincides with the uniform approximation of Eq. (8) for most of the range of $\Delta\nu$ except the region close to the classical satellite.

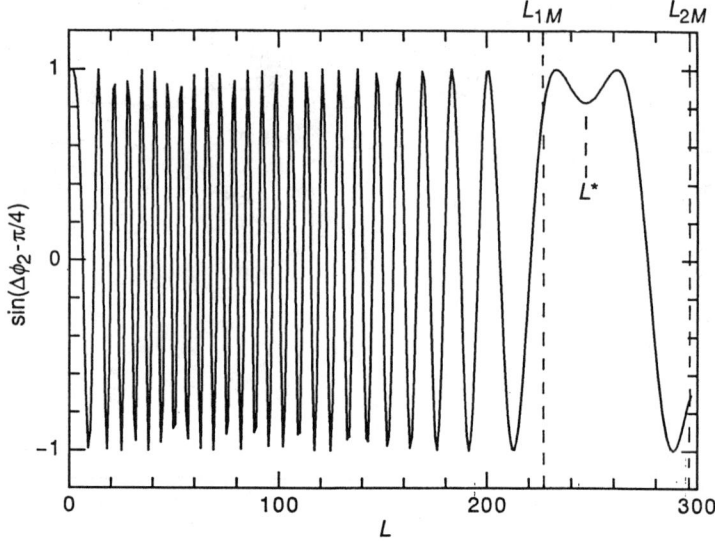

Fig. 10. L-dependence of $\sin[\Delta\phi_2 - \frac{\pi}{4}]$.

Figure 10 shows $\sin[\Delta\phi_2 - \frac{\pi}{4}]$ plotted against L. It is seen that the oscillation becomes stationary around L^* in the region between L_{1M} and L_{2M}. Equation (11) for $\Delta\phi_2$ may be roughly approximated as

$$\Delta\phi_2(L) \simeq \frac{1}{\hbar v} \int_{b_L}^{R_2} \frac{\Delta V(R) - hc\nu}{\sqrt{R^2 - b_L^2}} R dR, \quad (21)$$

where $b_L = (L + \frac{1}{2})/k_E$ is the impact parameter corresponding to L. Equation (21) implies that the nature of $\Delta V(R)$ as a function of R is transferred directly into the nature of $\Delta\phi_2(L)$ as a function of L. If $\Delta V(R)$ has an extremum within R_2, it is possible for $\Delta\phi_2(L)$ to have an extremum. In the present case of blue-wing satellite, $\Delta V(R)$ has a maximum at $R = R_0, (R_1 < R_0 < R_2)$. $\Delta\phi_2(L)$ then possibly has a maximum at $L = L^*, (L_{1M} < L^* <_{2M})$ as long as the turning point R^t (the larger one of R_i^t and R_f^t) is smaller than R_0. This is the reason why $\Delta\phi_2(L)$ is stationary between L_{1M} and L_{2M}, as shown in Fig. 10. One cannot use, therefore, the random-phase approximation in Eq. (14b) for the summation over L and the phase factor approximated by $\sin^2[\Delta\phi_2(L^*) - \frac{\pi}{4}]$ remains after the summation. This gives q_2 the oscillatory structure as shown in Fig. 8. This type of oscillation mechanism, which is different from the Rosenthal mechanism,[52] has been also discussed to interpret the oscillatory integrated cross-sections of inelastic atomic collisions. [53,54]

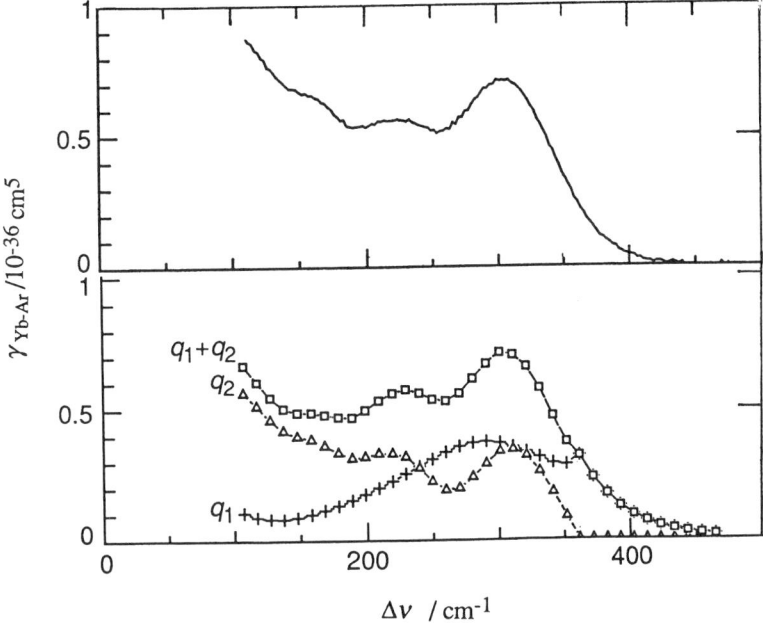

Fig. 11. Experimental and calculated RAC's in absolute scale for the blue-wing of Yb-Ar system at 970 K. Upper trace is from the experiment (Ref. 40). Lower traces are the q_1 and q_2 terms averaged over the Boltzmann distribution of velocity, and the calculated RAC is the sum of the two terms.

We have carried out the Boltzmann-average integration over E in Eq. (4) to calculate the RAC at a temperature T. The calculated RAC for the blue wing of Yb-Ar at $T = 970$ K is compared with the experimental one in Fig. 11. The Boltzmann-averaged q_1 and q_2 terms are also shown in Fig. 11. The calculated RAC is the sum of these two terms. The experimental profile is reproduced fairly well by the calculation. A better simulation will be obtained if we use more flexible potentials than those given by Eqs. (18) and (19). The oscillations seen in q_1 and q_2 before the Boltzmann average (Fig. 8) is not completely smeared out even after the average. The main satellite peak is formed due to a co-operative superposition of the outer-most peaks of the q_1 and q_2 terms. For small detunings, the q_2 term dominates over the q_1 term. This means that the contribution from the outer Condon point dominates over the interferential contribution from R_1 and R_2. It is seen that the oscillation between the line core and the main satellite peak comes from the q_2 term. The potential parameters of Eqs. (18) and (19) used in the calculation are $\epsilon = 357.0$ cm^{-1}, $\alpha = 0.91$ a_0^{-1}, $R_0 = 7.6$ a_0, $V_2 = 120.0$ cm^{-1}, and $\alpha_2 = 1.68$ a_0^{-1}. The maximum of ΔV ($= \epsilon$) is located at the detuning beyond the satellite peak where the intensity is about 45 % of the peak satellite intensity.

D. Absolute RAC's for far-wings of Hg-Diatom systems

Figure 12 gives the measured RAC's in absolute scale for the red and blue wings of Hg $6s6p$ $^3P_1$ − $6s^2$ $^1S_0$ transition at 254 nm broadened by N_2, CO, H_2 and D_2. The gas-cell temperature is in the range from 408 to 483 K. No significant change in the profiles is observed within the present temperature variation. The profiles are highly asymmetric with respect to the line center. The red wing is much more extended than the blue wing. The red wing profiles for Hg-H_2 and Hg-D_2 give no appreciable difference each other. The profiles for Hg-N_2 and Hg-CO show a prominent satellite (shoulder) in their blue wings. The satellite for Hg-N_2 begins to grow at smaller $\Delta \nu$ than the one for Hg-CO. The blue wing profiles for Hg-H_2 and Hg-D_2 show a weak and broad shoulder. There is an appreciable difference between the blue-wing profiles for Hg-H_2 and Hg-D_2 around the spectral region where a weak satellite (shoulder) exists.

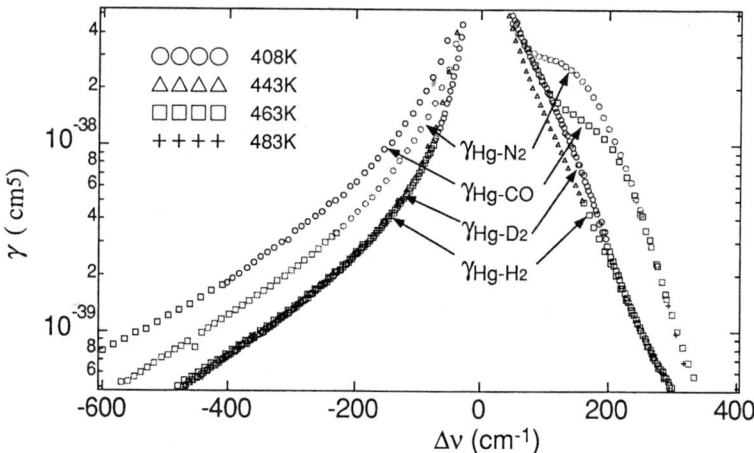

Fig. 12. $\Delta \nu$ dependence of the absolute reduced absorption coefficient for Hg-N_2, Hg-CO, Hg-H_2 and Hg-D_2.

Two excited states (A and B) are known to arise from the Hg*($^3P_1$) + N_2 (CO) (X $^1\Sigma$) limit. At larger separations between Hg and AB, (AB=H_2, CO,

H_2, and D_2) the A state is dominated by the character ($J = 1$, $\Omega^p = 0^+$) while the B state by $(J, \Omega) = (1,1)$, where J is the total electronic angular momentum and Ω is its projection onto the Hg–AB axis. The A state or the B state has generally a deeper or shallower well, respectively, than the ground X state. Thus the red wing may be assigned to the $A-X$ transition and the blue wing to the $B-X$ transition. Since the ground state has possibly a deeper well than the B state, the difference potential ΔV_{B-X} between the two state can have a maximum which may account for the blue-wing satellites. The difference in the far-wing profile between Hg-H_2 and Hg-D_2 systems cannot be explained within the quasi-static framework. We have seen in the previous session that the phase-interference effect is important in considering the blue-wing satellites. This effect depends on the kinetic wavenumber k_E of the system and, therefore, the relative velocity of the colliding particles. The observed difference in blue-wing profile between Hg-H_2 and Hg-D_2 may be explained in terms of the velocity effect or the phase-interference effect.

E. Decay channels of far-wing excited Hg-N_2 and Hg-H_2 systems

Figure 13(A) gives the excitation spectra for yielding Hg(6^3P_1) and Hg(6^3P_0) obtained for Hg-N_2 by scanning the pump-laser wavelength in the vicinity of the Hg $6\ ^3P_1$ - $6\ ^1S_0$ resonance line.

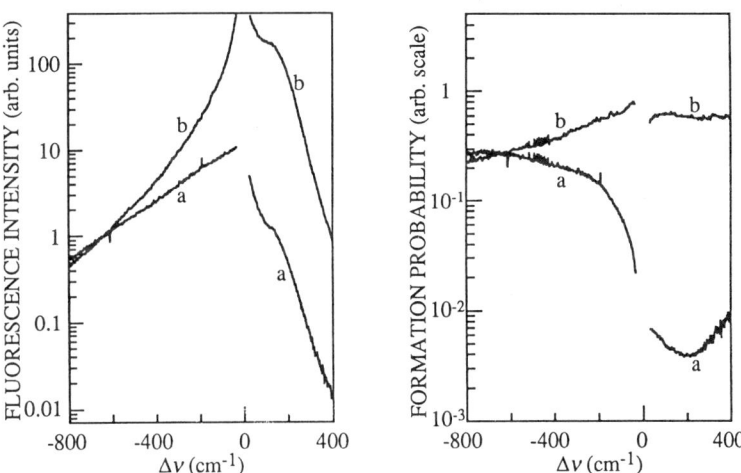

Fig. 13. (A) Excitation spectra for yielding Hg($^3P_1$) (plot a) and Hg($^3P_0$) (plot b) in Hg-N_2 collisions. (B) $\Delta\nu$ dependence of the probability for yielding Hg($^3P_1$) (plot a) and Hg($^3P_0$) (plot b) in Hg-N_2 collisions.

These spectra can be taken as the far-wing absorption spectra associated with following decay channels:

$$\text{Hg}^*\text{N}_2 \to \text{Hg}(^3P_1) + \text{N}_2, \qquad (22)$$

and

$$\text{Hg}^*\text{N}_2 \to \text{Hg}(^3P_0) + \text{N}_2, \qquad (23)$$

where Hg*N$_2$ is the collisional Hg-N$_2$ quasimolecule excited in the far wings of the Hg line. The far-wing-excited Hg*N$_2$ is, therefore, a transit state which exists during the collisional fine-structure transition:

$$\text{Hg}(^3P_1) + \text{N}_2 \to \text{Hg}(^3P_0) + \text{N}_2. \qquad (24)$$

In Fig. 13(A), the spectra are plotted against the excitation wavenumber given in terms of detuning from the Hg line center. The plots (a) and (b) correspond to the decay channels (23) and (22), respectively. The intensity ratio I_a/I_b of the two plots are normalized according to the population ratio of the nascent products $N(^3P_0)/N(^3P_1)$. The far-wing profiles shown in Fig. 13(A) is similar but not equal to the total absorption profiles shown in Fig. 12.

If the profiles shown in Fig. 13(A) is normalized for the total absorption profile shown in Fig. 12, then we have a probability that the excited state Hg*N$_2$ decays to yield the final state Hg($^3P_0$) or Hg($^3P_1$) as a function of the excitation wavenumber. The probability spectrum thus obtained are plotted in Fig. 13(B) on an arbitrary scale against the excitation wavenumber expressed in terms of detuning $\Delta\nu$ from the Hg line center. As seen in Fig. 13(B), the decay probability to $^3P_0$ is much higher for the $A-X$ excitation (negative $\Delta\nu$) than the $B-X$ excitation (positive $\Delta\nu$). This means that the A state serves as a more effective intermediate in decaying to $^3P_0$ than the B state. This is contrary to the case of bound-bound excitation for the Hg-N$_2$ van der Waals complex,[56,57] where the bound B state is found to be more effective than the bound A state in yielding $^3P_0$. In the present case, the blue-wing $B-X$ excitation is the free–free type and is mostly followed by the process (22). The $A-X$ excitation, on the other hand, tends to go from the free–free type to the bound–free type as $|\Delta\nu|$ increases in the red wing. This gives higher probability for the process (23) at larger $|\Delta\nu|$ in the red wing. We have observed essentially similar results on the Hg-CO system. A more detailed discussion will be given elsewhere.[48]

Figure 14(A) gives the far-wing excitation spectra due to the chemical reaction:

$$\text{Hg}^*\text{H}_2 \to \text{HgH}(X^2\Sigma^+, v=0, j=17.5) + \text{H}. \qquad (25)$$

The spectra is plotted as a function of the pump wavenumber expresses as the detuning $\Delta\nu$ from the Hg line center. The red and blue wing may be assigned to the $A-X$ and the $B-X$ transitions as in the case of Hg-N$_2$. Figure 14(B) gives the probability spectrum obtaind by normalizing the excitation spectrum in (A) against the total absorption spectrum of Hg-H$_2$ shown in Fig. 12. The probability is higher in the red wing than in the blue wing by a factor of 5. We have also measured the nascent rotational distributions of HgH($X^2\Sigma^+, v = 0$) at $\Delta \nu = -200$) and $+100$ cm^{-1} and found very similar distributions for these two distinct detunings. Discussions of these findings will be given elsewhere.

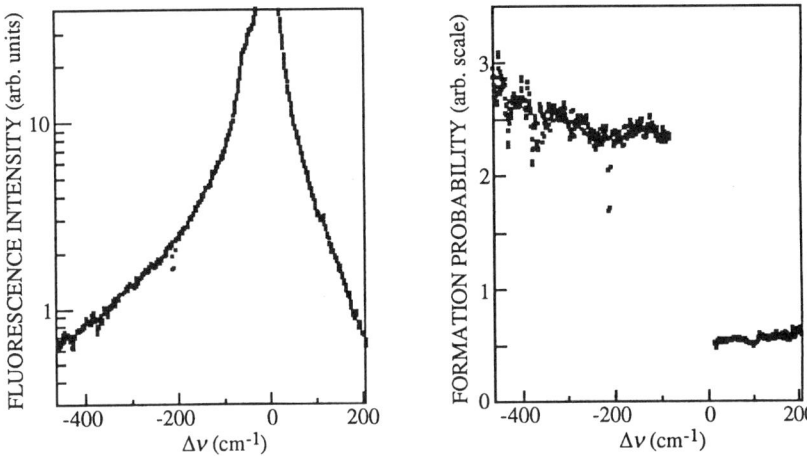

Fig. 14. (A) Excitation spectra for yielding HgH($X^2\Sigma^+, v = 0, j = 17.5$) in Hg-H$_2$ collisions. (B) $\Delta \nu$ dependence of the probability for yielding HgH($X^2\Sigma^+, v = 0, j = 17.5$) in Hg-H$_2$ collisions.

CONCLUSION

We have improved the classical double-beam absorption/dispersion method to determine absolute reduced absorption coefficients (RAC's) for the pressure-broadened far-wings of the metal resonance lines. With this improvement, the absolute RAC's for foreign-gas broadening and for self-broadening can be determined simultaneously over the spectral range up to 2000 cm^{-1}.

The origin of the oscillatory satellite structure is discussed in detail. The random-phase approximation in the summation over L is not valid for analyzing the blue-wing structure. Since the oscillatory blue-wing structure is very sensitive to the form of potential curves, it is possible to determine both of the upper and lower state potentials by the analysis of the observed structures. The absolute values of RAC's are very important in such an analysis.

Combination of the far-wing excitation and probe (FEP) approach and the double-beam absorption/dispersion approach is usefull to explore the dynamical aspect of the collisional quasimolecules.

ACKNOWLEDGEMENT

I am greatly indebted to all the members of our laboratory (laboratory of UV spectroscopy at RISM), especially to Dr. K. Ueda, Dr. K. Ohmori, Dr. M. Okunishi, Mr. H. Chiba. I am grateful to the students, both past and present, who have contributed material for this article, in particular Dr. T. Maeyama, O. Sonobe, H. Ito, S. Moriyama, T. Kurosawa, T. Nakamura. I am also grateful to Prof. Shimakura at Niigata University who has contributed the pseudo- potential calculation for the present work.

REFERENCES

1. R. E. M. Hedges, D. L. Drummond, and A. Gallagher, Phys. Rev. A **6**, 1519 (1972).

2. C. L. Chen and A. V. Phelps, Phys. Rev. A **7**, 470 (1973).

3. C. G. Carrington and A. Gallagher, Phys. Rev. A **10**, 1464 (1974).

4. D. L. Drummond and A. Gallagher, J. Chem. Phys. **60**, 3426 (1974).

5. G. York, R. Scheps, and A. Gallagher, J. Chem. Phys. **63**, 1052 (1975).

6. R. Scheps, C. Ottinger, G. York, and A. Gallagher, J. Chem. Phys. **63**, 2581 (1975).

7. J. F. Kielkopf and N. F. Allard, J. Phys. B **13**, 709 (1980).

8. J. F. Kielkopf, J. Phys. B **11**, 25 (1978).

9. H. Harima, Y. Fukuzo, K. Tachibana, and Y. Urano, J. Phys. B **14**, 3069 (1981).

10. H. Harima, K. Tachibana, and Y. Urano, J. Phys. B **15**, 3679 (1982).

11. H. Harima, T. Yanagisawa, K. Tachibana, and Y. Urano, J. Phys. B **16**, 4365 (1983).

12. H. Harima, T. Yanagisawa, K. Tachibana, and Y. Urano, J. Phys. B **16**, 4529 (1983).

13. J. L. Carlsten, A. Szoke, and M. G. Raymer, Phys. Rev. A **15**, 1029 (1977).

14. P. Thomann, K. Burnett, and J. Cooper, Phys. Rev. Lett. **45**, 1325 (1980).

15. A. Corney and J. V. M. McGinley, J. Phys. B **14**, 3047 (1981).

16. W. J. Alford, K. Burnett, and J. Cooper, Phys. Rev. A **27**, 1310 (1983).

17. W. J. Alford, N. Andersen, K. Burnett, and J. Cooper, Phys. Rev. A **30**, 2366 (1984).

18. W. J. Alford, N. Andersen, M. Belsley, J. Cooper, D. M. Warrington, and K. Burnett, Phys. Rev. A **31** 3012 (1985).

19. B. Cheron, R. Scheps, and A. Gallagher, J. Chem. Phys. **65**, 326 (1976); Phys. Rev. A **15**, 651 (1977).

20. M. G. Raymer, J. L. Carlsten, and G. Pichler, J. Phys. B **12**, L119 (1979).

21. J. F. Kielkopf and R. A. Miller, J. Chem. Phys. **61**, 3304 (1974).

22. H. C. Petzold and W. Behmenburg, Z. Naturforsch, **33a**, 1461 (1978).

23. See, for examples, Ref. 1 and the review by N. Allard and J. Kielkopf, Rev. Mod. Phys. **54**, 1103 (1982).

24. K. M. Sando and J. C. Wormhoudt, Phys. Rev. A **7**, 1889 (1973).

25. J. Szudy and W. E. Baylis, J. Quant. Spectrosc. Radiat. Transfer, **15**, 641 (1975).

26. R. J. Bieniek, Phys. Rev. A **15**, 1513 (1977); **23**, 2826 (1981); R. J.Bieniek and T. J. Streeter, Phys. Rev. A **28**, 3328 (1983).

27. K. Niemax and G. Pichler, J. Phys. B **7**, 1204 (1974); **8**, 179 (1975).

28. D. Gebhard and W. Behmenburg, Z. Naturforsch, **30a**, 445 (1975).

29. G. V. Zhuvikin, N. P. Penkin, and L. N. Shabanova, Opt. Spectrosc. **41**, 425 (1976).

30. K. Niemax, M. Movre, and G. Pichler, J. Phys. B **12**, 3503 (1979).

31. J. Heunennkens and A. Gallagher, Phys. Rev. A **27**, 1851 (1983).

32. C. G. Carrington, D. N. Stacey and J. Cooper, J. Phys. B **6**, 417 (1973).

33. M. Movre and G. Pichler, J. Phys. B **10**, 2631 (1977); **13**, 697(1980).

34. W. Kamke, B. Kamke, I. Hertel, and A. Gallagher, J. Chem. Phys. **80**, 4879 (1984).

35. P. D. Kleiber, A. M. Lyyra, K. M. Sando, V. Zafiropoulos, and A. Gallagher, J. Chem. Phys. **85**, 5493 (1986).

36. S. Bililign and P. D. Kleiber, J. Chem. Phys. **96**, 213 (1992); S. Bililign, P. D. Kleiber, W. R. Kearney, and K. M. Sando, J. Chem. Phys. **96**, 218 (1992).

37. Wm. R. Kearney and K. M. Sando, Phys. Rev. A **46**, 6977 (1992).

38. Y. Sato, in *The Physics of Electronic and Atomic Collisions*, editted by T. Andersen, B. Fastrup, F. Folkmann, H. Knudsen and N. Andersen, XVII ICPEAC, AIP Press, pp. 665 (1993).

39. T. Maeyama, H. Ito, H. Chiba, K. Ohmori, K. Ueda, and Y. Sato, J. Chem. Phys. **97**, 9492 (1992).

40. K. Ueda, O. Sonobe, H. Chiba, and Y. Sato, J. Chem. Phys. **95**, 8083 (1991).

41. K. Ohmori, T. Kurosawa, H. Chiba, M. Okunishi, and Y. Sato, J. Chem. Phys. **100**, 5381 (1994).

42. P. S. Julienne and F. H. Mies, Phys. Rev. A **34**, 3792 (1986).

43. K. Ueda and K. Fukuda, J. Phys. Chem. **86**, 678 (1982).

44. K. Ueda, T. Komatsu, and Y. Sato, J. Chem. Phys. **91**, 4495 (1989).

45. N. Shimakura, private communication.

46. K. Ueda, H. Sotome, and Y. Sato, J. Chem. Phys. **91**, 1907 (1991).

47. K. Ueda, O. Sonobe, H. Chiba, Y. Sato, and T. Namioka, Rev. Sci. Instrum. **63**, 1690 (1992).

48. K. Ohmori, T. Kurosawa, H. Chiba, M. Okunishi, K. Ueda, and Y. Sato, to be submitted.

49. W. L. Wiese, M. W. Smith and B. M. Miles, Atomic Transition Probabilities NSRDS-NBS22 (1969).

50. B. Sayer, M. Ferray, J. P. Visticot and J. Lozingot, J. Phys. B **13**, 177 (1980).

51. J. P. Visticot and J. Szudy and B. Sayer, J. Phys. B **14**, 2329 (1981).

52. H. Rosenthal and H. M. Foley, Phys. Rev. Lett. **23**, 1480 (1969); H. Rosenthal, Phys. Rev. A **4**, 1030 (1971).

53. F. J. Smith, Phys. Letter **20**, 271 (1966).

54. R. E. Olson, Phys. Rev. A **2**, 121 (1970).

55. In this approximation, the cosine and sine functions in Eqs. (21a,b) are replaced by a constant ($= 1/\sqrt{2}$) when Eqs. (21a,b) are used in Eq. (20). See Refs. 24 and 25.

56. C. Jouvet and B. Soep, J. Chem. Phys. **81**, 2229 (1984).

57. K. Yamanouchi, S. Isogai, S. Tsuchiya, M.-C. Duval, C.Jouvet, O. B. d'Azy, and B. Soep, J. Chem. Phys. **89**, 2975 (1988).

Line Shapes in the Far Wings of $Hg\,^3P_1$-$^1S_0$ Resonance-Line Broadened due to the $Hg\,^3P_1 \rightarrow\,^3P_0$ Fine-Structure Transitions in Collisions with N_2 and CO

Y.Sato, K.Ohmori, T.Kurosawa, H.Chiba, M.Okunishi, and K.Ueda
*Research Institute for Scientific Measurements, Tohoku University,
Katahira, Aoba-ku, Sendai 980, Japan*

Far-wing excitation and probe(FEP) technique gives a far-wing spectrum of an atomic-resonance line broadened due to a specific-exit channel such as the chemical reaction, the energy transfer, and the elastic scattering.[1] The FEP technique consists of laser excitation of the collisional quasimolecule in the far wings of a pressure-broadened atomic-resonance line and detection of the nascent products. We have recently reported preliminary results of our FEP experiments on the collision-induced fine-structure transition of Hg:[2]

$$Hg^*(^3P_1) + N_2 \rightarrow Hg^*(^3P_0) + N_2. \qquad (1)$$

We present, at this conference, more detailed and extended findings for the transit-region dynamics of process (1) as well as the results of the FEP experiment on the identical fine-structure transition of Hg induced by collisions with CO.

The experimental system has already been described in the accompanying paper.[1] The HgAB(AB=N_2 or CO) collisional quasimolecule is excited with the pump pulse, and the nascent products, $Hg^*(^3P_0)$ and $Hg^*(^3P_1)$, are probed by the laser-induced fluorescence method. Moreover, in this experiment, the absolute ratio of the yields of $Hg^*(^3P_0)$ and $Hg^*(^3P_1)$, Q_0/Q_1, is determined at a given pump wavelength from the fluorescence intensity.

We have measured the excitation spectra for the formation of $Hg^*(^3P_0)$ as well as $Hg^*(^3P_1)$ scanning the pump-laser wavelength in the vicinity of the $Hg\,^3P_1$-$^1S_0$ resonance line. The results are presented in Fig.1 for the HgN_2 quasimolecule as a function of the pump wavenumber expressed as the detuning Δv from the resonance-line center. The Δv relates to the Hg-N_2 distance for which the photoabsorption occurs. There exist two excited states, A and B, arising from the $Hg^*(^3P_1)+N_2$ limit. The red and blue wings, corresponding to the lower and higher frequency side of the line-center, are assigned as the A-X and B-X excitation, respectively.[2,3] A satellite structure is seen as a shoulder in the blue wing at $\Delta v \cong 150 cm^{-1}$. It is attributable to the maximum of the difference potential between the B and X states.

In order to obtain probability functions for the formation of $Hg^*(^3P_0)$ as well as $Hg^*(^3P_1)$, we have normalized the excitation spectra for the total absorption spectrum of the HgN_2 quasimolecule which we have obtained using a different apparatus. The results are presented in Fig.2. One can see that the B-X excitation results in exclusively the elastic scattering yielding $Hg^*(^3P_1)$. For the A-X excitation, on the other hand, the $Hg^*(^3P_1)$ formation probability becomes lower gradually with the increasing detuning; while the $Hg^*(^3P_0)$ formation probability becomes higher rapidly and finally surpasses

© 1995 American Institute of Physics

the Hg*($^3P_1$) formation probability.

When the quasimolecule is excited to the upper-state potentials, the elastic scattering and the fluorescence decay compete with the nonadiabatic transition to the a state arising from the Hg*($^3P_0$)+N$_2$ limit. The Hg*($^3P_1$)N$_2$ populated in the free states predominantly undergo the elastic scattering in a very short time($\cong$2ps under the present conditions) to produce Hg*($^3P_1$); while those in the bound states are energetically inaccessible to the Hg*($^3P_1$)+N$_2$ limit and then decay in time through the nonadiabatic transitions to the a state and a radiative transition to the X state. The B-X excitation is mainly the free-free type since the B state has a shallow well($D_e \cong 40 cm^{-1}$).[4] The A-X excitation, on the contrary, tends to go from the free-free type to the bound-free type as $|\Delta v|$ increases in the red wing since the A state has a much deeper well ($D_e \cong 660 cm^{-1}$) than the X state.[4]

The measured values for the ratio Q_0/Q_1 allow us to estimate time constants for the nonadiabatic transitions to the a state from the A state (τ_A^a) and from the B state (τ_B^a), assuming the elastic scattering completed in 2ps and fluorescence-life time to be 100ns: $\tau_A^a \cong$ 5-15ns and $\tau_B^a \cong$ 300ps. The $B \rightarrow a$ transition is much faster than the $A \rightarrow a$ transition. The B state is Coriolis' coupled to the a state directly at close Hg-N$_2$ configuration; while the A state is not directly coupled to the a state, and the B state serves as an intermediate in the coupling between these states. This difference in the coupling scheme can be the reason why τ_A^a is by far larger than τ_B^a.

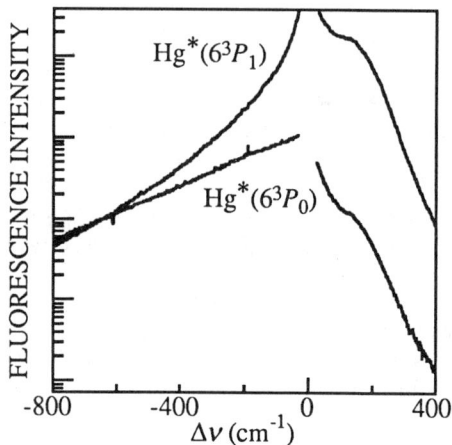

FIG.1. Excitation spectra of the HgN$_2$ collisional quasimolecules for the formation of Hg*(6^3P_0) and Hg*(6^3P_1). Δv = 0cm^{-1} denotes the Hg 6^3P_1-6^1S_0 resonance - line center.

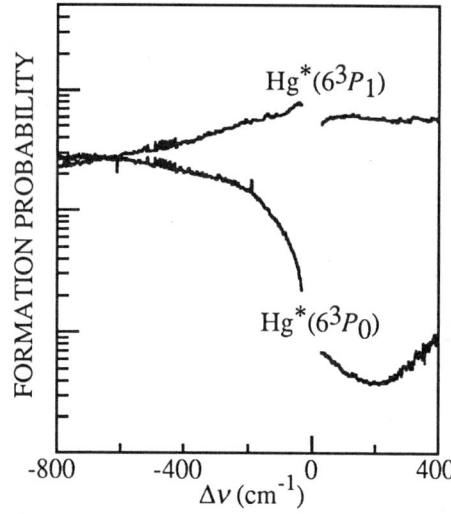

FIG.2. Detuning dependence of the Hg*(6^3P_0), Hg*(6^3P_1) formation probability. for the HgN$_2$ quasimolecules.

1. K.Ohmori et al., abstract of this conference.
2. K.Ohmori et al., J.Chem.Phys.**100**, 5381(1994).
3. K.Ohmori et al., unpublished.
4. K.Yamanouchi et al., J.Chem.Phys.**89**, 2975(1988).

Line Shapes in the Far Wings of Hg$^3P_1$-$^1S_0$ Resonance Line Broadened due to the Chemical Reactions: Hg*($^3P_1$) + H$_2$, D$_2$ → HgH($X^2\Sigma^+$,v,j) + H, D

K.Ohmori, H.Chiba, T.Kurosawa, M.Okunishi, K.Ueda, and Y.Sato,
*Research Institute for Scientific Measurements, Tohoku University,
Katahira, Aoba-ku, Sendai 980, Japan*

Continuum spectra are observed in the far wings of Hg$^3P_1$-$^1S_0$ resonance line broadened due to collisions with H$_2$ and D$_2$. In the classical Franck-Condon limit, the photoabsorption at a given wavelength in the far wings takes place at a fixed internuclear configuration between the relevant upper and lower electronic states of the transient HgH$_2$ or HgD$_2$ quasimolecule. Based on this principle, we have investigated the chemical reactions:

$$\text{Hg*}(^3P_1) + \text{H}_2, \text{D}_2 \rightarrow \text{HgH}, \text{HgD}(X^2\Sigma^+, v, j) + \text{H, D} \tag{1}$$

using "far-wing excitation and probe"(FEP) technique in an attempt to know the dynamics in the transit regions of those processes. The FEP technique consists of laser excitation of the colliding pair in the far wings and detection of the specific quantum state of the nascent product.

Our experiments have been performed by a pump-probe method in a photon-beam-gas-cell arrangement. The outputs of two dye lasers pumped by a pulsed Nd:YAG laser are frequency-converted to generate pump and probe pulses. The HgH$_2$(HgD$_2$) collisional quasimolecule is excited with the pump pulse; its wavelength covers from 256.7 to 252.4nm in the vicinity of the Hg$^3P_1$-$^1S_0$ resonance line(253.7nm). The nascent products, HgH(HgD) and Hg*($^3P_1$), are probed by the laser-induced fluorescence method. Reactions of the ground-state Hg atom with H$_2$(D$_2$) are highly endothermic; therefore, HgH is not formed without irradiation of the pump laser. Secondary collisions are negligible under the present conditions.

We have measured the excitation spectra for the formation of the products scanning the pump-laser wavelength in the vicinity of the Hg resonance line. The result is presented in Fig.1 for HgH($X^2\Sigma^+$, v=0, j=17.5) as a function of the pump wavenumber expressed as the detuning Δv(in cm^{-1}) from the resonance-line center. The red and blue wings correspond to the negative and positive detunings. From the classical Franck-Condon principle, $hc(v_0+\Delta v)$ equals the difference in the potential energy between the upper and lower states at the Hg-H$_2$ internuclear distance $R_{\Delta v}$ for which the photoabsorption occurs, where hcv_0 is the excitation energy at the Hg resonance-line center. In going to the larger |Δv| regions in the red and blue wings, $R_{\Delta v}$ becomes smaller on the attractive and repulsive upper potentials, respectively.

There exist two excited states(A and B) arising from the Hg*($^3P_1$)+H$_2$ limit; the A and B states have the characters designated as |$J_a \Omega_a p$>=|1 0$^-$> and |1 1>, respectively, where J_a is the total electronic-angular momentum of the quasimolecule, and Ω_a is its projection onto the collision axis.

© 1995 American Institute of Physics

In the Franck-Condon regions for transitions from the ground X state, the potential curves may be quite similar for HgH_2 and HgNe since the polarizabilities and hard-sphere diameters of H_2 and Ne are comparable. By analogy with the HgNe quasimolecule,[1] the A state may have a deeper well than the X state; while the B state may have a shallower one than the X state. The red and blue wings may then be assigned as the A-X and the B-X transitions, respectively.

We have normalized the excitation spectrum against a total absorption spectrum of the HgH_2 quasimolecule, which we have obtained using a different apparatus. Thus, a probability for the formation of the HgH is obtained as a function of the detuning, and it is presented in Fig.2.

The probability is higher in the red wing than in the blue wing by a factor of 5. This means that the A state serves as an intermediate in the formation of the HgH much more effectively than the B state.

With increasing the detuning, the mean translational energy of the final states of photoabsorption decreases in the red wing and increases in the blue wing; nevertheless, the probability exhibits almost flat profiles in both wings. This suggests that the HgH is formed without any activation barrier in both pathways via the A and B states.

We have also measured the nascent rotational distributions of $HgH(X^2\Sigma^+, v=0)$ at Δv=-200 and +100 cm^{-1} and found very similar distributions for these two distinct detunings. This indicates that the HgH is formed through the identical transition state for both of the A-X and B-X excitation

The B state is Coriolis' coupled to the A state at closer Hg-H_2 distances The present results suggest that the HgH is formed via the B-X excitation followed by the nonadiabatic transition to the A state.

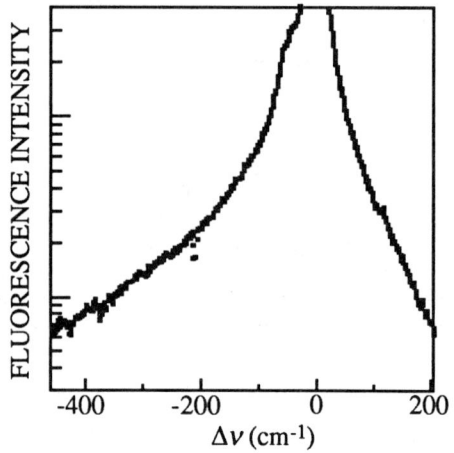

FIG. 1. Excitation spectra of the HgH_2 collisional quasimolecule for the formation of $HgH(X^2\Sigma^+, v=0, j=17.5)$. $\Delta v = 0$ cm^{-1} denotes the Hg 6^3P_1-6^1S_0 resonance-line center.

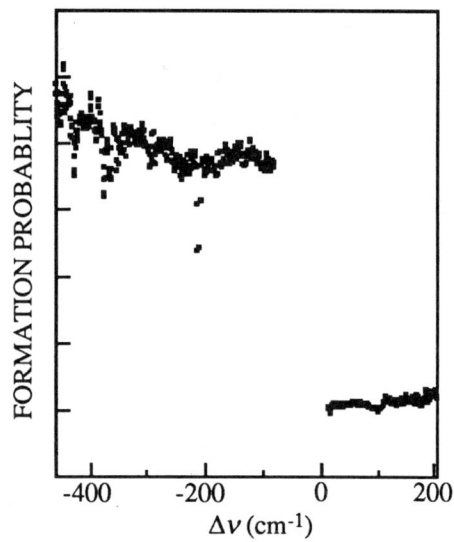

FIG.2. Detuning dependence of the HgH $(X^2\Sigma^+, v=0, j=17.5)$ formation probability.

1. K. Fuke et al., J.Chem.Phys. **81**, 2591(1984); K.Yamanouchi et al., J.Chem.Phys. **88**, 205(1988).

INTERATOMIC POTENTIALS FOR Cd-Xe AND Cd-Ar FROM THE Cd 326.1 nm LINE WINGS MEASUREMENTS

G.D. Roston[+] and T. Grycuk

Institute of Experimental Physics, University of Warsaw,
Hoża 69, 00-681 Warszawa, Poland

Although vibronic spectra for van der Waals diatomics are now available the measurements of wings of line broadened by neutral perturbers are still of very interest for determination of interatomic potentials. Analysis of far wings of the spectral line allows to derive the potential curves in the wide range of separations: from a repulsive wall up to the van der Waals attraction region, in contrary to the method involved vibronic spectra of supercooled molecules which in the most cases yield only some part of the potential well.

We have studied the absorption profile of the 326.1 nm Cd resonance line broadened by pressure effects of xenon and argon. The temperature dependent spectra were carefully measured with the 6 m concave grating spectrometer in the double-beam configuration, similarly as it has been done previously[1] for Cd-Kr. Fig. 1 shows the example of such spectra in comparison to that observed for the pure Cd vapour of the same density. The effects of Cd-Ar(Xe) interactions appear in the spectral range of ~ 1000 cm^{-1} to the red side of the line center and for xenon they are here stronger than for argon. In contrary to this the blue wing spectrum, showing the distinct satellite band, is more extended for Cd-Ar. Spectral features observed are attributed to the transitions from the ground state $X^1 0^+$ to the excited A and B states by way of that established in Ref. 1.

Using the well known quasistatic inversion method[2] it is possible to derive the shape of the potential curve from the spectrum. In particular, the van der Waals coefficients ΔC_6^0 and ΔC_6^1 for the difference potentials between the ground state – $X^1 0^+$ and the excited ones – $A^3 0^+$ and $B^3 1$, respectively, can be determined from logarithmic plot of the near red wing spectrum which in the van der Waals region shows the linear dependence with a slope of -1.5.

The near red wing profile for Cd-Ar and Cd-Xe, corrected for the Cd-Cd contribution and extrapolated to $T \to \infty$ were analysed accordingly to such a procedure yielding ΔC_6 parameters collected in table I.

Because of rather weak temperature dependence of the spectra considered it was impossible to determine the ground state potential weel with reasonable accuracy. Therefore, the far red wing spectra and the blue ones were analysed

---

[+]Permanent address: Department of Physics, University of Alexandria, Alexandria, Egypt

using the ground state potentials derived from molecular beams spectroscopy[3] (for Cd–Ar) and from some estimation (for Cd–Xe).

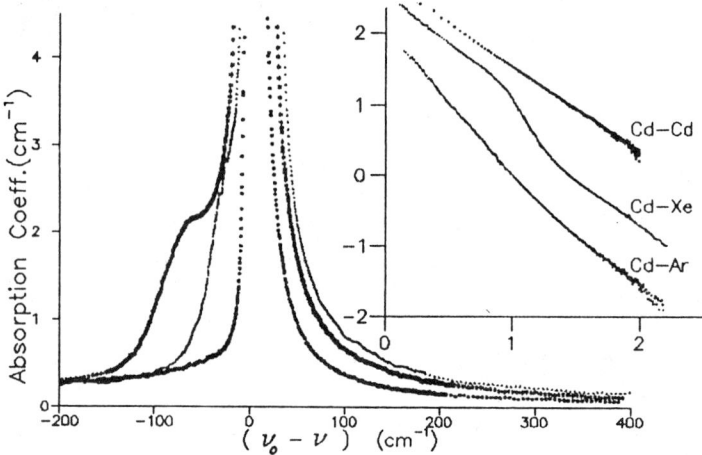

Fig. 1: Absorption profiles of the Cd 326.1 nm line broadened by 1.367×10^{19} cm^{-3} of Ar and by 1.325×10^{19} cm^{-3} of Xe in comparison with that for a pure Cd (the narrowest profile). Density of Cd is the same for all cases: 3.62×10^{18} cm^{-3}. The red wing spectra (the absolute absorption coefficients in 10^{-40} cm^5) plotted in logarithmic scale are inserted. The Cd–Xe profile is here multiplied by 10 and the Cd–Cd one is multiplied by 100.

Table I. Values of ΔC_6 parameters (eV Å^6)

	ΔC_6^1	ΔC_6^0
Cd–Cd	168.8 ± 3	not observed
Cd–Xe	85.7 ± 2	43.04 ± 0.5
Cd–Kr[a]	58.1 ± 5	37.2 ± 2
Cd–Ar	not observed	28.7 ± 1

[a]Reference 1.

This work is supported by the State Committee for Scientific Research within the project 2 0261 91 01.

REFERENCES

1. T. Grycuk, M. Findeisen and A. Śniecińska, AIP Conf. Proc. 216, Spectral Line Shapes 6, ed. L. Frommhold and W. Keto, (1990), p. 174.
2. R.E. Hedges, D.L. Drummond and A. Gallagher, Phys. Rev. A 6, 1519, (1972).
3. R. Bobkowski, M. Czajkowski and L. Krause, Phys. Rev. A 41, 243, (1990).

SEMICLASSICAL COHERENCES IN COLLISIONAL REDISTRIBUTION

Ronald J. Bieniek
Physics Department, University of Missouri-Rolla, Rolla, MO 65401-0249, U.S.A.

Ian M. Bell
Clarendon Laboratory, Oxford University, Oxford OX1 3PU, U.K.

If detuned polarized laser light is absorbed during an atomic collision, the polarization of the fluorescence depends upon the detuning. Quantum mechanical descriptions have been developed to describe such collisional redistribution.[1] Popular semiclassical theories have been proposed for collisional redistribution[2], but have faltered under the harsh trial of numerical tests. Although a full quasiclassical theory has explained many features in the red-wing of the Sr-Ar collision system, but had serious deficiencies in the red-wing of the Sr-He system.[3] We report here that the introduction of appropriate, but simple phases into the evolution of excited states significantly improve agreement between a semiclassical model and brute-force quantal computations.

The temporal evolution in the semiclassical model of redistribution proceeds as follows[3]: 1) polarized radiation can be absorbed at the (vertical) Condon points R^c for transitions; 2) orbitals both parallel and perpendicular to the incident radiation will be excited; 3) collision continues, switched to the excited potential; 4) excited orbitals rotate with the internuclear axis because of molecular locking; 5) rotation continues until the separation of the nuclei reaches a decoupling radius R_{dc}, after which the orbitals suddenly become space-fixed. The major coherence effect is from the phase difference between the excited Σ and Π orbitals when they reach the decoupling radius on the outgoing trajectory. We can account for the difference in the phase that accumulates from the initiation of the collision on the ground state, and then evolution on the excited state from a Condon point to the decoupling point. Based on a quantal formulation of alignment effects,[4] the relevant phase difference is

$$\Delta\phi(\ell,R_\Pi^x,R_\Sigma^x)^{p,q} = \phi_\Pi(\ell,R_\Pi^x)^p - \phi_\Sigma(\ell,R_\Sigma^x)^q$$

where p and q indicate either an incoming (i) or outgoing (o) excitation, and:

$$\phi_s(\ell, R_s^x)^i = \int_{R_s^x}^{\infty} k_g^\ell(r)\,dr + \int_{R_s^t}^{R_s^x} k_s^\ell(r)\,dr + \int_{R_s^t}^{R_{dc}} k_s^\ell(r)\,dr$$

$$\phi_s(\ell, R_s^x)^o = \int_{R_s^t}^{\infty} k_g^\ell(r)\,dr + \int_{R_s^t}^{R_s^x} k_g^\ell(r)\,dr + \int_{R_s^x}^{R_{dc}} k_s^\ell(r)\,dr$$

where where g indicates the ground state, s either the excited Σ or the Π state, R_s^t is a turning point, and $R_s^x = \min(R_s^c, R_{dc})$. For detunings δ, the Condon points R_s^c satisfy the equation $\Delta V_s(R_s^c)-V_s(\infty)=hc\delta$, where $\Delta V_s(R)=V_s(R)-V_g(R)$. After lengthy mathematical manipulations of the locking/decoupling model, but with retained phase terms, one can achieve formulae for polarizations that are no more complicated (or computationally difficult) than quasiclassical expressions previously derived without coherences.[3]

To investigate the importance of phase-coherence in real polarization spectra, we applied this semiclassical model to Sr-Ar(He) collision systems because numerical quantum mechanical results on analytic potentials were available for comparison.[1] The figure below compares "exact" quantal results to those obtained by the semiclassical model with and without phase-coherence terms. For red detunings within the classical rainbow satellite ($\delta \leq 30$ cm^{-1}), there are three Condon transition points, producing six possible points for semiclassical excitation (three incoming and three outgoing). This generates a rather strenuous test of the semiclassical model. In the Sr-Ar case, the inclusion of phase coherence terms substantially raises the polarization for incident detunings inside of the classical satellite, bringing the semiclassical model into much better agreement with quantal computations. The effect is even more pronounced in the Sr-He case, a much more quantal system. The standard quasiclassical model underestimates the residual polarization at small detuning, and even gives an incorrect trend in the polarization curve.[3] As can be seen, the locking/decoupling model with phase-coherence terms is in excellent agreement with the quantum mechanical results.

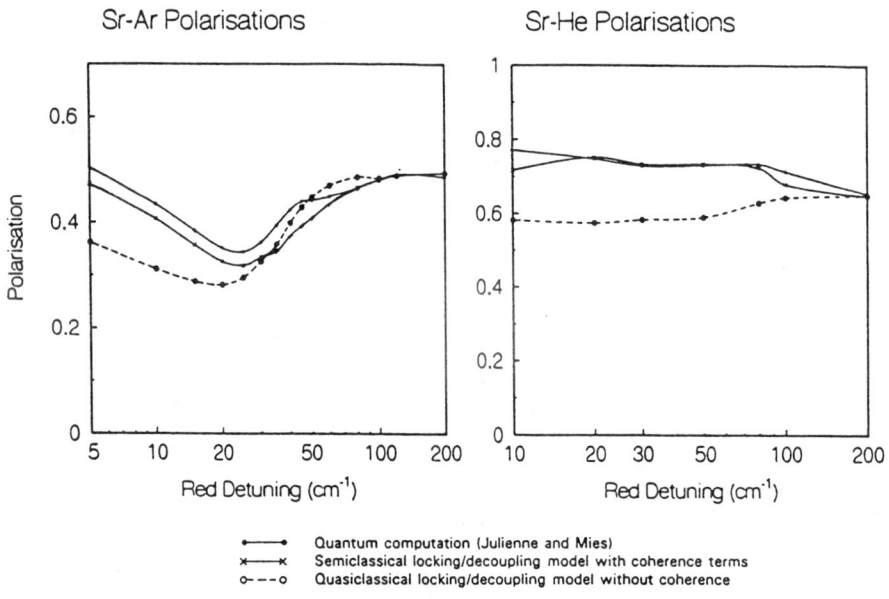

•——• Quantum computation (Julienne and Mies)
×——× Semiclassical locking/decoupling model with coherence terms
o---o Quasiclassical locking/decoupling model without coherence

1. P.S. Julienne and F.H. Mies, *Phys. Rev A* **34**, 3792 (1986).
2. E.L. Lewis, J.M. Slater, and M. Harris, *J. Phys. B* **14**, L173 (1981); E.L. Lewis, M. Harris, W.J. Alford, J. Cooper, and K. Burnett, *J. Phys. B* **16**, 553 (1983).
3. R.J. Bieniek, P.S. Julienne, and F. Rebentrost, *J. Phys. B* **24**, 5103 (1991).
4. M. Glass-Maujean and J.A. Beswick, *Phys. Rev. A* **36**, 1170 (1987).

EFFECTS OF INTERACTION Hg-C_{60} OBSERVED ON THE FAR RED WING OF THE Hg 253.7 nm LINE

T. Grycuk and M. Tchaplyguine[+]
Institute of Experimental Physics, University of Warsaw,
Hoża 69, 00-681 Warszawa, Poland

E. Czerwosz and P. Byszewski
Institute of Vacuum Technology, Długa 44/50, 00-241 Warszawa, Poland

Since the first report presenting the spectroscopic evidence for the existence of fullerenes[1] many articles dealing with spectroscopy of this new material appeared. However, only a few papers were related to spectroscopic studies of a single fullerene molecule in the gas phase. A situation is even worse in the case of mixtures of fullerenes with metals which e.g. for alkalies are extensively studied in condensed media.

In this paper we present the first measurements of spectrum of the Hg+C_{60} mixture in the gas phase. The fullerene molecule plays here a role of perturber for the resonance line of mercury at 253.7 nm. The Hg-C_{60} spectrum was registered in absorption over the spectral region 230 – 360 nm for the temperatures between $\sim$ 300 – 800°C at which the C_{60} density[2] was varied between $\sim$ 3 x 10^{11} – 4.4 x 10^{15} cm^{-3}. The great effort was made in preparation of the absorption cell (UV grade quartz) because C_{60} is degraded on heating in oxygen.

For temperatures greater than 335°C The Hg density was constant (dry vapour) and in this stage of our work it was rather high, ammounting to 8.1 x 10^{18} cm^{-3}.

Effect of interaction Hg-C_{60} is observed for the temperatures higher than about 500°C (densities of fullerenes greater than 6 x 10^{12} cm^{-3}) and for densities lower than $\sim$ 5 x 10^{14} cm^{-3} a perturbation of the line profile is very similar to that observed in the presence of rare gases. This is demonstrated as a monotonous enhancement of the red wing of line. It seems that the source of this effect can be the long-distance van der Waals interaction between the Hg atom and the total fullerene molecule.

For the greater densities of fullerene a qualitatively new spectral feature appear: a distinct diffuse band at $\sim$ 280 nm which is not observed in the spectrum of pure fullerene. Under such conditions the C_{60} band at 330 nm is also seen. The new band we try to interpret taking into consideration the stronger intermolecular forces. One of the serious possibilities is a charge-transfer interaction between the Hg atom and fullerene, because the latter is a

---

[+]Permanent address: Institute of Physics, St. Petersburg State University, St.Petergof 198904, Ulyanovskaya 1, Russia

quite strong acceptor of electrons. Further work in this direction is in progress.

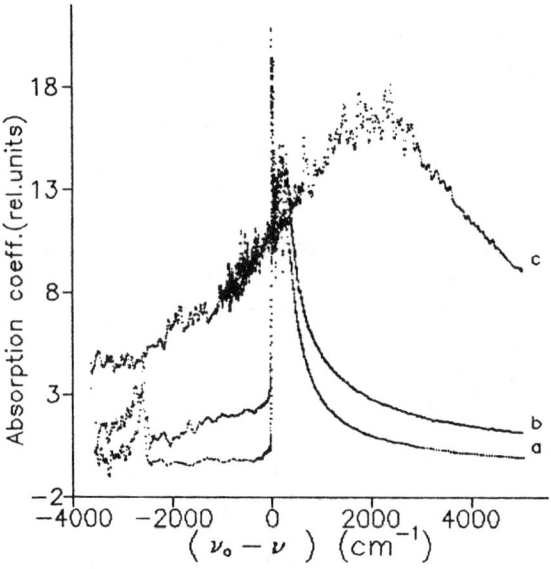

Fig. 1: Absorption spectrum of the Hg+C_{60} mixture.
(a): T = 481°C, $[C_{60}]$ = 2.56 x 10^{12} cm^{-3}; (b): T = 589°C, $[C_{60}]$ = 7.84 x 10^{13}; (c): T = 729°C, $[C_{60}]$ = 1.73 x 10^{15}; Density of Hg was 8.1 x 10^{18} for the all cases. A position of the Hg 253.654 nn line is marked with a narrow emission line from a RF Hg lamp.
The structured band on the blue wing of line is attributed to the Hg_2 molecule ($^3 1_u$ ($^3P_2 + {}^1S_0$) ← X transition).

REFERENCES

1. H.W. Kroto, J.R. Heath, S.C. O'Brien, R.F. Curl, R.E. Smalley, Nature **318**, 162 (1985).
2. J. Abrefach, D.R. Olander, M. Balooch and W.J. Siekhaus, Appl. Phys. Lett. **60** (11), 1313 (1992).

LINE SHAPES IN DENSE MEDIA

RAMAN SPECTRA OF FORMAMIDE IN DMSO.

A. Mortensen and O. Faurskov Nielsen
University of Copenhagen, Copenhagen, Denmark

J. Yarwood
Sheffield Hallam University, Sheffield, England

V. Shelley
University of Durham, Durham, England

The Raman spectrum of the carbonyl band of formamide is asymmetric. Diluting formamide with DMSO causes this asymmetry to change. From this it is concluded that the asymmetry is due to two different "sites".

Formamide ($HCONH_2$) is the simplest molecule containing the biologically important peptide group (-CONH-) and is thus a model compound for studying interactions in peptides.

The isotropic Raman spectrum of the carbonyl band of formamide is asymmetric. The observation of the asymmetry is made difficult by the overlapping NH_2 bending mode at 1600 cm^{-1}. The Raman spectra shown in Fig. 1 are therefore of the deuterated isotopomer ($HCOND_2$).

In the neat liquid $HCOND_2$ the asymmetry is seen as a wing on the high-frequency side of the carbonyl band. Addition of DMSO to $HCOND_2$ causes the frequncy of the carbonyl band to increase. At the same time, the band gets broader and the relative intensity of the wing increases. When the mole fraction of $HCOND_2$ is around 0.3, the band becomes asymmetric to the low-frequency side. At this point, further addition of DMSO causes a narrowing of the carbonyl band but only a minor shift of the frequency. At very low concentrations the band finally becomes symmetric (Fig. 1).

The behavior of the carbonyl band of $HCOND_2$ is interpreted in terms of two "sites". One "site" is a hydrogen-bonded aggregate (a linear "chain" of molecules), which is responsible for the carbonyl band in the neat liquid, and the other is a species with no hydrogen bonds to the carbonyl group ("monomer")[1,2] giving rise to the wing on the high-frequency side of the carbonyl band. Diluting $HCOND_2$ with the DMSO causes the chains to break up, thus increasing the proportion of "monomer" and thereby the intensity of the wing. At very low concentrations of $HCOND_2$ in DMSO there are no hydrogen bonds to the carbonyl groups of $HCOND_2$ and only one "site" is present (the nonhydrogen-bonded "monomer") and the carbonyl band is thus symmetric.

The anisotropic Raman carbonyl band of $HCOND_2$ is asymmetric as well. This band is broader and at higher frequency than the isotropic component[1,2]. This last phenomenon is called the noncoincidence effect. This band also shifts upwards in frequency but the shift is less than the corresponding shift of the isotropic component. At very low concentrations the isotropic and anisotropic components coincide.

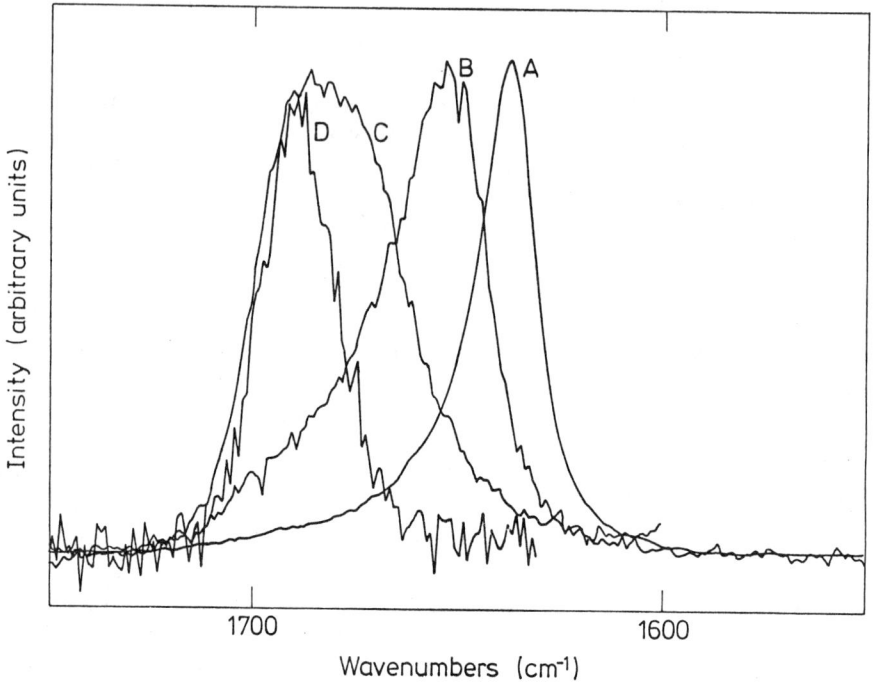

Fig. 1. Isotropic Raman spectra of $HCOND_2$ in DMSO. Mole fraction of $HCOND_2$: A) 1, B) 0.64, C) 0.31 and D) 0.09.

1. A. Mortensen, O. F. Nielsen, J. Yarwood and V. Shelley, J. Phys. Chem. **98**, 5221 (1994).
2. A. Mortensen, O. F. Nielsen, J. Yarwood and V. Shelley, Vib. Spectrosc. (submitted).

Field Ionization of High Rydberg States of CH3I in Liquid Argon

R. Reininger and A. Al-Omari

Synchrotron Radiation Center, University of Wisconsin - Madison,
3731 Schneider Dr., Stoughton, WI 53589, USA.

The effect of high medium densities on low-n Rydberg states has attracted renewed interest both experimentally[1] and theoretically[2]. Due to their rapid broadening with increasing density, high-n Rydberg states cannot be spectrally separated. As well known, high-n Rydberg states within $2\sqrt{F}$ (a.u.) below the ionization limit can field ionize in the presence of an electric field F. By studying the electric field dependence of the photoionization in a dense Ar gas doped with CH3I, Meyer and Reininger[3] observed an increase in the photoionization current with increasing electric field. This was attributed to the field ionization of high Rydberg states of the dopant.

We have extended the density range of the previous measurements[3] up to the triple point liquid in argon. The field effects in the photoionization of pure CH3I and in the photoionization of CH3I doped in liquid argon are shown in Fig. 1a and 2a, respectively. The field ionization in pure CH3I can be easily recognized below the two ionization thresholds[4]. As expected, the spectra recorded in the liquid differ considerably from those in the gas. The high density induces a considerable energy shift and a smearing of the rich autoionization structure present in Fig. 1a between the two ionization limits, 9.538 and 10.164 eV. As seen in Fig. 2a, field ionization also occurs in the liquid in the threshold region and at about 9.25 eV. This is due to the field ionization of high-n Rydberg states of CH3I converging to the first and second ionization limits.

Figure 1b shows several field signals in pure CH3I. They were obtained by subtracting the spectra recorded at 833 V/cm from those measured at higher electric fields. The field signals obtained in the liquid are displayed in Fig. 2b. As seen in the figures, there are two main peaks in each spectra. The energy separation between them, i.e., the spin-orbit splitting, does not change with the sample density. This is, however, not the case for the width of the peaks, which are considerably broader in the liquid. It is worth noting that all the peaks shown for the liquid sample have, practically, the same width.

The field signals were calculated using the first and second moments of the polarization energy between the positive ion and the medium. Good agreement was obtained between the experimental and calculated field signals in the whole density range from the gas to the triple point liquid. Moreover, the energy positions of the experimental field signals were obtained by shifting the calculated signals by the quasi-free electron energy in Ar[5] at each density.

1. E. Morikawa, A.M. Köhler, R. Reininger, V. Saile, and P. Laporte, J. Chem. Phys. **89**, 2729 (1988).
2. R.M. Stratt and J.E. Adams, J. Chem. Phys. **99**, 775 (1993).
3. J. Meyer and R. Reininger, Phys. Rev. A **47**, R3491 (1993).
4. A. Al-Omari and R. Reininger, Chem. Phys. Lett. **220**, 437 (1994).
5. B. Plenkiewicz, P. Plenkiewicz, and J.-P. Jay-Gerin, Phys. Rev. A **40**, 4113 (1989).

This work is based upon research conducted at SRC, Univ. of Wisconsin, which is supported by the NSF under award No. DMR-9212658

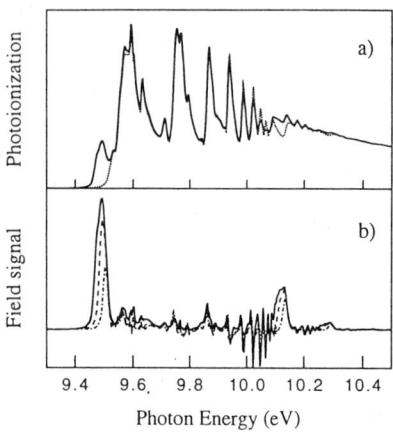

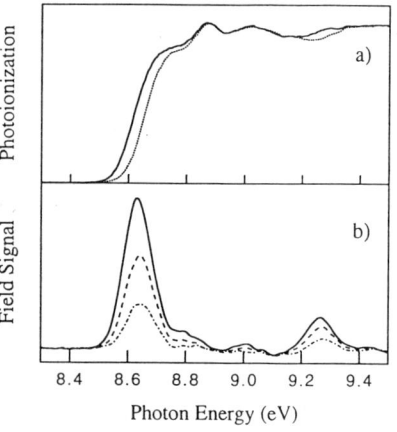

Fig. 1: a) Photoionization of pure CH_3I recorded with 9633 V/cm (solid line) and 833 V/cm (dotted line). b) Field signals (see text) for 9633 (solid line), 5633 (dashed line), and 2700 V/cm (dot dashed line).

Fig. 2: a) Photoionization of CH_3I doped in liquid argon ($\rho = 16.5 \times 10^{21}$ cm^{-3}) recorded with 9633 V/cm (solid line) and 833 V/cm (dotted line). b) As in Fig. 1b.

LINE SHAPES OF THE FUNDAMENTAL VIBRATION-ROTATION-PHONON AND PURE ROTATION-PHONON SPECTRA IN HD

R. M. Herman, B. Weiner and P. B. Shaw
The Pennsylvania State University, University Park, PA 16802 USA

ABSTRACT

A density matrix description of the HD vibration-rotation-phonon and pure rotation-phonon spectra is presented. Because the instantaneous force on any HD molecule of interest gives both the phonon-induced dipole and the phonon-rotation coupling, the problem can be completely solved by using a force spectrum inferred from the corresponding phonon spectrum in D_2.

DISCUSSION

Following the highly successful theory[1] for collisionally induced dipole interferences in gaseous HD vibration-rotation spectra, which led to pressure dependent line shapes of mixed Lorentzian-anomalous dispersion character, we employ our present concepts to explain the observed[2] solid and liquid state profiles in the fundamental band (3620-3740 cm^{-1} region, encompassing the $Q_1(0)$ and $R_1(0)$ features) and in the pure rotation spectrum (70-130 cm^{-1}, encompassing the $R_0(0)$ line). For this purpose, we have developed a quantum mechanical density matrix approach which incorporates many of the data and conceptual relationships of the gas phase problem.

The phonon-induced transition dipole for any HD molecule is assumed proportional to the instantaneous force, $\vec{F}$, exerted by all of its surroundings. For the "uninterfered" phonon (Q_R) branch of the $Q_1(0)$ line (the sharp $Q_1(0)$ feature is present only through induction by neighboring quadrupolar states) we infer its shape and intensity through a straightforward transformation of the corresponding vibration-phonon spectrum[3] in D_2. We label this spectrum, by itself, $L_{ph}(\omega)$. The coupling of the $(v,J)=(1,1)$ state to the ground $(0,0)$ state of HD occurs through two interfering routes: a direct excitation through the permanent transition dipole of HD, augmented by a static solid-induced dipole term; and through a coupling to the phonon Q_R-branch through the interaction potential $-\vec{F}\cdot\vec{d}$ where, $\vec{d}$ is the displacement of the geometric center from the center of mass in HD. Because the force spectrum is known, the spectral content of the $-\vec{F}\cdot\vec{d}$ interaction is entirely understood in terms of $L_{ph}(\omega)$ and its Kramers-Kronig conjugate function, $A_{ph}(\omega)$. With the above assumptions, the complete spectrum can be obtained in the closed form

$$L(\omega) = \text{Re} \left\{ \left[L_{ph}(\omega) + iA_{ph}(\omega) \right] \right.$$
$$\left. \times \left[\left(1 + \frac{\Omega}{\omega - \omega_1 + i\gamma_1}\right)\left(1 + \frac{\Omega + \Delta(\omega)}{\omega - \omega_1 + i\gamma_{ph} - \Delta(\omega)}\right) - \frac{i\Omega^2(\gamma_1 + i(\omega - \omega_1))}{\Delta(\omega)((\omega - \omega_1)^2 + \gamma_1^2)} \right] \right\}$$

Here, ω_1 is the $R_1(0)$ line frequency displacement from the $Q_1(0)$ feature, as statically perturbed by the lattice, γ_1 is the "uninterfered" $R_1(0)$ line (half) width, set equal to that of the sharpest $Q_1(0)$ feature and $\Delta(\omega)$ is a dynamical phonon induced (complex) energy shift function which is proportional to $A_{ph}(\omega) - iL_{ph}(\omega)$ and can be completely calculated. The only adjustable parameters are Ω, which essentially sets the strength of the uninterfered transition dipole moment for the $R_1(0)$ line and the phonon states effective line width, γ_{ph}. The imaginary part represents the negative real part of the dielectric response, which has not yet been observed. The term $L_{ph}(\omega)$ by itself represents the uninterfered phonon spectrum while the last term (γ_1-dependent part) gives the uninterfered $R_1(0)$ line. The remaining terms describe three principal interference effects: a pronounced interference minimum to the low frequency size of the sharp $R_1(0)$ feature; a substantial alteration in the shape and strength of the $R_1(0)$ feature; and an interference peak on the high frequency side of the $R_1(0)$ feature.

Choices of Ω and γ_{ph} which give reasonable agreement with the experimentally observed spectra have been made, leading to somewhat different total transition moments for the uninterfered R(0) features from those presently accepted[4].

REFERENCES

1. R. M. Herman, R. H. Tipping and J. D. Poll, Phys. Rev. A20, 2006 (1979).
2. A. R. W. McKellar and M. J. Clouter, Can. J. Phys. 68, 422 (1990).
3. A. Crane and H. P. Gush, Can. J. Phys. 44, 373 (1966).
4. J. D. Poll, R. H. Tipping, S-Y. Lee, S. Lee, T. W. Noh and J. R. Gaines, Phys. Rev. B39, 11372 (1989).

Many-body Correlations in the Far Infrared Absorption in Liquid OCS

H. Stassen and W. A. Steele

Pennsylvania State University, University Park, Pennsylvania 16802

At liquid densities the far infrared (FIR) absorption spectrum of OCS contains remarkable contributions from interaction-induced processes in addition to the allowed orientational contribution arising from the small permanent dipole moments ($\mu_o = 0.72$ D), especially in the spectral wing.[1] In this study, we calculated the induced pair dipoles $\vec{\mu}_{ij}$ for a sample of 256 molecules of OCS described by atom-atom (12/6)-Lennard-Jones potentials[2] in the NVE ensemble at 298 K and a molar volume of 61.70 cm^3/mole. The spectral time correlation function was evaluated from MD simulations

$$C(t) = \sum_{i=1} \sum_{j \neq i} \sum_{k=1} \sum_{l \neq k} \langle \vec{\mu}_{ij}(0) \cdot \vec{\mu}_{kl}(t) \rangle \tag{1}$$

and separated into its component terms. These are two two-body CFs:

$$C_{2A}(t) = \sum_{i=1} \sum_{j \neq i} \langle \vec{\mu}_{ij}(0) \cdot \vec{\mu}_{ij}(t) \rangle \; ; \; C_{2B}(t) = \sum_{i=1} \sum_{j \neq i} \langle \vec{\mu}_{ij}(0) \cdot \vec{\mu}_{ji}(t) \rangle, \tag{2}$$

three three-body CFs:

$$C_{3A}(t) = \sum_{i=1} \sum_{j \neq i} \sum_{k \neq i,j} \langle \vec{\mu}_{ij}(0) \cdot \vec{\mu}_{ik}(t) \rangle \; ; \; C_{3B}(t) = \sum_{i=1} \sum_{j \neq i} \sum_{k \neq i,j} \langle \vec{\mu}_{ij}(0) \cdot \vec{\mu}_{ki}(t) \rangle \; ;$$

$$C_{3C}(t) = \sum_{i=1} \sum_{j \neq i} \sum_{k \neq i,j} \langle \vec{\mu}_{ji}(0) \cdot \vec{\mu}_{ki}(t) \rangle \tag{3}$$

and a four-body contribution $C_4(t)$ (where all indices in eq.(1) differ from each other). We expanded this analysis to the auto CFs of dipole-induced dipoles (DID-DID), quadrupole-induced dipoles (QID-QID) and octopole-induced dipoles (OID-OID) as well as the cross CFs (DID-QID, DID-OID, QID-OID). The pair dipoles $\vec{\mu}_{ij}$ were obtained from the multipole expansion for the electric fields emanating from the molecular dipoles, quadrupoles ($Q_o = -0.79$ DÅ) and octopoles ($\Omega_o > 10$ DÅ^2) and the interaction of these moments with the anisotropic molecular polarizabilities.

In table I we summarize the C(0)-values of the many-body CFs for the various induction mechanisms. It becomes evident that the large C_{2A}-amplitudes are partially canceled by the next largest CF which is C_{3C}. The QID containing CFs in addition exhibit interferences due to correlations described by the C_{2B}.

Table I. Amplitudes of the many-body CFs in units of their corresponding total CF from eq.(1). The values for the total CF are in 10^{-62} C^2m^2.

CF	DID-DID	DID-QID	DID-OID	QID-QID	QID-OID	OID-OID
C_{2A}	1.49	-0.09	1.27	3.11	-0.88	1.22
C_{2B}	-0.01	0.33	-0.07	-1.20	1.70	-0.09
C_{3A}	0.15	0.01	0.02	0.11	0.10	0.02
C_{3B}	-0.01	-0.07	-0.01	0.25	-0.46	0.02
C_{3C}	-0.50	0.82	-0.22	-1.08	0.49	-0.13
C_4	-0.11	0.00	0.02	-0.18	0.04	-0.03
total	18.4	-1.8	-20.0	0.8	1.6	26.8

Comparing the three auto CFs we find the strongest cancellation phenomena of the C_{2A}-amplitude in the case of the QID, whereas the short-range induction of OID shows only small cancellations of the C_{2A} by the other component many-body CFs. The dynamics of the cancellation effect as depicted in fig. 1 indicates the expected tendency that more short-range inductions give more rapidly decaying CFs.

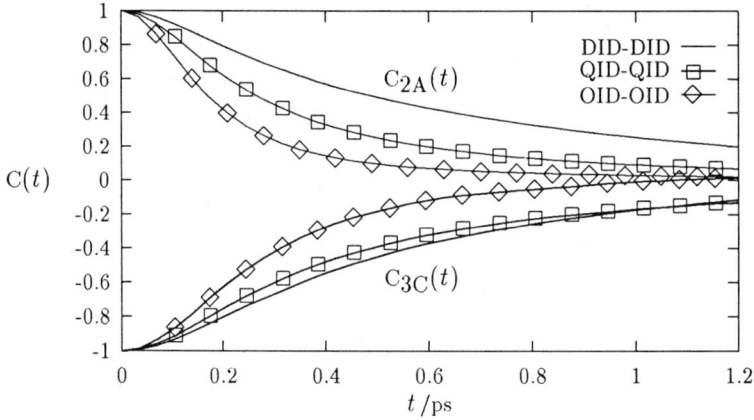

Fig. 1. Normalized $C_{2A}(t)$ and $C_{3C}(t)$ for the DID-DID, QID-QID and OID-OID.

1. H. Stassen and Th. Dorfmüller, Chem. Phys., in press.
2. J. Samios, H. Stassen, and Th. Dorfmüller, Chem. Phys. 160, 33 1992.

An analysis of the radiation line shape
from a multiple quantum well structure in four-wave-mixing

C. J. Hsu
Department of Physical Sciences,
University of New Brunswick, Saint John Campus, N.B. E2L 4L5

ABSTRACT

The optical line shape of multiple quantum well (MQW) structures has been analyzed based on the ordinary optical Bloch equation. This result shows that the dominating non-Lorentzian peak is from light hole exciton and a weak dispersive-like shape is due to the heavy hole exciton. The shapes given in the figures show an excellent agreement between the experimental observation and this theory.

SUMMARY

In recent wave mixing experiments with multiple quantum well (MQW) structures, a combination of a strong non-Lorentzian line shape and a very narrow and weak dispersive-like one has been observed [1]. This line shape was attributed to spectral diffusion, and was treated by using a redistribution of the sites with a modified optical Bloch equation (MOBE)[2,3]. The main feature of this treatment is the consideration of some sort of collective sites among the excitons, but in such collective sites, consideration was not given to whether, and how this property might be affected by the applied fields and their coherence.

It has been found recently that field-induced extraresonances [4] give rise to both the above line shapes in a two-wave mixing, in which an instantaneous population inversion is responsible for a stimulated scattering. In this report, I shall extend yhe use of population fluctuations to the four-wave-mixing situation, since such fluctuations are particularly effective in accounting for the collective properties through the involvement of the coherence and intensities of the applied fields.

In this work, the energy transitions of both types of the hole have wide inhomogeneous widths. When such a large energy distribution of the light hole exciton is pumped, it gives rise to the strong component of the non-Lorentzian peak; and the weak dispersive-like component is mainly due to the far off-resonant heavy hole exciton. Details of this result will be presented elsewhere.

FIGURE

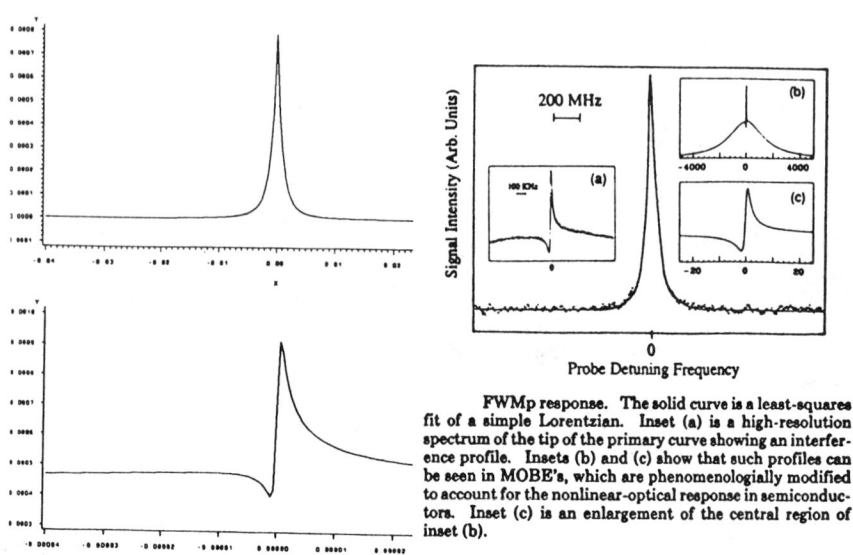

FWMp response. The solid curve is a least-squares fit of a simple Lorentzian. Inset (a) is a high-resolution spectrum of the tip of the primary curve showing an interference profile. Insets (b) and (c) show that such profiles can be seen in MOBE's, which are phenomenologically modified to account for the nonlinear-optical response in semiconductors. Inset (c) is an enlargement of the central region of inset (b).

On the left hand side, the result of this analysis is presented. In this result, the transverse relaxation rate for both type of exciton is 1 unit which corresponds to 1 cm^{-1} (30 GHz). The widths of the two distributions used are 140 and separated by 110 units (approximately the separation of the spectral peaks arising from the two types of hole). The longitudinal relaxation rates are $1.3*10^{-3}$ and $5*10^{-6}$ unit for the light holes and heavy holes respectively. the pump Rabi-frequency is 0.01 unit.

The right hand side shows the experimental observations and the result based on MOBE [1]. Notice that a direct measurement of the widths from the figure does not give any of the above relaxation rate due to dephasing- and field induced effect [4].

REFERENCES

1. J. T. Remillard, H. Wang, M.D. Webb, D.G. Steel, J.Oh, J. Pamulapati, and P.K. Bhattacharya, Opt. Lett., 14, 1131(1989)
2. Hailin Wang and D. G. Steel, Phys. Rev. A43,3823(1991)
3. P. R. Berman and R. G. Brewer, Phys. Rev. A32, 2784 (1985)
4. C. J. Hsu, Opt. Soc.AM. B7, 2155 (1990)
   C. J. Hsu and C. H. Leung, to appear in Physica (1994)

ABSORPTION SPECTRA SHAPES OF SILVER COLLOID AGGREGATES

V.P.Safonov, Yu.E.Danilova, V.A.Markel
Institute of Automation & Electrometry SB RAS,
Novosibirsk, 630090, Russia

It is known, that an aggregation of microparticles (monomers) in colloid solution leads to appearence of a long wavelength wing in absorption spectrum. This fact was discussed many times (see, for example, [1]), but an adequate understanding of the problem was not gained yet.

Our aim in this work was to correlate experimental characteristics of silver colloids, as determined by optical absorption and electron microscopy, and to compare them with theoretical models.

Three different colloids were studied: Ag(PVP)[2], collargol[3] and Ag(NaBH$_4$)[4]. Results, gained with two former colloids, stabilised by polymers, are similar. Their micrographs are much alike. Freshly prepared, these colloids have a single resonance in the absorption spectra, associated with excitation of electron plasma in metal microparticles. For both sols its occurs in the vicinity of $\lambda = 400$ nm, FWHM being $\Delta\lambda_m = 70 \div 90$ nm (curve 1, Fig. 1) Micrographs show particles of $2R = 5 \div 10$ nm in diameter, being separated by distances several times more than the diameters. We computed the extinction of monomers in a frame of Mie theory with including size effect. Calculated spectrum is consistent with experimental rather well.

Addition of the alkali leads to substantial changes in colloids spectra, which take place in few days. The plasmon resonance is widened, so the FWHM becomes $100 \div 130$ nm and the second resonance appears with peak position at 570 nm (curve 2, Fig. 1). On micrographs the main part of monomers (90%) in our samples are aggregated in clusters. Similar spectra were reported in[1].

Other shape of spectrum is typical for Ag(NaBH$_4$) colloid in age of more than several hours after preparation (curve 3, Fig. 1). There is a slow decrease in the long wavelenght absorption. Note, that the freshly prepared Ag(NaBH$_4$) has spectrum, similar to the same for Ag(PVP) colloid.

Electron microscopic study shows that there are clusters of varios sizes with number of the monomers N from a few units to $N \approx 1000$ in all studied colloids. Large clusters ($N > 200$) have loose structure and one can describe them as fractals, having the relation of particles number in cluster to its radius as $N \propto R^D$, where D is fractal dimension. From micrographs, using "box counting" method we found for our colloids $D = 1.70 \pm 0.05$.

There was observed the distinction between Ag(PVP) and Ag(NaBH$_4$) colloids. In Ag(PVP) clusters the monomers are separated on distances $0.5 \div 5$ nm, whereas in Ag(NaBH$_4$) some fraction of monomers looks as a sticked together.

© 1995 American Institute of Physics

We have carried out the computer calculations of absorption spectra of small clusters in the model of dipole-dipole interactions of monomers. Clusters were generated in "random walker" model. For $N = 15 \div 30$ the well-pronounced second peak was obtained. However the second peak detuning from the monomer's resonance was less than on curve 2, Fig. 1.

It is very interesting to note that the shape of the long-wavelength wing of the spectrum 3 Fig.1 (Ag(NaBH$_4$)) is closed to predicted by the scale-invariant theory of fractal clusters[5]. According to the theory, the long-wavelength absorption of a fractal cluster decrease by a power law $\lambda \cdot A \propto |X|^{d-1}$, when a light frequency is characterised by the variable $X = -R^{-3} Re[(\varepsilon + 2\varepsilon_0)/(\varepsilon - \varepsilon_0)]$, d is optical spectral dimension, ε, ε_0 — dielectric constants of the monomers and the host medium (water). ε and a light frequency are related by Drude formula. Fig.2 gives evidence the power-law dependence of $\lambda \cdot A$ on $|X|$ in some spectral region. Measured $d = 0.59$, the theory[5,6] gives $d = 0.4 \div 0.5$. Scaling region in the spectrum is $\lambda = 470 \div 670$ nm.

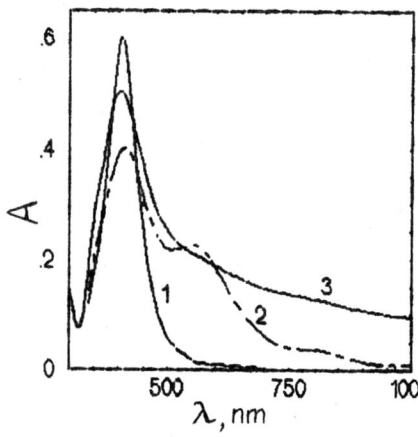
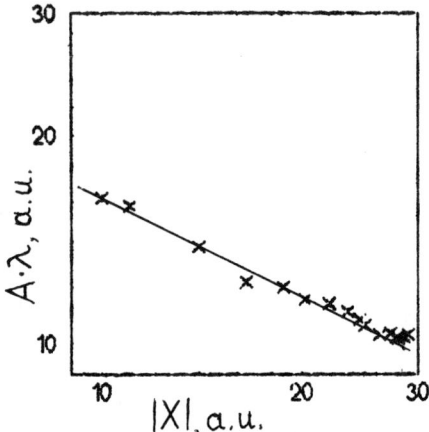

Fig. 1. Extinction spectra of silver sols.

Fig. 2. Shape of wing of Ag(NaBH$_4$) spectrum in log-log scale.

1. Surface Enhanced Raman Scattering. Ed.R.K.Chang and T.E.Furtak (Plenum Press, N.Y. 1982)
2. H.Hirai, J.Macromol.Sci.-Chem., A13, 633 (1979)
3. A.V.Butenko, et al. Z.Phys. D17, 283 (1990)
4. J.A.Creighton, et al. J.Chem.Soc.Faraday Trans. II, 75, 790 (1979)
5. V.A.Markel, L.S.Muratov, M.I.Stockman, T.F.George, Phys.Rev. B43, 8183 (1990)
6. V.Shalaev, V.A.Markel et al, Fractals, 2, 201 (1994)

BROADENING AND SHIFTING IN NEUTRALS AND IONS

ABILITY TO PREDICT MOLECULAR ROTATION-VIBRATION LINE SHAPES

Sheldon Green
NASA, Goddard Space Flight Center
Institute for Space Studies, New York, NY 10025

ABSTRACT

Within the "impact approximation," calculation of line shapes requires accurate molecular scattering matrices which, in turn, requires solving for molecular collision dynamics on an accurate intermolecular potential energy surface. For some simple, but nontrivial molecular systems the interaction potentials are now well determined from ab initio electronic structure calculations and/or from analysis of spectra of the van der Waals complexes, and it is feasible to solve numerically the coupled channel scattering equations as well. In such cases predicted line widths and shifts are generally in satisfying accord with experiment. Recent results for line widths, line shifts, collisional transfer of intensity in overlapping lines, and Dicke narrowing are presented as examples of current capabilities.

INTRODUCTION

The theory of line shapes in molecular vibration-rotation spectra has been derived from several different starting points. The first treatment by Anderson[1] assumed semiclassical (classical path) collision dynamics, solved the collision dynamics using second-order perturbation theory, and assumed that only long-range electrostatic forces were important. Until recently virtually all calculations were done within this framework. Although this method gives moderately accurate linewidths for systems dominated by strong long-range forces (e.g., dipole forces) it suffers from many well known shortcomings; in particular, it does not give satisfactory results for nonpolar systems. Barranger[2] provided a fully quantal framework which was subsequently rederived by Fano[3] within a generalized theory of relaxation in many-body systems. This formalism was specialized to the case of molecular vibration-rotation spectra by Ben-Reuven.[4] Most contemporary theoretical studies of line shapes follow the formulation of Ben-Rueven.

Hess[5] noted that all of these methods ignore motion of the spectral molecule in the laboratory frame and hence cannot account for Doppler broadening or Dicke narrowing; he suggested that the Boltzmann equation provides a natural framework for including these effects. Line shape theories starting with a quantum molecular version of the Boltzmann equation, the Waldman-Snider equation, have been derived.[6] To date, however, calculations within this framework have been done for only one system, D_2 in He.

To make calculations tractable all the methods described above introduce both the binary collision approximation and the impact approximation, limiting validity to densities which are not too large and to frequencies which are not too far from line centers. Interestingly all the methods then yield essentially the same formula: if the two-body collisional S-matrices are known one can calculate generalized line shape cross sections which describe the width and shift of isolated (Lorentzian) lines and also collisional transfer of intensity between overlapping lines. (Anderson theory also fits into this framework, obtaining the collisional S-matrices with a perturbation approximation to a semiclassical description of the collision.) Calculation of accurate S-matrices, of course, requires detailed knowledge of the intermolecular potential. The last ten years have seen much progress in obtaining potentials for non-polar systems from either quantum molecular structure calculations or from analysis of spectra of weakly bound van der Waals complexes. Similar progress has been made in solving collision dynamics within an accurate quantum coupled channel framework. Such calculations have been applied to line shapes for several simple, but nontrivial molecular systems, and ability to predict reliable values is illustrated here by recent results for CO-He, HCl-Ar, HF-Ar, and D_2-He.

CO - He

CO-He collision dynamics have been extensively studied because rotational excitation rates are quite important for understanding radioastronomical observations of CO in the interstellar gas.[7] An extensive ab initio potential energy surface has been computed for this system.[8] Molecular scattering calculations on this surface were shown to accurately predict microwave (pure rotational) line widths for the lowest few rotational levels from liquid nitrogen to room temperature.[9] Even more impressively, subsequent microwave line shapes at cryogenic temperatures[10] (below 4 K) were found to agree quite well with theoretical predictions.[11] Calculations also quantitatively reproduce infrared[12] and CARS[13] line shapes for a range of rotational levels. Further, theory[14] was able to predict collisional transfer of intensity at higher pressures in both infrared[15] and CARS[13] spectra.

Despite the apparent successes of this work, very recent analysis[16] of spectra of the CO-He van der Waals complex suggests that the theoretical potential may be inaccurate in the region of the minimum and raises questions about this good agreement, especially with the cryogenic measurements. The spectroscopically determined potential agrees well with the theoretical potential at higher energies and predicted line shape cross sections at liquid nitrogen temperatures and above are changed only slightly; however, below 4 K the new potential predicts linewidths much larger than measured. It is suggested in Ref. 16 that the validity of the impact approximation at these extremely low temperatures needs to be reexamined.

HCl - Ar

Analyses of the HCl-Ar van der Waals complex have provided

successively more accurate (rigid rotor) potential energy surfaces.[17] Accurate molecular scattering calculations[18] using the best of these are in good accord with the far IR (pure rotational) line width data from 125 K to room temperature. A very recent analysis of van der Waal spectra provided the vibrational dependence of this surface as well,[19] and molecular scattering calculations using this surface are in accord with linewidths and shifts for the vibrational bands.[20]

HF - Ar

Line shapes in HF perturbed by Ar show a strong dependence on the vibrational band suggesting the importance of vibrational dephasing effects. Analysis of spectra of the HF-Ar van der Waals complex provided an accurate potential energy surface which also includes dependence on the HF vibrational coordinate.[21] Accurate quantum molecular scattering calculations for this system which treat HF as a vibrating rotor are in harmony with earlier width and shift data for the pure rotational, fundamental, and first overtone bands; in fact, the experimental data contain significant discrepancies and it was suggested that the theory is probably more accurate than these experimental data.[22] New, precise measurements on the fundamental band, made subsequent to these calculations, are in very good (1%) accord with predictions.[23]

D_2 - He

An accurate ab initio potential energy surface, which includes dependence on the D_2 vibrational coordinate, is available for this system.[24] Molecular scattering calculations on this surface which treated D_2 as a vibrating rotor[25] were able to reproduce measured line widths and shifts in the CARS vibrational Q-branch.[26] Dicke narrowing has also been measured for this system,[26] and this is the only case for which theories based on the Waldman-Snider formalism have yet to be tested;[27] agreement between theory and experiment was quite satisfactory, suggesting the viability of this theoretical approach.

DISCUSSION

The fact that the current paradigm, when used with accurate intermolecular potentials and converged coupled channel molecular scattering calculations, can correctly predict line shapes for several systems is an indication of the fundamental correctness of the theory, at least near line centers and at densities which are not too high. It is well known that this theory fails in the far wings, for example, and development of a more general theory is still needed.

It should be admitted in closing that the examples provided here were not chosen at random. Successful prediction of line shapes requires an accurate intermolecular potential, and these are still not well determined for many systems of interest. H_2-Ar can be cited as one example of a system for which more work is required. A potential energy surface which was thought to be quite accurate was, in fact, determined for this system from spectra of the van der Waals complex.[28] However, line shape calculations using this surface predicted CARS linewidths in only moderate agreement with experiment and line shifts which were nearly a factor of two too large.[29] More recently a potential for this system was obtained from large-scale molecular structure calculations and was further adjusted to agree with the spectrally determined potential in the region where the latter is thought to be most accurate.[30] Line shape calculations with this surface predict line shifts in much better accord with experiment but the linewidths are now too small.[31] It is clear that line shape parameters can be quite sensitive to details of the potential and that an accurate potential is a prerequisite for useful predictions.

ACKNOWLEDGMENTS

This work was supported by NASA Headquarters, Office of Space Science and Applications, Astrophysics Division, Infrared and Radio Astrophysics Program. I thank Bruce Haas for logistical support which made possible presentation of this work at the 12th International Line Shape Conference.

REFERENCES

1. P. W. Anderson, Phys. Rev. **76**, 647 (1949).
2. M. Barranger, Phys. Rev. **111**, 481 (1958); **111**, 494 (1958); **112**, 855 (1958).
3. U. Fano, Phys. Rev. **131**, 259 (1963).
4. A. Ben-Reuven, Phys. Rev. **145**, 7 (1966).
5. S. Hess, Physica **61**, 80 (1972).
6. L. Waldmann, Z. Naturforsch. **13a**, 609 (1958); R. F. Snider, J. Chem. Phys. **32**, 1051 (1960); L. Monchick and L. Hunter, J. Chem. Phys. **85**, 713 (1987); R. Blackmore, J. Chem. Phys. **87**, 791 (1987).
7. S. Green and P. Thaddeus, Astrophys. J. **205**, 766 (1976).
8. L. D. Thomas, W. P. Kraemer, and G. H. F. Diercksen, Chem. Phys. **51**, 131 (1980).
9. L. D. Thomas and S. Green, J. Chem. Phys. **73**, 5391 (1980).
10. D. R. Willey, R. L. Crownover, D. N. Bittner, and F. C. DeLucia, J. Chem. Phys. **89**, 1923 (1988); D. R. Willey, T. M. Goyette, W. L. Ebenstein, D. N. Bittner, and F. C. DeLucia, J. Chem. Phys. **91**, 122 (1989).
11. S. Green, J. Chem. Phys. **82**, 4548 (1985); A. Palma and S. Green, J. Chem. Phys. **85**, 1333 (1986).
12. S. Green and A. J. Mannucci, J. Chem. Phys. **97**, 1610 (1992).
13. W. S. Hurst, G. J. Rosasco, and S. Green, in preparation.
14. S. Green, J. Boissoles, and C. Boulet, J. Quant. Spectrosc. Rad. Transf. **39**, 33 (1988).
15. F. Thibault, J. Boissoles, R. LeDoucen, R. Farrenq, M. Morillon-Chapey, and C. Boulet, J. Chem. Phys. **97**, 4623 (1992).
16. C. E. Chuaqui, R. J. Le Roy, and A. R. W. McKellar, J. Chem. Phys. **101**, 39 (1994).
17. J. M. Hutson and B. J. Howard, Molec. Phys. **43**, 493 (1981); J. M. Hutson and B. J. Howard, Molec. Phys. **45**, 769 (1982); J. M.

Hutson, J. Chem. Phys. **89**, 4550 (1988).
18. S. Green, J. Chem. Phys. **92**, 4679 (1990).
19. J. M. Hutson, J. Chem. Phys. **96**, 4237 (1992).
20. C. F. Roche, J. M. Hutson, and A. S. Dickinson, J. Quant. Spectrosc. Rad. Transf. (in press).
21. J. M. Hutson, J. Chem. Phys. **96**, 6752 (1992).
22. S. Green and J. M. Hutson, J. Chem. Phys. **100**, 891 (1994).
23. A. Pine. J. Chem. Phys. **101**, 3444 (1994).
24. W. Meyer, P. C. Hariharan, and W. Kutzelnig, J. Chem. Phys. **73**, 1880 (1980).
25. S. Green, R. Blackmore, and L. Monchick, J. Chem. Phys. **91**, 52 (1989).
26. K. C. Smyth, G. J. Rosasco, and W. S. Hurst, J. Chem. Phys. **87**, 1001 (1987).
27. R. Blackmore, S. Green, and L. Monchick, J. Chem. Phys. **91**, 3846 (1989).
28. R. J. LeRoy and J. M. Hutson, J. Chem. Phys. **86**, 854 (1987).
29. S. Green, J. Chem. Phys. **95**, 3888 (1991).
30. D. W. Schwenke, S. P. Walch, and P. R. Taylor, J. Chem. Phys. **98**, 4738 (1993).
31. S. Green, D. W. Schwenke, and W. M. Huo, J. Chem. Phys. **101**, 15 (1994).

COLLISIONAL BROADENING AND COUPLING IN THE ROTATIONAL SPECTRUM OF THE ASYMMETRIC ROTOR CHF2Cl

G. Buffa and O. Tarrini

Dipartimento di Fisica dell'Università, Piazza Torricelli 2, I-56126 Pisa, Italy

G. Cazzoli, L. Cludi, G. Cotti, and C. Degli Esposti

Dipartimento di Chimica dell'Università, Via Selmi 2, I-40126 Bologna, Italy

The line-shape of 15 selected CHF_2Cl rotational transitions is studied in the 59 – 1049 GHz frequency region using two different spectrometers. The broadening parameters are measured for self-collisions and for collisions with N_2 and O_2. For some pair of lines a large collisional coupling effect is observed and analyzed. The dependence of the collisional parameters on the numbers J, K_a and K_c and on the perturbing molecule is studied. Within the framework of Ref. 1, we extend to line coupling the semiclassical approximations usually used for line broadening.[2]

Figs. 1 and 2 summarize some results. Fig. 1 reports, versus J, the self-pressure broadening and coupling for $K_c=J-1$, $K_a=2\leftarrow 1$. Units are MHz/Torr. Solid line is the calculated broadening coefficient γ of the single transition $J+1_{2,K_c} \leftarrow J_{1,K_c}$. Dots are the calculated broadening coefficients $\bar\gamma$ of the overlapped pair $J+1_{2,K_c} \leftarrow J_{1,K_c}$, $J+1_{3,K_c} \leftarrow J_{2,K_c}$. The coupling parameter ζ is given by $\zeta = \gamma - \bar\gamma$. Crosses are experimental data. The apparent discontinuity at $J \simeq 20$ is explained as a transition from resolved to unresolved regime.

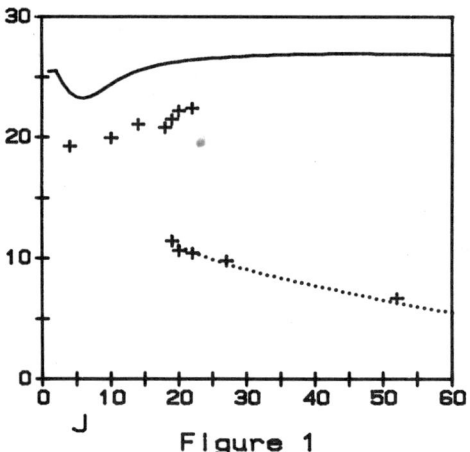

Figure 1

In Fig. 2 the self–pressure broadening and coupling coefficients are reported vs K_a for $J=53\leftarrow 52$, $K_c=52-K_a$. Solid line is the calculated broadening coefficient γ of the single transition $53_{K_a+1,K_c} \leftarrow 52_{K_a,K_c}$. Dots are the the calculated broadening coefficients $\bar\gamma$ for the overlapping pairs. When $K_a<10$ the

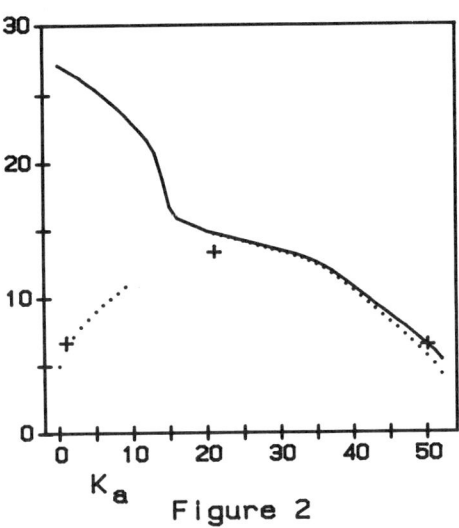

Figure 2

pair is $53_{K_a+1,K_c} \leftarrow 52_{K_a,K_c}$ and $53_{K_a+2,K_c} \leftarrow 52_{K_a+1,K_c}$, while it is done by $53_{K_a+1,K_c} \leftarrow 52_{K_a,K_c}$ and $53_{K_a+1,K_c+1} \leftarrow 52_{K_a,K_c+1}$ if $K_a > 20$. Line coupling (difference between solin and dotted line) is large for some cases and negligible for others. The agreement between theory and experiment is good. In Fig. 1, at low J, the calculated broadening γ of isolated lines is 10–20% in excess, but this is a usual consequence of the approximations of Ref. 2.

On the whole, from our study some conclusions about line coupling are obtained of general validity for the spectra of asymmetric top molecules with large dipole moment:

(i) Many pairs of coupling lines are usually present.

(ii) Line coupling may be very large for some pairs of lines but negligible for other apparently similar pairs.

(iii) Empirical models, such as energy-gap rate laws, are not useful for this case. In order to explain the effect one must consider not only energy levels, but also the dipole moment matrix elements of transitions between initial and final states of the coupling lines.

(iv) Line coupling is due mainly to long range collisional interaction. Hard collisions with small impact parameter and randomization of the rotational state have no coupling effect.

REFERENCES

1. G. Buffa and O. Tarrini, Phys. Rev. A **16**, 1612 (1977).

2. P. W. Anderson, Phys. Rev. **76**, 647 (1949); C. T. Tsao and I. Curnutte, J. Quant. Spectrosc. Radiat. Transfer **2**, 41 (1962).

A Study of Collisional Lineshapes of Ammonia Transitions

A. Ciucci[+], G. Baldacchini, F. D'Amato
ENEA, Dip. Innovazione, Settore Elettroottica e Laser, C.R.E. Frascati
C.P. 65, 00044 Frascati (Roma) Italy

G. Buffa and O. Tarrini
Dipartimento di Fisica, Università di Pisa
Piazza Torricelli 2, 56100 Pisa, Italy

The interest of the spectroscopy community on molecular parameters as transition frequencies, strengths, pressure shift and pressure broadening coefficients has been always well alive in the past, but it has particularly grown lately during the last decade due to the availability of high resolution laser sources. Among them a special place is occupied by the tunable diode lasers (TDL) which possess a linewidth emission much smaller than the Doppler width of the most common molecular species. So TDL's coupled to optical cells, which reproduce the physical conditions of interest, can provide high resolution spectra from which the line parameters may be obtained.

Of all the parameters the pressure broadening and shift measurements are essential for determining intermolecular forces, for modeling planetary atmospheres, and for quantitative analytical studies. However it is imperative that the experimental errors be kept as low as possible, and this aim has been reached both by upgrading the apparatus as a whole and by using an automatic analysis of the data. Indeed we have developed a software called LINEFIT 2.0, which is dedicated to the reduction and fitting of spectroscopical data. Moreover it offers to the operator the option of an active interaction which allows, among other things, a strong reduction of computational time. By using this computer program and a TDL system expecially assembled for lineshape measurements as a function of temperature, we started to study a small region of the ν_2 band of the ammonia molecule around ≈ 938 cm^{-1} from -100 to $+100$ °C.

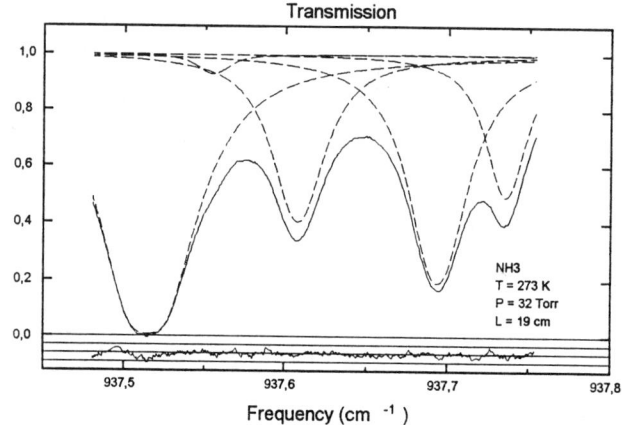

Fig. 1

Figure 1 shows the experimental transmission spectrum, normalized in amplitude and linearized in frequency, as a continuous curve containing, from left to right, the aQ(8,3), 2sQ(11,9), aQ(10,5), aQ(9,4) and the aQ(7,1) transition lines. The aQ(8,3) is well saturated, and remains so in almost all the experiments except those at very low pressures. In spite of this serious limitation it has been possible to obtain its linewidth by using the software mentioned above. The dashed curves represent the transmissions of the single lines as calculated in order to obtain the best fit. The result of this procedure is shown at the bottom of the figure, where the residual, i.e. the difference between experimental and fitted curve, is shown, and it is satisfactory if we take into account that the distance of two horizontal lines corresponds to 1%.

Figure 2 shows the linewidth the aQ(8,3) so obtained as a function of pressure and with error bars. The straight line represents the best fit, having excluded the last point at ≈ 60 Torr, which is clearly outside the common linear behaviour. Anyway Fig. 2 gives a broadening coefficient of 13.4±0.5 MHz/Torr, which compares fairly well with the value of 13.6 MHz/Torr calculated by using the semiclassical impact approximation.

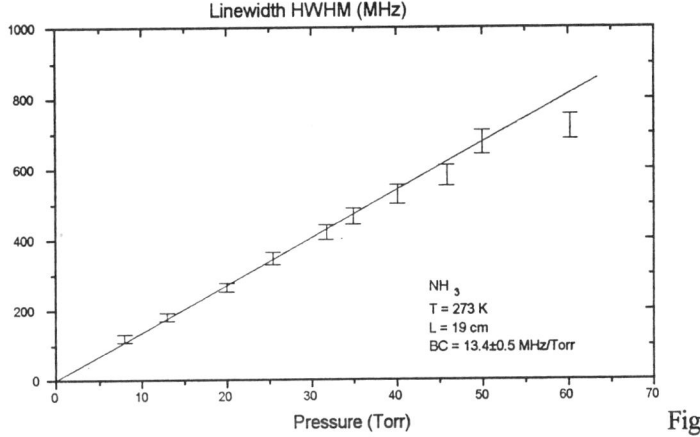

Fig. 2

Measurements have been extended at various temperature, with a general agreement between experiments and theory only when all the nearby lines are taken into account in the fitting procedure, i.e. both the weak 2sQ(11,9) line and the strong aQ(11,6) one at $v_0 = 937.268$ cm^{-1}. This last line is not shown in Fig. 1 and not measured directly by the TDL spectrometer.

Reference:
- Baldacchini, J. Baltussen, A. Bellatreccia, P. Cardoni, F. D'Amato, R. Giovagnoli, I. Cenciarelli, **RT/INN/90/52**
- A. Ciucci and F. D'Amato; **RT/INN/94/01**
- Baldacchini, G. Buffa and O. Tarrini; Il Nuovo Cimento **13D**, 719 (1991)

[+] ENEA guest through a INFM fellowship.

FOREIGN-GAS BROADENING OF STARK-TUNED IR RESONANCES IN AMMONIA AT LOW PRESSURES

R. A. Gordon
Physics Department, The Technical University of Denmark,
Building 309, DK-2800 Lyngby, Denmark

ABSTRACT

Several different Stark-tuned rotational-vibrational resonance transitions in the ν2-band of ammonia have been studied as a function of buffer gas pressure using mixtures of helium, argon, freon 22, and sulphur hexafluoride in ammonia at room temperature. The primary purpose of the measurements has been to provide additional information about collisional line broadening in ammonia for several qualitatively different collision mechanisms at infrared frequencies at low gas pressures (10 mTorr - 2 Torr) where such measurements are scarce or non-existent.

The infrared measurements reported here were carried out under strong saturation conditions ($I \geq I_o$) using a 24-cm long Stark absorption cell with the entrance Gaussian laser beam width much larger (by a factor ~ 2-5) than the Stark electrode gap. Here I is the laser intensity and I_o is the relaxation-time dependent saturation parameter. Under such conditions, the Stark-tuned infrared absorption line will be a sensitive function of the collision relaxation times appearing in the plasma dispersion function line shape in contrast to more conventional Lambert-Beer absorption line measurements at low pressures [1,2]. The collision broadening experimentally observed using this method will be given by an average over the velocity distributions of both target and incident molecules in contrast to 2-photon Doppler-free techniques.

We report here the results of buffer gas collisional line broadening observed for the saQ(5,3) ν2 ammonia infrared resonance line ($\Delta m = 0$) which was Stark-tuned through the fixed frequency (P32) of a carbon dioxide laser field propagating through the Stark absorption cell. The laser intensity (0.5 - 6.0 W) and ammonia gas partial pressures (50 - 300 mTorr) were chosen to maximize the sensitivity of the ammonia resonance line shape and line width to changes in the buffer gas pressure. Ample time, varying from 10 - 30 minutes, was allowed for pressure stabilization within the Stark absorption cell after changing each buffer gas pressure. In addition, care was taken to ensure that the carbon dioxide laser was centered at the P(32) transition maximum.

The measured pressure dependence of the ammonia self-broadening was consistent with the well-established value (20 MHz/Torr ± 15%) [3] and more than an order of magnitude larger than that obtained with helium and argon as buffer gases in agreement with the work of others at higher pressures [4,5]. The heavier Freon 22 and sulphur hexafluoride gases both exhibited strong interactions with ammonia leading to an absorption resonance line broadening equal to (0,8 ± 0,15) and (0,7 ± 0,15), respectively, of that corresponding to ammonia self broadening itself. Measurements of two other ammonia v2 infrared resonance lines exhibited similar buffer gas broadening effects although the measurements are less quantitative. Infrared measurements of Freon 22 and sulphur hexafluoride buffer gas broadening in ammonia have not, to the authors knowledge, been reported before but are believed to be accurate within the experimental uncertainty quoted above. It is worth noting that it is not possible to distinguish between different ammonia-buffer gas resonance line broadening mechanisms for Freon 22 and sulphur hexafluoride solely on the basis of the velocity-averaged collision broadening measurements reported here over the gas-discharge limited range of gas pressures attainable using Stark cell tuning.

Finally, it is worth noting that the experimental method of measuring buffer gas line broadening employed here under strong absorption conditions can be expected to provide a useful supplement to standard laser-diode infrared absorption line measurements which are much less sensitive to pressure-dependent line broadering at pressures less than 1 - 2 Torr.

The author would also like to thank Jørgen Heintz for assistance in carrying out the measurements reported.

REFERENCES

1. R. A. Gordon, M. Owner-Petersen, and E. Dalsgaard, Spectral Line Shapes (de Gruyter & Co., Berlin 1985) Vol. 3, p. 741.

2. R. A. Gordon, 11th Int. Conf. on Infrared and Millimeter Waves (Conf. Digest, Pisa, Oct. 20-24, 1986) p. 623.

3. H. J. Clar, R. Schieder, G. Winnewisser, and Koichi M. T. Yamada, J. Mol. Structure 190, 447 (1988).

4. G. Baldacchini, A. Bizzarri, L. Nencinci, G. Buffa, and O. Tarrini, J. Quant. Spectrosc. Radiat Transfer 43, 371 (1990).

5. P. M. Beckwith, D. J. Danagher, and J. Reid, J. Mol. Spectrosc. 121, 209 (1987).

LINESHAPE ANALYSIS OF SPEED DEPENDENT COLLISIONAL WIDTH INHOMOGENEITIES IN CO BROADENED BY He, N$_2$, AND XE

P. Duggan, P.M. Sinclair, A.D. May, and J.R. Drummond
Department of Physics, University of Toronto, Toronto, Canada.

ABSTRACT

We present high resolution (10^{-4} cm^{-1}) and high signal to noise ratio (2000) infrared lineshapes of CO perturbed by Xe, N$_2$, and He in the intermediate regime from Doppler to collision broadening. It is found that the pressure broadened component of the lineshape cannot be fit by a Lorentzian lineshape. A speed dependent collisional width is required to fit the spectra, particularly when broadened by the larger Xe perturber. Based on the theory of Robert et al. [1992], a collision broadening lineshape model is derived, and we also include the effect of Dicke narrowing.

RESULTS

Speed dependent broadening has been seen in atomic emission spectra by Shannon et al.[1986], and in Raman CARS spectra by Farrow et al. [1989]. In this experiment, it is observed in simple absorption spectra at room temperature, and at pressures of less than 1 atmosphere. Essentially, speed dependence manifests itself when the collision rate of the active molecule, hence the broadening, depends upon its speed. It is particularly noticeable for a light active atom in a heavy perturber bath, such as CO in Xe. The collisional width associated with the different speeds is dependent upon the intermolecular potential, and is plotted in figure 1, assuming straight line collisions and r^{-n} type potentials. As extremes, we also plot a "hard core" r$^{-\infty}$ potential and a speed independent Lorentzian model. Speed distributions for CO in He and Xe are also shown. The total spectrum is obtained by integrating Lorentzians of the appropriate width over the speed distribution, the result being non-Lorentzian.

In order to compute a lineshape, folding in the Doppler effect and Dicke narrowing, we Fourier transform a speed dependent correlation function of the form

$$C_{sd}(t) = X_{sc}(t) \int_0^\infty M(v) \exp(-\Gamma(v)t) dv$$

where M(v) is the Maxwellian speed distribution, and $X_{sc}(t)$ is the soft collision translational (or Doppler) correlation function. The degree of Dicke narrowing of this component is determined by the value of the mass diffusion coefficient at the relevant pressure. The speed dependence lies in the choice of $\Gamma(v)$, which depends upon the potential. In figure 2 we show fits and residuals (expt - fit)×20 for CO in Xe (1:100 ratio) at 20 kPa and 298K for: 1) Voigt model; 2) Soft collision translation with Lorentzian broadening; 3) Soft collision translation and speed dependent broadening. Both the correct collision broadening and Doppler effect are required to fit the data. A more detailed description of these results is given in Duggan et al. [1994].

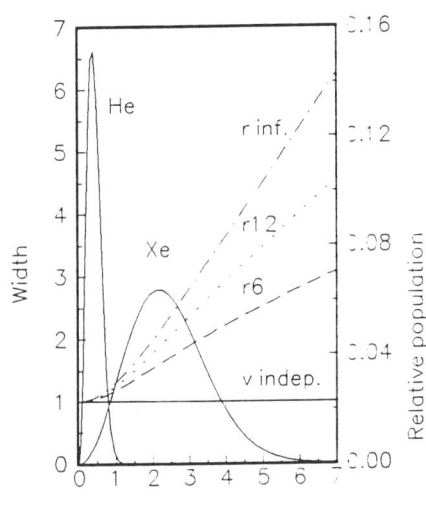

Figure 1

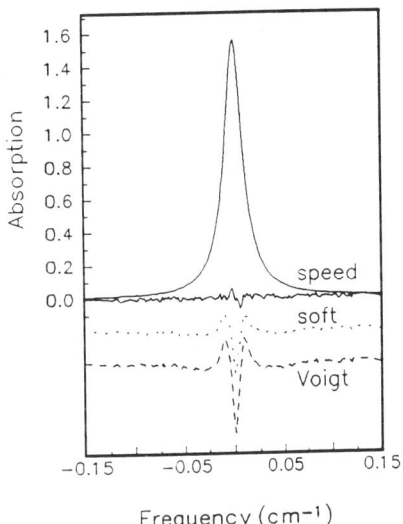

Figure 2

REFERENCES

[1] D. Robert, J.M. Thuet, J. Bonamy, and S. Temkin, Phys. Rev. A **47(2)**, R771 (1993).
[2] I. Shannon, M. Harris, D.R. McHugh, and E.L. Lewis, J. Phys. B **19**, 1409 (1986).
[3] R.L. Farrow, L.A. Rahn, G. O. Sitz, and G.J. Rosasco, Phys. Rev. Lett. **63(7)**, 746 (1989).
[4] P. Duggan, P.M. Sinclair, A.D. May, and J.R. Drummond, to be submitted to Phys. Rev. A.

BROADENING AND SHIFTING OF THE RAMAN Q BRANCH IN PURE D_2 AND D_2 - He MIXTURES

P. M. Sinclair, P. Duggan, M. P. Le Flohic, J. W. Forsman,
J. R. Drummond and A. D. May

Department of Physics, University of Toronto, Toronto, Canada, M5S 1A7

ABSTRACT

We report the broadening and shifting of the Raman Q branch lines in D_2 and D_2 - He mixtures, all at 305.2 K. From a linear fit of the widths and shifts as a function of density we have determined broadening and shifting coefficients with an uncertainty of ± 0.5%. None of the existing semi-classical or quantal coefficients agree with the measurements at this level of accuracy. As well we discuss nonlinear contributions to the widths and shifts. Measurements of the broadening of the depolarized part of the Q branch in D_2 are reported for the first time.

SUMMARY

We have measured the broadening and shifting coefficients in pure D_2 and in several mixtures of D_2 and He. From the intercepts of plots of the coefficients versus the molar fraction D_2 the infinite dilution values have been determined. These values, as well as the pure gas coefficients, can then be compared to theory. As an example, we show in Fig. 1 the shifting coefficients for D_2 infinitely diluted in He. Also shown are the theoretical values of Green et al.[1] An 11% difference is observed between the experiment and the theory. We also show the theoretical results shifted by a constant value such that there is agreement for our best point, J = 1. The theory comes close to passing through the other three points. We feel that the theory may be in error due to the difficulty of calculated the differential phase shift between the vibrational states, caused by the isotropic intermolecular interaction.

In the above discussion we have assumed that the individual lines are Lorentzian in shape and that they broaden and shift linearly with density. Because very high signal to noise ratios were obtained on some of our spectral profiles, several deviations from this simple model were evident. Some of the effects we have investigated are: (i) the depolarized component of the line shape, (ii) line mixing, (iii) the finite duration of collisions, and finally (iv) nonlinear broadening and shifting. This work has resulted in two benefits. Firstly, we have obtained new results such as the first measurements of line mixing[2] and depolarized broadening[3] in D_2. Secondly, we have been able to remove systematic errors from the linear broadening and shifting coefficients obtained using the simple Lorentzian model discussed above.

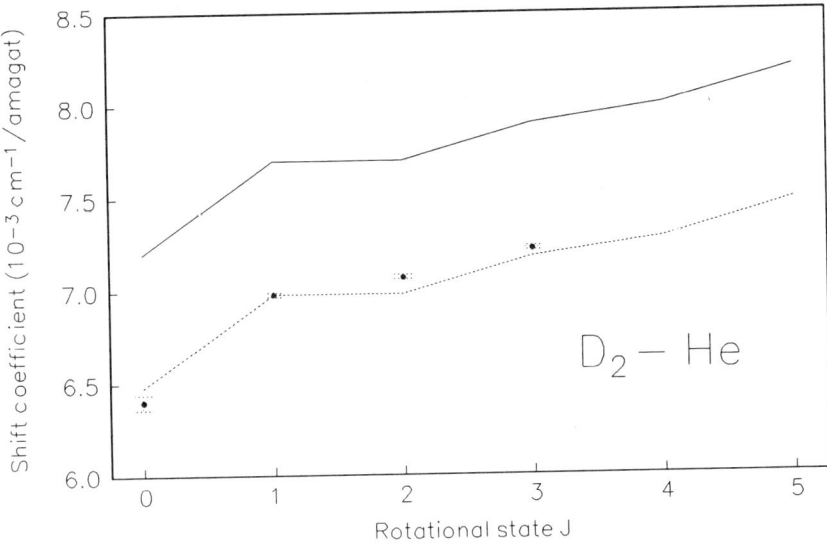

Fig. 1. Shifting coefficients of the Q branch lines of D_2 infinitely diluted in He as a function of the rotational quantum number J. The solid line represents calculated values taken from reference [1]. The dotted line is the same line but shifted to pass through our most precise value, that for J = 1.

REFERENCES

1. S. Green, R. Blackmore, and L. Monchick, J. Chem. Phys. **91**, 52(1989).
2. P. M. Sinclair, J. W. Forsman, J. R. Drummond, and A. D. May, Phys. Rev. A. **48**, 3030 (1993).
3. P. M. Sinclair, P. Duggan, M. P. Le Flohic, J. W. Forsman, J. R. Drummond and A. D. May, submitted to Can. J. Phys. (1994).

BROADENING AND SHIFT OF THE LINES OF MOLECULAR IONS BY COLLISIONS WITH NEUTRAL PERTURBERS

O. Tarrini and G. Buffa
Dipartimento di Fisica dell'Università, Piazza Torricelli 2, I-56126 Pisa, Italy

G. Cazzoli and L. Dore
Dipartimento di Chimica dell'Università, Via Selmi 2, I-40126 Bologna, Italy

The collisional broadening of the rotational spectral lines of neutral molecules has recently received much attention both experimentally and theoretically. Also the spectral line shapes in highly ionized plasma were extensively studied. By contrast, only few studies can be found dealing with the broadening of the spectra of molecular ions due to collisions with neutral perturbers,[1] in spite of the importance of this topic, particularly in the interpretation of radio astronomical data.

We studied HCO^+, produced in a magnetically enhanced negative glow, perturbed by argon at 77 K. The collisional parameters were determined for three lines:[2] $J = 1 \leftarrow 0$, $2 \leftarrow 1$, and $4 \leftarrow 3$. An additional result obtained with a FIR laser sideband spectrometer is also given for the broadening of the $8 \leftarrow 7$ transition, allowing a better investigation of the trend with J.

Our theoretical treatment shows that the interaction between the ionic monopole and the multipole moments, permanent or induced, of the neutral perturbers plays no role in the collisional relaxation. The effect of the ionic monopole is restricted to affecting the collisional trajectory and to inducing a polarization in the perturber. In the case of HCO^+ perturbed by argon, by far the largest relaxation effect comes from the interaction between the dipole of the ion and the dipole induced in argon by the ionic charge.

We extend to the case of ions the methods already used for neutral molecules,[3] and perform calculations of the pressure broadening and shift.

Results of measurements and calculations are reported in Table I. Numbers in parenthesis are 1-σ statistical errors. For the shift of two lines only the sign could be determined. As far as the $8 \leftarrow 7$ transition is concerned, the unsatisfactory long time stability of the free running sideband FIR laser spectrometer prevents any pressure shift determination.

Table I. Measured and calculated broadening (HWHM) and shift parameters of HCO$^+$ rotational lines $J+1 \leftarrow J$ by argon pressure at 77 K. Units are mHz/Torr.

line J	frequency (MHz)	measured broadening	calculated broadening	measured shift	calculated shift
0	89 188.53	21.49 (21)	21.02	+2.27 (6)	+2.91
1	178 375.06	17.51 (19)	17.22	+	+0.85
2	267 557.619		13.17		+0.01
3	356 734.25	14.56 (7)	10.25	−	−0.28
7	713 340.5	10.42 (40)	4.49		−0.65

On the whole, in view of the experimental errors and of the approximate character of the theory, the agreement between measurements and calculations is satisfactory. The largest difference is found for the broadening of the high J transition. This may be ascribed to the break down of two of the approximation used in the calculations. The semiclassical approximation becomes less reliable when a larger energy amount is transferred in the collisions from internal degrees to translation. Moreover, the assumption of straight lines collisional trajectories becomes less realistic when the collisional broadening cross section is reduced. In order to clarify this point, calculations with a more realistic translational motion are in progress.

REFERENCES

1. R. T. G. Anderson, C. S. Gudeman, T. A. Dixon, and R. C. Woods, J. Chem. Phys. **72**, 1332 (1980); C. S. Gudeman, PhD. Thesis, University of Wisconsin, Madison (1982); C. Woods, in *Molecular Ions: Spectroscopy, Structure, and Chemistry*, T. A. Miller and V. E. Bondybey (Editors), North Holland, Amsterdam (1983).

2. G. Buffa, O. Tarrini, G. Cazzoli, and L. Dore, Phys. Rev. A in press.

3. P. W. Anderson, Phys. Rev. **76**, 647 (1949); C. T. Tsao and I. Curnutte, J. Quant. Spectrosc. Radiat. Transfer **2**, 41 (1962).

Non-adiabatic effects in the broadening and shift of the K $7s\ ^2S_{1/2} - 4p\ ^2P_{3/2}$ transition perturbed by Ar

Warren Kreye
UCS/Research and Instructional Computer Center
Wright State University, Dayton, Ohio 45435

John Kielkopf
Department of Physics, University of Louisville
Louisville, KY 40292

Fully quantum mechanical spectral line shifts and widths were computed for the K $7s \to 4p$ transition perturbed by Ar. Non-adiabatic theory was used which included j-degeneracy and inelastic collisions corresponding to the $j = 3/2 \to j = 1/2$ transition in the $4p$ state. Recently Kreye and Kielkopf[1] reported adiabatic quantal calculations of the widths for this transition.

The bases for the theory are Baranger's papers[2,3] for a system of two electronic states. He derived an expression (see Eq. (77c) in Ref. 3) for the half width at half maximum intensity, w/n, which is both illustrative and informative:

$$w/n = \frac{v}{2c} \left[\sigma_{i,in} + \sigma_{f,in} + \int d\Omega \mid f_i(\Omega) - f_f(\Omega) \mid^2 \right] \quad (1)$$

where σ is the scattering cross section, i and f refer to the initial and final electronic states, *in* refers to the inelastic component which can occur in the $4p$ state, and $f(\Omega)$ is the scattering amplitude. The integral term contains all the elastic contributions to the width, and the cross product $f_i^*(\Omega) \times f_f(\Omega)$ which connects the initial and final electronic states. Baranger's result (Eq. (77a) in Ref. 3) for the shift has a dependence only on the scattering amplitudes, indicating that only elastic collisions contribute to the shift.

Expressions for the inelastic cross sections, σ_{in}, are given in terms of the transition matrix, **T**, in Mies' paper on fine structure transitions.[4] The scattering amplitudes, $f(\Omega)$, were evaluated from **T**, expressed as a reactance matrix **K** using the methods of Wilson and Shimoni[5]. The K-Ar interactions were computed with a two-parameter pseudopotential model[6,7,1] in which the electron density, ρ, was considered constant within a radius, r_0, about the Ar atom. These parameters were adjusted to make the theoretical shift and width agree with the experimental results over a temperature range from 443 to 760 K.

Approximately equal results were computed for initial states of either parity. For an initial $7s^-$ state at 400 K, the contributions to w/n in units of 10^{-20} cm^2 were 3.86 for the $7s$ state, 1.089 for the total of inelastic and elastic collisions in the $4p$ state, 0.038 for the inelastic collisions in the $4p$ state, and -0.291 for the $7s4p$ cross product. Figure 1 presents the computed values of w/n as a function of temperature T for two trial potentials which differ in their values of r_0 for the Ar radius. The experimental values were obtained by Kreye[8] and Kielkopf and Knollenberg[9] using Fabry-Perot interferometry. Similar results for the shift show equally good agreement between the computed and experimental values.

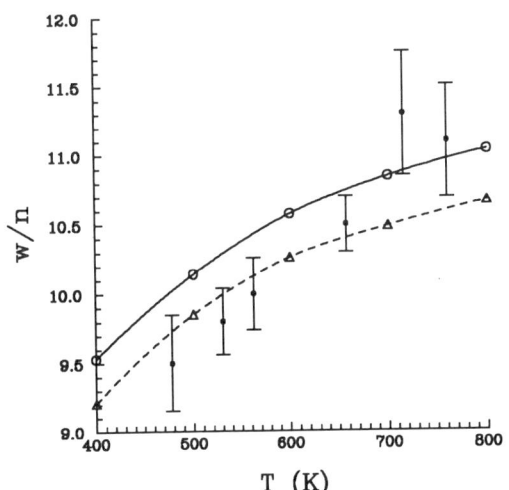

Figure 1
Theoretical collision broadening, w/n, in units of (10^{-20}) cm^{-2} versus temperature T for two potentials both with $\rho = 25$ e$^-$ Å^{-3}:
V36B (- -△- -) with $R_0 = 0.6085$ Å;
V35B (-○-) with $R_0 = 0.6080$ Å.
Experimental values are shown with errorbars.

REFERENCES

1. W.C. Kreye and J.F. Kielkopf, J. Phys. B **24**, 65 (1991).
2. M. Baranger, Phys. Rev. **111**, 481 (1958).
3. M. Baranger, Phys. Rev. **112**, 855 (1958).
4. F.H. Mies, Phys. Rev. **A7**, 942 (1973).
5. A.D. Wilson and Y. Shimoni, J. Phys. B **8**, 2415 (1975).
6. W.E. Baylis, J. Chem. Phys. **51**, 2665 (1969).
7. J. Pascale and J. Vandeplanque, J. Chem. Phys. **60** 2278, (1974).
8. W.C. Kreye, J. Phys. B **15**, 371 (1982).
9. J.F. Kielkopf and R.B. Knollenberg, J. Phys. B **14**, 1263 (1981).

Intensities and N_2-broadened Half-widths in the ν_3-Fundamental Band of CO_2 at Atmospheric Temperatures.

Zhenhua Li[†] and Prasad Varanasi[†]
Institute for Terrestrial and Planetary Atmospheres
The University at Stony Brook
Stony Brook, NY 11794-5000, U.S.A.

and

Mark Weber[‡]
Laboratory for Extraterrestrial Physics
Code 690, NASA-Goddard Space Flight Center
Greenbelt, MD 20771, U.S.A.

We have measured the absolute intensities and N_2-broadened half-widths of lines between $P(44)$ and $R(44)$ in the ν_3-fundamental band of $^{12}C^{16}O_2$ at 216, 250, and 300 K with a BRUKER IFS-120 HR Fourier transform spectrometer. Spectral resolution as high as 0.002 cm^{-1} has been employed. The line parameters of CO_2 were obtained using a non-linear least squares fitting program for the lineshapes.[1] The vibrational intensity[2] derived from all of the measured line intensities at 216, 250, and 300 K is 9.412 ($\pm$ 0.001) $\times$ 10^{-17} cm molecule^{-1} at 300 K. This value is 3 % higher than the value reported by Johns[2] and used in the HITRAN compilation.[3] The discrepancy, however small, might be due to systematic errors resulting from uncertainties in the ppm mixing ratio of the sample used in our laboratory and the small pathlengths elsewhere. While this apparent discrepancy is the subject of further study in our laboratory, it is gratifying to note that there is internal consistency in the measurements of the intensities as far as their dependence with temperature and variation with the rotational quantum number are concerned.

[†] Supported by the Atmospheric Radiation Measurements Program of the U.S. Department of Energy and the Upper Atmosphere Research Program of the National Aeronautics and Space Administration.

[‡] NAS/NRS Resident Research Associate.

The N_2-broadened half-widths $\gamma_m^o(T)$ (cm^{-1} atm^{-1}) of the lines of CO_2 measured at 216, 250, and 300 K have been fit to the following polynomial:

$$\gamma_m^o(T) = (296/T)^n \, \alpha \, (1 + a_1 m + a_2 m^2 + a_3 m^3),$$

where $\alpha = 0.0934 \pm 0.0003$, $n = 0.747 \pm 0.003$, $a_1 = -0.0179 \pm 0.0005$, $a_2 = (4.52 \pm 0.29) \times 10^{-4}$, and $a_3 = -(4.21 \pm 0.46) \times 10^{-6}$. m is the usual line index which is the rotational quantum number J for lines in the P-branch and $J+1$ for lines in the R-branch. The linewidths are in good agreement with the values measured by Johns[2] and Devi et al[4] at room temperature. The value of n obtained by us is consistent with the values reported by Cousin et al[5] and Rosenmann et al.[6]

REFERENCES

1. M. Weber, "*Non-linear Least Squares Fit of Lineshapes in High Resolution Fourier Transform Spectra: Manual for FTFIT.FOR and Accompanying Programs,*" (1994; unpublished).

2. J.W.C. Johns, *J. Mol. Spectrosc.* **125**, 442 (1987); **134**, 433 (1989).

3. L. S. Rothman, R. R. Gamache, R. H. Tipping, C. P. Rinsland, M. A. H. Smith, D. C. Benner, V. M. Devi, J. M. Flaud, C. Camy-Peyret, A. Perrin, A. Goldman, S. T. Massie, L. R. Brown, and R. A. Toth, *JQSRT* **48**, 469 (1992).

4. V. M. Devi, D. C. Benner, C. P. Rinsland, and M. A. H. Smith, *JQSRT* **48**, 581 (1992).

5. C. Cousin, R. Le Doucen, C. Boulet, and A. Henry, *Appl. Opt.* **24**, 3899 (1985).

6. L. Rosenmann, M. Y. Perrin, and J. Taine, *J. Chem. Phys.* **88**, 2995 (1988).

LINE SHAPES AND COLLISIONAL EFFECTS IN RESONANT DEGENERATE FOUR-WAVE MIXING

Larry A. Rahn
Sandia National Laboratories, Livermore, CA 94551-0969

Skip Williams
Los Alamos National Laboratory, Los Alamos, NM 87545

Richard N. Zare
Stanford University, Stanford, CA 94305

Resonant four-wave mixing techniques are finding increasing application as sensitive diagnostic probes of composition and temperature in highly collisional, luminous environments such as those found in combustion[1] and plasmas[2]. The interpretation of resonant four-wave mixing signals for diagnostic applications, however, requires a detailed understanding of the effects of different mechanisms, laser polarizations, and collisional processes on the signal strength. To this end we have developed perturbation theory expressions[3] that describe the resonant four-wave mixing signal polarization, collisional, and velocity effects in the weak-field limit. We have found that the line shapes, in particular, can be interpreted with this treatment to provide information about electronic dephasing rates, unequal relaxation of the molecular population, orientation, and alignment, and the effects of competing mechanisms, such as thermal gratings.[4]

In this paper we report high resolution measurements of degenerate four-wave mixing (DFWM) and nearly degenerate four-wave mixing (NDFWM) spectra in atmospheric pressure H_2-O_2 flat flames with and without dilution. These spectra are interpreted using least-squares fits to perturbation theory line-shape expressions. We find that, in the He diluted flame, the NDFWM line width depends on the laser polarizations and suggests the presence of unequal relaxation of the population, orientation and alignment for low values of J. These effects are not observable on the DFWM line shape, but they do affect the DFWM intensity and are an important consideration for the extraction of relative internal-state distributions. We also show, for undiluted flames, the effects of quenching-induced thermal gratings on the NDFWM lineshape. This mechanism is found to be most important at higher densities and is elucidated by time-delay scans of the probe laser.[5]

This research was funded by the U. S. Department of Energy, Basic Energy Sciences, Division of Chemical Sciences and the Air Force Office of Scientific Research.

REFERENCES

1. R. L. Farrow and D. J. Rakestraw, Science, **257**, p. 1894 (1992).
2. S. Williams, D. S. Green, S. Sethuraman, and R. N. Zare, J. Amer. Chem. Soc., **114**, p. 9122 (1992).
3. S. Williams, R. N. Zare, and L. A. Rahn, J. Chem. Phys., **in press** (1994).
4. L. A. Rahn and M. S. Brown, Opt. Lett., **in press** (1994).
5. S. Williams, L. A. Rahn, P. H. Paul, J. W. Forsman, and R. N. Zare, Opt. Lett., **submitted** (1994).

EXCIMERS

DIFFUSE BANDS IN INTERMETALLIC EXCIMERS

Slobodan Milošević
Institute of Physics, University of Zagreb, HR-41000 Zagreb, Croatia

ABSTRACT

The recent progress in experimental and theoretical study of the alkali- group IIB intermetallic excimers will be summarized. The main concern here is an understanding of interatomic interaction and collision dynamic in intermetallic systems. In the experiments, intermetallic vapors or amalgams are illuminated by a pulsed laser radiation. The reactive and non reactive energy transfer processes occur in the vapor or plasma, resulting with emission of the excited atoms and molecules. The analysis of the emission show a complexity of vapor and plasma conditions. Appropriate vapor mixture preparation, exciting laser wavelength selection and temporal spectral analysis allow discrimination among different collision processes. The photochemical reaction of excited alkali dimer with group IIB atoms, photoassociation of alkali and IIB atoms, quasi resonant and resonant alkali atom excitation with subsequent energy transfer to IIB atoms, are illustrated for Li-Zn and Li-Cd systems. Accompanying theoretical work provides interatomic potential energy curves necessary for interpretation of the observed emission. The emphasis is made on proper treatment of spin-orbit interaction which increases from Zn towards Hg. Recent calculation of potential energy surfaces for the Li_2Zn system shows that this exciplex can be formed in the excited state with considerable binding energy.

INTRODUCTION

An alkali- group IIB metal systems have drawn our attention due to many open questions on interatomic interactions and collision dynamics[1]. Intermetallic excimers are usually prepared in collisions of excited atoms or molecules. This determine the main characteristic of the present work which connects molecular spectroscopy and collision energy transfer studies. Both analysis of the molecular band emission or determination of the cross sections for energy transfer provide information about energy potentials of interaction.

The metal vapor mixture contains atoms and dimers of both atom groups. The alkali-group IIB molecules are less abudant due to lower binding energy and usually direct laser excitation is precluded with absorption of more abudant alkali dimers in the same spectral region. However upon laser excitation various processes occur. Among them photochemical reactions are used to prepare excited intermetallic molecules $(AB)^*$

$$A_2^* + B \rightarrow (AB)^* + A \qquad (1)$$
$$A_2 + B^* \rightarrow (AB)^* + A \qquad (2)$$

© 1995 American Institute of Physics

where A and A_2 are an alkali atom and dimer, respectively, and B stands for atoms of group IIB. A_2^* are prepared in various states by single or two photon excitation, and B^* is usually in the $^3P_1$ metastable state.

The $(AB)^*$ molecules are also prepared in photoassociation process

$$A + B + h\nu \to (AB)^* \qquad (3)$$

The energy transfer collisions were observed, such as

$$B^* + A \to A^* + B \qquad (4)$$
$$A^* + B \to A + B^* \qquad (5)$$
$$A^* + B + h\nu \to A + B^* \qquad (6)$$

where $h\nu$ is a resonance photon (e.g. 2P-2S transition at 670.8 nm for Li). There are several other contributions to the overall emission spectrum which arise from other competing collision energy transfer processes. Those which are changing intially prepared A_2^* state

$$A_2(\Lambda)^* + A \to A_2(\Lambda')^* + A \qquad (7)$$

or energy pooling

$$B^* + B^* \to B^{**} + B \qquad (8)$$
$$A^* + A^* \to A^{**} + A \qquad (9)$$

which involves radiatively trapped resonant excited alkali atoms and metastable B^* atoms, are among the most important. With pulsed laser excitation multiphoton ionization of atoms or dimers can also take place.

We found that termalization collisions affect spectra substantially at higher total vapour pressures in the heat-pipe oven

$$AB(\Lambda, v',J') + A \text{ or } B \to AB(\Lambda, v'\pm\Delta v, J'\pm\Delta J) + A \text{ or } B \qquad (10)$$

For reactions (1) and (2) the entrance and exit channels are monitored by observing the fluorescence emission of the A^* and $(AB)^*$ molecules. By the nature of $(AB)^*$ preparation, which gives broad rovibrational distribution, the emission is in the form of structured continua (diffuse bands or satellites in the far wings of resonance lines). The main spectral peaks of this continua are related to the extrema in relevant difference potentials. For complete understanding of spectral shapes, help from potential energy *ab initio* calculations is necessary. Recently, progress has been made on almost all systems considered, by quantum-chemistry groups in Bonn, Göttingen, Leiden and Pisa. With these results and detailed spectral simulations nascent rovibrational distribution from reactive collision can be extracted which in turn provide information about involved A_2+B potential energy surfaces.

With respect to the energy transfer (4)-(6) there are many cases with small energy defect between highly excited levels of alkalis and excited levels of group IIB atoms. Especially interesting are those collisions which involve metastable or radiatively trapped states where efficient energy transfer could be obtained due to the prolonged availability of excited species.

Processes (7)-(10) which complicate spectra from intermetallic mixtures, can not be avoided, but a certain selection could be achieved through temporal spectral analysis.

EXPERIMENTAL METHOD

One of the main questions in the present studies is how to prepare the vapor mixture which provides enough collisions of desired kind. A common approach is to create a hot and dense metal mixture by filling a standard heat-pipe oven (HPO) with a certain mass ratio of alkali and IIB metals. Generally, behaviour of metal mixtures (total pressure above the liquid or a melting point) is quite different from properties of isolated metal constituents. One has to rely on phase diagrams of metal alloys[2] and Raoult law in order to predict behaviour of amalgam and the ratio of partial vapor pressures in the mixture. Difficulties could arise since for certain mass ratios different intermetallic compounds are created in the liquid.

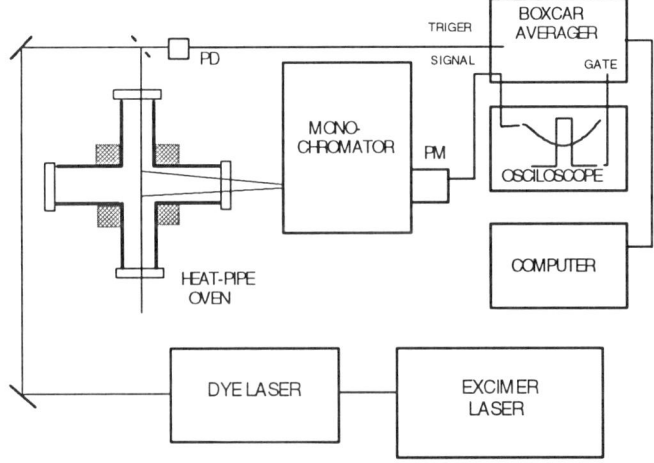

Fig. 1. Experimental setup for the preparation of alkali-IIB dimers in a heat pipe oven.

In addition, if vapor pressures differ very much a stream of vapor of the element with higher vapor pressure is introduced from one arm of the HPO into the vapor of the element with lower vapor pressure in the central part of the HPO[3,4]. In order to determine the cross sections for the collisions the knowledge of atom and molecule densities in ground and excited states is necessary. For alkali atoms and dimers in the ground state this could be done by measurements of absorption coefficient of the vapor mixture[5] whereas for excited states, in principle, an additional laser should be used. The use of HPO with gravitational heat-pipe as a heater is of great advantage, providing well defined column length, a homogeneous temperature profile and finaly direct measurement of the total pressure and temperature of the metal mixture.

A typical experimental set-up is shown in Fig. 1. The excitation is usually made by a pulsed laser with about 16 ns time duration and repetition rate of 10 Hz. Lasers excite different electronic states of alkali dimer or atom by a single or two photon absorption. Recently we used methods like OODR or PFOODR [6] to prepare a wide span of highly excited lithium dimer states. During their lifetime excited species suffer many collisions with other atoms. Additional selection of the desired process is achieved by temporal analysis of the observed spectra. In some cases correction of diffuse band spectral shapes due to the absorption of alkali dimer should be performed (for example potassium dimer B-X absorption overlap with K- group IIB emission).

ALKALI-GROUP IIB MOLECULES

In Fig. 2. we show a simplified potential energy diagram for Li,Na,K-group IIB atom combinations. In the left panel a typical example of molecular potentials is given. The AB molecules are much less bound compared to alkali dimers. However the excited states are substantialy bound. The well depths for the first four states are listed in Table 1. A lower lying excited states of AB molecules dissociate to the excited alkali atom and group IIB atom in the ground state. The exception is LiCd excimer where the $Cd(^3P_{0,1})$ levels lye bellow the Li(3P) level.

The AB molecules are prepared either in the $2^2\Pi$ or in $2^2\Sigma^+$ and $1^2\Pi$ states, and the emission to the $1^2\Sigma^+$ ground state is observed as a blue-green and red bands, respectively. The later are associated with quasistatic red wings of alkali resonance lines broadened with IIB atoms. The positions of the main emission peaks are shown in Table 2 and Table 3. For the blue-green emission two main peaks are observed in LiCd, LiHg and NaCd, NaHg cases. This is due to the spin-orbit splitting of the $2^2\Pi$ state which increases from Zn towards Hg. *Ab initio* calculations show that the wavefunction of the $2^2\Pi$ state is influenced by the (nsnp) ^3P configuration of the IIB atom[8]. Consequently molecular states are described in different Hund's coupling cases in combinations with Zn towards Hg (eg. $2^2\Pi \rightarrow 2^2\Pi_{1/2}$, $2^2\Pi_{3/2} \rightarrow 1/2, 3/2$). For the Hund's coupling case c (LiCd, LiHg, NaCd, NaHg), only Ω is a good quantum number and an interesting avoided crossings appear (e.g. between $3^2\Sigma^+(1/2)$ and $2^2\Pi(1/2)$). In the LiZn, NaZn and K-IIB cases the splitting is not sufficient to be observed in the present type of experiments.

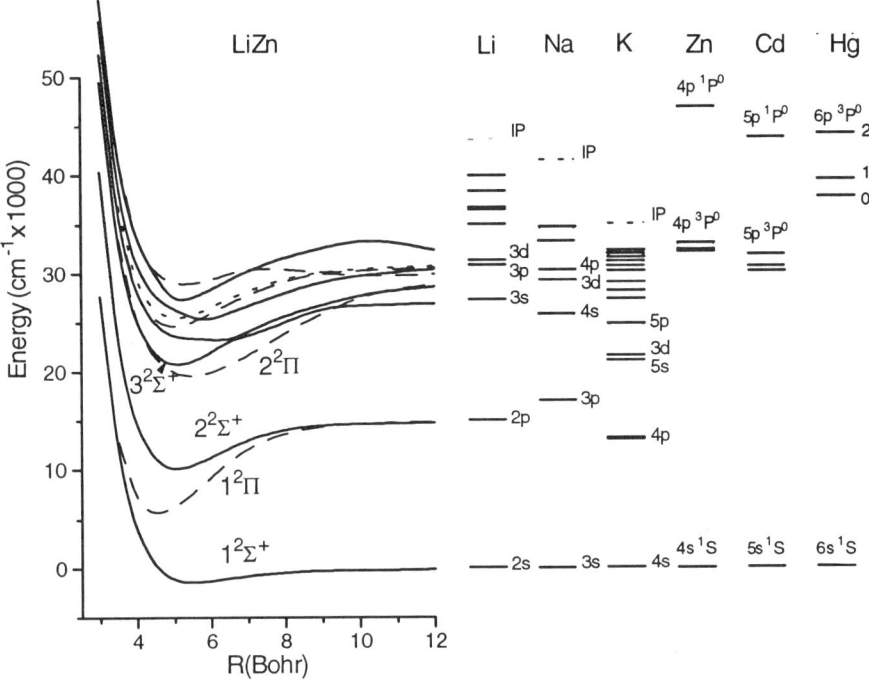

Fig. 2. Energy term diagram of alkali- group IIB atom combinations. In the left panel the non-relativistic *ab initio* potentials of the LiZn are shown[7].

DIFFUSE BANDS

Fig. 3. gives an example of the observed blue-green emission for LiZn, LiCd, LiHg. In all cases shown, AB($2^2\Pi$) molecules are created in the photochemical reaction of type (1). The $Li_2(2^1\Pi_u)$ state is prepared in a high rovibrational level by an excimer laser at 308 nm. The main peak position is indicated with black dots. Simple analysis yields an estimation of the spin-orbit splitting between $2^2\Pi_{1/2}$ and $2^2\Pi_{3/2}$ states to be 350 cm^{-1} and 1160 cm^{-1} for LiCd and LiHg, respectively. Analogus splitting in NaCd and NaHg has been found to be 230 cm^{-1} and 945 cm^{-1}, respectively. A detailed spectral simulation of the observed emission has been performed in LiZn and LiCd cases [17,18]. Generally one has to know the Franck-Condon factors and the rovibrational level distribution to be able to calculate the emission intensities. The FC's can be determined satisfactorily using *ab initio* potential energy curves and corresponding electronic transition dipole moment functions. The population distribution however depends on the collision process by which excimer is created.

		Zn		Cd		Hg	
		D_e	R_e	D_e	R_e	D_e	R_e
Li	$1^2\Sigma^+$	1314	5.5 [7]	1426	5.7 [8]	653	5.7 [4]
						851	5.67 [10]
						856	5.71 [11]
	$2^2\Sigma^+$	4742	5.1 [7]	4944	5.5 [8]		
	$1^2\Pi$	9176	4.6 [7]	9048	5.0 [8]		
	$2^2\Pi$	9322	5.5 [7]	8500	6.0 [8]		
Na	$1^2\Sigma^+$	830	6.5 [3]	481	6.3 [9]	593	6.5 [12]
						1208	6.4 [13]
						443	8.9 [14]
	$2^2\Sigma^+$	3439	5.5 [3]	1488	6.37 [9]		
	$1^2\Pi$	7712	5.05 [3]	6375	5.5 [9]		
				7170	5.5 [9]		
	$2^2\Pi$	8257	5.76 [3]	7950	6.67 [9]		
				7589	6.58 [9]	7830	6.0 [12]
K	$1^2\Sigma^+$	232	8.2 [15]	300	8.0 [15]	520	7.4 [15]
						423	9.3 [16]
	$2^2\Sigma^+$	1000	6.8 [15]	1370	7.0 [15]	1460	7.0 [15]
	$1^2\Pi$	5200	4.75 [15]	5400	5.95 [15]	5700	5.9 [15]
	$2^2\Pi$	5575	6.5 [15]	5791	7.0 [15]	5692	6.5 [15]

Table 1: The AB molecular parameters for the 1, 2 $^2\Sigma^+$ and 1, $2^2\Pi$ states. D_e is in cm^{-1} and R_e in Bohr.

		Zn	Cd	Hg
Li	exp.	477 [7]	482, 490 [8]	443, 467 [4]
	cal.	482 [17]	484, 493 [17]	447, 462 [18]
Na	exp.	478 [3]	479, 484 [21]	450, 470 [13]
	cal.	475 [5]	-	-
K	exp.	-	634 [5]	640 [5]
	cal.	636 [5]	635 [5]	625 [5]

Table 2. Wavelength of the main peaks (in nm) of the blue-green emission in Li,Na,K-dimers which is due to the $2^2\Pi \rightarrow 1^2\Sigma^+$ transitions.

Fig. 4. shows the effective potentials for J'=71/2 (which is the most probable at T= 1000 K) together with several Mulliken difference potentials (MDP's). The MDP's show how lower vibrational levels are sampled for the transition from a given upper vibrational level in terms of the classical Franck-Condon principle. As an example, the transition from level v' = 7 is indicated with

	Zn	Cd	Hg
Li exp.	-	-	-
cal.	940 [7]	904 [8]	800 [18]
Na exp.	-	691-727 [20]	643-669 [13]
cal.	720 [3]	775 [5]	690 [5]
K exp.	-	847-876 [5]	810-829 [21]
cal.	812-836 [5]	827-865 [5]	817-834 [5]

Table 3: Wavelength of the main peaks (in nm) of the red emission in Li,Na,K-dimers which is due to the $2^2\Sigma^+ \to 1^2\Sigma^+$ transitions.

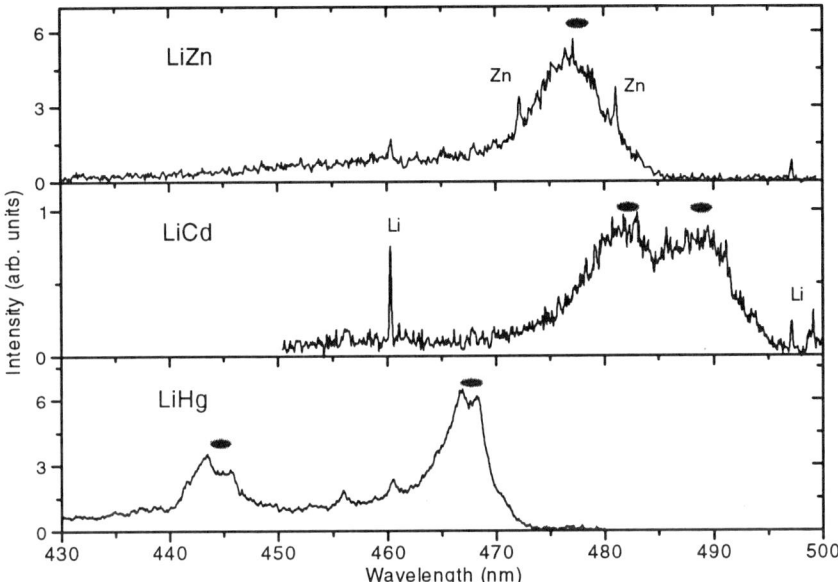

Fig. 3. Time-integrated Li-IIB blue-green emission. Total vapor pressures, temperatures and laser wavelength are: LiZn - 52 Torr, 1020 K, 308 nm, LiCd - 20 Torr, 1000 K, 308 nm, LiHg - 40 Torr, 1120 K, 351.1 nm (Ar$^+$ laser [19]).

vertical lines. For v' ≤ 4 vibrational levels of the LiZn state only transitions to the bound levels of the ground state are possible. The maxima of the MDP's, connected with a large increase of intensity, fall into the bound part of the ground state. For higher vibrational levels of the upper state these maxima fall into the free part of the ground state, consequently increasing the bound-free emission whereas only transitions to higher bound levels are sampled which are of negligible intensity.

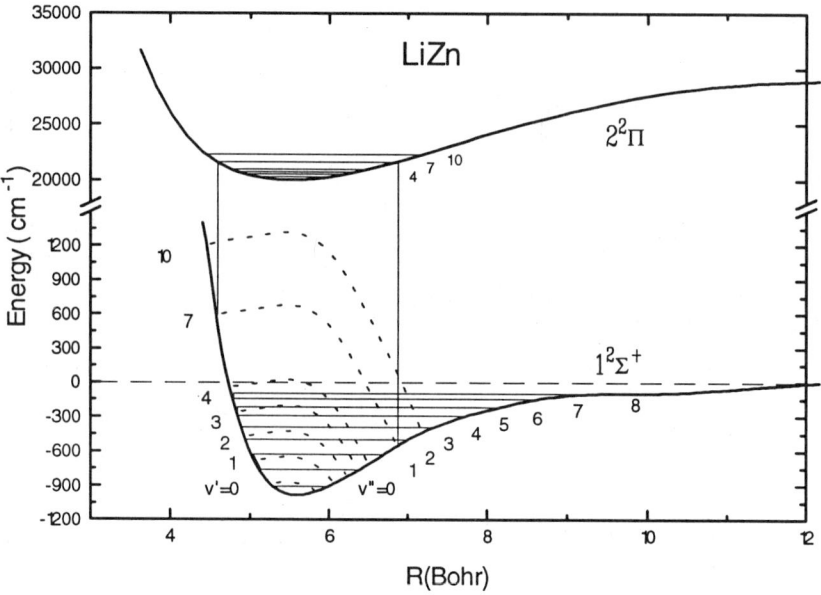

Fig. 4. The effective potentials of the $2^2\Pi \to 1^2\Sigma^+$ transition for J=71/2 and the Mulliken difference potentials. Note two different scales on vertical axis.

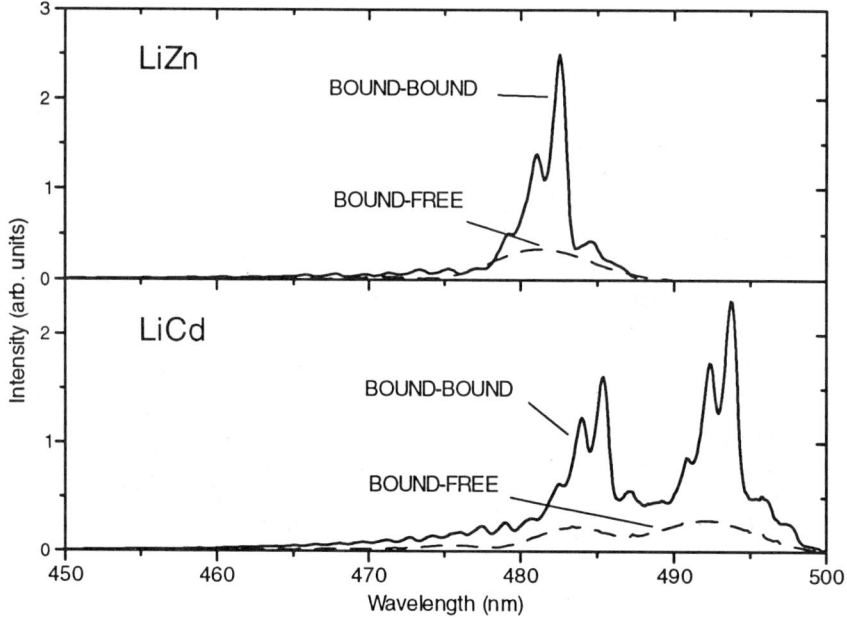

Fig. 5. Spectral simulation of the LiZn and LiCd blue-green bands.

Fig. 5. gives results of quantum mechanical simulation, separately for rovibrationaly averaged bound-bound and bound-free contributions. The Boltzmann distribution of rovibrational leves is assumed with effective temperature T= 1000 K. This results show that interplay between dominance of either bound-bound or bound-free transitions is possible, depending on rovibrational distribution of the upper state. Note that the width at half maximum of bound-free peaks is larger by a factor of about two than of the bound-bound peak.

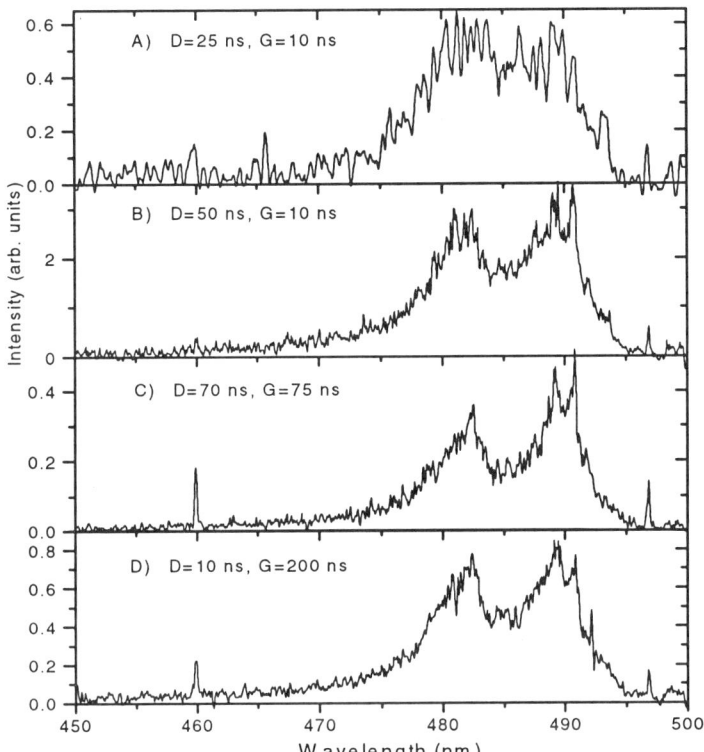

Fig. 6. Dependence of the LiCd emission on time delay and time gate of the boxcar averager. Temperature of vapor 1030 K and buffer gas pressure of argon 100 Torr.

For the case where $2^2\Pi$ states are created in the photochemical reaction we found that nascent rovibrational distribution can be substantially different from Boltzmann distribution. However collisions of kind (10) tend to smear out nascent distribution[19]. If recording of spectra is performed over large time after laser pulse, various distributions contribute to the overall spectra (from nascent to the almost thermalized). Therefore comparison of the experimental shapes with those from quantum simulations is not simple. Fig. 6 shows how LiCd blue-green emission evolves for different time delays from the beginning of laser pulse. Laser excitation wavelength is 308 nm and LiCd is created through the process (1). For larger time delays and integration times (Fig.6 c), additional sharp features appear, which are due to the bound-bound transitions. This is due to the process (10) and the thermalization of nascent distribution. At the beginning of laser pulse (Fig.6 a) signal to noise ratio is strongly affected by jitter of trigger pulse. However it can be seen that spectral shape is broad and mainly of bound-free character. The same effect have been observed in time-integrated measurements depending on total pressure[7,8,18,19]. Besides molecular emission, also atomic emission is observed which is due to other processes like collision energy transfer, ionization or energy pooling.

At high total pressures usually used to create excited excimers the effective lifetimes of different atomic and molecular levels excited through processes (1)-(6) are shorter than their spontaneous lifetimes[22]. However we found that for LiZn and LiCd mixtures (50% of atomic density) at total pressures bellow 5 Torr and temperature about 1000 K, the effective lifetimes for blue-green emission are between 20-30 ns, which is close to spontaneous lifetimes expected from calculations[15]. Under this conditions spectral shapes show dominance of bound-free emission, which proves that nascent distribution from photochemical reaction is preserved and that collisions of kind (10) are negligible.

FORMATION PROCESSES OF ALKALI-IIB MOLECULES

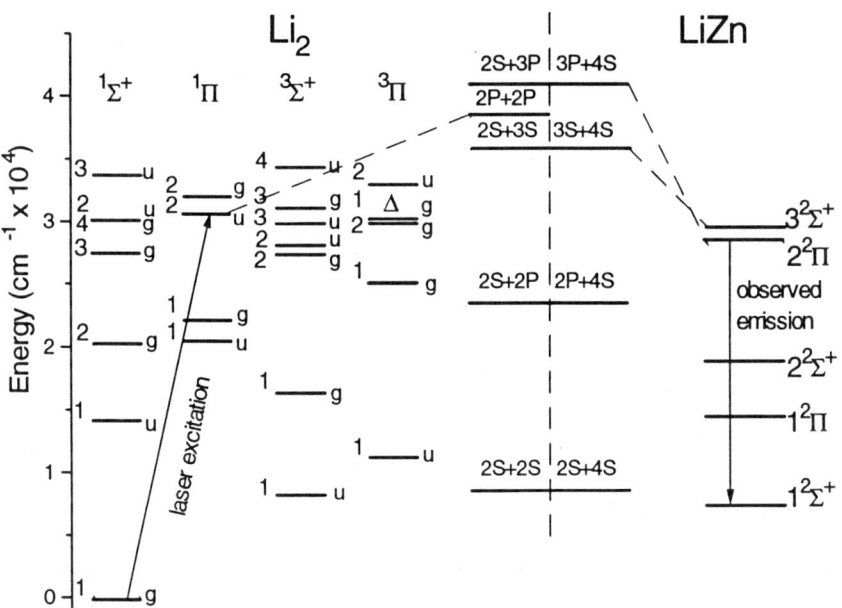

Fig. 7. Energy term diagram showing entrance and exit channels of the photochemical reaction

6. D.S. Chen, L.Li, X.T.Wang, Li Li, G.Hui, H.Ma, L.Q.Li, X.Y.Xu and D.Y.Chun, J.Mol.Spectrosc. 161, 7 (1993) and references therein
7. S. Milošević, X. Li, D. Azinović, G. Pichler, M. C. van Hemert and R. Düren, J. Chem. Phys. 96, 7364 (1992)
8. M. C. van Hemert, D. Azinović, X. Li, S. Milošević, G. Pichler and R. Düren, Chem. Phys. Lett. 200, 97 (1992)
9. C.Angeli and M.Persico, private communication
10. U.Buck, H.O.Hope, F.Huisken and H.Pauly, J.Chem.Phys. 60, 4925 (1974)
11. R.E. Olson, J.Chem.Phys. 49, 4499 (1968)
12. E.Czuchaj, F.Rebentrost, H.Stoll and H.Preuss, Chem.Phys.Lett., 178 246 (1991).
13. L.Windholz, G.Zerza, G.Pichler and B.Hess, Z.Phys. D.-At.Mol.Clust. 18, 373 (1991)
14. L.Hüwel, J.Maier and H.Pauly, J.Chem.Phys. 76, 4961 (1982).
15. E.Czuchaj, F.Rebentrost, H.Stoll and H.Preuss, Chem.Phys.Lett., 218, 454 (1994).
16. U.Lackschewitz, J.Maier and H.Pauly, J.Chem.Phys. 84, 181 (1986).
17. X. Li, Ph.D thesis, Spectroscopic studies of LiLi, LiZn and LiCd molecules in vapor mixtures, Inst. of Physics, Univ. of Zagreb, 1992
18. I.Vezmar, Degree thesis, Laser spectroscopy of LiHg excimer (Croatian) University of Zagreb 1993.
19. X. Li, S. Milošević, D. Azinović, G. Pichler, R. Düren and M. C. van Hemert, in press Z.Phys. D (1994).
20. G.Pichler, D.Veža and D.Fijan, Opt.Commun. 67, 45 (1988).
21. G.Pichler, D.Fijan, D.Veža, J.Rukavina and J.Schlejen, Chem.Phys.Lett. 147, 497 (1988).
22. M.Musso, L.Windholz, F.Fuso and M.Allegrini, J.Chem.Phys. 97, 7017 (1992).
23. R.Düren, B.Heumann and S.Milošević, to be published
24. M.Allegrini, G. de Filippo, F.Fuso, D.Gruber, L.Windholz and M.Musso, Chem.Phys. in press
25. T.B.Lucatorto and T.J.McIlrath, Appl.Opt. 19, 3948 (1980).
26. C.He and R.A. Bernheim, Chem.Phys.Lett. 190, 494 (1992).
27. H.Umemoto, J.Kikuma, A.Masaki, and S.Sato, Chem.Phys. 127, 227 (1988).
28. S.Milošević, D.Azinović and G.Pichler, to be published

Theoretical and Experimental Studies of the LiHg-Blue Green Bands

D. Gruber, L. Windholz, X. Li*
Institut für Experimentalphysik, Technische Universität Graz, A-8010 Graz

M. Gleichmann, B. Heß
Lehrstuhl für Theoretische Chemie, Universät Bonn, D-53115 Bonn

Representants of Ia-IIb excimers only have a very weak bound ground state. Since dissociation energy usually is even lower than thermal energy of these molecules in their gas phase, one has to produce Ia-IIb molecules by some photochemical reaction in one of their electronically excited bound states[1].

In our experiments we found the following mechanisms to be suited for production of electronically excited LiHg:

$$Li_2(C^1\Pi_u) + Hg(6s^2\ ^1S_0) + h\nu_{Laser} \rightarrow LiHg(III_{1/2}, II_{3/2}) + Li(2s\ ^2S) \quad (1)$$

and the photoassociative reaction

$$Li(2s\ ^2S) + Hg(6s^2\ ^1S_0) + h\nu_{Laser} \rightarrow LiHg(II_{3/2}). \quad (2)$$

The Li-Hg vapor mixture was produced in a crossed stainless steel heat pipe oven, operated at temperatures of about 1150 K using argon as buffer gas at pressures of about 10 kPa. Laser radiation was provided by an Ar^{++} pumped dye laser operated with stilbene III. Fluorescence light was collected by an imaging lens and focused onto the entrance slit of an 1100 mm scanning monochromator equipped with a 1200 grooves/mm holographic grating, detected by a Peltier-cooled photo multiplier tube, preamplified and processed by a lock-in amplification system.

A basis set of 142 contracted Gaussian functions is used to represent the one particle space of the molecule. The kinematical relativistic effects are taken into account by application of the spin-free no-pair hamiltonian with external field projectors. Complete active space self consistent field (CASSCF) orbitals[2], generated from an active space, comprising the 6s and 6p orbitals of Hg and the 2s orbitals of Li, are used as a basis of the multi-electron functions; all electrons are treated explicitly in these calculations. Dynamic correlation effects are introduced in a multi-reference double-excitation configuration interaction treatment (MRDCI) where thirteen electrons are correlated.

---

* presently at Max Planck Institut für Extraterrestrische Physik, Garching

Figure 1 shows fluorescence spectra for the spectral range studied. Results for photochemical production of LiHg according to (1) is shown in Fig.1a and for the photoassociative reaction (2) in Fig.1b. It clearly can be seen, that preparation of the Li_2 in its $Li_2(C^1\Pi_u)$ reagent's state results in several loss channels giving rise to molecular and atomic transitions of Li and Li_2 whereas these loss channels can be suppressed by the photoassociation (2).

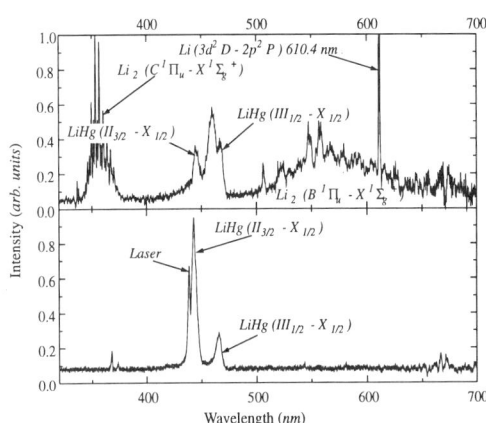

Applying laser induced fluorescence spectroscopy, we were able to determine the well depth of the LiHg ground state. Table I shows a comparison between values obtained by experiment, *ab-initio* calculations and scattering experiments.

Fig.1: LIF-spectra of the Li+Hg vapor mixture investigated for a) photochemical reaction (1), b) photoassociative reaction (2)

Table I: Well depth D_e and bond distance R_e for the LiHg ground state

R_e Å	D_e cm^{-1}	Potential	Source
3.00	851	inversion	Buck et al.[3]
3.03	653	*ab-initio*	present
	780	LIF	present

1. D. Gruber, M. Musso, L. Windholz, M. Gleichmann, B.A. Heß, F. Fuso, M. Allegrini, J. Chem. Phys. *in press* (1994)
2. B.J. Roos, in *Ab initio Methods in Quantum Chemistry - II*, K.P. Lawley (ed.) pp 399, John Wiley & Sons Ltd. (1987)
3. U. Buck, H.O. Hoppe, F. Huisken, H. Pauly, J. Chem. Phys. 60, 4925 (1974)

Temperature Dependence of the Kr*Ar Exciplex Emissions Lineshape

R. Reininger[1], J.L Subtil[2], C. Vincent-Donnet[2], P. Laporte[2], and P. Gürtler[3]

(1) Synchrotron Radiation Center, University of Wisconsin - Madison, 3731 Schneider Dr., Stoughton, WI 53589, USA.
(2) Laboratoire T.S.I., Université Jean Monnet, URA 842 CNRS, 23 rue du Docteur Paul Michelon, 42023 Saint-Etienne Cedex 2 - France
(3) Hamburger Synchrotronstrahlungslabor (HASYLAB) at DESY, Notkestr. 85, D-22603 Hamburg, FRG

Previously reported work on pure rare gas excimer luminescence have revealed a very weak environmental influence on these VUV emissions. In particular, Cheshnovsky *et al.*[1] demonstrated that the excimer bandwidth in xenon and argon was predominantly temperature dependent. Furthermore, they found no significant change in the width for liquid and gas samples at the same temperature. They interpreted their results using a simple theoretical model based solely on molecular effects. Recently, part of us reported on similar results on Kr and extended the previous measurements in Ar, Kr and Xe to densities ranging from the gas to the liquid[2].

In the present contribution we summarize the temperature dependence of the Kr*Ar exciplex emissions width between 300 and 90 K. The argon density was varied from the gas to the liquid. At some temperatures, measurements were performed both in the gas and in the liquid. The exciplex emissions are due to the transitions from the relaxed $0^+(^3P_1)$ and $1(^3P_2)$ to the $0^+(^1S_0)$ ground state. Their lifetimes at liquid densities[3] being ≈ 2 ns for the former and ≈ 200 ns for the latter. Since the energy of both transitions is ≈ 9.25 eV they were separated using time resolved techniques.

As shown in Fig. 1 these emissions have been obtained in very good experimental conditions, i.e., almost free from neighboring atomic $^3P_1$ (10.0 eV) and excimer Kr_2^* (8.4 eV) emissions. This was achieved by combining selective excitation provided by monochromatized synchrotron radiation and samples

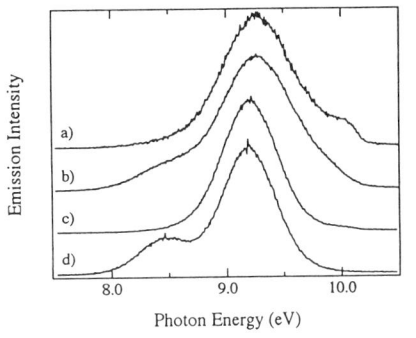
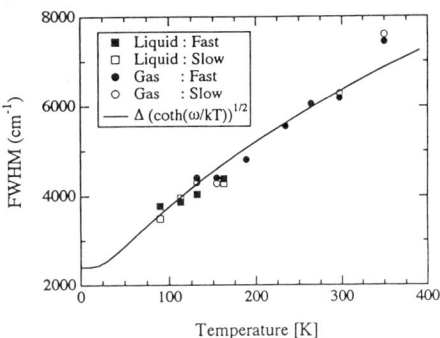

Fig. 1: Emission intensity recorded after $^3P_1$ excitation. a) Fast and b) slow components. T=300 K, and ρ=1.7x10^{21} cm^{-3}. c) Fast and d) slow components. T=162 K, and ρ=3.9x10^{21} cm^{-3}. The curves are displaced vertically.

Fig. 2: Temperature dependence of the exciplex FWHM.

containing only ppm of Kr according to previously reported techniques[3]. The rather large temperature effect on the line width is displayed in Fig. 2. As seen in the figure, there is no density effect within the experimental accuracy. The line shown in the figure was calculated based on the model by Cheshnovsky et al.[1] with ω=60 cm^{-1} for the vibrational constant and Δ=2400 cm^{-1} for the width at 0 Kelvin. The value obtained for ω is much smaller than those associated with the rare gas excimers[2]. This is expected based on the *ab initio* excited potentials[4] which show much shallower wells than in the excimer cases.

Financial support by NATO Grant No. CRG 920574 is gratefully acknowledged.

1. O. Cheshnovsky, B. Raz, and J. Jortner, Chem. Phys. Lett. **15**, 475 (1972).
2. E. Morikawa, R. Reininger, P. Gürtler, V. Saile, and P. Laporte, J. Chem. Phys. **91**, 1469 (1989).
3. P. Laporte, J.-L. Subtil, R. Reininger, and P. Gürtler, Chem. Phys. Lett. **174**, 61 (1990).
4. F. Spiegelmann, F.X. Gadea, and M.C. Castex, Chem. Phys. **145**, 173 (1990).

ULTRACOLD ATOMS

SLOW ATOM COLLISIONS: A REVIEW

Paul S. Julienne
Molecular Physics Division, National Institute of Standards and Technology,
Gaithersburg, Maryland 20899, USA

The experimental field of cooling and trapping of atoms has seen rapid development over the last decade with the consequence that several methods of trapping atoms in the microkelvin temperature range are now available. Laser cooling can be applied to alkali, alkaline earth, and rare gas metastable atoms, but temperature and trap density (generally $< 10^{12}$ atoms/cm^3) will depend on the atomic species and the methods used. The effects of collisions between trapped atoms are readily observed. The absorption or emission of one or more photons during a collision of two ultracold atoms induces a rich variety of processes that can be detected due either to heating and loss of trapped atoms or to formation of ionic products. The spectral signature of the signal that results from varying the frequency of the exciting light contains information about both the ultracold collision dynamics and the structure and potentials of the ground or excited molecular states involved.

Collisions of atoms with T < 1 mK are novel for two reasons: (1) quantum effects due the large de Broglie wavelength compared to the range of the ground state interaction potential, and (2) excited state spontaneous decay during the long time of interaction. The former dictates a quantum treatment of the collision, whereas the latter requires the inclusion of decay and dissipation. A number of theoretical tools have now been applied to describe the absorption of light during ultracold collisions: quasistatic models, semiclassical optical Bloch equations, complex potential models, Monte Carlo wavefunction methods, Franck-Condon models, and Landau-Zener models. Fortunately, a number of simplifications in our understanding of these cold collisions are now possible. This review talk will concentrate on describing how these simplifications allow us to think about the underlying physics of these collisions.

The basic physics and spectral line shapes for ultracold collisions in a light field can be understood in terms of Franck-Condon factors and their semiclassical interpretation. Two ground state atoms collide in state $|g>$ with potential $V_g(R)$, and are excited by light of frequency ω to excited state $|e>$ with potential $V_e(R)$. The nature of the collision and the spectral signal are strongly influenced by the long range potentials. The ground state van der Waals potential, varying as R^{-6}, is of much shorter range and is nearly flat at distances where R in on the order of the De Broglie wavelength. By contrast, the excited state resonant dipole potential, varying as R^{-3}, is very long range, with a magnitude comparable to the natural decay linewidth where $R \approx \omega/c = \lambda/2\pi$ is on the order of the wavelength of the atomic S→P transition, typically near R = 1000 Å. In the quantum picture, the amplitude for a process that involves the absorption of a photon will be proportional to the Franck-Condon factor: $<g|d(R)|e>$, where $d(R)$ is a transition dipole. In a semiclassical picture, the photon is "absorbed" near the Condon point R_C where the photon energy matches the difference between excited and ground state potentials: $\hbar\omega = V_e(R_C) - V_g(R_C)$. The usual stationary phase approximation for the Franck-Condon integral is equivalent to a Landau-Zener treatment of the curve crossing at R_C of field-dressed molecular potentials.

When the detuning to the red of atomic resonance is very large, the vibrational spacing is large compared to the natural linewidth, the excited molecular state can vibrate many times before decaying, and the spectral signature is a narrow feature that describes the free-bound transition. Tuning across many such features gives a high resolution photoassociation spectrum that effectively maps out the bound state structure of levels with large outer turning points in the excited molecular state. The Franck-Condon factors giving the line shapes are strongly influenced by quantum effects in the ground state collision that are very sensitive to the details of the ground state scattering. On the other hand, when the detuning is very close to atomic resonance, only a few atomic linewidths to the red, the Condon point is at very long range, and the excited molecule only evolves a short distance on the excited state before decaying. In this regime, quantized bound state structure is not resolved, but fully quantum mechanical methods have shown that a simple semiclassical picture is remarkably accurate for the very long range excitation process, even at the extremely low temperatures in atom traps. In this picture, the transition from the ground to the excited state is described by the Landau-Zener curve hopping probability evaluated at the Condon point, and the reduction in excited state population is described by calculating the classical probability of survival along the classical trajectory from the Condon point to the short range region of chemical size, where the process occurs that gives the observed signal.

References to work in this field are too numerous to include an inclusive list of specific references here. Reviews of work on cold collisions are available from the theoretical[1] and experimental[2] viewpoints and recent theoretical work is summarized in reference 3.

1. P. S. Julienne, A. M. Smith, and K. Burnett, Adv. At., Mol, Opt. Physics, **30**, 141(1993).
2. T. Walker and P. Feng, Adv. At. Mol. Opt. Physics **34**, to be published.
3. P. S. Julienne, K.-A. Suominen, and Y. B. Band, Phys. Rev. A **49**, 3890(1994); K.-A. Suominen, M. Holland, K. Burnett, and P. S. Julienne, Phys. Rev. A **49**, 3897(1994).

THEORY OF LINE SHAPES FOR COLD ATOM COLLISIONS

Carl J. Williams[†]
Molecular Physics Division, National Institute of Standards and Technology,
Gaithersburg, MD 20899

High resolution molecular spectra obtained from the photoassociation of two colliding ultracold atoms contains detailed information on ground and excited state interactions. High resolution is achieved in such spectra, even though the initial state is a continuum state, because the thermal energy k_BT of optically trapped atoms is typically close to the natural linewidth of the rovibrational molecular states. Below we discuss general features of the photoassociation spectra of alkali atoms along with some details of the line shapes produced in the photoexcitation of colliding ultracold atoms. The theory is presented in terms of the collisions of alkali atoms since these are the systems of current experimental interest.[1-4]

Photoassociation spectroscopy has several novel features that makes it a very useful new experimental tool: 1) it probes molecular structure right up to the dissociation limit including vibrational states with outer turning points greater than 100 Å, 2) it excites only the lowest few rotational levels because of the extremely small initial collision energies, 3) it populates states that are normally inaccessible to conventional spectroscopy, 4) it exhibits hyperfine structure that in conjunction with theory provides an unambiguous signature for assigning the electronic character of the spectra, and 5) it produces asymmetric line shapes that illustrate the Wigner law behavior characteristic of the ground state continuum wavefunction in the quantum threshold (T → 0 K) limit. This latter behavior makes photoassociation spectroscopy an extremely sensitive probe of long range ground state interactions.

When the photon energy $\hbar\omega$ of a tunable laser plus the asymptotic kinetic energy ε of two colliding ground state atoms matches the energy difference between an excited rovibrational bound level b and the asymptotic atomic ground state energies, photoexcitation results in level b appearing as a scattering resonance. Detection of a product signal p versus a probe frequency ω, that is red detuned from the alkali atomic $^2S \to {}^2P$ transition, measures the positions of bound rovibrational levels b of electronically excited molecular states correlating to an excited ^2P alkali atom and a ground ^2S alkali atom. Probe frequencies blue of atomic resonance also produce a photoexcitation spectrum that can be interpreted in terms of continuum - continuum transitions.[4]

The rate coefficient $K_p(T,\omega)$ for the inelastic process yielding product p from two colliding ground state ^2S alkali atoms is given by:

$$K_p(T,\omega) = \left\langle \frac{\pi v}{k^2} \sum_{\ell=0}^{\infty} (2\ell+1) \left|S_p(\varepsilon,\ell,\omega)\right|^2 \right\rangle, \quad (1)$$

where $\varepsilon = \hbar k^2/2\mu$ is the asymptotic kinetic energy of the two ground state alkali atoms, μ is the reduced mass, ℓ is the relative angular momentum quantum number, ω is the photon frequency, and $S_p(\varepsilon,\ell,\omega)$ is the S-matrix element for the process that forms product p from the initial ground state channel. The brackets <...> imply an average over the distribution of initial velocities v, which is assumed to be

---

[†]Permanent Address: James Franck Institute, Univ. of Chicago, Chicago, IL 60637

Maxwellian. In the weak field limit our full close coupling results[5] shows that the following simple resonant scattering expression for the bound state b:

$$|S_p(\varepsilon,\ell,\omega)|^2 = \frac{\gamma_p \gamma_s(\varepsilon,\ell)}{(\varepsilon + \hbar\omega - E_b)^2 + (\gamma/2)^2}, \quad (2)$$

is an ideal approximation. In Eq.(2), E_b is the position of the bound state; $\gamma = \gamma_p + \gamma_s(\varepsilon,\ell) + \gamma_o$ is the total width (in energy units) of the excited bound state, $\gamma_p/\hbar$ is the rate for the bound state resonance b to decay to the detected product p, $\gamma_s(\varepsilon,\ell)/\hbar$ is the stimulated emission rate back to the ground state, and $\gamma_o/\hbar$ is the decay rate due to other processes such as spin-orbit or hyperfine predissociation. Finally, Fermi's golden rule gives $\gamma_s(\varepsilon,\ell) = 2\pi I |\langle \varepsilon,\ell | V_{rad}(R) | b \rangle|^2$, where $|\varepsilon,\ell\rangle$ is the ground state energy normalized continuum wavefunction, and $V_{rad}(R) = (2\pi I/c)^{1/2} d(R)$ is the radiative coupling matrix element proportional to the molecular transition dipole $d(R)$ and the square root of the probe laser intensity I.

Although two state models can provide insight into the physics of these systems, comparison of theory with experiments requires more detailed models. Because the asymptotic energy is typically larger than the atomic hyperfine splitting it is essential to include hyperfine structure in the calculations. One way that this is seen is in the experimental photoassociation spectroscopy of the $Na_2(1_g)$ state that correlates at short internuclear separation to the $^1\Pi_g$ state.[1] The $Na_2(1_g)$ photoassociation spectra spans from v=38 to v=99, or from a 170 cm^{-1} below dissociation to 0.2 cm^{-1} from dissociation, where the photoassociation spectra becomes so congested that the series can no longer be easily followed. Very near dissociation the vibrational spectra consists of clusters of lines, each cluster approximately 1 GHz in width, that are clearly a result of hyperfine structure.[6] However, the $Na_2(1_g)$ for v=48 (80 cm^{-1} below dissociation) shows only four isolated features corresponding to J = 1, 2, 3, and 4. At higher resolution these individual rotational features still show some residual hyperfine structure. The v=48 state basically lacks hyperfine structure since the $Na_2(1_g)$ is better described as a $^1\Pi_g$ state for this binding energy. However, the transition dipole moment for the $^1\Pi_g$ state is nonzero only as a result of the residual triplet character. In addition numerous other states have been observed, including the purely long range 0_g^- state of both Na_2[1b] and Rb_2.[2b] In Na_2 this state has a binding energy of 2 cm^{-1} and an inner turning point of 30 Å.

1. a) P. D. Lett, K. Helmerson, W. D. Phillips, L. P. Ratliff, S. L. Rolston, and M. E. Wagshul, Phys. Rev. Lett. **71**, 2200, (1993); b) L. P. Ratliff, M. E. Wagshul, P. D. Lett, S. L. Rolston, and W. D. Phillips, J. Chem. Phys. (in press), 1994.
2. a) J. D. Miller, R. A. Cline, and D. J. Heinzen, Phys. Rev. Lett. **71**, 2204 (1993); b) R. A. Cline, J. D. Miller, and D. J. Heinzen, unpublished.
3. E.R.I. Abraham, N.W.M. Ritchie, W.I. McAlexander, and R.G. Hulet, unpublished (1994).
4. V. Bagnato, L. Marcassa, C. Tsao, Y. Wang, and J. Weiner, Phys. Rev. Lett. **70**, 3225 (1993).
5. R. Napolitano, J. Weiner, C. J. Williams, and P. S. Weiner, Phys. Rev. Lett. submitted.
6. C. J. Willams and P. S. Julienne, J. Chem. Phys. (in press), 1994.

ULTRACOLD COLLISION STUDIES: A NEW PHYSICAL REGIME WHERE LINESHAPE ANALYSIS INFORMS DYNAMICS

J. Weiner and R. Napolitano
Department of Chemistry and Biochemistry
University of Maryland
College Park, MD 20742
USA

V. Bagnato and L. Marcassa
Optical Group
University of Sao Paulo
Sao Carlos, Brazil

P. S. Julienne
Physics Laboratory
National Institute of Standards and Technology
Gaithersburg, MD 20899

ABSTRACT

We present results from photoassociative ionization (PAI) spectroscopy in which analysis of the free-bound transition line shapes reveal Wigner threshold contributions from individual partial waves as well as the temperature of colliding atoms confined in a magneto-optical trap (MOT).

Photoassociation spectroscopy was first proposed by Thorsheim, Weiner, and Julienne [1] and recently observed by several groups [2-5]. It is a high resolution probe of molecular states formed by photoexciting two colliding cold atoms to an excited molecular dimer state. Because the thermal energy $k_B T$ of trapped atoms is typically close to the natural linewidth of the atomic cooling transition, the high resolution photoassociation spectrum exhibits only slight thermal broadening. Recently, 1 MHz resolution line shapes have been reported for both Na [2] and Rb [5] photoassociative spectra. The ultracold experiments confirm the prediction that detailed information of the ground state potential and scattering wave function can be obtained from line shapes and intensity patterns. Thus, experimental photoassociation spectra in conjunction with theoretical analysis should permit an accurate determination of scattering lengths and other ground state $T \to 0$ collision properties relevant to achieving Bose-Einstein condensation in alkali systems [6].

The photoassociation spectrum is proportional to the rate coefficient $K_p(T,\omega)$ for the inelastic process yielding product p from two colliding ground state ^2S atoms:

$$K_p(T,\omega) = \left\langle \frac{\pi v}{k^2} \sum_{l=0}^{\infty} (2l+1) |S_p(\varepsilon,l,\omega)|^2 \right\rangle$$

where $\varepsilon = \hbar k^2/2\mu$ is the asymptotic kinetic energy of the ground state $^2S+^2S$ atoms, μ is the reduced mass, l is the relative angular momentum quantum number, and $S_p(\varepsilon,l,\omega)$ is the S-matrix element for the process that forms product p from the initial ground state channel. The brackets $<...>$ denote an average of the distribution of initial velocities v. An excellent approximation to the full quantum close coupling calculation is the following simple resonant scattering expression for an isolated resonance,

$$|S_p(\varepsilon,l,\Delta_b)|^2 = \frac{\gamma_p \gamma_s(\varepsilon,l)}{(\varepsilon-\Delta_b)^2 + (\gamma/2)^2}$$

where $\Delta_b(\omega) = E_b - \hbar\omega$ is the detuning relative to the position E_b of the bound state. Results for s, p, and d wave contributions to the v=48 and v=85 vibrational levels of the 1_g excited state of the collision intermediate are shown in Fig. 1.

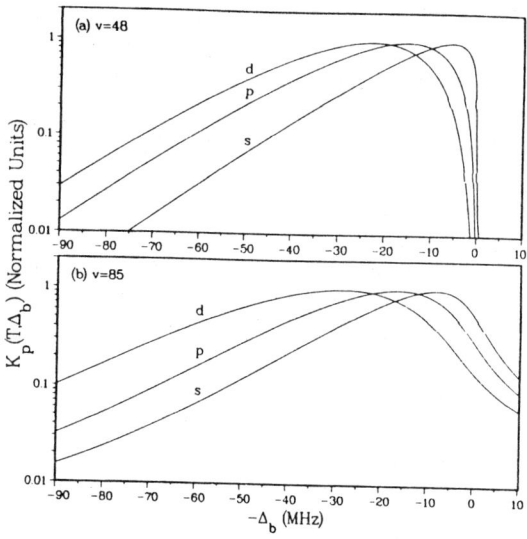

Fig. 1 Calculated $K_p(T, \Delta_b)$, normalized to unity at the peaks, vs. $-\Delta_b$ for (a) v=48 and (b) v=85

REFERENCES

1. H. R. Thorsheim, J. Weiner, and P. S. Julienne, Phys. Rev. Lett. **58**, 2420, (1987)
2. L. P. Ratliff, M. E. Wagshull, P. D. Lett, S. L. Rolston, and W. D. Phillips, J. Chem. Phys. **101**, 2638 (1994)
3. E. R. I. Abraham, N. W. M. Ritchie, W. I. McAlexander, and R. G. Hulet (to be published)
4. V. Bagnato, L. Marcassa, C. Tsao, Y. Wang, and J. Weiner, Phys. Rev. Lett. **70**, 3225 (1993)
5. R. A. Cline, J. D. Miller, and D. J. Heinzen, Phys. Rev. Lett. **73**, 632 (1994)
6. B. J. Verhaar, and H. T. C. Stoof, Phys. Rev. A **47**, 4114 (1993)

PHOTOASSOCIATION OF ULTRACOLD ATOMS

D. J. Heinzen, J. D. Miller, and R. A. Cline
The University of Texas, Austin, TX 78712

ABSTRACT

Cold-atom photoassociation is important as a novel molecular bond formation mechanism, as a loss mechanism in optical atom traps, as a possible source of ultra-cold molecules, and as a new spectroscopic method.[1-3] In this process, a colliding pair of laser-cooled atoms A and B absorbs a photon of frequency ν to produce a bound, excited molecule AB*. Photoassociation resonances occur that are associated with the excitation of specific ro-vibrational states of AB*; these resonances can be very narrow because the kinetic energy spread of the atoms is small at low temperature (e.g. k_BT/h = 21 MHz at T = 1 mK). Unique, long-range molecular states are readily studied because the initial collisional state has a large amplitude at long-range.

We have carried out photoassociation studies of trapped, laser-cooled Rb atoms, and explored Rb_2^* states with binding energies from less than 0.3 cm^{-1} to more than 1000 cm^{-1}, with outer turning points from 17 a_0 to 250 a_0, and with a resolution of order 1 MHz. The Rb atoms are held in a far-off resonance optical dipole force trap, have a density of order 10^{12}/cm^3, and a temperature of about 1 mK. We detect the photoassociation resonances by monitoring the loss of atoms from the trap, which occurs because the excited Rb_2^* dimers decay predominantly to free states with an energy much greater than the trap depth. These studies have revealed many new features of long-range molecular states. Several vibrational series are observed, with intensity oscillations that reflect the structure of the initial collisional state wavefunction. We observe the bound levels of the unique 0_g^- molecular "pure long-range state", which has an inner turning point beyond 25 Bohr. We also observe well-resolved rotational structure and predissociation broadening of the photoassociation resonances.

Long range interaction potentials can be derived from the photoassociation spectra. We can extract the coefficient C_3 of the dominant C_3/R^3 resonant dipole interaction between the atoms to an accuracy of about 1%, and may eventually be able to realize 0.1% accuracy. This analysis indirectly yields the excited atomic lifetime to comparable accuracy. Information on the ground state interaction potential is contained in the free-bound Franck-Condon factors, and in the intensities and lineshapes of rotationally resolved lines. These lines exhibit asymmetric lineshapes that reflect the distribution of initial collisional state energies, as modified by transmission of the colliding atoms through the ground state centrifugal barrier.

Analysis of the photoassociation spectra is simplified if the atoms are prepared in the spin-polarized (F=3, M_F=3) state, because to a good approximation the collisions then occur on a single, triplet potential curve. As illustrated in Fig. 1, we have obtained photoassociation spectra with spin-polarized atoms. An analysis of this spectrum, in progress, should yield the ground state triplet scattering length and coefficient C_6 of the long range $-C_6/R^6$ atomic interaction.

In summary, we have obtained highly resolved cold atom photoassociation spectra which provide a wealth of information on the long-range interactions between the atoms. With further analysis, an almost complete characterization of both excited and ground state interactions should be possible. We gratefully acknowledge the support of the National Science Foundation, the A. P. Sloan Foundation, and the R. A. Welch Foundation.

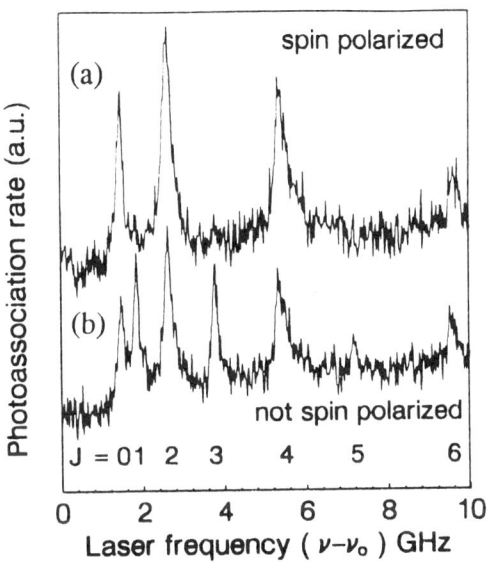

Fig. 1. High-resolution photoassociation spectrum of ^{85}Rb atoms, showing the rotationally resolved structure of a single 0_g^- vibrational line. a) Atoms are optically pumped into the F = 3, M_F = 3 ground state sublevel. b) Atoms are pumped into F = 3, but are in a random distribution of Zeeman sublevels.

1. H. R. Thorsheim, J. Weiner, and P. S. Julienne, Phys. Rev. Lett. 58, 2420 (1987).
2. J. D. Miller, R. A. Cline, and D. J. Heinzen, Phys. Rev. Lett. 71, 2204 (1993).
3. P. D. Lett *et al.*, Phys. Rev. Lett. 71, 2200 (1993).

LIQUID HELIUM LINE BROADENING AND SHIFTS OF ALKALINE EARTH AND ALKALI METAL ATOMS

Y. Takahashi, K. Fukuda, T. Kinoshita, and T. Yabuzaki
Department of Physics, Faculty of Science,
Kyoto University, Kyoto 606-01, Japan

ABSTRACT

With newly developed laser sputtering method we implanted various kinds of atoms in superfluid helium. An atomic bubble model could rather quantitatively explain the results of laser and rf spectroscopy of these atoms.

OPTICAL SPECTRA

New methods for implantation of foreign neutral atoms into superfluid helium (He II) have recently been developed.[1,2] We could succesfully creat alkaline earth and alkali metal atoms in He II by laser sputtering method[2] (density 10^8–10^{10}).

We investigated the optical properties of these atoms in He II using dye and Ti:sapphire lasers[3]. The experiments were performed at the helium temperature of 1.6 K at the saturated vapor pressure. The obtained emission spectra were broadened by about 1nm FWHM and blueshifted by about 1nm compared with the corresponding spectra for free atoms. On the other hand, the excitation spectra were very broad (5–10 nm) and considerably blue shifted (10–20 nm). These results were explained by a spherical atomic bubble model where a spherical vacancy is assumed to be formed around the atom. In this model the optical excitation occurs at a smaller bubble or vacancy radius determined by the ground state wave function, whereas the emission occurs after the bubble is expanded due to the larger extent of the excited state wave function. The quantitative explanation of the observed spectra could be done by using a configuration coordinate diagram where the bubble radius is taken as the coordinate.

Recently we performed laser spectroscopy of atoms in pressurized liquid helium. The excitation and emission spectra shifted toward the shorter wavelength sides as the pressure was increased, which was consistent with a simple atomic bubble model.

We also found that the excitation spectrum of the Cs D2 line had a doubly shaped profile. This cannot be explained by a simple spherical bubble model. The nonspherical density distribution of surrounding helium associated with a dynamical Jahn-Teller effect should be responsible for the observed structure.

ZEEMAN AND HYPERFINE SPECTRA

We also performed optical pumping and optical detection of magnetic resonance of alkali-metal atoms (Cs and Rb) in He II[4]. Using the circularly polarized pumping beam tuned to the D1 line (778nm for Rb and 876 nm for Cs), we achieved large spin polarization of the atoms in the ground states, more than 50 % for both electron and nuclear spins, especially for the Cs atoms. The fact that the D1-and D2-excitation spectra are resolved in He II makes the optical pumping possible for this system.

We observed the magnetic resonances between the ground-state Zeeman sublevels and also hyperfine levels through monitoring the D1 fluorescence by means of the optical-rf double resonance technique. The ground state g values in He II were the same as in vacuum within the experimental error. However, the hyperfine constant of the ground state of the Cs atom in He II was slightly larger than in vacuum. This is the direct evidence of the compression of the ground state wave function due to the surrounding helium. The strength of the shift was consistent with the prediction based on an atomic bubble model. We also observed the sublevel coherent transient signal (transient nutation) of alkali atoms in He II. The deduced sublevel linewidth for the Cs atom was smaller than 10 mG. The narrow linewidth is advantageous for many applications, which need high resolution, of the spin polarized atoms in He II.

REFERENCES

1. H. Bauer *et al.*, Phys. Lett. **A 137**, 217(1989); ibid **A 146**, 134(1990)
2. A. Fujisaki *et al.*, Phys. Rev. Lett. **71**, 1039(1993)
3. Y. Takahashi *et al.*, Phys. Rev. Lett. **71**, 1035(1993)
4. T. Kinoshita *et al.*, Phys. Rev. **B 49**. 3648(1994)

"SLOW" MOLECULES ENSEMBLE CHOICE IN ROOM TEMPERATURE GAS FOR HIGH RESOLUTION COHERENT TRANSIENT SPECTROSCOPY

Rubtsova N.N., Vasilenko L.S., Hvorostov E.B.
Quantum Electronics Department,
Institute of Semiconductor Physics,
acad. Lavrentyev prospect, 13, 630090, Russia

ABSTRACT

Coherent transient method was applied to demonstrate experimentally the possibility to choose "slow" molecules sub-ensemble in room temperature gas for its use in high resolution spectroscopy and in "slow" molecules collisional interaction investigation.

INTRODUCTION

High resolution spectroscopy in gases supposes employment of Doppler-free methods, and coherent radiation in time separated fields [2], based on photon echo in standing waves, is one of them. Moreover, such coherent transient spectroscopy has advantages over saturated absorption spectroscopy [2], being able to provide signal increase by several orders of magnitude at spectral resolution controlled by homogeneous broadening. The last one depends in its turn on collision broadening and transit-time effect. Further improvement of spectral resolution in a gas needs to overcome transit-time effect and second order Doppler effect, which can be realized by employment of "slow" particles. Aim of this work is to investigate experimentally the possibility to choose sub-ensemble of molecules with low translational motion velocities in room temperature gas by using coherent transient spectroscopy.

EXPERIMENTAL PROCEDURE

Photon echo in running waves which arises as a media response to excitation by two resonant radiation pulses at time moment t=2T was used to select ensemble of slow gas particles (here T is the time delay between two exciting pulses, time moment t=0 coincides with first of them). Excitation of photon echo by pulses of narrow line frequency tunable laser radiation with controlled pulse length allows to work at "wide spectral line" echo conditions and provides possibility to choose molecular sub-ensemble with its width in the space of longitudinal (relative direction of light propagation) velocities controlled by exciting pulse spectral width [3]. Selection of low radial velocity particles can be performed by choosing of large time delay T [3]. In this case high velocity (in radial direction) particles escape the laser field region, while coherent media response (photon echo) is generated mainly by "slow" particles.

Experiments were performed in SF_6 gas at P(33) A_2^1 vibrational-rotational transition of ν_3 0-1 vibrational mode. Frequency tunable stable CW CO_2 laser at generation line 10P(18) served as a source of radiation. Exciting pulses were formed by electrooptic shutter with time delay T and pulse widths controlled in a wide range. Coherent response (photon echo) was detected by fast response photoresistor. Photon echo sensitivity detection was improved by utilizing of heterodyne CW waveguide CO_2 laser frequency detuned 20 MHz relative to exciting radiation.

RESULTS

Appropriate choice of exciting radiation pulse lengths allowed us to detect echo from molecular ensemble of longitudinal velocities no higher than one tenth of average thermal velocity. Effective temperature of such sub-ensemble is about 3 ^{o}K. Photon echo decay rate versus time delay T proved to be pressure independent at gas pressures lower than 0.5 mTorr and shows good correlation with the particles inverse flight time across laser beam (transit-time effect). Photon echo decay kinetics differs from exponential one at high time delay T. This part of kinetic curve can be described in a good approximation by T to minus fourth power dependence which confirms dominating contribution of "slow" particles to the echo response. Echo signal is generated by molecules of radial velocities no higher than R/2T (R is laser beam radius) which accounts approximately one fourth part of average thermal velocity in our experimental conditions and corresponds to effective temperature of sub-ensemble lower than 20°K.

CONCLUSION

Experimental results have shown possibility to detect coherent responses of "slow" molecules in low pressure room temperature gas. Revealed at high time delay reduction in photon echo decay rate seems to be promising for "slow" molecules employment in high resolution coherent spectroscopy.

REFERENCES

1. Vasilenko L.S., Matveyenko I.D., Rubtsova N.N., Optics communications, 1985, v.53, p.371-374.
2. Vasilenko L.S., Rubtsova N.N., Izv. AN SSSR, ser. fiz., 1989, v.53, p.2329-2333.
3. Vasilenko L.S., Rubtsova N.N., Izv. RAN, ser. fiz., 1994, No.9 (to appear).

Appendix

Minutes of the meeting of the International Committee for the International Conference on Spectral Line Shapes, held in Toronto on June 16 1994.

1) The resignation of Dr. Daniel Kelleher from the International Committee was accepted. The committee expresses its deep appreciation to Dan for his contribution to the line shape community.

2) The committee accepted Dr. Eugene Oks of Auburn University and Dr. George Tabisz of the University of Manitoba as new members of the International Committee.

3) It was agreed that all appointments to the International Committee would be, in the future, for a period of four years.

4) It was agreed that the responsibilities of the members of the International Committee were;
 (i) To assist the local conference committee by suggesting names for invited speakers and topics for discussion.
 (ii) To assist the local committee by serving as chairs for sessions during the meetings.
 (iii) To serve as referees for articles in the conference proceedings.
 (iv) To promote participation in the conference, both nationally and internationally.

5) It was agreed that multiple submissions for the scientific program (abstracts) from one group would normally (but not always) imply separate presenters at the conference. Exceptions to these guidelines will be at the discretion of the local organizers. The presenters name was to be underlined on the abstract as opposed to being necessarily, the first listed.

6) It was agreed, that in the case presenters were unable to attend the meeting (for legitimate reasons) that another attendee be permitted to present the poster on their behalf.

7) It was agreed that if none of the authors listed on the abstract paid the conference registration fee, then the article would not be included in the conference proceedings.

8) It was agreed to pursue the question of publishing the proceedings through a recognized journal.

9) Through this item, the chairman reports to the International Committee the consensus reached at the discussion group concerning including the liquid state at future meetings. The discussion group agreed that the conference not necessarily be restricted to plasmas and neutral gases and that the local committee have the right to invite any speaker on any subject judged to be of general interest to the line shape community. It was noted, that in practice, all poster submissions dealing with line shapes have been accepted without restriction in the past.

Toronto, June 21, 1994   A. David May

Author Index

A

Alexiou, S., 81, 91
Al–Omari, A., 355
Angelo, P., 122, 215
Aparicio, J. A., 30, 62
Arranz, J.P., 213
Astapenko, V.A., 83
Azinovic, D., 256

B

Babin, S.A., 85
Back, C., 215
Bagnato, V., 417
Baldacchini, G., 374
Bauer, A., 310
Behery, M.M., 254
Bell, I.M., 347
Ben Lakhdar, Z., 213
Ben Nessib, N., 213
Berman, R., 258
Bielski, A., 89
Bieniek, R.J. 347
Birnbaum, G., 288
Blagojević, B. 75
Bliman, S., 70
Bloch, D., 241, 249
Boissolles, J., 265
Borysow, A., 221, 227, 237
Boulet, C., 265
Brezina, R., 300
Buffa, G., 372, 374, 382
Byszewski, P., 349

C

Calisti, A., 105, 153
Carlier, J., 310
Cazzoli, G., 372, 382
Chaker, M., 158
Chiba, H., 341, 343
Ciucci, A., 374
Ciurylo, R., 89
Cline, R.A., 420
Cludi, L., 372
Coe, S., 158
Cooper, J., 225
Cornille, M., 70
Coté, C.Y., 155
Çotti, G., 372
Ćuk, M., 79
Czerwosz, E., 349

D

D'Amato, F., 374
Danilova, Yu.E., 363
Degli Esposti, C., 372
de la Rosa, I., 30, 62
Demura, A.V., 100, 156, 177
Depiesse, M., 93
Derevianko, A., 34
Derfoul, H., 122
Devdariani, A.Z., 101, 235
Dimitrijević, M.S., 75, 87
Djurović, S., 77
Döhrn, A., 68
Domyslawska, J., 89
Dore, L., 382
Drummond, J.R., 258, 308, 378, 380
Ducloy, M., 241, 249
Duggan, P., 258, 378, 380

F

Fichet, M., 241
Filippov, N.N., 298, 302
Forsman, J.W., 308, 380
Fukada, K., 422

G

Galal, A.A., 254
Gao, B., 225
Gauthier, P., 122, 215
Gigosos, M.A., 30, 62
Gilles, D., 155, 156, 158
Glaz, W., 223
Gleichmann, M., 406
Glenzer, S., 100, 134, 153
Godbert, L., 105, 153
Godon, M., 310
Goldhar, J., 211
Goly, A., 95
Gondal, M.A., 229
Gordon, R.A., 376
Grabowski, B., 209
Green, S., 300, 367
Griem, H.R., 99, 101, 211
Grigorchuk, N.I., 260
Grigoriev, I.M., 302
Gruber, D., 406
Gruszka, M., 227
Grützmacher, K., 32
Günter, S., 64, 68, 151, 217
Gürtler, P., 408
Grycuk, T., 345, 349

H

Halenka, J., 151, 209, 217
Hammel, B.A., 105
Hannon, S., 306
Heinzen, D.J., 420
Helbig, V., 68, 99
Herman, R.M. 357
Heß, B., 406
Himmel, G., 28
Hirsch, S., 28
Hsu, C.J., 361
Huynh, B.C., 156
Hvorostov, E.B., 424

I

Ivković, M., 58, 60

J

Jiang, Z., 155, 158
Johns, J.W.C., 306
Julienne, P.S., 413, 417

K

Kablukov, S.I., 85
Katsonis, K., 70
Keane, C.J., 105
Keitel, C.H., 251
Khan, M.A., 229
Khorev, S.V., 85
Kieffer, J.C., 155, 158
Kielkopf, J., 384
Kinoshita, T., 422
Kobilarov, N., 60, 77
Könies, A., 64, 68, 217
Konjević, N., 58, 60, 75, 77, 100
Korn, G., 158
Kreye, W., 384
Kukushkin, A.B., 83
Kurosawa, T., 341, 343

L

Langer, S.H., 105
Laporte, P., 408
Leboucher–Dalimier, E., 122, 215
Lednev, M.G., 231, 235
Le Doucen, R., 304
Lee, R.W., 105, 211
Le Flohic, M.P., 258, 380

Leng, Y., 211
Lesage, A., 93
Li, J., 252
Li, X., 406
Li, Z., 386
Liu, W.-K., 300

M

Ma, Q., 274, 310, 314
Mar, S., 30, 62
Marcassa, L., 417
Markel, V.A., 363
Maron, Y., 81
May, A.D., 258, 308, 378, 380
Moraldi, M., 237
McQuarrie, B., 225
Meftah, T., 153
Mijatović, Z., 60, 77
Miller, J.D., 420
Milošević, S., 256, 391
Mortensen, A., 353
Mossé, C., 153
Mourou, G., 158

N

Napolitano, R., 417
Narducci, L., 251
Nesmelova, L.I., 312
Nguyen, H., 213
Nielsen, O.F., 353

O

Ohmori, K., 341, 343
Oks, E., 34, 98, 101, 102
Okunishi, M., 341, 343
Olchawa, W., 151, 209
Oppo, G.L., 251

P

Papagiorgiou, N., 241
Pérez, C., 30, 62
Peyrusse, O., 155, 158
Pichler, G., 256
Podivilov, E.V., 160
Popović, M.V., 75
Poquérusse, A., 122, 215
Purić, J., 79

R

Rahn, L.A., 388
Rais, M.H., 229
Reininger, R., 355, 408
Richou, J., 93
Rodimova, O.B., 312
Roston, G.D., 345
Rozanov, A.V., 302
Rubtsova, N.N., 424

S

Safonov, V.P., 363
Sahal–Bréchot, S., 87
Sato, Y., 316, 341, 343
Schlüter, H., 28
Schöning, T., 66
Schuller, F., 241
Scully, M., 251
Seidel, J., 32, 102
Serapinas, P., 97
Shapiro, D.A., 85, 160
Shaw, P.B., 357
Shelley, V., 353
Sinclair, P.M., 308, 378, 380
Stamm, R., 101, 105, 153
Stassen, H., 359
Steele, W.A., 359
Stefanović, I., 58
Stehlé, C., 36, 156, 177
Steiger, A., 32
Stobbe, M., 217
Strow, L., 306
Subtil, J.L., 408
Szudy, J., 89

T

Tabisz, G.C., 223, 225
Takahashi, Y., 422
Talin, B., 105, 153
Tarrini, O., 372, 374, 382
Tchaplyguine, M., 349
Thibault, F., 304
Tipping, R.H., 274, 314
Tobin, D., 306
Tonkov, M.V., 298, 302, 304
Trawinski, R.S., 89
Tserkovnyi, S.I., 233
Tvorogov, S.D., 312

U

Ueda, K., 341, 343

V

van Wijngaarden, W.A., 252
Varanasi, P., 386
Vartanyan, T.A., 249
Vasilenko, N.N., 424
Vincent–Donnet, C., 408
Voslamber, D., 3, 101

W

Weber, M., 386
Weiner, B., 357
Weiner, J., 417
Williams, C.J., 415
Williams, S., 388
Windholz, L., 406
Wujec, T., 95

Y

Yabuzaki, T., 422
Yarwood, J., 353
Youssef, N.H., 254

Z

Zagrebin, A.L., 231, 233, 235
Zare, R.N., 388
Zheng, K.C., 221

Subject Index

For brevity, xx(ion) refers to various states of ionization or to the parent atom or molecule from which ions were created.

A

adiabatic, 81
Ag, 363
Al(ion), 97, 128, 213, 215
alignment relaxation, 388
anisotropic potential, 222, 223, 231, 237, 274, 288ff, 314
application of spectroscopy, 258
Ar, 225, 235, 237, 289, 299, 302, 314, 316, 345, 347, 355, 368, 376, 384, 408
Ar(ion), 68, 85, 105
astrophysics, 36ff
asymmetry, 122, 156
atmospheric windows, 310

B

Ba, 316
Ba(ion), 5
Balmer, 9, 29
beams, 5, 252
Bethe–Salpeter, 64
Boltzmann eq., 367
Born approximation, 83
Bose–Einstein condensation, 417

C

C(ion), 4, 53, 70, 77, 95, 135ff, 211
C_{60}, 349
Ca, 229, 316
CARS, 368
Cd, 256, 345, 391
CF_2(ion), 129, 215
charge exchange, 6ff
chemical reactions, 343
CH_4(ion), 136
CHF_2Cl, 372
CH_3I, 355
Cl(ion), 105
close coupling, 67, 416, 418
CO, 266, 316, 341, 368, 378
CO_2, 227, 237, 265, 274, 288, 299, 304, 306, 314, 386
CO_2(ion), 95, 136
collision induced spectra, 122, 221, 223, 229, 233, 235, 274, 318, 357, 359
colloid, 363
combustion, 388
cooling of atoms, 413
correlation function, 289, 359
Coulomb–Bethe approximation, 213
Coulomb–B–O approximation, 213
coupled channel scattering, 367
continuum lowering, 113
Cs, 243, 422

D

D, 343
D(ion), 3ff, 28,
D_2, 300, 308, 316, 343, 357, 367, 380
Debye length, 45, 142
deconvolution, 93
density diagnostic, 107ff
depolarized scattering, 380
Dicke narrowing, 28, 160ff, 367, 378
diffusion tensor, 163ff
dimers, 274
dissociative recombination, 89
DMSO, 353
Doppler free, 3ff, 32, 241, 252, 424
Doppler width, 89, 152, 160ff, 218
duration of collision, 306, 380
Dyson integral, 91

E

ECS, 265, 373
energy pooling, 256, 392
excimers, 391, 406, 408
excitons, 260

F

F(ion), 136
far wings, 178, 231, 274, 288, 312, 314, 316, 341, 343, 345, 349, 392
Fermi's golden rule, 416
field autocorrelation function, 46
field ionization, 38, 355
fine structure, 42, 217
fluorescence, 16
foil ablation, 123, 215
fractal, 364
Frank–Condon factor, 395, 413, 420
Freon–22, 376
FWM, 361, 388

G

gas pinch, 135, 153
Gaunt factor, 142
Green's function, 91, 151, 217

H

H, 129, 343
H(ion), 3ff, 39ff, 58, 64, 68, 70, 129, 151
H_2, 221, 316, 343, 388
H_2(ion), 95,
HCO^+, 382
$HCOND_2$, 353
$HCONH_2$, 353
HCl, 368
HD, 225, 288, 357
He, 225, 233, 235, 256, 265, 289, 299, 300, 302, 304, 316, 347, 367, 376, 378, 380, 388, 422
He(ion), 4ff, 29, 30, 39ff, 52, 58, 60, 62, 105ff, 158, 209,
heat pipe, 393
hexadecapole, 227
HF, 302, 368
Hg, 235, 316, 341, 343, 349, 406
H_2O, 274ff, 310, 314

I

ICF, 3, 105ff, 123, 153, 158
induced dipole, 223, 225, 237, 242, 318
inelastic collisions, 223, 256, 317, 384, 391
intermolecular potential, 221, 227, 237, 249, 316, 393, 413, 415, 421
interstellar gas, 368
ion broadening, 34
ion dynamics, 39ff, 60, 64, 77, 91ff, 111ff, 151, 153

J

Jahn–Teller effect, 422

K

K, 384
Kr, 235, 246, 316, 408

L

laser plasma, 122, 155, 158, 211, 215,
Li, 256, 391, 406
Li(ion), 5, 108ff, 135, 153, 155, 158
LiF (ion), 129
line asymmetry, 39
line interference, 38, 292, 298, 300, 302, 304, 306, 308, 312, 368, 372, 380
line strengths, 275, 386

liquids, 353, 355, 257, 359, 422
London–van der Waals, 243
Lorentzian, 134, 368
LTE, 42ff, 95, 159, 211
Lyman α, 3ff, 32, 151, 211
Lyman β, 8

M

magnetic resonance, 423
many body effects, 359
MEG, 308
memory function, 289
microfield, 91, 109ff, 156, 177, 209, 217
microwaves, 28
Mie theory, 363
model microfield method (MMM) 42ff
molecular dynamics, 359
molecular ions, 382
multiphoton, 83, 392

N

natural line width, 113, 251, 252
NLTE, 50, 63, 66, 89, 109, 254
N(ion), 79, 95, 135, 211
N_2, 227, 274ff, 289, 304, 310, 314, 316, 341, 372, 378, 386
N_2(ion), 95, 136
Na, 417
Na_2, 416
Ne, 233, 316
Ne(ion), 68, 89, 105, 134ff, 289
NH_3, 374, 376
noise, 93
nonlinear broadening and shifting, 380
nutation, 423

O

O, 254
O(ion), 4, 75, 136ff
O_2, 372, 388
optical pumping, 423
orientation relaxation, 388
oscillator strength, 87, 214
OCS, 359

P

phase coherence, 348
phonon spectra, 357
photoassociation, 414, 415, 417, 420

Subject Index 433

photochemical reaction, 391, 406
photon echo, 424
planetary atmosphere, 274
plasma instabilities, 97
plasmon, 363
polarizability, 300
polarization, 4, 347, 388
polarization(atomic), 231
polarization induction, 83
predissociation, 420

Q

quadrupole, 227
quadrupole induction, 226
quantum well, 361
quasi−molecules, 316, 341
quasistatic, 91, 178, 314, 316, 343, 345, 394, 413

R

radiation trapping, 392
radiative acceleration, 41
radiative pressure, 159
radiative transfer, 114, 134
Rayleigh scattering, 137
Rb, 417, 420, 423
Rb(ion), 87
Rb_2, 416, 420
Rydberg states, 355

S

satellites, 109
selective reflection, 241, 249
SF_6, 376, 425
SF_6(ion), 136
shock wave, 221
S−matrix, 300, 368, 415, 418
solids, 357
spectral moments, 221, 227
spectral simulation, 106ff
speed dependent widths, 378
spin relaxation, 294
Sr, 316, 347
stars, 36, 87, 221, 254
Stark broadening and shifting, 5ff, 28, 34, 58, 60, 66, 75, 79, 81, 87, 95, 97, 134, 152, 155, 158, 209, 211, 217
Stark tuning, 376
state to state rates, 308
stellar winds, 36ff
stimulated emission, 83, 85
synchrotron radiation, 408

T

T(ion), 3ff
T matrix, 384
temperature diagnostic, 107ff
Thompson scattering, 135
T matrix, 69
tokomak, 3, 34
torques, 288
transit time, 424
translational motion, 221, 237
trapping of atoms, 413, 417, 420
two photon, 3ff, 32, 403

V

Voigt profile, 34, 89
vibrational dephasing, 308

W

Waldmann−Snider eq., 367
wall interaction, 242, 249
Wigner function, 162
Wigner law, 415, 417

X

Xe, 235, 289, 302, 316, 345, 378
Xe(ion), 105
X−ray diagnostics, 105, 122

Y

Yb, 316

Z

Z_{eff}, 3, 34
Zeeman, 4ff
Zn, 391